AVIATION AND THE ENVIRONMENT

Aviation and the Environment

Jon C. Goodman
Editor

Nova Science Publishers, Inc.
New York

LIBRARY OF CONGRESS CATALOGING-IN-PUBLICATION DATA

Available upon request

ISBN: 978-1-60692-320-7

Published by Nova Science Publishers, Inc. ✢ New York

CONTENTS

PREFACE

Aviation contributes a modest but growing proportion of total U.S. emissions, and these emissions contribute to adverse health and environmental effects. Aircraft and airport operations, including those of service and passenger vehicles, emit ozone and other substances that contribute to local air pollution, as well as carbon dioxide and other greenhouse gases that contribute to climate change. EPA estimates that aviation emissions account for less than 1 percent of local air pollution nationwide and about 2.7 percent of U.S. greenhouse gas emissions, but these emissions are expected to grow as air traffic increases.

Chapter 1 - Exhaust emissions from aircraft contribute to degradation of urban air quality near airports [Carslaw et al., 2006; Farias and Simon, 2006; Peace et al., 2006, and Pison and Menut, 2004] and can influence background atmospheric chemistry in major flight corridors [Klemm et al., 1998]. They may also impact global climate directly by enhancing the greenhouse effect and indirectly by altering the properties of background atmospheric aerosol and cloud particles in the upper troposphere and lower stratosphere (UT/LS), thereby affecting absorption, emission, and transmission of both visible and infrared radiation [IPCC, 1999]. In order to accurately attribute the atmospheric impacts of current aviation operations, and reliably predict future impacts, it is necessary to have a good understanding of the gaseous and particulate emissions of different aircraft types, as well as an understanding of the fundamental chemical and dynamical processes that occur in the relevant regions of the atmosphere.

The goals of this White Paper are to summarize the ways in which aircraft emissions impact atmospheric chemistry in the UT/LS, to examine what has been learned since the last major assessments, and to prioritize future scientific studies that can reduce the most important uncertainties that remain and that address new problems that have arisen.

Chapter 2 - The global commercial aircraft fleet currently numbers about 10,000 and flies several billion kilometres per year while burning more than 100 MT of fuel per year at high temperatures producing mostly water and CO_2. However, NOx (= $NO+NO_2$), other minor gaseous species, organic aerosols from unburnt fuel and soot and ions are also injected at cruise altitudes located in upper troposphere and lower stratosphere (UT/LS), a region particularly sensitive to atmospheric climate change.

The demand for air transportation in the US is projected to grow three fold by 2025 while similar growth is projected for the aviation industry world wide. Future climate impacts are expected to increase based on this projected aviation growth and resulting changing atmospheric conditions. These impacts relate to the impact of tripling aviation system capacity and the resulting global impact of these additional engine emissions which are estimated to be approximately twice as large as at the turn of the last century. However, if current economic projections obtain for this period, boundary layer (BL) NOx emissions may also double and hence their contribution to the UT region. In addition to global climate impacts there is also potential for even greater regional or local effects. The growth of emissions of both BL and aircraft NOx will likely lead to an increased production of ozone in the UTLS. This increase in UT/LS ozone will cause a significant increase in the radiative forcing, which in turn will contribute to global warming.

Chapter 3 - Generally, the climatic impact of air traffic (of which a substantial part may be due to contrails and contrail cirrus) today (year 2000) amounts to 2-8% of the global radiative forcing associated with climate change. Due to the projected increase in air traffic [ICAO, 2007] the relative importance of air traffic is going to increase drastically. In the long term it may well be, that the most serious threat to the continued growth of air travel is its impact on climate [Green, 2005]. In view of the societal relevance and economic importance of sustainable growth of global aviation, it would be appropriate that the climate science community received sufficient funding, allowing significant progress estimating climate impacts, in order to ensure that political decisions are based on increasingly sound scientific knowledge. Aircraft-induced cloudiness, which comprises contrail cirrus and modification of cirrus by aircraft exhaust soot emissions are the most uncertain component in aviation climate impact assessments [IPCC, 2007]. Since they may be the largest component in aviation radiative forcing aircraft-induced cloudiness and contrail cirrus in particular requ ire a largeresearch effort.

Chapter 4 - Theme 4 of the ACCRI, "Contrails and Contrail-Specific Microphysics", reviews the current state of understanding of the science of contrails: 1) how they are formed, 2) their microphysical properties as they evolve, 3) how they develop into contrail cirrus and if their microphysical properties can be distinguished from natural cirrus, 4) their radiative properties and how they are treated in global models and 5) the ice nucleating properties of soot aerosols and whether these aerosols can nucleate cirrus crystals.. Key gaps and underlying uncertainties in our understanding of contrails and their effect on local, regional and global climate are identified and recommendations are provided for research activities that will remove or decrease these uncertainties.

Contrail formation is described by a simple equation that is a function of atmospheric temperature and pressure, specific fuel energy content, specific emission of water vapor and the overall propulsion efficiency. Thermodynamics is the controlling factor for contrail *formation* whereas the physico-chemistry of the emitted particles acts in a secondary role. The criteria for contrail formation determine whether a contrail will form but does not predict whether the contrail will persist or spread into an extensive cirrus-like cloud.

Chapter 5 - In this subject-specific white paper, we present a literature survey of past and current developments regarding the impact of contrails and contrail cirrus on the radiation field of the Earth's atmosphere and climate. A number of recommendations for future long-term and short-term actions that are required to comprehend and quantify this important subject are subsequently outlined.

We first present a survey on the background of the basic problem of aviation's impacts on climate and climate change, followed by a discussion of perspectives based on conclusions of the 1999 Intergovernmental Panel on Climate Change (IPCC) Special Report, and the doubling and tripling growths of aviation industry in the next 20 to 40 years as projected by the Next Generation Air Transportation System, United Nation International Civil Aviation Organization, European Union Nations, and the United Kingdom. In response to the pressing need for further study of the potential impact of aircraft emission on climate and environment, a "Workshop on the Impacts of Aviation on Climate Change" was organized and held in Boston, MA on June 7-9, 2006, and a report on the findings during this workshop was later published.

Chapter 6 - The effect of aircraft emissions on the climate of Earth is one of the most serious long-term environmental issues facing the aviation industry (IPCC, 1999; Aviation and the Environment – Report to the United States Congress, 2004). Aviation emissions, including gases and particles in the upper troposphere and lower stratosphere, have both direct and indirect climate effects. The direct effect is principally the emission of carbon dioxide, a powerful greenhouse gas. The indirect effects include the changes in ozone due to emissions in nitrogen oxides, the effects of aerosol emissions and water vapor on clouds, and the effects associated with contrails and contrail-induced cirrus clouds.

As stated in *the Executive Summary of the Workshop on the Impacts of Aviation on Climate Change,* June 7-9, 2006, Boston, MA (hereafter, the *Workshop Executive Summary,* http://web.mit.edu/aeroastro/partner/reports/climatewrksp-rpt-0806.pdf), "The effects of aircraft emissions on the current and projected climate of our planet may be the most serious long-term environmental issue facing the aviation industry... The only way to ensure that policymakers fully understand trade-offs from actions resulting from implementing engine and fuel technological advances, airspace operational management practices, and policy actions imposed by national and international bodies is *to provide them with metrics that correctly capture the climate impacts of aviation emissions.*"

Chapter 7 - The United Nations Framework Convention on Climate Change (UNFCC) entered into force in 1994 with the objective for 'stabilization of greenhouse gas concentrations in the atmosphere at a level that would prevent dangerous anthropogenic interference with the climate system'. The Kyoto Protocol (1997) set out to reduce emissions of most long-lived greenhouse gases in developed countries to below their 1990 levels. Probably as a result of convenience and simplicity, the chosen metric to compare the climate impact of these greenhouse gases was the 100-year Global Warming Potential (GWP), as calculated by the Intergovernmental Panel of Climate Change Second Assessment Report (IPCC, 1995).

As an integral and growing part of the global economy and transportation sector, aviation has the potential to significantly contribute to changes in the Earth's climate. However, the impact of short-lived species (e.g. nitrogen oxides (NOx), an ozone precursor which in turn impacts on methane) and effects (e.g. aviation induced contrails) on the climate system depends upon geographical and altitudinal location, season, time of the day and the background meteorology and chemistry during their release (Rogers et al., 2000; Sausen et al., 2005). Such short-lived species therefore require an appropriate metric which takes into consideration these dependencies (Rogers et al., 2002a). For the aviation sector the potential climate impact is dependent upon both long-lived and short-lived emissions and effects, making the choice of a suitable metric that integrates over all effects more difficult.

Chapter 8 - The impact of climate-altering agents on the atmospheric system is a result of a complex system of interactions and feedbacks within the atmosphere, and with the oceans, the land surface, the biosphere and the cryosphere. Climate metrics are used as a proxy to simplify interpretation of the complex science and associated feedbacks to indicate the ultimate effect of constituent changes in the atmosphere. Aviation is just one contributor to these constituent changes in the atmosphere but the potential impact of aviation on climate is expected to grow over the coming decades as demand for air travel increases. It is necessary to quantify the impact of aviation so that appropriate policy actions may be defined. The objective of this report is to examine the capabilities and limitations of current climate metrics in the context of the aviation impact on climate change, to analyze key uncertainties associated with these metrics and, to the extent possible, to make recommendations on future research and about how best to use metrics currently to gauge aviation-induced climate change.

Chapter 9 - Aviation contributes a modest but growing proportion of total U.S. emissions, and these emissions contribute to adverse health and environmental effects. Aircraft and airport operations, including those of service and passenger vehicles, emit ozone and other substances that contribute to local air pollution, as well as carbon dioxide and other greenhouse gases that contribute to climate change. EPA estimates that aviation emissions account for less than 1 percent of local air pollution nationwide and about 2.7 percent of U.S. greenhouse gas emissions, but these emissions are expected to grow as air traffic increases.

Two key federal efforts, if implemented effectively, can help to reduce aviation emissions—NextGen initiatives in the near term and research and development over the longer term. For example, NextGen technologies and procedures, such as satellite-based navigation systems, should allow for more direct routing, which could improve fuel efficiency and reduce carbon dioxide emissions. Federal research and development efforts—led by FAA and NASA in collaboration with industry and academia—have achieved significant reductions in aircraft emissions through improved aircraft and engine technologies, and federal officials and aviation experts agree that such efforts are the most effective means of achieving further reductions in the longer term. Federal R&D on aviation emissions also focuses on improving the scientific understanding of aviation emissions and developing lower-emitting aviation fuels.

In: Aviation and the Environment
Editor: Jon C. Goodman,
ISBN: 978-1-60692-320-7

Chapter 1

AVIATION-CLIMATE CHANGE RESEARCH INITIATIVE (ACCRI) SUBJECT SPECIFIC WHITE PAPER (SSWP) ON UT/LS CHEMISTRY AND TRANSPORT SSWP # I, JANUARY 25, 2008

Darin Toohey[1], Linnea Avallone[2] and Martin Ross[3]
[1,2] University of Colorado-Boulder, Colorado, USA
[3] The Aerospace Corporation, El Segundo, California, USA

EXECUTIVE SUMMARY

Aircraft emissions of particles, particle precursors, NOx, and water vapor, can have significant impacts on chemistry in the upper troposphere and lower stratosphere (UT/LS). Previous groups have assessed the important terms involving UT/LS chemistry and noted the following issues that limit the ability to reduce uncertainties in assessments of aircraft impacts:

Incomplete knowledge of exhaust emissions of gases (primarily sulfur oxides) and particles (e.g., soot) and their geographic and altitudinal distributions.
Important discrepancies between modeled and measured distributions of key HOx and NOx radical species involved in ozone formation and destruction.
Poor understanding of the sources of NOx in the upper troposphere, especially lightning.
Incomplete knowledge of the evolution of NOx and NOy in aircraft plumes during the first ~24 hours following emission.
Incomplete understanding of, and potential non-linearities in, the coupling among CH_4, CO, OH and O_3 in the troposphere.
Potential scavenging and removal of NOx by aerosols and cirrus.
Limited understanding of atmospheric transport, especially that between the stratosphere and troposphere.

This SSWP summarizes important results in key areas since the last major aircraft impacts assessment [IPCC 1999]. Significant progress has been made in the areas of:

Measurements of emissions of chemi-ions, NOx, and trace organic species from aircraft engines.
Observations constraining the lightning and convective fluxes of NOx to the upper troposphere.
Measurements of HOx, its precursors, and coupled NOx/HOx chemistries in the UT.
Rates rate and extent of conversion of NOx to NOy in the UT.
New observations of water vapor and particles that help to constrain important processes that determine stability of cirrus clouds and persistent contrails.
Model studies of the impact of aircraft emissions of particles on ozone in the UT/LS.
Model studies of the potential role for destruction of ozone in the UT by heterogeneous reactions involving halogen species.

In addition to studies that can lead to improvements in our understanding of the impacts of aircraft emissions, there are longstanding issues and new observations that raise important new questions about our understanding of UT/LS chemistry that may have significant, including:

Ongoing discrepancies of upwards of 30% between observations of water vapour in the cold, dry upper troposphere and lower stratosphere that limit our ability to predict formation and persistence of cirrus clouds and, hence, their impact on the budgets of trace species that control ozone abundances in the UT/LS.
Important discrepancies between modeled and observed HOx species (primarily HO_2) at high NO values in the region where subsonic aircraft emissions represent the most significant perturbation to chemistry.
New observations of heterogeneous activation of chlorine in the tropopause region.
Observations that indicate greater abundances of inorganic bromine than previously believed, presumably due to more efficient transport of short-lived bromine sources to the UT.
Observations of significant uptake of nitric acid in ice particles and an increased role for HNO_3 in the stability of ice in the UT/LS.

Perhaps the most significant new result related to the impacts of some of these new findings is that of Sovde et al. [2007] that shows a reversal in the sign of ozone response to increased aircraft emissions in the UT, primarily as a result of heterogeneous chemistry on particles. If confirmed, this result could have important implications for the sign and magnitude of climate impacts due to aircraft.

These results, if studied with the best modeling tools available, should help constrain the role of aircraft emissions on chemistry in the UT/LS. It is expected that the new result will imply a diminished enhancement of ozone due to NOx/hydrocarbon chemistry in the UT, and possibly ozone losses in some regions where aircraft emissions enhance the production of particulate surfaces areas or the lifetimes of cirrus clouds. Constraints on OH abundances throughout the troposphere should reduce the uncertainties in modelled impacts of aircraft emissions on the lifetime of methane, which is currently believed to have a negative forcing on climate. Finally, modeling studies of the sensitivity of ozone and HOx to heterogeneous processes, including sedimentation of particles that contain HNO_3 and halogen activation, should help to define the range of possible impacts these processes, which are currently poorly understood, could have.

Ideally, to make the best use of the new results in a future aircraft impacts assessment, the following issues will need to be better understood. Progress in all areas is likely to take the concerted efforts of a number of research groups involved in atmospheric measurements (both in situ and from satellites) and modeling programs designed to explore the new results in great detail. Among the issues identified in this SSWP are:

Resolving discrepancies in water vapor measurements should be the highest priority for addressing remaining uncertainties in UT/LS chemistry. It would also be desirable to develop a standard for water vapor measurements under cold, dry conditions so that more costly large-scale intercomparisons and validations can be infrequent. This top priority cannot be overlooked – anything less, and it is likely that in a few years' time, a similar group will be making the same recommendation. Validations of temperature should be a nearly equal priority, and should be feasible with a small augmentation to a water vapor program.

Addressing gaps in measurement capabilities for species that are important in assessing the impact of heterogeneous reactions and plume dispersion processes. Programs should be started very soon, even with limited funds, so that investigators have confidence that in a few years' time they will be able to participate in missions of opportunity. Priority should be placed on instrumentation with a heritage, even if from other platforms, so that development of calibrations and standards does not take up a significant fraction of the available resources. Instruments using new techniques would be desirable in a few cases for corroboration of the most critical measurements.

Developing a strategy for model simulations to assess the range of possible impacts and that incorporate new results, especially those relating to plume dispersion and non-linear effects. The program should focus on assessing the range of impacts over a wide set of boundary conditions for those processes that are currently unconstrained by observations (e.g., redistribution of nitric acid by sedimentation, chlorine and bromine chemistry, unknown coupled HOx/NOx chemistry, errors in water vapor and supersaturation).

Guided by results from studies of the above issues, new questions should be developed to help guide measurement programs (dedicated or flights of opportunity).

Convene annual meetings of investigators participating in aviation impacts-related activities to foster frequent exchange of ideas. Rather than a comprehensive meeting, discussion of presentations and discussions should focus on results of studies that reduce the critical uncertainties in aircraft impacts or studies that highlight new and important processes that could result in a major shift in understanding of those processes. The community should be conditioned to respond quickly and productively to new developments and shifting priorities, much like the atmospheric chemistry community responded to the ozone hole and methyl bromide issues.

1. Introduction and Background

Exhaust emissions from aircraft contribute to degradation of urban air quality near airports [Carslaw et al., 2006; Farias and Simon, 2006; Peace et al., 2006, and Pison and Menut, 2004] and can influence background atmospheric chemistry in major flight corridors [Klemm et al., 1998]. They may also impact global climate directly by enhancing the greenhouse effect and indirectly by altering the properties of background atmospheric aerosol and cloud particles in the upper troposphere and lower stratosphere (UT/LS), thereby affecting absorption, emission, and transmission of both visible and infrared radiation [IPCC, 1999]. In order to accurately attribute the atmospheric impacts of current aviation operations, and reliably predict future impacts, it is necessary to have a good understanding of the gaseous and particulate emissions of different aircraft types, as well as an understanding of the fundamental chemical and dynamical processes that occur in the relevant regions of the atmosphere.

The goals of this White Paper are to summarize the ways in which aircraft emissions impact atmospheric chemistry in the UT/LS, to examine what has been learned since the last major assessments, and to prioritize future scientific studies that can reduce the most important uncertainties that remain and that address new problems that have arisen.

2. Processes that Impact Climate

2.a. Current State of the Science

Two previous assessments have thoroughly reviewed the important properties of emission products that are thought to be the most relevant to atmospheric chemistry [IPCC, 1999; Brasseur et al., 1998]. Based on these reports, the most important products of combustion of aircraft fuel (e.g., kerosene) are CO_2, H_2O, NOx, soot, and oxides of sulfur. All of these species interact strongly with infrared or visible light, serving to directly warm or cool the planet. Some can alter the nature and radiative properties of particulate matter (e.g., aerosols and clouds) or can promote formation of new particles by changing the extent of supersaturation through influence on temperature and water vapor abundances. Some, such as NOx and soot, can also have important indirect impacts on the atmosphere, including subtle shifts in chemical balance that can alter the natural abundances of radiatively important gases such as O_3 and CH_4, or cause the redistribution of naturally occurring species such as H_2O and HNO_3 via sedimentation of large particles. Finally, through influences on radiation balance, these emissions can impact atmospheric transport, especially between the troposphere and stratosphere.

These different, and in some cases offsetting, effects have been studied before in some detail. IPCC [1999] identified warming due to enhancements of CO_2, contrails and cirrus, and O_3 (which is thought to be increased by NOx chemistry), and cooling by CH4 (which is thought to decrease as a result of enhancements of OH by NOx chemistry), as the most likely to have significant impacts on climate. It was believed that only one of these processes, warming by CO_2, was well understood, whereas the relative scientific understanding of the others was listed as fair to poor. An update of this assessment by Sausen et al. [2005], recognized that work published since the turn of the century reduced some of the key uncertainties. Nevertheless, the limited understanding of those processes continues to represent a major hurdle to reducing the overall uncertainties in aviation impacts [Wuebbles et al., 2006]. Of particular interest are impacts of NOx on the chemistry of ozone and on the budget of methane, which together could represent more than half of the total impact of aircraft emissions on climate. If aviation transport continues to grow, it is estimated that the number of flights will double from present rates by about 2025 [Cox, 2007]. Unless major changes to combustion systems can be implemented, aircraft emissions can also be expected to nearly double by 2025. Consequently, the impacts of aviation operations on climate and the oxidative capacity of the atmosphere are of great interest.

Both the IPCC [1999] and the Workshop on the Impacts of Aviation on Climate Change [Wuebbles et al., 2006, hereafter called the "2006 Workshop"] concluded that the following processes that influence NOx chemistry contributed most to uncertainties in assessments of the impact of the chemistry of aircraft exhaust on Earth's climate:

1. Incomplete knowledge of exhaust emissions of gases (primarily sulfur oxides) and particles (e.g., soot) and their geographic and altitudinal distributions.
2. Important discrepancies between modeled and measured distributions of key HOx and NOx radical species involved in ozone formation and destruction.
3. Poor understanding of the sources of NOx in the upper troposphere, especially lightning.
4. Incomplete knowledge of the evolution of NOx and NOy in aircraft plumes during the first ~24 hours following emission.
5. Incomplete understanding of, and potential non-linearities in, the coupling among CH_4, CO, OH and O_3 in the troposphere.
6. Potential scavenging and removal of NOx by aerosols and cirrus.
7. Limited understanding of atmospheric transport, especially that between the stratosphere and troposphere.

In addition, we note the critical nature of understanding the processes controlling water vapor in the UT/LS [see IPCC 1999]. Water vapor is important not only because it is a greenhouse gas that is directly emitted by aircraft but also because it is a significant source of odd-hydrogen (HOx) in the UT/LS. Species in the HOx family produce and destroy ozone, largely determine the lifetimes of CH4 and CO, and also influence NOx chemistry under the conditions that prevail in the UT/LS. Finally, H_2O is the major condensable species, playing a key role in the formation of ice particles and polar stratospheric clouds in the UT/LS (see SSWPs III and IV). As discussed in detail in a separate SSWP, the relative humidity variable, RHi, is the critical quantity for understanding formation, growth, and evaporation of ice-containing particles in the UT/LS. Therefore, direct emissions of water vapor to the atmosphere, as well as indirect influences of other trace combustion products on water vapor distributions and temperatures in the UT/LS, can have major impacts on the chemistry of the atmosphere.

Due to the strong non-linear coupling between NOy, particles, and water/ice precipitation, all of these factors are influenced by processes discussed in other SSWPs, most importantly, that on clouds and aerosols. Thus, the discussion here will overlap strongly with other SSWP topics that address uncertainties in water vapor measurements and parameterizations of aerosol properties and clouds. Of particular interest to UT/LS chemistry are factors that limit the ability to predict the presence of ice and the extent of uptake of nitric acid. The rates of heterogeneous reactions that repartition NOx into NOy and that release active forms of chlorine vary by several orders of magnitude, depending on the abundances of condensed HNO_3, a quantity that itself is non-linear with respect to temperature and relative humidity (essentially a threshold with temperature or RHi) [e.g., see WMO 2006 and references therein]. In addition, a significant confounding factor is that heterogeneous reactions between halogens and temporary NOx reservoirs can release photolytic sources of HOx, which, in turn, destroy methane and accelerate the gas-phase formation of HNO_3. Enhancements of reactive chlorine also alter methane abundances. It is safe to say that highly accurate measurements of water vapor are critical for any assessment of atmospheric chemistry that is influenced by heterogeneous chemistry.

These issues are explored in detail in the following two major sections. The remainder of Section 2 will summarize studies that have led to significant improvements in our understanding of aircraft impacts on chemistry in the UT/LS. Section 3 will report on recent observations that raise important new questions about chemical processes in the UT/LS; new modeling efforts will be necessary to determine their proper roles in future aviation impacts assessments.

2.b. The Role of UT/LS Chemistry in Aviation Impacts on Climate

The 2006 Workshop considered the combined impacts of NOx emissions on ozone abundances and, through perturbations to HOx chemistry, on methane abundances, to comprise the bulk of the total uncertainty in climate forcing due to aviation [Wuebbles et al., 2006]. This SSWP examines recent results that address the various aspects of UT/LS chemistry that were identified in the 1999 IPCC and 2006 Workshop reports and listed in the previous section. Figure 1, reproduced from Sausen et al. [2005], updates a similar figure from IPCC [1999]. It shows the Global Radiative Forcing (RF) framework that has largely informed the bulk of recent scientific research into the impacts of aviation on climate. As is clear from figure 1, terms relating to chemistries of NOx and HOx are among the three largest contributors to the aircraft RF, and, as will be shown in Section 3 below, the third term related to contrails is itself influenced by NOx chemistry via the role of HNO_3 in ice stability and contrail evolution. Consequently, uncertainties in the chemistry of aircraft emissions in the UT/LS dominate the overall uncertainty in climate forcing due to aviation.

A key result of research conducted in the 1990s and summarized in Chapter 2 of IPCC [1999] was that the response of ozone to changes in NOx reverses sign in the lower stratosphere. Formation of ozone by photochemistry initiated by oxidation of volatile organic compounds dominates in the upper troposphere, whereas catalytic destruction of ozone by NOx dominates in the middle stratosphere. The discovery in the early 1990s of a shift in the relative roles of halogens and NOx in the lower stratosphere due to heterogeneous conversion of N2O5 to HNO_3, lead to reexamination of the impacts of emissions from supersonic aircraft. Model studies soon found that NOx enhancements near 20 km due to supersonic aircraft (or upward transport of subsonic aircraft emissions) would lead to increases in ozone, thereby reducing reactive halogens [e.g., Weisenstein et al., 1993].

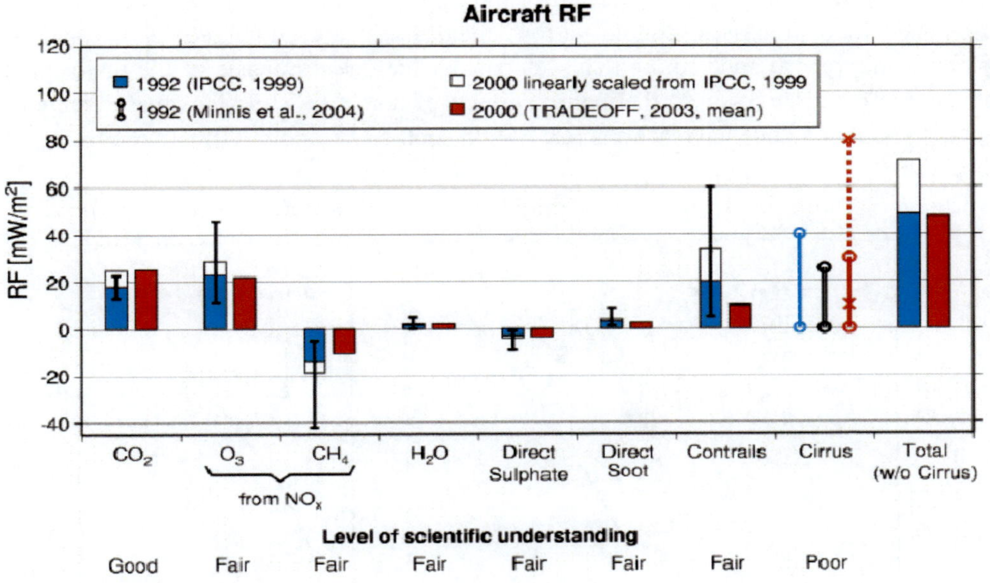

Figure 1. Global radiative forcing (RF) [mW m-2] from aviation for 1992 and 2000, based on IPCC (1999) and TRADEOFF results. The whiskers denote the 2/3 confidence intervals of the IPCC (1999) values. The lines with the circles at the end display different estimates for the possible range of RF from aviation-induced cirrus clouds. In addition the dashed line with the crosses at the end denotes an estimate of the range for RF from Sausen et al., [2005].

Figure 2, taken from the 1999 IPCC Report, reveals this dual nature, and illustrates why transport and mixing processes are critical in determining the response of ozone to aircraft NOx emissions. Although the simulation shown in figure 2 was designed simply to illustrate the sensitivity of ozone to a change in NOx, and not to predict the true response of ozone to a specific perturbation due to aviation, it still serves to frame the discussion of impacts and uncertainties that follows. For example, it is easy see that emissions that remain in the upper troposphere will lead to an increase in ozone, whereas those that reach the stratosphere will increase ozone below 24 km, but decrease it above. The net impact of NOx emissions thus depends strongly upon the vertical distribution of the resultant perturbation to background levels. Consequently, the impact of NOx on ozone will differ for subsonic and supersonic aircraft, which deposit their exhaust mainly in the UT and LS, respectively [IPCC, 1999]. Thus, in order to assess the impacts of aviation, the proportion of stratospheric (e.g., supersonic) and tropospheric (e.g., subsonic) emissions from a future fleet of aircraft (the so-called mixed fleet) must be known [Gauss et al., 2006]. What is important to note here is that assessments of the impact of emissions of a particular assumed fleet of aircraft on ozone have relied explicitly on the ability to accurately model this altitude dependence of the ozone response to changes in NOx, the vertical distribution of which depends not only on the flight altitude, but also upon knowledge of the vertical transport of NOx and possible redistribution by cloud and aerosol processes. These themes will become important later in this SSWP, as the implications are explored of new observations in the UT/LS that show a more important role for heterogeneous chemistry and possible redistribution of HNO_3 than was known at the time of the previous assessment.

The strong linkages between these three topics, especially heterogeneous chemistry and aerosol and cloud processes, couple various themes that are addressed in this and other chapters of this report, and require that we consider the direct impacts of the major aircraft combustion products, as well as the indirect effects of non- CO_2 emissions that participate in gas-phase and heterogeneous reactions (e.g., SOx, soot, NOx, and H_2O) with the background atmosphere.

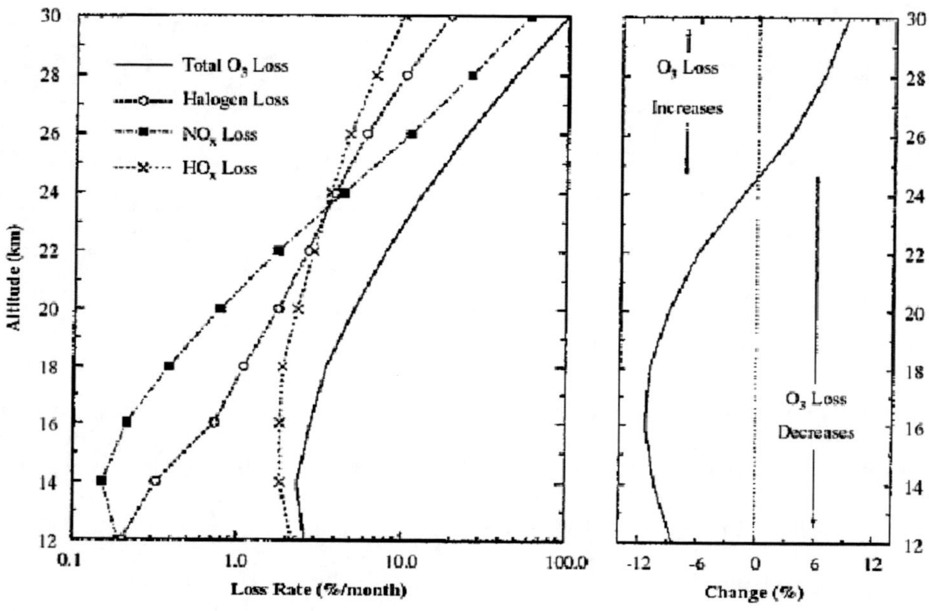

Figure 2. One-dimensional model results for the month of March at northern midlatitudes used to illustrate the relative roles of ozone-destroying radicals (left panel) and percentage change in the ozone destruction rate for a uniform 20% increase in NOx (right panel) as functions of altitude [IPCC, 1999].

2.c. Advancements since the 1999 IPCC Report

Since the publication of the 1999 IPCC report, there have been more than several hundred studies that address important issues raised in that report. While it is not possible to do justice to all of these studies in this SSWP, we summarize here where significant advances have been made.

To help define the range of species and concentrations of important engine exhaust emissions, new measurements have been obtained of soot and particle precursor gases [Dakhel et al., 2007; Hays and Vander Wal, 2007; Karcher et al., 2007; Sorokin and Arnold, 2004] such as chemi-ions [Arnold et al., 2000; Eichkorn et al., 2002; Haverkamp et al., 2004; Miller et al., 2005; Sorokin and Arnold, 2006], sulfur and NOx [Herndon et al, 2004; Schroder et al., 2000; Schumann et al., 2002; Tsague et al., 2006, 2007; Wormhoudt et al.,

2007], and volatile organic compounds (VOCs) and particles [Anderson, et al., 2006; Herndon et al., 2006; Lobo et al., 2007; Knighton et al., 2007; Nyeki et al., 2004; Sorokin et al., 2001; Wey et al., 2007; Wilson et al., 2004; Yelvington et al., 2007], in the exhaust of engines or aircraft on the ground and at cruise altitudes. In addition, new laboratory studies have further defined the reactivity of engine- emitted soot, most importantly regarding uptake of water and reactivity to NOx, NOy, and O_3 [Popovicheva et al., 2000, 2003, 2004, 2007; Shonija et al., 2007; Talukdar et al., 2006; Wei et al., 2001]. These new studies help to constrain parameters that are critically important for modeling the perturbations of reactive species (e.g., NOx and VOCs) and particle evolution (e.g., chemi-ions, VOCs, and soot) emitted by aircraft in the UT/LS [Ma and Zhao 2000; Petzold et al., 2005; Wei and Liu 2007]. Key new results and implications of these studies are summarized in Section 2.c.I.

Evidence is mounting from more than a decade of in situ measurements and from new satellite observations that air in the UT/LS is influenced considerably by convective transport from the surface. In fact, there are more recent studies reporting on this issue than for any of the other issues of this SSWP. In Section 2.C.II. some new results are highlighted, in particular those that address some key uncertainties in NOx and HOx budgets. Of particular interest to this SSWP are efforts to quantify lightning, biomass burning, and convective PBL (planetary boundary layer) pollution sources of NOx to the upper troposphere [Brunner et al, 2001; Decaria et al., 2005; Fehr et al., 2004; Hudman et al., 2007; Koike et al., 2002; Lange et al., 2001; Leue et al., 2001; Levy et al., 1999; Ma et al., 2002; Martin et al., 2006, 2007; Muhle et al., 2002; Parrish et al., 2004; Pierce et al., 2003; Ridley et al., 2005; Sauvage et al., 2007; Schumann and Huntrieser, 2007; Sioris et al., 2007; Smyshlyaev et al., 2003; Stohl et al., 2002; Thakur et al., 1999; van Noije et al., 2006; Wang et al., 2000; Zhang et al., 2000; Ziereis et al., 1999, 2000], fluxes that were highlighted in previous assessments as being poorly constrained. Not only do these sources of NOx (and, hence, NOy) dominate the odd nitrogen budget in the UT, thereby setting the background conditions upon which aircraft emissions represent a small, but potentially significant, perturbation, incomplete knowledge of their magnitudes and seasonal and geographic distributions make it difficult to directly attribute NOx enhancements to aircraft operations except in highly localized plumes or heavily travelled flight corridors [Brunner et al., 2005; Colette and Ancellet, 2005; Colette et al., 2005; Grewe et al., 2002; Koike et al., 2000; Marecal et al., 2006; Mari et al., 2002; Meijer et al., 2000; Park et al., 2004; Schlager et al., 1999; Tsai et al., 2001; Wang and Prinn, 2000]. New in situ observations with a larger suite of measurements of tracers for biomass burning, human activities, lightning, and stratospheric fluxes [Bertram et al., 2007; Singh et al., 2007], not only provide for attribution of sources other than aircraft emissions, but also provide new clues into photochemical processes that transform reactive NOx into species that serve as reservoirs or that can redistribute NOy (hence, NOx) by condensation onto particles followed by sedimentation [Neuman et al., 2006].

The interactions of NOy species with particles [Gao et al., 2004; Popp et al., 2006; Karcher and Voigt, 2006; Voigt et al., 2006, 2007] raise important new questions that rely on the ability to model formation, composition, and reactivity of particles [Considine et al., 2000; Meier and Hendricks, 2002; Meilinger et al., 2001; von Kuhlmann and Lawrence; 2006]. Several key new modeling studies have shown that heterogeneous chemistry involving NOx, HOx, and halogens, is extremely important in particle-rich exhaust plumes and persistent contrails, and, depending on the subsequent behavior of these species as these plumes and contrails disperse, can even have important implications on the sign of ozone response to aircraft exhaust on hemispheric scales [Meilinger et al., 2005; Sovde et al., 2007]. These results and their implications are discussed in Section 2.c.III.

The importance of convective sources of HOx in the upper troposphere has been known for many years [Collins et al., 1999; Crawford et al., 1999; Muller and Brasseur, 1999; Reiner et al., 1999; Singh et al., 2000]. New observations of HOx and volatile organic compounds in conjunction with modeling studies, continue to reinforce this view [Colomb et al., 2006, Mari et al., 2002; Olson et al., 2004; Ravetta et al., 2001; Singh et al., 2004; Snow et al., 2003, 2004; Stickler et al., 2006; Wang and Chen, 2006], and they provide some important insights into the nature of previous disagreements between modeled and measured HOx that seem to depend on NOx [Ren et al., 2008] (the previously referenced "coupled HOx/NOx discrepancy" [e.g., Faloona et al., 2000]). New measurements of HO2NO2 [Murphy et al., 2004; Kim et al., 2007] could help to identify important missing chemistry, while issues of resolution have been shown to be important under some conditions [Olson et al., 2006].

Measurements of water vapor in the upper troposphere and the stratosphere, where the naturally occurring humidities are the lowest found on Earth, have always been a source of controversy [e.g., Kley et al., 2000]. Not only are emissions of water vapor from aircraft critical for understanding radiative impacts of exhaust, accurate knowledge of background water vapor distributions and temperatures, and the microphysics of water-containing particles, are essential in order to accurately model heterogeneous chemistry, HOx distributions, and possible redistribution of reactive species in the UT/LS by sedimentation. Ongoing studies by a number of groups [Bencherif et al., 2006; Bortz et al., 2006; Ferrare et al., 2004; Folkins et al., 2006; Gao et al., 2005; Gulstad and Isaksen, 2007; Helten et al., 1999; Kley et al., 2000; Luo et al., 2007; Marecal et al., 2007; Miloshevich et al., 2006; Nedoluha et al., 2002; Park et al., 2004; Ramaswamy et al., 2001; Spichtinger et al., 2002; Troller et al., 2006; Vaughan et al., 2005; Vay et al., 2000 that have improved our understanding of water vapor and supersaturation are summarized in Section 2.c.IV. New studies addressing temperatures in the UT/LS are summarized in Section 2.c.V.

In addition to results that have improved our understanding of key uncertainties outlined in previous assessments, there have been some observations, some controversial, that raise important new questions about our basic understanding of chemistry in the UT/LS that could have major implications for the impacts of aviation. These will be presented in Section 3 of this SSWP, and include new studies related to the bromine budget [Dorf, et al., 2006a, 2006b; Salawitch, et al., 2005; Schauffler, et al., 1999; Sioris, et al., 2006; Theys, et al., 2007], the unusual impacts of bromine on NOx chemistry [Sinnhuber and Folkins, 2006; Hendricks, et al., 2000; Yang, et al., 2005], and new observations of chlorine activation in the UT/LS [Thornton, et al., 2003, 2005, 2007] that call for a fresh look at the potential impacts of heterogeneous reactions in the UT/LS, especially in persistent contrails [Borrmann, et al., 1996; Lelieveld, et al., 1999; Bregman, et al., 2002].

2.c.I. Engine Emissions

Although knowledge of the emissions of sulfate was identified as a key uncertainty in previous assessments, the main issue was not so much the sulfate itself, as the impact of fuel sulfur on particle nucleation. Since then, a number of studies have characterized particulate emissions from a variety of aircraft engines. The most significant new result is that particle production does not closely track fuel sulfur content [Wey et al., 2006; Yelvington et al., 2007]. While studies have shown that ion nucleation is the probable mechanism for volatile aerosol production in aircraft exhaust [e.g., Miller et al., 2005], measurements of positive and negative chemiions have revealed a greater role for LVOCs (low volatility VOCs) than previously believed [Eichkorn et al., 2002; Sorokin and Arnold, 2006; Miller et al., 2005].

In a study of an on-wing commercial gas turbine engine, Lobo et al. [2007] recently found little dependence of particulate emissions with varying fuel sulfur content, although they did observe that the soluble mass fraction of particles increased with distance from the engine exit plane and with increasing aromatic and sulfur content of the fuel, consistent with increased uptake of water by hygroscopic particles. Recent measurements of engine-generated soot [Shonija et al., 2007] found significant water uptake due to the existence of impurities within the engine, with amounts of absorbed water increasing with decreasing temperatures in the exhaust plume (reaching 18% by weight at threshold conditions for contrail formation). In light of previous observations of significant uptake of water by soot, these authors have inferred that to be hygroscopic, soot does not have to be processed by reactions with sulfuric or nitric acids, as was previously believed, and that impurities in engine-generated soot will play key roles in the formation of CCN in aircraft plumes. These results are consistent with a laboratory study of Talukdar et al. [2006], who found that uptake of nitric acid on aviation kerosene soot is reversible, and not a significant source of NOx, as had been suggested previously. They are also consistent with another study that found the characteristics of soot emitted by engines are determined largely by combustor processes, and not by subsequent reactions in the turbine/nozzle.

It is important to recognize that measurements of soot from combustors must be considered carefully, as it may be chemically and physical unstable, as shown in a recent study by Popovichava et al. [2003]. In addition, it is unclear whether ground level measurements will apply under cruise conditions, where combustion is more complete and LVOC emissions are likely to be significantly smaller. But from the majority of new studies, it does appear that aircraft-generated particles are relatively hygroscopic, and therefore are likely to be good CCN. A new particulate emission inventory developed under the European PartEmis program should help reduce uncertainties in modelled impacts of particulate emissions by aircraft [Petzold et al., 2005].

Important new measurements of the emissions of hydrocarbons and NOx, including speciation, have been obtained in the exhaust plumes of a variety of aircraft types during the APEX campaign [Herndon et al., 2004; Herndon et al. 2007; Knighton et al., 2007; Wormhoudt et al., 2007]. To first order, the results are in good agreement with previous studies, increasing confidence in the emissions databases used for modeling aircraft impacts. Additional insights from these studies include the finding that fuel type and plume age appear to have only minor effects on the emissions of hydrocarbons, including speciation, whereas temperature appears to be an important factor. NOx emissions were found to increase with thrust, while the fraction of NO_2/NOx decreased from 80% at lowest thrust to below 7% at highest thrust. Nitrous acid (HONO) was found to be a minor species (~7%) that increased with thrust, and also served as a good indicator for predicting abundances of other trace species, such as oxides of sulfur.

In summary, new results indicate an increased role for hydrocarbons in formation of particles in aircraft exhaust, a decreased tendency for reduction of HNO_3 to NOx on soot, and, as will be discussed in a separate chapter, a general increase ice-forming activity for aircraft emissions. This raises the importance of heterogeneous chemistry to reduce NOx, and increase the importance of HOx and halogens, in persistent contrails.

2.c.II. Sources of NOx and HOx in the Upper Troposphere

Motivated by the dominant role placed on NOx and HOx by previous aircraft assessments [Brasseur et al., 1998; IPCC 1999], the past decade has been witness to a multitude of studies to attribute sources of these species in the upper troposphere, especially those that could potentially be due to aircraft. A brief review of some important new results is presented below.

Sources of NOx

The main source of NOx in the stratosphere is oxidation of N_2O, and based on tight correlations that have been observed between NOy (the sum of reactive nitrogen species) and N_2O, it is relatively straightforward to simulate the impact of an additional source of NOx from direct injection of aircraft exhaust or parameterized transport from the troposphere [IPCC 1999]. However, there are a number of potentially significant sources of NOx to the upper troposphere, not just those from aircraft emissions, all of which must be reasonably well understood in order to determine the perturbation of NOx due to aircraft [IPCC 1999]. Of these non-aircraft sources, lightning and convective transport from the boundary layer have stood out as dominant sources of NOx in the UT [Grewe et al., 2002]. The studies are too numerous to describe here, but we summarize a few key results that have emerged from these studies that significantly improve our understanding of NOx sources.

Around the time of the 1999 IPCC assessment, lightning was estimated to represent a source strength of about 3-5 Tg(N) yr-1. In a comprehesive review of three decades of research on this topic, Schumann and Huntrieser [2007] have concluded that the best estimate for the annual lightning NOx source is 5±3 Tg(N) yr^{-1}. Consistent with this, in a recent study using a combination of space-based NO2 observations from SCIAMACHY, O_3 observations from OMI and MLS, and HN O_3 observations from ACE-FTS, Martin et al. [2007] determine a range of 6±2 Tg(N) yr^{-1} for the lightning NOx sources. For reference, such a source-strength is about 8-10 times larger than the estimated NOx source from aircraft emissions [Kraabol et al., 2002] but only about $1/8^{th}$ of the total NOx source strength assumed in state-of-the-art aircraft NOx emissions impacts studies [e.g., Gauss et al., 2006].

It is important to note that aircraft emissions are more confined in altitude and to heavily traveled corridors than these other sources, so they can still represent a large local perturbation. What makes assessing aircraft contributions so difficult, then, is not only the quantification of these larger global sources, but specifying their geographic distributions with sufficient precision so that the contributions due to the highly localized aircraft emissions can be quantified. In other words, the large, distributed sources determine the broader background abundances of NOx into which the aircraft emissions represent a highly localized perturbation. Thus, studies addressing the contributions of various sources of NOx (or NOy) to the UT are critical for evaluating the significance of that due to aircraft.

Source Attribution of NOx in the Upper Troposphere

Singh et al. [2007] analyzed observations of reactive nitrogen species in the UT over North America in the summer of 2004, reporting that ~30% of the NOy in the UT is in the form of NOx. PAN and HNO_3 were the dominant reservoirs of reactive nitrogen in the UT and LS, respectively. Relying on tracers for biomass burning emissions (e.g., HCN) and anthropogenic pollution, they concluded that lightning represents a larger source of NOx to that region than was believed previously. Model simulations based on these observations

[Hudson et al., 2005] imply that lightning was responsible for approximately 75% of the NOx observed in this region. These results suggest that the NOx observed in this region is relatively 'fresh', that is, it is undergoing photochemical aging (e.g., oxidizing). Consistent with this, Sioris et al. [2007] reported large local NO_2 enhancements at ~10 km that they attributed to lightning, estimating that it is responsible for 60% of the upper tropospheric NO_2 in the tropics.

Bertram et al. [2007] develop the idea of a 'photochemical clock', using the ratio of observed NOx to that determined with a photochemical model with similar total NOy - 16 - (i.e., NOx_{obs}/NOx_{ss}) to estimate that ~17% of the air in the UT under the conditions sampled was transported from the planetary boundary layer. Furthermore, they estimate a turnover rate by convection of 0.1 day^{-1} for air in the UT (although it should be noted that this is includes altitudes somewhat below typical aircraft cruise altitudes).

These results suggest that non-aircraft sources of NOx to the upper troposphere are more important than previously believed, consistent with the observations of Klemm et al., [1998], who found that clear perturbations due to aircraft in the northest Atlantic corridor were difficult to identify on scales larger than a few km due to natural variability, whereas in 'fresh' plumes between 15 and 90 minutes in age, enhancements of up to 10 ppb were observed. Based on NOy/O_3 correlations, Koike et al. [2000] estimated that the mean NOy enhancement in the North Atlantic corridor is of order 70 ppt at 11 km, implying NOx enhancements of about 40% above backgrounds. They also found the NOy enhancements to increase with increasing ozone (e.g., closer to the chemical tropopause). Given the more recent observations of Singh et al. [2007] of significant transport from the surface, Koike et al. [2000] may have significantly overestimated the NOx contributions from aircraft.

Sources of HOx

Not only does OH largely determine the lifetime of methane, a greenhouse gas that plays a key role in the Aircraft RF uncertainties framework (figure 1, [Sausen et al., 2005]), both OH and HO_2 participate in catalytic cycles that destroy ozone and are necessary for ozone production. Therefore, models must be able to reproduce both total HOx abundances and the partitioning within the HOx family (the generally preferred indicator being the OH/HO_2 ratio) over a wide range of conditions found in the UT/LS.

Measurements of HOx carried out in the 1990s revealed significantly larger abundances of this critical oxidizer than could be modeled with assumed sources [e.g., see Faloona et al., 2000]. By the time of the 1999 IPCC assessment, it was well known that sources of HOx in addition to H_2O/O_3 photochemistry were required to resolve this discrepancy, especially in the upper troposphere [Collins et al., 1999; Crawford et al., 1999; Muller and Brasseur, 1999; Reiner et al., 1999; Singh et al., 2000]. Since then, a number of ongoing studies related to sources of HOx have been published, and models for assessing aircraft impacts have used any available in situ observations to constrain parameterizations of HOx, including measurements of species such as H_2O_2, whose abundances serve as sensitive indicators of HOx chemistry [Brunner et al., 2005]. The basic understanding of HOx chemistry seems to be relatively sound, in that it is widely acknowledged that additional sources, generally gases transported from the PBL by convection (in agreement with the conclusions based on NOx partitioning described above), are required to fully explain HOx abundances. The partitioning between OH and HO_2 varies with NOx in a fashion that can be reproduced reasonably well by models [for example, see Brunner et al., 2005, Ren et al., 2008, and references therein]. Figure 3 shows comparison of OH measurements from recent missions with modeled OH abundances, indicating good agreement over a wide altitude range [Ren et al., 2008].

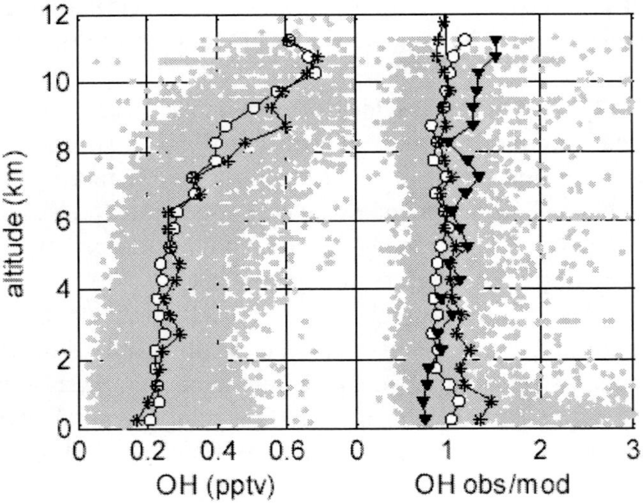

Figure 3. (left panel) Comparison of the median vertical profiles of measured (circles) and modeled (stars) of OH for INTEX-A. (right panel) Measured-to-modeled OH in INTEX-A (circles), TRACE-P (stars) and PEM Tropics B (triangles). Individual 1- minute measurements from INTEX-A are shown (gray dots) [from Ren et al., 2008].

The results shown in figure 3 indicate that there should be a firm basis for model simulations of OH distributions over a wide range of conditions, as is required to predict the lifetime of CH_4 to a reasonable degree of accuracy. However, important model-measurement discrepancies remain in modeling the partitioning of OH and HO_2 that are not well understood, as will be discussed in Section 3 [Hudman et al., 2006; Ren et al., 2008]. One of the challenges in comparing modeled and measured HOx is the inherent non-linearities in HOx chemistry; in essence, unless the photochemical conditions are highly uniform during sampling, some differences in modeled and measured total HOx or OH/HO_2 can be due simply to the coarse temporal resolution of the model. As shown by Olson et al. [2006], such errors are most problematic at high solar zenith angles and at high and variable NOx conditions. In light of the significant role that heterogeneous chemistry plays in the effect of NOx on ozone in the UT, this type of issue could become very important in future assessments of aircraft impacts.

There are several implications of the results highlighted above that are worth noting here. First, the increased role of convection from the PBL to sources of NOx and HOx to the upper troposphere reduces the significance of aircraft perturbations of these species or their precursors. Thus, it is likely that model simulations used in prior assessments, updated to reflect these new observations, would find the impacts of aircraft emissions to ozone and methane in the UT/LS to be diminished. However, increased transport of short-lived species from the PBL also implies increased production of aerosols in the UT due to oxidation of these gases into less volatile products. Second, increased 'aging' of UT air results in a shift in the partitioning from NOx to NOy. As discussed in the following sections, this has important implications for the role of long-lived reservoirs of nitrogen oxides in particle stability. Heterogeneous reactions are effective in denoxifying cold, particle rich regions of the atmosphere, such as where persistent contrails are formed. Thus, increased transport from the PBL implies a greater role for ozone-destroying reactions of HOx and halogen radical species that are normally kept in lower abundances by NOx.

2.c.III. Conversion of NOx to NOy

The laboratory finding that uptake of nitric acid on aircraft kerosene soot is reversible [Talukdar et al., 2006] implies that emissions of soot will not shift the partitioning of NOy to NOx in aircraft plumes, as was believed previously. This result, together with new measurements of the hygroscopicity of soot and the subsequent formation of CCN and emissions of particles from engines (e.g., see Section 2.c.I. and SSWPs III and IV), implies, rather, that in plumes, contrails, and potentially even in heavily traveled flight corridors, there will be more rapid conversion of NOx to NOy. Although the impacts of these new findings have yet to be fully explored, results from recent modeling efforts provide clues as to what might be the tendencies.

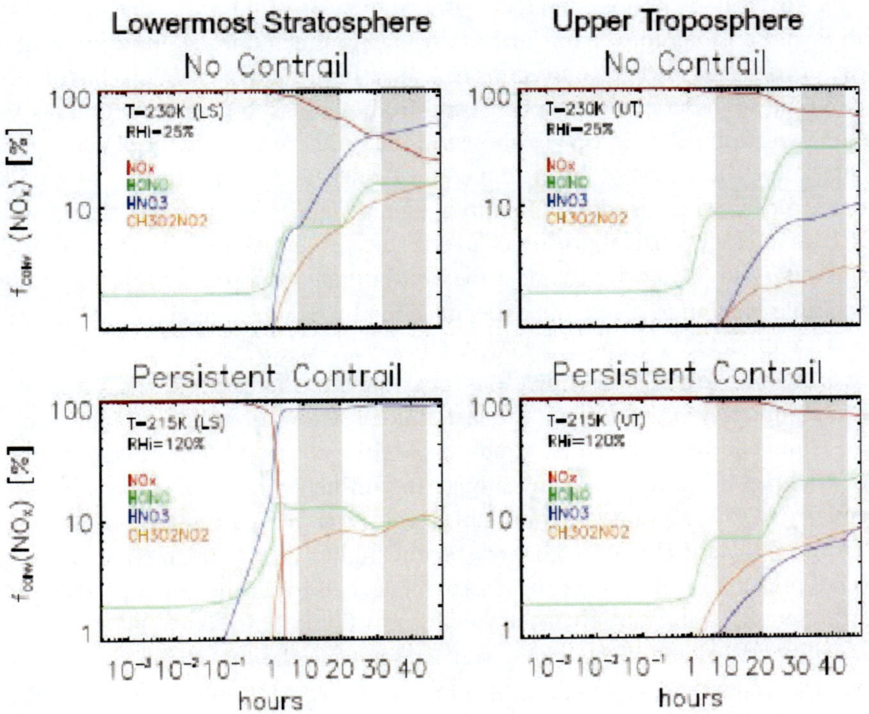

Figure 4. Model results from Meilinger et al. [2005] showing the impact of heterogeneous processing of NOx in a persistent contrail in the lower stratosphere (left panels) and in the upper troposphere (right panels). Shaded regions refer to nighttime.

A modeling study by Kraabol et al. [2002] found that reactions that form odd-nitrogen reservoirs in aircraft plumes and persistent contrails reduce the magnitude of changes in ozone as a result of the conversion of ~25-35% of the aircraft NOx to NOy. A subsequent study by Meilinger et al. [2005] found that NOy formation depends very strongly on heterogeneous reactions, especially in the lowermost stratosphere. Figure shows the shift in NOy partitioning due to heterogeneous chemistry in a persistent contrail. In the lowermost stratosphere, NOx is completely converted to HNO_3 in a matter of hours, whereas without a contrail, even after a few days, conversion of NOx to NOy is only 50%. According to Meilinger et al., in the lower stratosphere, ozone destruction by chlorine and bromine enhances that due to $NO+O_3$ in the early plume and dominates over NOx-induced ozone production in the aged plume. This is the result of combined effects of halogen activation and denoxification by heterogeneous reactions on contrail ice particles. The situation in the upper troposphere is less clear, and the tendency of ozone depends strongly on temperatures in the initial plume and persistent contrail. However, reductions in net ozone production or shifts from ozone production to loss result from the more complete treatment of heterogeneous chemistry. The recent modeling study of Sovde et al. [2007] examines the global implications of heterogeneous reactions on the ozone changes induced by aircraft exhaust products. Although they focus on the impacts of a mixed fleet for the year 2050, there are some important new conclusions that extend the results of Meilinger et al. [2005] to hemispheric scales. (It is also important to note that even in a mixed fleet, operations of subsonic aircraft dominate the overall emissions budget). As shown in figure 5, the most significant implication of more rapid conversion of NOx to NOy is the complete reversal in the sign of the response of ozone to nitrogen emissions (e.g., see figure 2) from net production to net loss below 18 km (i.e., in the upper troposphere) and from net loss to net production above 24 km. Although the two ozone change curves shown in the right panel of figure 2 and figure 5 have similar shapes, they are nearly mirror images of one another, as figure 2 deals with the quantity ozone loss, whereas figure 5 shows ozone gain, with altitude. Using reasonable estimates for an average vertical profile of ozone, the percent change in ozone near 25 km in figure 5 is about +2 to +4%, whereas near the mid-latitude tropopause (12-16 km) the change is of comparable magnitude, but opposite in sign. In essence, one could achieve similar changes to those modeled in figure 2 by *decreasing* NOx by ~10%.

It is worthwhile to consider how it is possible for the sign of the impacts of NOx emissions to completely reverse since the last major reviews of aviation (and even the 2006 Workshop). Hints can be found in the study by Meilinger et al. [2005] discussed above and one by Hendricks et al. [2000] who investigate the influence of naturally occurring bromine on the chemistry of aircraft emissions in the UT/LS. First, the partitioning of NOx emissions is shifted far more toward HNO_3 in the more recent studies than in the model used to generate figure 2 (and presumably the state-of-the-art models used at the time of the 1999 IPCC Assessment). Second, (and largely a consequence of this shift from NOx to HNO_3) the relative contributions of the NOx, HOx, and halogen families to ozone loss in the UT/LS differ in the more recent model simulations from those used for previous assessments.

Hendricks et al. [2000] found the somewhat surprising result that bromine radicals, even at the minor abundances that are thought to be present in the UT/LS, efficiently convert NOx to NOy by heterogeneous hydrolysis of $BrONO_2$ on background and aircraft-produced aerosols. They showed that this process can even be an important pathway for denoxification in the lowermost stratosphere. .As noted by Meilinger et al. [2005], such halogen chemistry becomes significantly more important in exhaust-influenced air in the plumes of aircraft, in cirrus, and in persistent contrails.

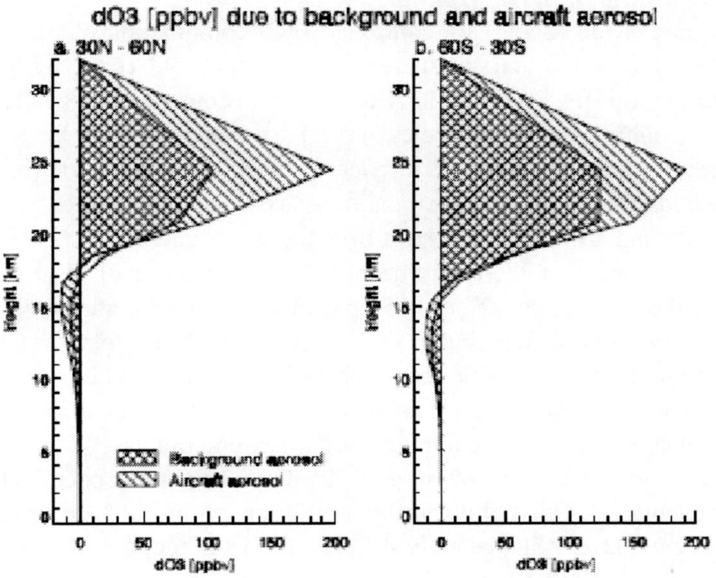

Figure 5. Vertical profile of the zonally averaged response of ozone to aircraft emissions of NOx assuming background aerosols and aircraft aerosol perturbations for a 2050 Mixed Fleet, as described in Sovde et al. [2007].

This issue will be addressed in more detail in Section 3, since the role of halogens in aviation impacts has received little attention and remains one of the major uncertainties in UT/LS chemistry.

Halogen chemistry may not be dominant throughout the UT/LS, but it is important to note that even a few tens of parts per trillion, background abundances of halogens are sufficient to compete with (and even dominate in some regions) HOx- and NOx-catalyzed destruction of ozone in the UT/LS. The non-linear coupling between HOx, NOx, and halogen oxides makes the assessment of the impacts of emissions of any specie that influences abundances of just one of these families very difficult to assess unless we have a solid quantitative understanding of each of the major ozone-destroying radical's response to changes in the abundances of the others. Although such an understanding has been achieved for the middle-to-upper stratosphere, the situation is less clear for the lowermost stratosphere and upper troposphere, especially for the reactive halogen species, abundances of which are so strongly modulated by heterogeneous processes. Given the additional complication of non-linearities in particle formation, composition, and heterogeneous reaction rates with respect to relative humidity, temperature, and abundances of H_2O and HNO_3, the details of plume formation and dispersion, particle growth, composition, and sedimentation, and the ability to predict the presence of ice crystals in the UT/LS all become essential factors in assessing the chemistry of aircraft exhaust. In light of the clearly dominant role played by water vapor in all of these issues, the next section will examine progress in understanding water vapor in the UT/LS.

2.c.IV. Water Vapor and Supersaturation

H_2O abundances in the UT/LS are controlled by a combination of transport processes. Both large- (e.g., Brewer-Dobson circulation) and small-scale (e.g., waves, convection) processes are important [IPCC 1999; SPARC 2000]. Temperature, chemistry (e.g., CH_4 oxidation) and microphysics also play roles. Transport phenomena are key elements in UT water distribution; these include such occurrences as horizontal transport from the tropics to sub-tropics and midlatitudes and vertical motions associated with mesoscale convection, midlatitude cyclones and downward transport from the stratosphere.

SPARC [2000] noted that there has been a 2 ppm increase of H_2O (~1%/yr) in the stratosphere since the mid-1950s, about 0.55 ppm of which can be attributed to increases in CH_4, while the source of the remaining ~1.5 ppm (75% of the total) remains unknown. Trends in relative humidity in the upper troposphere have been found in some latitude bands, but there is no apparent global trend; variability from ENSO, large-scale circulation modes and temperature all contribute to the complexity of attributing trends.

Agreement amongst measurements of H_2O in the lower stratosphere (60-100 mb) has always been problematic. Although typically clustering within 10% of each other, some individual instruments have systematically differed from the mode of the measurements by 25-30%. The source of this disagreement is under investigation.

Water measurements in the upper troposphere are less numerous than those in the stratosphere, and they are less reliable overall. Radiosonde data are not sufficiently accurate for determining trends at the level of importance for understanding perturbations by aircraft. Measurements from TOVS are reasonable, on average, but very difficult to validate because of the high temporal and spatial variability of H_2O vapor in the UT. The measurement of tropospheric water vapor amounts via radio occultation of Global Positioning Satellite (GPS) signals has become a fairly mature technique, and methods for determining vertical profiles of water with high vertical resolution (a few hundred meters) are under development [e.g., Troller et al., 2006].

Since the last water vapor assessment [SPARC 2000], a number of uncertainties relevant to aircraft impacts have been addressed in some detail, as described below: Intercomparison experiments and laboratory work for stratospheric water vapour instruments have been ongoing; validation of satellite H_2O retrievals and numerous correlative measurements have been conducted; improvements in radiosonde H_2O measurements have been made; a number of process studies have been conducted to investigate the role of convection and cloud microphysical properties in UT/LS H_2O distributions and studies of stratosphere-troposphere exchange mechanisms.

Intercomparison and Validation

Detailed intercomparisons of lidar, radiosondes, and frost-point sensors (AFWEX) revealed that the frost-point/chilled mirror measurements are "drier" (i.e., lower water vapor) than the others by 10-25% in the UTLS [Ferrare et al., 2004]. During the 2003 AWEX-G campaign, (designed to validate the AIRS measurements from the A-train satellites), six radiosonde-type sensors were flown against the University of Colorado Cryogenic Frostpoint Hygrometer (CFH). With appropriate corrections for solar heating, data from the Vaisala RS-90 sensor was found to be suitably accurate for use in validation studies [Miloshevich et al., 2006].

Intercomparisons between the satellite-based POAM measurement (solar occulation) and the in situ MOZAIC data set showed that POAM water vapor values are about 10% higher than those determined with capacitive humidity sensors flown on several in-service aircraft [Nedoluha et al., 2002]. Finally, based on comparisons made during the SONEX and POLINAT campaigns in 1997, Tunable Diode Laser (TDL) and cryogenic hygrometers were found to agree to within their stated instrumental accuracies of 10% [Vay et al., 2000], whereas a similar intercomparison conducted between the POLINAT and MOZAIC datasets found water vapor measurements to agree within 5% [Helten et al., 1999]. However the agreement between measured values of relative humidity was worse, potentially pointing to temperature measurement problems.

Perhaps of most significance for this White Paper will be the upcoming results from the AquaVIT blind intercomparison that was carried out at the AIDA chamber in Karlsruhe in Fall 2007 (http://imk-aida.fzk.de/campaigns/RH01/Water-Intercomparison-www.htm). This formal program brought together more than twenty instruments that measure water vapor and/or condensed water for a two-week measurement campaign. The results of a formal blind intercomparison among a subset of the instruments are due out Spring 2008, and should elucidate some of the reasons why water vapor measurements in the cold, dry UTLS have disagreed to a level that is greater than their reported uncertainties.

Observations in UT

Observations of relative humidity over ice (RHi) and supersaturation in the upper troposphere have been analyzed in detail, and both radiosonde measurements and those derived from the chilled-mirror "SnowWhite" frost point hygrometer show frequent supersaturation with respect to ice during wintertime (24% of time) [Vaughan et al., 2005]. Data from MLS show occurrences of high supersaturations in only about 0.5% of observations overall, with considerably larger frequencies of occurrence found over Antarctica [Spichtinger et al., 2002]. Only one direct observation of RHi relevant for assessing supersaturation in an aircraft-related contrail has been reported. Gao et al. [2005] argued that the high supersaturations they observed might be due to co-condensation of other species (e.g., HNO_3) in cloud particles.

Climatology/Mechanistic Studies

Ten years of MOZAIC data have been compiled to relate UT water to deep convection and moisture transport [Luo et al., 2007]. Interannual variability is observed to correlate in some cases with average temperature and/or ENSO, but is not fully explained by either. Regional differences are well-explained by convective frequency. However, no trend in H_2O abundances has been found in the MOZAIC data over the period Aug 1994 to Dec 2003 [Bortz et al., 2006].

Comparison of global or mechanistic model results with observations can also provide insight into the significance of various transport processes for determining the water vapor distribution. For example, MOZART model results and HALOE water vapor data are in good agreement with respect to the seasonal cycle of vertical transport (the so-called "tape recorder"), but some significant differences exist in distributions around the tropopause [Park et al., 2004]. Much of this difference is attributed to the model's treatment of moisture transport in the monsoon regions, as well as stratosphere-troposphere exchange in those areas.

Similar results were obtained when comparing simulations from the NCAR Community Atmosphere Model (CAM 3.0) to HALOE observations and reanalyses by ECMWF. Deficiencies in the calculation of stratospheric water vapor are attributed to weaknesses in the model's core stratospheric dynamics, in particular, the lack of a QBO and crude representation of planetary waves [Gulstad and Isaksen, 2007]. The authors also note the importance of the model's temperature fields, which continue to show a polar cold bias; this particularly affects water vapour distributions in the southern hemisphere.

To date, mechanistic model simulations have focused on the representation of water vapor in the tropics. For example, tropical climatologies of H_2O, CO, HNO_3 and O_3 are compared to calculations of vertical profiles of the same species obtained from four models with differing parameterizations of convection [Folkins et al., 2006]. No single model/parameterization emerged as "best", with each having some failings in its ability to reproduce observations. Comparisons of balloon-borne water vapor observations over Brazil with profiles calculated by the Brazilian Regional Atmospheric Modeling System (BRAMS) and ECMWF global analyses illustrate the importance of both model vertical resolution and the treatment of microphysics in the ability to calculate realistic water vapor profiles [Marécal et al., 2007].

2.c.V. UT/LS Temperatures

Atmospheric temperature is a fundamental quantity in all areas that this SSWP considers – gas-phase and heterogeneous chemistry, the formation and persistence of condensed matter (e.g., cirrus, contrails, polar stratospheric clouds), and transport processes. Thus uncertainties in our knowledge of the mean temperature in the UT/LS, as well as its natural variability, impact a wide range of processes important for understanding the impacts of aircraft emissions on climate. Furthermore, the inability of models to adequately simulate the temperatures in the atmospheric regions of interest may have significant impacts on their treatment of heterogeneous processes and parameterizations of microphysics (in addition to the role of temperature in model dynamics, such as the classic GCM "cold pole" problem). A review of the temperature trends associated with the broader climate change issue is beyond the scope of this document and controversies surrounding the temperature record for the surface and mid-troposphere will not be discussed.

A comprehensive review of temperature trends in the stratosphere was published in 2001 [Ramaswamy et al., 2001]. This work indicated that temperature trends in the lower stratosphere were negative (-0.5 ± 0.25 $_o$C/decade) and consistent with known trends in stratospheric ozone as well as other greenhouse gases. These authors noted, however, that better knowledge of the vertical profiles of ozone and water vapor, and their changes, throughout the upper troposphere and lower stratosphere were critical for proper attribution of the observed temperature changes. Stratospheric temperature trends updated through 2005 are presented in Chapter 5 of WMO [2006] and are consistent with those reported earlier.

Similar exhaustive trend studies for the UT/LS have not been carried out, although data for this region do exist (from radiosondes, satellites and even in-service aircraft). One regional study [Bencherif et al., 2006] uses radiosonde data gathered over South Africa to show that temperatures are decreasing throughout the UT/LS (200 hPa and altitudes above) between 1980 and 2001. In that region, upper tropospheric temperatures have decreased at a rate of -0.10 ± 0.18 $_o$C/decade, a value similar to that reported by Parker et al. [1997] for an analysis based on globally gridded radiosonde observations.

Sensitivity of the rates of chemical processes to temperature can be significant. In general an error of a degree or two makes little difference in the rate of a gas-phase process; however, the same cannot be said for heterogeneous chemical transformations. The composition of condensed phases is often a strong function of temperature, as is the threshold for condensation. For example, at 200 K, a 1-K change in temperature changes the saturation vapor pressure of water over ice by approximately 15%. When coupled to uncertainties in water vapor measurements, errors in temperature observations or calculations can have dramatic impacts on the determination of conditions such as supersaturation or the presence of polar stratospheric clouds, and hence, chlorine activation.

2.d. Present State of Measurements and Data Analysis

To understand the photochemistry of ozone in the UT/LS, it is important to know the distributions of the major species that produce ozone (HOx, NOx, and hydrocarbons) and those that destroy it (HOx, ClOx, BrOx, and NOx). Due to the strong coupling between species within the radical families and between species from different families, it is not necessary to measure all of the important species simultaneously. However, it is important to have a good understanding of interrelationships between the major ozone-forming/destroying radicals under the wide range of conditions that prevail where aircraft emissions can be found. This includes temperatures that can range from ~190-240 K, solar zenith angles from 0 degrees to greater than 90 degrees, and ozone abundances that range from tens to thousands of ppb.

Not only is it a primary emission product of combustion, NOx has a controlling influence on partitioning within the HOx and halogen families. Therefore, measurements of NOx in the UT/LS are important for defining the range of variation of the other ozone-controlling radicals. Results from a number of major aircraft campaigns, some designed to validate new orbiting platforms, as well as routine measurements from commercial airliners equipped with instrumentation, have provided a wealth of information relevant for understanding oxidation, as well as ozone formation and loss, in the UT/LS. The results summarized in section 2.c for UT NOx and HOx chemistries have provided a strong foundation for new modeling studies to address the impacts of NOx emissions on ozone and methane in the broader upper troposphere and lower stratosphere. However, new results pointing to a reversal in the impacts on ozone in aircraft contrails and cirrus clouds raises important questions about the completeness of the measurements. Unfortunately, observations in regions of low NOx have not been a major priority of recent aircraft campaigns, and key satellite instruments do not have sufficient vertical or horizontal resolution to examine these kinds of issues in narrow regions where heterogeneous chemistry could play a dominant role .

Our understanding of the distribution of water vapor and the processes that control it remains problematic. In the regions where heterogeneous chemistry would be most important (i.e., at or near the tropopause), long-standing discrepancies between measurements makes it extremely difficult to predict the chemical response to any perturbation, let alone one that includes potential ice nuclei, water vapor and important co-condensable species such as nitric acid, plus species that can inhibit ice formation (such as volatile organic compounds). Although this issue is addressed in detail in another SSWP in the context of cirrus and persistent contrail formation, the critical role that these observations play in allowing for the prediction of the reactivities of particles and, hence, their importance to this SSWP, cannot be understated. Resolution of this problem is critical for assessing the impacts of aircraft emissions on particle formation, heterogeneous chemistry, redistribution of condensable species, transport of emissions to the stratosphere, and production of HOx. Currently, the reported differences of up to 30% between widely respected measurements is unacceptable, especially when they imply strange behavior for particles that could change our fundamental view of the nature of aerosols and clouds [e.g., cubic ice, nitric acid antifreeze, and very large supersaturations].

Important results on water vapor measurements are expected in 2008 from the recent AquaVIT intercomparison discussed in Section 2.c.IV.; however, it is important to note that laboratory intercomparisons of the same or similar instruments have been carried out before, and while they have answered some questions, they have largely been unsuccessful at resolving the major discrepancies in the atmospheric measurements themselves. Consequently, the state of *agreement* among water vapor measurements remains inadequate for assessing the key remaining aviation impacts issues, even though the instruments themselves may be in a mature state.

A new approach to water vapor intercomparisons would be welcome. One approach that could be promising - dedicated flights into the combustion plumes of rockets and aircraft - is described in more detail below. In 2008, potentially important results will be forthcoming from a small pilot program called "PUMA" (Plume Ultrafast Mesaurements Acquisition) that explore the nature of the discrepancy between water vapour measurements in the UT/LS and the implications of heterogeneity on interpretations of non-linear processes (such as threshold behaviors for condensation and evaporation of ice, HOx and halogen photochemistry, and redistribution of major species, such as H_2O and NOx). Preliminary analyses of H_2O and particulate water data in evaporating plumes are quite promising, and indicate that future measurements in these environments could play a critical role in validating the accuracies of water vapor measurements. An interesting question raised by these studies is whether the highly perturbed plumes represent a realistic environment for investigating fundamental photochemical and dynamical issues important in the UT/LS. From the point of view of the assessment of aircraft emissions, it would seem that such environments, especially the plumes and persistent contrails produced by aircraft themselves, would be ideal natural 'laboratories' for studying important processes identified in these SSWPs. In addition, there are some who argue that pushing measurements outside their normal dynamic range is one sure way to find problems that might help in identifying those issues that are important under more normal conditions.

Finally, it is important to note here that satellite observations, with a few noteworthy exceptions, have not yet been a major driving force in refining our understanding of aircraft impacts. However, following completion of validation activities, new results from the AURA platform, as well as those from SCIAMACHY, ACE, etc., will be analyzed in light of the issues raised here and in previous assessments. It is very likely that significant new insights into convective sources of NOx and NOy, HOx, and aerosols will be forthcoming from analyses of observations made from numerous satellite platforms. Such results will be especially important in defining the basic state of the UT/LS into which aircraft emissions represent a small, but important, perturbation.

2.f. Current estimates of climate impacts and uncertainties

Since the IPPC [1999] Assessment and the Sausen [2005] of the Brasseur et al. [1998] European Report, there are no direct comparative model studies that address current estimates of climate impacts and uncertainties. However, on the basis of the new results presented above, some general conclusions can be drawn. First, on the basis of improved understanding of upper tropospheric sources of NOx, in particular, due to lightning and convection from the PBL, it can be interred that the climate impact of aircraft emissions on regional and global scales will be reduced. Second, on the basis of studies showing an increased sensitivity of NOx and NOy to heterogeneous chemistry, it is likely that for subsonic emissions there will be regions of the atmosphere where aircraft NOx and particles may, in fact, result in ozone losses, especially in the tropopause and LS regions. On the basis of this result, one would expect the climate impacts of subsonic aircraft emissions to be smaller than previously believed, and possibly reversed in sign relative to previous evaluations (e.g., negative instead of positive), whereas the impacts of supersonic emissions would be greater than previously believed, and positive instead of negative. Third, the observation of nitric acid-containing particles in the UT/LS, along with measurements indicating more vigorous transport of NOx from the surface, raises the possibility that NOx and NOy are processed more rapidly in the UT/LS than previously believed. Finally, the presence of reactive halogens in the UT/LS, species that, at the abundances that have been observed, can only coexist with NOx if there is rapid heterogeneous processing, raises the possibility for highly non-linear photochemistry that can result in a net positive or net negative change in ozone with aircraft emissions of NOx and particles.

It is likely that future studies of the climate impacts of subsonic aircraft emissions that have more realistic treatments of lightning and convective sources of NOx, more complete treatments of redistribution of NOy, especially in persistent contrails, and heterogeneous halogen chemistry will find that the climate impacts are reduced, or even reversed in sign (i.e. ozone losses due to aircraft) in the UT/LS. This possibility, calls into question the uncertainties ascribed to the chemistry of NOx emissions by aircraft. Our understanding probably remains as "fair", until new CTM studies can be carried out, but the magnitudes of the error bar placed on the RF terms in figure 1 may be too small, and may need to accommodate a reverse in sign, at least until the implications of these new results can be properly assessed with new model studies.

The growing body of HOx observations in the UT indicates that OH abundances are at the high end of most model predictions, resulting in a lower lifetime for methane in the UT. This implies that, at least in these regions, methane will have a greater sensitivity to perturbations on NOx and aerosols due to aircraft emissions. In addition, the potential for an increased role in halogen chemistry in cirrus and persistent contrails raises the possibility that aircraft perturbations to methane may currently be underestimated, as the reaction of methane with chlorine atoms is likely to be more important in the UT/LS than is currently believed.

2.g. Interconnectivity with other SSWP theme areas

As has been discussed earlier, the chemistry of aircraft emissions is highly non-linear and strongly coupled with important processes dealt with in other SSWPs, including formation of persistent contrails and cirrus. Furthermore, and potentially more problematic for assessing impacts, emissions of NOx could alter redistribution of NOy and water, not only from aircraft exhaust, but from the background atmosphere as well if the addition of NOx results in enhanced large-particle stability and sedimentation. It is also possible for NOx influences to impact transport of NOy and H_2O (although the latter may be too small to matter) from the UT into the LS. As noted in Section 2.d., the greatest uncertainty for this SSWP is due to the implications of continuing discrepancies in water vapor measurements in the cold and dry regions of the UT/LS. Thus, there is a strong interconnection between this SSWP and those on particle microphysics and contrail and cirrus cloud formation.

3. Outstanding Issues

Progress made in areas highlighted in Section 2.c., especially that relating to the importance of heterogeneous chemistry, raise new questions about the fundamental chemistries of NOx, HOx, and halogens, and the interactions of ice and nitric acid in the UT/LS, all which can have important consequences in future assessments of aviation impacts. Key new findings in these areas are summarized in Section 3.a. Although their impacts have not yet been adequately assessed, their tendency to push the effects of aviation emissions in the same general direction that has been found in model studies summarized in Section 2.c.III. is somewhat troublesome, in that they have the ability to offset some of the advances that have occurred over the past decade.

3.a. Science

The key developments in UT chemistry summarized in Section 2.c. place considerably more emphasis on the role of heterogeneous chemistry of non-aircraft species, such as the halogens, on understanding the distributions of background H_2O and nitrogen oxides, and on the need for new studies that address chemical heterogeneities of the UT/LS. One of the interesting consequences of the increased importance of heterogeneous processes is the change in sign of ozone response with NOx perturbation described earlier. This section will highlight important issues listed in the 2006 Workshop [Wuebbles et al., 2006] that remain unresolved, and new findings that raise new questions about chemistry in the UT that must be understood before uncertainties in the impacts of aircraft emissions on chemistry in the UT/LS can be reduced further.

3.a.I Discrepancies in Coupled HOx and NOx Chemistry

The ability to realistically simulate ozone production and loss and the coupling between CH4, CO, OH, and O_3 relies upon an accurate model representation of the response of HOx (and, to a lesser extent, halogen radicals) to variations in NOx. There have been a significant number of campaigns where NO, NO2, OH, HO2, and ozone have been measured simultaneously, and the first-order linkages between the NOx and HOx families have been demonstrated. However, model comparisons with HOx observations have been somewhat problematic [Faloona et al., 2000]. Olson et al. [2006] show that most of the previous model-measurements discrepancies at high NOx (e.g. during SONEX) can be explained by non-linearities of HOx chemistry under highly variable conditions for NOx (i.e., the model timescales are too long, relative to the measurements, such that averages of derived quantities do not represent quantities derived from averages of the individual measurements – see also Wild and Prather [2006]).

Despite considerable progress that has been made in the area of tropospheric HOx chemistry, as noted in two very recent papers [Hudman et al., 2007; Ren et al., 2008], observations continue to highlight important discrepancies between models and measurements. Figure 6 taken from Ren et al. [2008] shows how well models agree with measurements of HOx during three recent major field campaigns for which there were comprehensive suites of measurements of sources of HOx. The agreement between modeled and measured OH is quite good over most of the range, except, perhaps, at the very highest NO where a slight underprediction develops for INTEX-A (where the highest NO values were observed). However, at high NO, measured HOx exceeds that from the model by as must as an order of magnitude at highest NO. Further insight into this issue is gained by examining the altitude dependence of the discrepancy, as shown in figure 7. Clearly these results are problematic for assessments of the impacts of aviation, since high NOx abundances can develop in heavily traveled flight corridors [e.g., see IPCC 1999]. The reasons for these discrepancies remain elusive. However, new observations of a critical species, pernitric acid (HO2NO2), whose abundance is determined by the coupled photochemistry of HOx and NOx, may help provide some answers [Murphy et al., 2002; Kim et al., 2007]. In a new report of simultaneous in situ observations of HO2NO2, NO2, and HO2, at aircraft cruise altitudes, Kim et al. [2007] found that abundances of HO2NO2 were about a factor-of-two low than those calculated with assumed photochemistry and observed abundances of HO2 and NO2. This discrepancy can be reconciled if one of the measurements (most likely HO2NO2 or HO2) were in error (too small or too large, respectively). However, it is interesting to note that the trend in this discrepancy with altitude is similar to that of figure 7, raising the possibility of missing or poorly understood chemistry coupling HOx to NOx in the relatively cold and dry upper troposphere. It is particularly problematic for assessments of aircraft emissions that the discrepancy is largest at cruising altitudes for most large subsonic aircraft.

3.a.II. Halogen Chemisty

In any modeling study of the impacts of a perturbation, it is important to start with a correct description of the composition of the background atmosphere. In previous aircraft assessments [Brasseur et al., 1998; IPCC 1999] it has been assumed that reactive halogens are not present in sufficient abundances to significantly impact ozone chemistry.

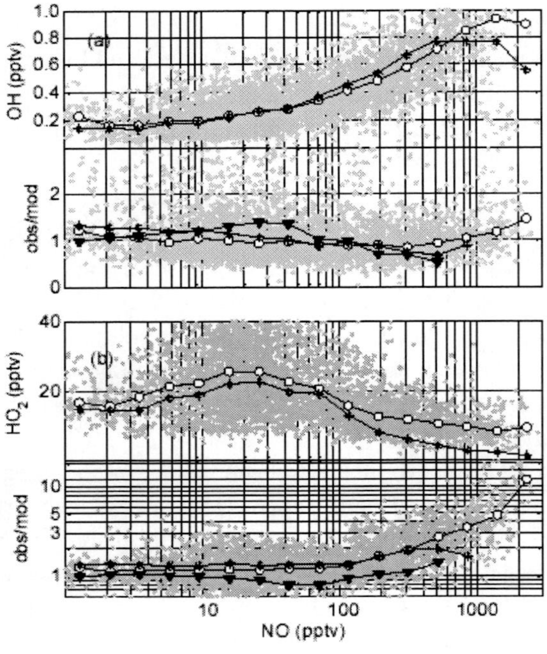

Figure 6. (a) Comparison of NO dependence for observations of OH (upper panel) and the ratio of measured-to-measured OH (lower panel). (b) Comparison of NO dependence for observations of HO2 (upper panel) values and the ratio of measured-to-modeled HO2 from INTEX-A (circles), TRACE-P (stars) and PEM Tropics B (triangles). Individual INTEX-A 1-minute measurements are shown (gray dots). All lines show the median profiles [from Ren et al., 2008].

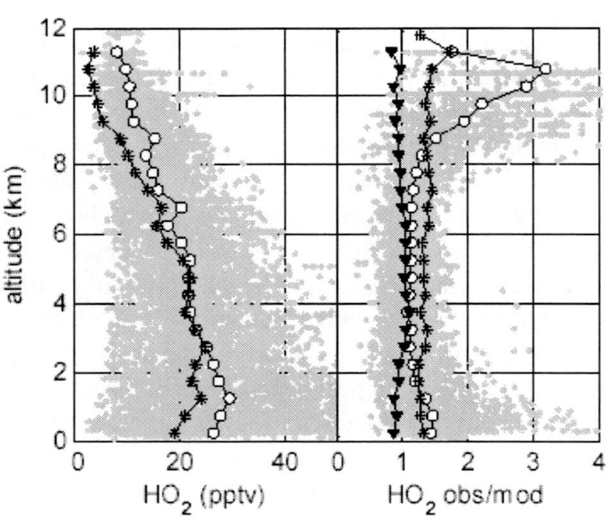

Figure 7. Similar to figure 3, but for HO2. (left panel) Comparison of the median vertical profiles of measured (circles) and modeled (stars) of OH for INTEX-A. (right panel) Measured-to-modeled OH in INTEX-A (circles), TRACE-P (stars) and PEM Tropics B (triangles). Individual 1-minute measurements from INTEX-A are shown (gray dots) [from Ren et al., 2008].

Such a view was not based on observations, as there were few reliable observations of ClO and BrO in the UT/LS. Following the first observations of enhanced ClO in the lowermost stratosphere in 1991 [e.g., Avallone et al., 1993], ozone loss due to heterogeneous chemistry on cirrus clouds was proposed as a way to explain a gap between modeled and measured ozone trends in the midlatitude LS [e.g., Borrmann et al., 1996; Solomon et al., 1997]. A detailed examination of water vapor and ClO measurements in the UT/LS found no evidence for heterogeneous activation of chlorine [Smith et al., 2001]. However, subsequent measurements of ClO in the Arctic and examination of measurements over the continental US, both near the tropopause, found evidence for widespread chlorine activation in regions of high particulate loading [Thornton et al., 2003, Thornton et al., 2007]. The diurnal behavior of reactive chlorine was very suggestive of rapid in situ processing by aerosols [Thornton, 2005].

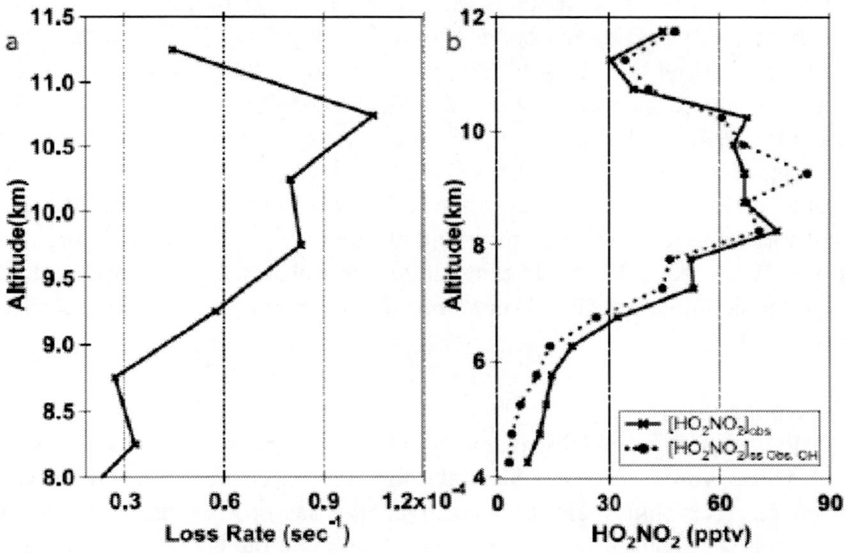

Figure 8. Implications of new observations reported by Kim et al. [2007] revealed an imbalance of production minus loss representing 50% of the magnitude of the production rate calculated from observed abundances of HO2 and NO2 (left panel). HO2NO2 abundances were in good agreement with steady-state calculations based on observed abundances of OH (right panel), suggesting a problem with coupled HO2/NO2 chemistry or one of the observations.

As noted in Section 2.c.III., modeling studies that included heterogeneous processing of NOx found significant changes in the response of ozone to aircraft emissions. In one case [Meilinger et al., 2005], it was the consideration of heterogeneous reactions on ice in persistent contrails that led to important changes in ozone response. In the case of the study by Hendricks et al. [2000], simply including heterogeneous reactions of bromine nitrate, significant denoxification occurred in some regions with important consequences on ozone. Finally, in the very recent study by Sovde et al. [2007], properly accounting for known heterogeneous reactions on aircraft-perturbed aerosol particles resulted in a complete reversal in sign of the ozone response to increased emissions in the UT. Based on these results alone, a reexamination of the role of heterogeneous reactions on background aerosols and in persistent contrails and cirrus using updated photochemical parameters is warranted.

Adding to the complexity of this issue, over the past decade there have been a number of reports (more than will be referred to here – see Salawitch et al. [2005]) of larger-than-expected abundances of BrO in the upper. Salawitch et al. [2005] make a strong case for the need to add upwards of 2-4 ppt of bromine to the stratospheric budget, either by transport of inorganic species (such as BrO, $BrONO_3$, and HOBr) or short-lived organic sources. In light of the increased number of surface sources required to explain recent NOx and HOx measurements in the UT, it seems reasonable that both types of species could contribute to this ~10-20% enhancement in the total bromine budget by short-lived species [e.g., Sinnhuber and Folkins, 2006]. However, there are some important caveats. First, it is only the remotely sensed observations of BrO that point to a need to increase the bromine budget beyond what measurements of organic source gases seem to suggest – in other words, beyond about 4 ppt of bromine from short-lived compounds [e.g., Dorf et al. 2006a]. Second, even the remote sensing observations of BrO do not agree; they split roughly 50/50 in number between those that agree [Schofield et al., 2004]; Sinnhuber et al., 2005] with a budget based on measurements of source gases and some short-lived compounds near the tropopause [e.g., Schauffler et al., 1999] and those that suggest missing nearly double those short-lived sources of bromine [e.g., Sioris et al., 2006; Theys et al., 2007].

This issue is treated in great detail in the recent WMO Ozone Report [2006], so will not be discussed further here, other than to note that due to the importance of bromine in some regions of the UT/LS (e.g., Hendricks et al., 2000), new observations of BrO with high spatial resolution, and in conjunction with observations of NOx and HOx, may be required to resolve this issue.

3.a.III. Potential Surprises

"Our vision is often more obstructed by what we think we know than by our lack of knowledge." These words of Krister Stehdahl, the Harvard Professor of Divinity, apply well to this problem. It is important to remember the lessons of the 1985 WMO Ozone Assessment, where the consensus view at the time was that the ClO dimer and heterogeneous reactions would not play important roles in stratospheric ozone chemistry. This lesson seems relevant to this White Paper, and the authors view several issues that fall in this category as the most important in terms of limiting our ability to accurately assess the current impacts of aviation on UT/LS chemistry and predict future impacts.

Scavenging of NOy

Another important series of new observations are those related to the formation of nitric-acid containing ice particles in the UT/LS [Voigt et al. 2007, Voigt et al. 2007, and Popp et al. 2006]. The fact that such particles are larger, and less abundant than other particles suggests that their sedimentation could impact distributions of reactive nitrogen and water in the UT/LS. Redistribution and/or removal of NOy and H_2O from the UT/LS could result in important non-linearities that are presently not treated adequately in models. For example, it is possible that addition of aircraft NOx, followed by enhanced sedimentation of nitric acid-containing particles, could denitrify a narrow layer centered about the flight corridor. In the tropics, such a process could even mean that aircraft emissions 'seed' the removal of NOy and water, thereby decreasing transport of these species to the stratosphere (i.e. a negative feedback loop). Recent observations of significant chlorine activation in broad region near the polar tropopause where NOy- containing particles were also observed [Thornton et al., 2003] suggest that such a feedback is possible. Thus, it is important to understand better uptake of NOy species on ice particles and the role of temperature and water vapor (i.e. RHi) on such processes. Key to such an understanding will be the accuracies of measurements of water vapor and condensed water in the UT/LS.

Non-linear Processes – Feedbacks and Plume Dispersion

The issue of potential surprises due to a lack of understanding of plume dispersion must be examined in greater detail. One of the ubiquitous features of in situ measurements of many types is their high degree of heterogeneity to very small scales [Richard et al. 2006, and Lovejoy et al. 2007]. In fact, for reactive species, this can translate down to sub- meter scales [unpublished results from the PUMA campaign]. Therefore, it is insufficient to assume simple gaussian plume dispersion when it is known that constituents exhibit a high degree of variability, even hours after they are emitted. This is especially the case when differences between vertical mixing and horizontal shear forces result in filamentary structures [e.g Fairley et al., 2007] that are difficult to describe with a simple gaussian parameterization.

We also lack a basic understanding of non-linear processes that can occur in the heterogeneous environment of an aircraft plume and persistent contrail. With the likely addition presence of solid or liquid mixtures of HNO_3 and H_2O (e.g. nitric acid trihydrate), in which the stability is proportional to the density of the plume raised to a power as large as four, and where heterogeneous reaction rates are strong non-linear functions of relative humidity and composition, this problem has only become more difficult to handle following observations of nitric acid-containing particles in the UT/LS. In a sense, this issue, along with the non-linear coupling between HOx, ClOx, BrOx, and NOx, is reminiscent of the ozone hole. While the effects will not be as severe, their role in the aircraft emissions assessment process is only now being addressed in sufficient detail.

3.b. Measurements and analysis

New and improved measurements and analysis of existing data should help to address some of the outstanding issues highlighted above. As noted previously, reanalysis of HOx measurements may help to resolve some discrepancies between models and measurements that have been noted previously. It is also possible, perhaps likely, that such analyses will raise new questions. In addition, ongoing observations of HOx, along with NOx, source gases, and tracers of transport from the PBL and stratosphere, as are planned for major campaigns such as ARCTAS in 2008 are critical for efforts to map out seasonal and regional variations of this critical oxidizer. Such observations will provide important constraints for models used to assess the role of HOx chemistry, especially tat related to methane oxidation.

New, fast response, in situ measurements in aircraft plumes, including particles, water vapor, several good tracers of combustion and mixing (e.g. CO_2 and CO), ice water content, HOx, NOx, and at least one halogen radical would go far toward reducing uncertainties resulting from non-linear processes. The capability exists for such measurements, although to date, they have not been carried out downstream of an aircraft or in aircraft flight corridors (the potential to rectify this situation exists during ARCTAS).

Continued analyses of satellite data, particularly those with sufficient horizontal resolution to identify regions of interesting chemistry (e.g. in persistent contrails, the North Atlantic Flight Corridor, or in the tropopause region), may shed light on the importance of potential non-linearities that may be difficult to examine by in situ methods. Of particular value would be studies of correlative measurements of clouds and trace constituents (e.g. TES, MLS, SCIAMACHY, OMI, AIRS, MODIS) that might reveal linkages between cloud occurrences and constituent abundances.

Efforts should continue to understand bromine and chlorine chemistry in the UT/LS, in particular the variations of abundances of BrO and ClO. Of particular interest would be high-resolution correlative measurements of these species with HO_2, OH, and NOx, along with their respective source gases. Observations in aircraft plumes and flight corridors would be especially helpful for constraining plume dispersion models. Finally, it will likely be necessary to carry out frequent water vapor measurements intercomparisons to continue to refine our understanding of the factors that influence the discrepancies that have been observed between various techniques.

4.a. Prioritization of Issues Based on Impact

The outstanding issues identified above can be prioritized on the basis of the level of scientific understanding and the magnitude of the terms each represents in the most recent IPCC "Radiative Forcing"-like representation of aviation effects on climate. Referring to figure 1, this would suggest that improvements in understanding of the processes that impact the distribution of ozone (28 mW m^{-2} and "fair") and the lifetime of methane (ŕ {20 mW m^{-2} and "fair") will be most significant. Of lesser importance are the impacts on direct radiative forcings due to emissions of CO_2, H_2O, sulfate and soot. Finally, of least importance would be investigation of issues that were not considered in detail in previous impacts assessments.

However, it is also worthwhile to consider prioritization of issues on the basis of the extent to which they may represent a dramatic shift in our basic understanding of the impacts of aviation. In this case, those issues deemed of least importance using the present framework of the IPCC Forcings, as outlined above, could be considered of highest priority from the point of view of uncertainty or "surprise". For example, if a proper treatment of heterogeneous chemistry on aircraft-produced particles or of aircraft emissions of NOx and H_2O on background aerosols results in a reversal of the sign of ozone change in the UT/LS, this would essentially render as moot the prioritization of issues based on the previous IPCC-like forcings. That is, because the sign for the radiative impact of aircraft-induced ozone changes could, in fact, be negative, a result that is outside the present estimate of uncertainty for that particular term in figure 1. While the possibility of this type of "surprise" is relatively small, given recent observations that raise questions about our understanding of heterogeneous chemistry in the UT/LS, it is prudent to examine the potential consequences of previously unknown processes before expending much effort toward reducing the uncertainties of processes that were previously believed to be the most important.

In the section that follows, we approach the prioritization from these different perspectives, beginning first with the conventional approach of prioritizing the issues on the basis of reducing the current list of uncertainties. We then follow with a prioritization of issues based on the potential for a major shift in our understanding of the impacts.

Priority 1 – Water Vapor Measurements

Long-standing discrepancies among water vapor measurements (both in situ and remote) in the coldest and driest regions in the UT/LS continue to limit efforts to accurately quantify the role of heterogeneous chemistry in conversion of NOx to NOy, to model HOx production and loss, to predict the frequency and extent of halogen activation, and to model the distribution of exhaust emissions (in particular, sedimentation of NOx and H_2O) in the UT/LS. Of critical importance is the characterization of the role of supersaturation (i.e., RHi) in particle formation and growth, both highly non-linear processes.

One method for assessing the accuracy of water vapor measurements is to examine observations from different pairs of instruments in a series of informal intercomparisons. From such opportunities, it is known that particular instruments report data that is consistently as much as 40% larger than all other techniques under the driest conditions in the UT/LS. These data have led researchers to conclude that large supersaturations (well over 150% in some cases) exist. Because all of the in situ instruments have been characterized separately in the laboratory, it has been argued that carefully designed and executed laboratory intercomparisons will help to resolve outstanding differences. A recent formal (double-blind) intercomparison (AquaVIT) has revealed some issues that may help to reduce the discrepancy among instruments. However, it will still be necessary to demonstrate consistent agreement amongst instruments under a wide range of conditions in actual atmospheric observations before this problem can be considered to be resolved.

Unfortunately, few, if any, dedicated intercomparison campaigns are being planned that will adequately address this critical issue. In part, this is due to the high costs that would be associated with a multi-platform, multi-instrument campaign which would be required to demonstrate good agreement over the wide range of conditions found in the UT/LS. For example, a month-long dedicated WB57F campaign based in Houston, designed to sample across a wide range of latitudes in order to encounter a reasonable dynamic range of water vapor values would involve over $1 million in aircraft operating costs and adequate funds for participant travel and post-mission analysis. In addition, it is unclear how new measurements obtained in this manner would resolve outstanding issues from previous campaign involving similar flight tactics. From many perspectives, a new approach aimed at clearly identifying instrument performance issues is required to make significant progress in this area, and to lend credibility to the results.

A promising new approach that could be taken to identify key areas of disagreement between instruments is to deploy them into combustion plumes in the UT/LS, both those laid down by aircraft and those laid down by rockets. The validity of this approach has been demonstrated recently in a pilot mission called PUMA (Plume Ultrafast Measurements Acquisition). In 2004, 2005, and 2006, exhaust emissions from three rockets (Atlas IIAS and two Space Shuttles) were sampled for particle size distributions, ice water content, water vapor, temperature, and carbon dioxide. The advantage to this approach is that a significant range of abundances of H_2O (from ambient levels near 4 ppm to over 30 ppm) are encountered at each altitude where the plumes are sampled, providing for a slope/intercept analysis for each instrument. Such an approach can reveal whether measurement differences are due to differences in calibration or to offsets, the latter of which can be significant for water vapor in the dry UT/LS. One of the interesting results from PUMA is the demonstration that the contrail evaporation point (when RHi drops below 100%) serves as an important independent validation of the H_2O vapor pressure measurement – that is, independent of the CO_2/H_2O emission index, which constrains the slope of a calibration (the "span" or response function), the instant when RHi drops below 100%, which can be identified unambiguously by an enhanced total water measurement such as CLH, is a powerful constraint on the accuracy of a total water measurement to a level that cannot be achieved in any laboratory calibration based on water vapor alone.

Priority 2 – Temperature measurements

As shown above, in the context of defining RHi, measurements of temperature on most platforms agree to a level that is better than the agreement amongst water vapour measurements. However, making accurate temperature measurements is a non-trivial process, especially on a fast-moving platform, such as the WB57F, ER-2, or HIAPER. For example, near 200 K, a difference of 1 $_{o}$C translates into an uncertainty of 10% in RHi. Thus, any program designed to address water vapor accuracies (especially one that relies on the vapor-ice transition such as that described above) must also address the accuracy of temperature measurements. It is the correction from observed to static temperatures using the "recovery temperature" equation that is most uncertain, as the correction involves quadratic terms for air speed that rely on highly accurate measurements of static and dynamic pressure. Consequently, accurate knowledge of the air flow around the aircraft surface where temperature probes are mounted is critical in order to determine recovery temperature to better than 1 $^{\circ}$C.

One issue that has been raised when different temperature measurements from the WB57F aircraft have been compared is that placement of inlets can have profound effects on water vapor measurements in clouds (or at RHi near 100%) due to possible inertial enhancement of particulate water. It is recommended here that to avoid ambiguities (such as pressure perturbations near blunt surfaces or under wings), it would be quite useful to install temperature probes in various locations around the aircraft, especially on wing pods or under the wings near where water vapor instruments are deployed in any campaign that has a focus on accuracies of water vapor measurements. Good agreement between such measurements (say one located on a wing pod and one on the nose) serves to provide increased confidence that differences between measurements of water vapour are not due to perturbations of the temperature/pressure field around an instrument. This approach was used successfully during the PUMA campaign. As shown below, such measurements represent a very small cost compared to the time that could be lost in post-mission analyses that must account for potential consequences due to placement of temperature and pressure measurements.

Priority 3. HOx Measurements

Critical to the modeling effort that is required to determine the impact of aircraft emissions on the global methane budget (and hence the radiative forcing term that is labeled by "CH_4" in figure 1) is the ability for the models to accurately simulate global OH distributions. Not only does the abundance of OH determine the tropospheric lifetime of methane and the rate of conversion of NOx to NOy, OH and HO_2 are important ozone destroying radicals. In addition, the ability to model the sources of HOx in the UT/LS improves knowledge of the surface convective sources that also contribute the budget of NOx in the UT/LS. Finally, measurements of OH and the OH/HO2 ratio provide constraints on NOx and halogen chemistries.

A substantial heritage of measurements of OH and HO_2 in the UT/LS has been established as a result of numerous campaigns involving the ER-2 and DC-8 aircraft. Because HOx abundances are fundamental to a number of important processes in models used to assess aircraft impacts, continuing to add to the current database of HOx measurements will serve to reduce important uncertainties in those models. Frequent intercomparisons between measurements of OH and HO2 using different techniques will also help investigators reduce their measurement uncertainties, and should be encouraged.

Priority 4 – Coupled HOx/NOx Chemistry

Possible discrepancies between modeled and measured HO_2NO_2, a compound that provides a critical link between the photochemistries of HOx and NOx families, should be investigated further. The current discrepancy points out a potential problem with the new measurements of HO_2NO_2 or one or more of the species that produces it, an error in a critical photochemical parameter, or missing chemistry that could be important in determining abundances of NOx or NOy in the UT/LS. Efforts to reduce uncertainties in the measurements of HO_2NO_2 and modeling investigations of potential errors in sources or sinks of HO_2NO_2 should be encouraged.

Prioritization Based on Potential Impacts that Are Currently Unknown

Although important uncertainties remain in the processes listed in the section above, for all of these it is possible to estimate the likely bounds of their impacts with investigations that are constrained by known uncertainties in existing measurements. For example, impacts could be assessed with a model that assumes ice particle formation in the UT/LS at supersaturations consistent with the low end (i.e., driest) of the water vapour measurements and with those consistent with the high end of the measurements. Based on the resulting range of impacts, the need to resolve the discrepancies in water vapour measurements could be quantified (for example, a range of 10%, rather than 30%, is required for adequate assessment of this term). However, for several processes, the observations may be too limited to provide a reliable estimate of the impacts of aircraft emissions. In this section, these processes are given high priority based on the possibility that they could be significant, but reasonable bounds cannot yet be placed on their potential impacts due to lacking observational constraints (e.g., the situation, although probably not as dramatic, can be likened to that of 1985 when it was believed that heterogeneous reactions were not significant for ozone balance and that CFC-related ozone loss would occur in the middle stratosphere at mid-latitudes).

Priority 1 – Investigations of Non-Linear Effects

Recent observations of nitric acid-containing particles [Popp, et al.,] and enhancements in reactive chlorine [Thornton et al.,] in the UT/LS outside the polar regions have raised the possibility that heterogeneous reactions could lead to conversion of NOx to NOy, and activation of chlorine, in persistent contrails or cirrus occurring in flight corridors. It is even possible that NOy could be redistributed by sedimentation of particles if they grow large enough in these regions. Such processes are strongly non-linear in plumes or exhaust-influenced regions, due to the threshold nature of particle formation and strong water dependence of heterogeneous reactions involving halogens and NOy. To understand the role of such processes in UT/LS chemistry, details of the dispersion of exhaust become extremely important.

At the present time, there are few observations of the variability of constituents in and subsequent dispersion to the background atmosphere of exhaust plumes. In addition, the chemical composition of particles in exhaust plumes has only recently begun to be studied, and measurements of reactive halogens in the UT/LS with instruments sensitive enough to observe their small-scale (e.g., plume scale) variability have been ignored. Given the recent model results shown in Sections 2.C.II and 2.C.IV above, it is important to investigate the potential impacts of dispersion processes on the chemistry of plumes. Significant progress toward setting possible limits on the importance of such processes would be possible with modeling efforts that consider extreme cases, such as complete removal of NOy by sedimentation in persistent contrails, slow dispersion of plumes, and rapid heterogeneous reactions. Such studies could then serve to guide observations of species such as HNO_3, particles, ClO, and BrO that would constrain the impacts of these processes on the chemistry of ozone in the UT/LS.

Priority 2 – The Role of Halogen Oxides in Background UT/LS Ozone Chemistry

Although it is believed that the importance of halogen oxides is limited by excess abundances of NOx in the UT/LS, recent observations of widespread, low levels (~1-2 ppt) of BrO throughout the UT/LS and narrow regions with significant enhancements of ClO raise important questions about our understanding of halogen chemistry in the altitude region where aircraft emissions have the greatest impact on ozone abundances. Because coupled NOx/HC/HOx (i.e. "smog") chemistry tends to produce ozone in the upper troposphere, whereas halogens solely (and rapidly) destroy ozone, a better understanding of the distributions of halogen radicals is necessary to accurately simulate the impact of aircraft NOx and H_2O emissions on ozone in the UT/LS. Of particular concern is the possibility that NOx serves as a catalyst for production of halogen oxides via rapid heterogeneous reactions in the presence of sunlight. This situation is somewhat the reverse of that in the winter polar stratosphere, where NOx serves to deactivate the halogen radicals via formation of relatively stable reservoirs. In the UT/LS at lower latitudes, however, rapid heterogeneous conversion of inorganic halogen acids (e.g., HOBr, HBr, HOCl, and HCl) is limited by availability of oxidants such as $ClNO_3$ and $BrNO_3$, such that addition of NOx serves as a catalyst for halogen activation, so long as particulate surface areas are sufficient. With recent studies showing a reversal in sign of the impact of aircraft emissions on ozone abundances due to more rapid heterogeneous chemistry and halogen activation, it is important that the issue of distributions of halogen oxides be revisited.

There are several cost-effective ways that this issue could be approached. First, because abundances of ClO and BrO are quite small in this region, it would be useful for a team of investigators composed of modelers and measurements experts to model the impact on ozone of extreme scenarios involving halogen radicals in the UT/LS using the few existing observations. The calculated ranges of ozone could then be used to re-examine the radiative impacts of aircraft emissions. In addition, new high-resolution in situ measurements of halogen oxides in the UT/LS could be obtained in conjunction with measurements of NOx and HOx as part of larger campaigns designed to study the oxidative state of the UT/LS. Such measurements in the upper troposphere have had a very low priority on previous missions, except for the 1998 WB-57F Aerosol Mission (WAM) and the 2000 SOLVE campaign, results of which have shown that active forms of chlorine are more prevalent than was believed previously, provided that ample aerosol surface area abundances (> 3 ŕЭm cm-3) are available. There are cost-effective ways to pursue this line of investigation, such as redeploying atomic resonance fluorescence (RF) instruments that have been used for over two decades for stratospheric measurements and that were previously flown on the WB-57F and DC-8 aircraft, in this case reconfigured for improved sensitivity under tropospheric conditions, or by adapting instruments that use an alternative detection technique (e.g., chemical ionization mass spectrometry - CIMS). In either case, there will be modest costs (see below) associated with the laboratory efforts required to optimize the existing stratospheric instruments for use in the UT/LS or those required to develop new calibrations and to develop a heritage of reliable observations, in the case of a new measurement technique, such as CIMS.

Laboratory measurements of key rate parameters at low temperatures of the UT/LS will continue to refine our understanding the sensitivities of NOx and heterogeneous chemistries to temperature, relative humidity and pressure, variables that can be important in the UT/LS.

4.b. Ability to Reduce Uncertainties

Given the wealth of new information regarding UT/LS chemistry that has become available in recent years, the ability to reduce uncertainties in estimates of the climate impacts of aviation is quite good. Significant progress can be made on nearly all of the topics presented in this SSWP within 3 to 5 years. The most problematic of the issues, those involving accuracies of water vapor measurements, plume dispersion, and heterogeneous chemistry, may require a longer timeframe to achieve the level of confidence that is associated with attribution of cause-and-effect for ozone destruction in the stratosphere, but given the level of knowledge already attained in the atmospheric chemistry community, it is not unreasonable to expect that an effort that is more focused on resolving the key issues outlined above can see significant progress within the time frame of 2 three-year grant cycles.

First, and most critical, will be detailed studies with models that can treat plume chemistry and dispersion to scope out the range of possible impacts of non-linear particle formation processes and heterogeneous chemistry. Coupled with this knowledge, field and laboratory studies can be carried out to reduce the uncertainties in the most critical parameters that are revealed in these model studies. Of particular significance will be those fields studies that can address plume processes directly with the powerful suite of instruments and platforms that are currently in the atmospheric sciences arsenal. With few exceptions (such as better instruments for measuring halogens at part-per-trillion abundances and new or improved instruments to measure oxygenated source gases for HOx), the instruments and platforms required to provide critical observations to constrain these process models already exist, and the investment in the investigations needed to answer the critical questions will be valuable for issues that reach beyond the impacts of aircraft (for example, alternative energy production, changing climate, new technologies, etc.).

4.c. Practical Use

Addressing all of the key issues above will have important practical applications, including improvements in measurements that address a broad range of atmospheric issues. Additional model development, especially an accurate and validated plume dispersion model can be quite useful for studying a number of issues related to climate change, including source apportionment of CO_2, an issue that will be of major importance in the future if CO_2 trading schemes become prevalent.

4.d. Achievability

As noted in Section 4.b., important results are clearly achievable in all areas outlined in this SSWP. In most cases, cost will be the primary limiting issue, as some instruments or platforms that may be required for the most definitive studies will require significant modifications or deployment costs. Improvements in models that will be necessary to assimilate the results from new observations may require the development of new codes (for example, a high-resolution plume dispersion model). However, to date technology does not seem to be what has limited the development of such a model.

4.e. Cost

Addressing the water vapor measurements issue will probably be the most productive use of funds at this point in time in terms of reducing uncertainties in aircraft climate impacts. However, due to the high level of interest for other programs (e.g., satellite validation and climate change studies in general), significant leveraging of funds should be possible, and should immediately be pursued. However, a business-as-usual approach is very likely not going to foster significant progress in this area, such that a new and creative program will be required. It would be helpful to develop clear milestones with broad community support, with implications for failure of PIs to meet stated accuracies. New and innovative approaches to validating water vapor (and condensed water) measurements in the cold and dry UT/LS, such as periodic direct flights in exhaust plumes to calibrate individual instruments, to reveal discrepancies between instruments, and to monitor instrumental drift, would be particularly useful. Such efforts that could also build on recent efforts, such as AquaVIT, to maintain a traceable set of intercomparisons, should be monitored regularly by a group of scientists who are both knowledgeable in the field, and outsiders who have an expertise in measurement intercomparisons and validations. It would be particularly helpful to develop a water vapor standard for calibrations and traceability, just as was done for ozone measurements, thereby reducing the reliance on costly large-scale laboratory intercomparisons.

It would be very useful to carry out an in-flight intercomparison of water vapour measurements in the UT/LS from a common platform, such as the DC-8 or WB-57, one that involves frequent sampling in aircraft plumes (both wet and dry). Not all instruments would have to participate in such an intercomparison, but it would be essential to have sufficient variety of existing instruments that span the range of current measurements (e.g., from those that are on the low side of the intercomparisons, such as frost point hygrometers, to those that are on the high side, such as the JPL TDL). Results in dry plumes can be traced to an absolute value using simultaneous measurements of CO_2, since the stoichiometry of combustion of aviation fuel is well known.

Overall, a ~$1-2 million program over five years, with funds provided from a variety of sources, would catalyze significant progress on this issue, and get away from the business-as-usual approach of providing limited funding for smaller, term efforts that piggy-back off larger projects, and end up suffering from too little funding without a guarantee of continued funds to thoroughly investigate the causes for discrepancies. A Water Vapor Campaign, whose chief focus is on reducing the uncertainties in measurements and maintaining a long-term, traceable record, should be a top priority for an aircraft impacts program, as well as a general world-wide program to monitor climate change.

With a clear focus on water vapor, other issues can be dealt with on an 'add-value' basis. For example, studies of non-linear processes in plumes would be a natural add-on to missions that use combustion plumes as a way to investigate instrument differences and, potentially, as a way to maintain a long-term calibration standard (assuming that combustion of kerosene will remain the method of choice for aircraft propulsion for many decades. Issues that require some instrument development (e.g., halogen and oxygenated organic compound measurements) should be initiated as soon as possible to reduce the long lead times that are associated with integration and demonstration of new instruments on research aircraft. Funding for these developments could be leveraged with funding agencies like NSF and DOE, insofar as other programs will benefit from the use of such instruments in other environments (e.g., halogens in the polar boundary layer, oxygenated compounds in urban pollution/source attribution studies, etc.). International cooperation would also help to reduce development time and cost, especially where there are common interests for measurement capabilities (i.e., it is cheaper per unit to build more than one).

Addition of increments of ~$300-500 K in a few key areas would likely result in important progress for most of the issues highlighted in this SSWP. A total program of $5 million, including the water vapor project mentioned above, would probably reduce most of the remaining climate uncertainties in aviation operations by half, and change the level of understanding from poor or fair to good for most, if not all, chemical terms in the climate forcing framework.

4.f. Timeline

Significant progress could be made on all of the issues discussed above within 3-5 years with an adequately resourced project. The expertise exists in the community and there would be limited need for development of new techniques. In fact, waiting longer could inadvertently result in significant additional expenses to carry out similar work, as experts in some areas retire or become involved in other projects. In the worst case, it is possible for an opportunity to be lost altogether. Because time is a factor, heritage should be a major factor in consideration of projects to fund. The cost of missed opportunities is difficult to estimate, but it vastly exceeds the cost of starting from scratch. instrument to service. (for example, it would be highly desirable to bring the NOAA-lyman alpha water instrument back into service, and waiting much longer may preclude this, and resurrecting this capability from scratch would be prohibitively costly, especially given the extraordinarily long record of measurements for this instrument) .

Immediate

A water vapor program should be developed immediately. This issue will be around for a long time, and waiting longer will only serve to up the overall cost. Development and integration of new (or modified) instruments designed to address key 'missing terms' or resolve discrepancies between measurements should also begin as soon as possible. Far too often, such measurements are missing from major campaigns due to lack of planning and preparation.

5. Best Way to Assess Uncertainties with Current Knowledge

In the absence of improvements in our understanding of the outstanding issues presented in Section 4.a., there are studies that can be undertaken now to assess the impacts of aviation on chemistry of the UT/LS that will represent a significant advance since the 1999 IPCC Report. Before recommending such studies, it is important to note that such an advance does not necessarily imply that all of the specific uncertainties reported in previous assessments will be improved. It is possible that new observations reported above may reveal gaps in our understanding that were not foreseen a decade ago.

As noted above, resolving the water vapor measurements discrepancy in the cold, dry UT/LS is crucial in order to improve our understanding of the climate impacts of aviation that are linked to chemistry. Therefore, it would be extremely useful to use the best available 3D global chemical transport models to study the sensitivity of climate impacts to the two extreme possibilities that are represented in the literature. Based on uncertainties described in SSWPs dealing with clouds and aerosols, it is unclear whether the models sufficiently capture the complexities of condensation and dehydration, so it may not be straightforward to study these extreme cases from 'first principles.' That is, a realistic treatment of particle formation, composition, reactivity, and sedimentation, as a function of supersaturation on the scales of individual plumes and persistent contrails may not yet be possible. In this case, it would still be very useful to use some statistical representation of occurrences of cirrus, contrails, and persistent contrails as a basis for estimating the frequency of heterogeneous chemistry events [e.g., Bregman et al., 2002; Meilinger et al., 2005] and their contribution to the d[O_3] and d[CH_3] terms in the radiative forcing framework (e.g., figure 1).

It is imperative that the recent results of Sovde et al. [2007] be examined in detail over the possible ranges of critical parameters such as lightning and convective fluxes of NOx, sources of HOx, microphysics of mixtures of HNO_3 and H_2O, and background abundances of halogens. Sensitivity tests of regional and global ozone and methane responses to aircraft emissions would help to narrow down the list of parameters to those that contribute to the bulk of the uncertainty in the aircraft RF terms. (This approach is similar to one taken several decades ago to define which rate parameters were most critical in determining ozone loss due to chlorine buildup, for example.)

New sensitivity studies should be carried out to address the role of processes that are highly scale-dependent, such as denoxification, sedimentation, and mixing. Processes that are important in persistent contrails, for example, may have very different impacts if they are modeled as being severe, but highly localized, versus moderate and more widespread. Effects such as redistribution of NOy by sedimentation are likely to be more severe, whereas those such as ozone loss due to chlorine activation may be less severe, in the former case (i.e., highly localized assumption).

Due to the large and growing body of HOx observations, it would be extremely useful to reevaluate the "CH_4' radiative forcing term with a CTM that is either constrained by or validated with observed OH fields.

Finally, it could be useful to carry out a series of focused observational studies to quantify the uncertainties in temperature and pressure measurements from aircraft. Not only will such studies improve our understanding of the uncertainties in past determinations of supersaturation, they will serve as the basis for much improved measurements of temperature in the UT/LS for future studies. Of particular value will be the development of ultra-fast (~100 Hz or faster) temperature probe for research aircraft such as HIAPER, the ER-2, the WB57F and Global Hawk, all of which can play important roles in defining thermodynamic variables in the UT/LS, but also for commercial aircraft that could be used to carry out long-term measurements in the UT/LS.

6. Summary

Aircraft emit a variety of species that can alter climate and the chemistry of Earth's atmosphere. In this context, the most important are emissions of NOx, particles, and water vapor, all of which interact to determine ozone distributions in the UT/LS, a region where radiatively active gases have a strong influence on temperature and dynamics. Previous assessments pointed to increases in ozone columns and reductions in methane (from the influence of NO on the OH/HO2) as the two chemical impacts that were likely to have the largest impact on climate (aircraft radiative forcing, RF). It was found that these two terms were of roughly equal magnitude, but opposite sign, so that the net climate impact of aircraft emissions chemistry was approximately neutral. However, the understanding of the processes that determine these quantities was considered poor to fair. In the view that UT/LS chemistry is controlled by NOx, these two terms will always cancel, because the processes that result in ozone production will lead to methane destruction.

New observations and modeling efforts undertaken over the past decade have raised important questions about the basis for earlier assessments. In particular, NOx in the UT/LS is found to be partitioned in long-lived reservoirs to a larger extent than previously believed, presumably by heterogeneous reactions. Convective and lightning sources of NOx to the upper troposphere have also been found to be more important than previously believed. In addition, reactive bromine and chlorine radicals have been observed in the UT or LS, implying a greater role for these species in partitioning of HOx. Finally, large particles containing nitric acid have been observed in the UT/LS. Models that include more vigorous heterogeneous chemistry in the UT/LS indicate that emissions of particles from aircraft may actually reduce ozone in the UT and increase ozone in the lower stratosphere, the opposite of what was reported in the previous assessments.

Given that the climate impacts from ozone changes are partially offset by those of methane changes (assuming that the inverse relationship between NOx and OH is maintained under these new conditions), the impact to climate overall may not change dramatically with this sign reversal in ozone changes. However, if these changes are confirmed, strategies for reducing the impacts of aircraft emissions on atmospheric chemistry and climate would be very different than those based on work summarized in previous assessments. Therefore, it is important that these new findings and their implications be explored in more detail before designing mitigation strategies.

Significant progress toward reducing the uncertainties in UT/LS chemistry identified here can be made with modest investments in key areas. The observational and modeling tools are largely available, thanks to the high priority that has been placed on understanding UT/LS chemistry. Several high priority studies are recommended here. Of greatest priority would be supporting efforts to resolve long-standing discrepancies among measurements of water vapor, including establishment of a water vapor standard that is appropriate for UT/LS conditions, and carrying out high-resolution measurements of water vapor, particles, and CO_2 in and around aircraft plumes with a platform such as the DC-8, WB-57, or HIAPER. Augmentations of measurements of key species to address coupled radical chemistry to the payloads for major campaigns could reduce uncertainties in basic ozone loss chemistry.

With added importance of aerosols and clouds to ozone chemistry in the UT/LS, it will be important to assess the importance of heterogeneous chemistry and aerosol formation and evolution in aircraft plumes, persistent contrails, and cirrus clouds. Models that treat plume dispersion with some realism may be necessary, although our knowledge of the potential range of impacts of plume processes can probably be improved by simple sensitivity tests that assume extreme bounds for processes such as denoxification and redistribution of species such as NOy and H_2O. It would be reasonable to expect that significant new results to improve our understanding of the impacts of aircraft exhaust on atmospheric chemistry and climate would be forthcoming within three to five years of formulation of a focused program to address the major uncertainties presented in this White Paper for a total expenditure of under $10 million, including funds from all sources. There are significant opportunities for synergistic studies that are currently in the planning stages or underway, with strategic placement of new funds to target particular elements that are critical for specifically assessing the impacts of aircraft.

REFERENCES

Anderson, B. E., et al. (2006), Hydrocarbon emissions from a modern commercial airliner, *Atmospheric Environment*, *40*(19), 3601-3612.

Arnold, F., A. Kiendler, et al. (2000), Chemiion concentration measurements in jet engine exhaust at the ground: Implications for ion chemistry and aerosol formation in the wake of a jet aircraft, *Geophysical Research Letters,* 27(12), 1723-1726.

Avallone, L. M., D. W. Toohey, et al. (1993), Balloon-Borne In-Situ Measurements of ClO and Ozone - Implications for Heterogeneous Chemistry and Midlatitude Ozone Loss, *Geophysical Research Letters,* 20(17), 1795-1798.

Bertram, T. H., A. E. Perring, et al. (2007), Direct measurements of the convective recycling of the upper troposphere, *Science,* 315(5813), 816-820.

Bencherif, H., R. D. Diab, T. Portafaix, B. Morel, P. Keckhut, and A. Moorgawa (2006), Temperature climatology and trend estimates in the UTLS region as observed over a southern subtropical site, Durban, South Africa, *Atmospheric Chemistry and Physics*, 6, 5121–5128.

Borrmann, S., S. Solomon, et al. (1996), The potential of cirrus clouds for heterogeneous chlorine activation, *Geophysical Research Letters,* 23(16), 2133-2136.

Bortz, S. E., et al. (2006), Ozone, water vapor, and temperature in the upper tropical troposphere: Variations over a decade of MOZAIC measurements, *Journal of Geophysical Research-Atmospheres*, *111*(D5), doi:10.1029/2005JD006512

Brasseur, G.P., R.A. Cox, et al. (1998), European scientific assessment of the atmospheric effects of aircraft emissions, *Atmospheric Environment*, 32 (13), 2329-2418.

Bregman, B., Wang, P. H., and Lelieveld, J. (2002), Chemical ozone loss in the tropopause region on subvisible ice clouds, calculated with a chemistry-transport model, *Journal of Geophysical Research-Atmospheres,* 107(D3), doi:10.1029/2001JD000761.

Brunner, D., et al. (2001), Nitrogen oxides and ozone in the tropopause region of the Northern Hemisphere: Measurements from commercial aircraft in 1995/1996 and 1997, *Journal of Geophysical Research-Atmospheres*, *106*(D21), 27673-27699.

Brunner, D., et al. (2005), An evaluation of the performance of chemistry transport models - Part 2: Detailed comparison with two selected campaigns, *Atmospheric Chemistry and Physics, 5*, 107-129.

Carslaw, D. C., et al. (2006), Detecting and quantifying aircraft and other on-airport contributions to ambient nitrogen oxides in the vicinity of a large international airport, *Atmospheric Environment, 40*(28), 5424-5434.

Chughtai, A. R., et al. (2003), The effect of temperature and humidity on the reaction of ozone with combustion soot: Implications for reactivity near the tropopause, *Journal of Atmospheric Chemistry*, *45*(3), 231-243.

Colette, A., and G. Ancellet (2005), Impact of vertical transport processes on the tropospheric ozone layering above Europe. Part II: Climatological analysis of the past 30 years, *Atmospheric Environment*, *39*(29), 5423-5435.

Colette, A., et al. (2005), Impact of vertical transport processes on the tropospheric ozone layering above Europe. Part I: Study of air mass origin using multivariate analysis, clustering and trajectories, *Atmospheric Environment*, *39*(29), 5409-5422.

Colette, A., and G. Ancellet (2006), Variability of the tropospheric mixing and of streamer formation and their impact on the lifetime of observed ozone layers, *Geophysical Research Letters*, *33*(9), doi:10.1029/2006GL025793.

Collins, W. J., et al. (1999), Role of convection in determining the budget of odd hydrogen in the upper troposphere, *Journal of Geophysical Research-Atmospheres*, *104*(D21), 26927-26941.

Colomb, A., et al. (2006), Airborne measurements of trace organic species in the upper troposphere over Europe: the impact of deep convection, *Environmental Chemistry*, *3*(4), 244-259.

Considine, D. B., et al. (2000), A polar stratospheric cloud parameterization for the global modeling initiative three-dimensional model and its response to stratospheric aircraft, *Journal of Geophysical Research-Atmospheres*, *105*(D3), 3955-3973.

Cooper, O. R., et al. (2006), Large upper tropospheric ozone enhancements above midlatitude North America during summer: In situ evidence from the IONS and MOZAIC ozone measurement network, *Journal of Geophysical Research- Atmospheres*, *111*(D24), doi:10.1029/2006JD007306.

Cox, V. (2007), *NextGen and Its Impact on Performance Worldwide Symposium on Performance of the Air Navigation System*, Presentation to the ICAO, Montreal, Quebec, Canada, March 26-30, 2007.

Crawford, J., et al. (1999), Assessment of upper tropospheric HOx sources over the tropical Pacific based on NASA GTE/PEM data: Net effect on HOx and other photochemical parameters, *Journal of Geophysical Research-Atmospheres*, *104*(D13), 16255-16273.

Dakhel, P. M., et al. (2007), Postcombustion evolution of soot properties in an aircraft engine, *Journal of Propulsion and Power 23*(5), 942-948.

DeCaria, A. J., et al. (2005), Lightning-generated NOx and its impact on tropospheric ozone production: A three-dimensional modeling study of a Stratosphere- Troposphere Experiment: Radiation, Aerosols and Ozone (STERAO-A) thunderstorm, *Journal of Geophysical Research-Atmospheres*, *110*(D14), doi:10.1029/2004JD005556.

Dessens, O. and P. Simon (2002), The importance of dynamics/chemistry coupling in the evaluation of aircraft emission impact studies, *Meteorologische Zeitschrift 11*(3), 161-175.

Dorf, M., et al. (2006), Balloon-borne stratospheric BrO measurements: comparison with Envisat/SCIAMACHY BrO limb profiles, *Atmospheric Chemistry and Physics 6*, 2483-2501.

Dorf, M., et al. (2006), Long-term observations of stratospheric bromine reveal slow down in growth, *Geophysical Research Letters 33*(24), doi:10.1029/2006GL027714.

Eichkorn, S., et al. (2002), Massive positive and negative chemiions in the exhaust of an aircraft jet engine at ground-level: mass distribution measurements and implications for aerosol formation, *Atmospheric Environment 36*(11), 1821-1825.

Esler, J. G., Roelofs, G. J., Kohler, M. O., and O'Connor, F. M., (2004), A quantitative analysis of grid-related systematic errors in oxidising capacity and ozone production rates in chemistry transport models. *Atmospheric Chemistry and Physics 4*, 1781.

Fairlie, T. D., et al. (2007), Impact of multiscale dynamical processes and mixing on the chemical composition of the upper troposphere and lower stratosphere during the Intercontinental Chemical Transport Experiment-North America, *Journal of Geophysical Research-Atmospheres 112*(D16), doi:10.1029/2006JD007923.

Faloona, I., et al. (2000), Observations of HOx and its relationship with NOx in the upper troposphere during SONEX, *Journal of Geophysical Research-Atmospheres 105*(D3): 3771-3783.

Farias, F. and H. ApSimon (2006), Relative contributions from traffic and aircraft NOx emissions to exposure in West London, *Environmental Modelling & Software 21*(4): 477-485.

Fehr, T., et al. (2004), Model study on production and transport of lightning-produced NOx in a EULINOX supercell storm, *Journal of Geophysical Research-Atmospheres*, *109*(D9), doi:10.1029/2003JD003935.

Ferrare, R. A., et al. (2004), Characterization of upper-troposphere water vapour measurements during AFWEX using LASE, *Journal of Atmospheric and Oceanic Technology*, *21*(12), 1790-1808.

Folkins, I., et al. (2006), Testing convective parameterizations with tropical measurements of HN O_3, CO, H_2O, and O_3: Implications for the water vapor budget, *Journal of Geophysical Research-Atmospheres*, *111*(D23), 10.1029/2006JD007325

Forster, C., et al. (2003), The residence times of aircraft emissions in the stratosphere using a mean emission inventory and emissions along actual flight tracks, *Journal of Geophysical Research-Atmospheres* *108*(D12), doi:10.1029/2002JD002515.

Gao, R. S., et al. (2004), Evidence that nitric acid increases relative humidity in low-temperature cirrus clouds, *Science 303*(5657): 516-520.

Gao, R. S., et al. (2006), Measurements of relative humidity in a persistent contrail, *Atmospheric Environment*, *40*(9), 1590-1600.

Gauss, M., et al. (2003), Impact of H_2O emissions from cryoplanes and kerosene aircraft on the atmosphere, *Journal of Geophysical Research-Atmospheres*, *108*(D10), doi:10.1029/2002JD002623.

Gauss, M., et al. (2006), Impact of aircraft NOx emissions on the atmosphere – tradeoffs to reduce the impact, *Atmospheric Chemistry and Physics 6*, 1529-1548.

Grewe, V., et al. (2002), Impact of aircraft NOx emissions. Part 1: Interactively coupled climate-chemistry simulations and sensitivities to climate-chemistry feedback, lightning and model resolution, *Meteorologische Zeitschrift 11*(3): 177-186.

Giannakopoulos, C., et al. (2003), Modelling the impacts of aircraft traffic on the chemical composition of the upper troposphere, *Proceedings of the Institution of Mechanical Engineers Part G-Journal of Aerospace Engineering 217*(G5): 237-243.

Gulstad, L. and I. S.A. Isaksen, (2007), Modeling water vapor in the upper troposphere and lower stratosphere, *Terrestrial, Atmospheric and Oceanic Science*, *18,* 415-436.

Haag, W., et al. (2003), Freezing thresholds and cirrus cloud formation mechanisms inferred from in situ measurements of relative humidity, *Atmospheric Chemistry and Physics*, *3*, 1791-1806.

Haverkamp, H., et al. (2004), Positive and negative ion measurements in jet aircraft engine exhaust: concentrations, sizes and implications for aerosol formation, *Atmospheric Environment 38*(18), 2879-2884.

Hays, M. D. and R. L. Vander Wal (2007), Heterogeneous soot nanostructure in atmospheric and combustion source aerosols, *Energy & Fuels 21*(2), 801-811.

Helten, M., et al. (1999), In-flight comparison of MOZAIC and POLINAT water vapour measurements, *Journal of Geophysical Research-Atmospheres*, *104*(D21), 26087-26096.

Hendricks, J., et al. (2000), Implications of subsonic aircraft NOx emissions for the chemistry of the lowermost stratosphere: Model studies on the role of bromine, *Journal of Geophysical Research-Atmospheres 105*(D5), 6745-6759.

Herndon, S. C., et al. (2004), NO and NO2 emission ratios measured from in-use commercial aircraft during taxi and takeoff, *Environmental Science & Technology 38*(22), 6078-6084.

Herndon, S. C., et al. (2006), Hydrocarbon emissions from in-use commercial aircraft during airport operations, *Environmental Science & Technology 40*(14), 4406-4413.

Hudman, R. C., et al. (2007), Surface and lightning sources of nitrogen oxides over the United States: Magnitudes, chemical evolution, and outflow, *Journal of Geophysical Research-Atmospheres 112*(D12), doi:10.1029/2006JD007912.

Immler, F., et al. (2007), Cirrus clouds, humidity, and dehydration in the tropical tropopause layer observed at Paramaribo, Suriname (5.8 oN, 55.2 oW), *Journal of Geophysical Research-Atmospheres 112*(D3), doi:10.1029/2006JD007440.

Intergovernmental Panel on Climate Change (1999). Aviation and the Global Atmosphere, Special Report of IPCC Working Groups I and III, Cambridge University Press, Cambridge, UK.

Jaegle, L., et al. (1999), Ozone production in the upper troposphere and the influence of aircraft during SONEX: Approach of NOx-saturated conditions, *Geophysical Research Letters*, *26*(20), 3081-3084.

Karcher, B., et al. (2007), Insights into the role of soot aerosols in cirrus cloud formation, *Atmospheric Chemistry and Physics*, *7*(16), 4203-4227.

Karcher, B. and C. Voigt (2006), Formation of nitric acid/water ice particles in cirrus clouds, *Geophysical Research Letters 33*(8), doi:10.1029/2006GL025927.

Kim, S., L. et al., (2007), Measurement of HO2NO2 in the free troposphere during the Intercontinental Chemical Transport Experiment–North America 2004, *Journal of Geophysical Research - Atmospheres, 112*, doi:10.1029/2006JD007676.

Klemm, O., et al. (1998), Measurements of nitrogen oxides from aircraft in the northeast Atlantic flight corridor, *Journal of Geophysical Research-Atmospheres 103*(D23): 31217-31229.

Kley, D., J. M. Russell III, and C. Phillips, eds., SPARC Assessment of Upper Tropospheric and Stratospheric Water Vapour, World Climate Research Programme, WCRP-113, Dec 2000.

Knighton, W. B., et al. (2007), Quantification of aircraft engine hydrocarbon emissions using proton transfer reaction mass spectrometry, *Journal of Propulsion and Power 23*(5): 949-958.

Koike, M., et al. (2000), Impact of aircraft emissions on reactive nitrogen over the North Atlantic Flight Corridor region, *Journal of Geophysical Research-Atmospheres 105*(D3): 3665-3677.

Koike, M., et al. (2002), Reactive nitrogen over the tropical western Pacific: Influence from lightning and biomass burning during BIBLE A, *Journal of Geophysical Research-Atmospheres*, *108*(D3), doi:10.1029/2001JD000823.

Kraabol, A. G., et al. (2002), Impacts of NOx emissions from subsonic aircraft in a global three-dimensional chemistry transport model including plume processes, *Journal of Geophysical Research-Atmospheres 107*(D22), .

Labrador, L. J., et al. (2004), Strong sensitivity of the global mean OH concentration and the tropospheric oxidizing efficiency to the source of NOx from lightning, *Geophysical Research Letters*, *31*(6), doi:10.1029/2003GL019229.

Lange, L., et al. (2001), Detection of lightning-produced NO in the midlatitude upper troposphere during STREAM 1998, *Journal of Geophysical Research-Atmospheres*, *106*(D21), 27777-27785.

Law, K. S., et al. (2000), Comparison between global chemistry transport model results and Measurement of Ozone and Water Vapor by Airbus In-Service Aircraft (MOZAIC) data, *Journal of Geophysical Research-Atmospheres*, *105*(D1), 1503-1525.

Lelieveld, J., et al. (1999), Chlorine activation and ozone destruction in the northern lowermost stratosphere, *Journal of Geophysical Research-Atmospheres*, *104*(D7), 8201-8213.

Leue, C., et al. (2001), Quantitative analysis of NOx emissions from Global Ozone Monitoring Experiment satellite image sequences, *Journal of Geophysical Research-Atmospheres*, *106*(D6), 5493-5505.

Levy, H., et al. (1999), Simulated tropospheric NOx: Its evaluation, global distribution and individual source contributions, *Journal of Geophysical Research-Atmospheres*, *104*(D21), 26279-26306.

Li, Q. B., et al. (2005), North American pollution outflow and the trapping of convectively lifted pollution by upper-level anticyclone, *Journal of Geophysical Research-Atmospheres*, *110*(D10), doi:10.1029/2004JD005039.

Lobo, P., et al. (2007), Physical characterization of aerosol emissions from a commercial gas turbine engine, *Journal of Propulsion and Power 23*(5): 919-929.

Longfellow, C. A., et al. (1999), Reactive uptake on hydrocarbon soot: Focus on NO2, *Journal of Geophysical Research-Atmospheres*, *104*(D11), 13833-13840.

Lopez, J. P., et al. (2006), CO signatures in subtropical convective clouds and anvils during CRYSTAL-FACE: An analysis of convective transport and entrainment using observations and a cloud-resolving model, *Journal of Geophysical Research-Atmospheres*, *111*(D9), doi:10.1029/2005JD006104.

Lovejoy, S., et al. (2007), Is isotropic turbulence relevant in the atmosphere?, *Geophysical Research Letters 34*(15), doi:10.1029/2007GL029359.

Luo, Z. Z., et al. (2007), Ten years of measurements of tropical upper-tropospheric water vapor by MOZAIC. Part I: Climatology, variability, transport, and relation to deep convection, *Journal of Climate*, *20*(3), 418-435.

Ma, J. Z., and X. J. Zhou (2000), Development of a three-dimensional inventory of aircraft NOx emissions over China, *Atmospheric Environment*, *34*(3), 389-396.

Ma, J. Z., et al. (2002), Summertime tropospheric ozone over China simulated with a regional chemical transport model. 2. Source contributions and budget, *Journal of Geophysical Research-Atmospheres*, *107*(D22), doi:10.1029/2001JD001355.

Marecal, V., et al. (2006), Modelling study of the impact of deep convection on the utls air composition - Part I: Analysis of ozone precursors, *Atmospheric Chemistry and Physics*, *6*, 1567-1584.

Marécal, V., G. Durry, K. Longo, S. Freitas, E. D. Rivière, and M. Pirre (2007), Mesoscale modelling of water vapour in the tropical UTLS: two case studies from the HIBISCUS campaign, *Atmospheric Chemistry and Physics* 7, 1471-1489.

Mari, C., et al. (2002), Sources of upper tropospheric HOx over the South Pacific Convergence Zone: A case study, *Journal of Geophysical Research-Atmospheres*, *108*(D2), doi:10.1029/2000JD000304.

Mari, C., et al. (2002), On the relative role of convection, chemistry, and transport over the South Pacific Convergence Zone during PEM-Tropics B: A case study, *Journal of Geophysical Research-Atmospheres*, *108*(D2), doi:10.1029/2001JD001466.

Martin, R. V., et al. (2006), Evaluation of space-based constraints on global nitrogen oxide emissions with regional aircraft measurements over and downwind of eastern North America, *Journal of Geophysical Research-Atmospheres*, *111*(D15), doi:10.1029/2005JD006680.

Martin, R. V., et al. (2007), Space-based constraints on the production of nitric oxide by lightning, *Journal of Geophysical Research-Atmospheres*, *112*(D9), doi:10.1029/2006JD007831.

Meijer, E. W., et al. (2000), Model calculations of the impact of NOx from air traffic, lightning, and surface emissions, compared with measurements, *Journal of Geophysical Research-Atmospheres*, *105*(D3), 3833-3850.

Meier, A. and J. Hendricks (2002), Model studies on the sensitivity of upper tropospheric chemistry to heterogeneous uptake of HNO_3 on cirrus ice particles, *Journal of Geophysical Research-Atmospheres 107*(D23), doi:10.1029/2001JD000735.

Meilinger, S. K., et al. (2001), On the impact of heterogeneous chemistry on ozone in the tropopause region, *Geophysical Research Letters*, *28*(3), 515-518.

Meilinger, S. K., et al. (2005), Microphysics and heterogeneous chemistry in aircraft plumes - high sensitivity on local meteorology and atmospheric composition, *Atmospheric Chemistry and Physics 5:* 533-545.

Miller, T. M., et al. (2005), Mass distribution and concentrations of negative chemiions in the exhaust of a jet engine: Sulfuric acid concentrations and observation of particle growth, *Atmospheric Environment 39*(17): 3069-3079.

Miloshevich, L. M., et al. (2006), Absolute accuracy of water vapor measurements from six operational radiosonde types launched during AWEX-G and implications for AIRS validation, *Journal of Geophysical Research-Atmospheres*, *111*(D9), doi:10.1029/2005JD006083.

Morris, G. A., et al. (2003), Potential impact of subsonic and supersonic aircraft exhaust on water vapor in the lower stratosphere assessed via a trajectory model, *Journal of Geophysical Research-Atmospheres*, *108*(D3), doi:10.1029/2002JD002614.

Muhle, J., et al. (2002), Biomass burning and fossil fuel signatures in the upper troposphere observed during a CARIBIC flight from Namibia to Germany, *Geophysical Research Letters*, *29*(19).

Muller, J. F., and G. Brasseur (1999), Sources of upper tropospheric HOx: A three-dimensional study, *Journal of Geophysical Research-Atmospheres*, *104*(D1), 1705- 1715.

Murphy, J. G., et al. (2004), Measurements of the sum of HO2NO2 and CH3O2NO2 in the remote troposphere, *Atmospheric Chemistry and Physics*, *4*, 377-384.

Nedoluha, G. E., et al. (2002), Polar Ozone and Aerosol Measurement III measurements of water vapor in the upper troposphere and lowermost stratosphere, *Journal of Geophysical Research-Atmospheres*, *107*(D10), doi: 10.1029/2001JD000793.

Neuman, J. A., et al. (2006), Reactive nitrogen transport and photochemistry in urban plumes over the North Atlantic Ocean, *Journal of Geophysical Research-Atmospheres*, *111*(D23), doi:10.1029/2005JD007010.

Nyeki, S., et al. (2004), Properties of jet engine combustion particles during the PartEmis experiment: Particle size spectra ($d > 15$ nm) and volatility, *Geophysical Research Letters*, *31*(18), doi:10.1029/2004GL020569.

Olson, J. R., et al. (2004), Testing fast photochemical theory during TRACE-P based on measurements of OH, HO2, and C H_2O, *Journal of Geophysical Research-Atmospheres*, *109*(D15), doi:10.1029/2003JD004278.

Olson, J. R., et al. (2006), A reevaluation of airborne HOx observations from NASA field campaigns, *Journal of Geophysical Research-Atmospheres* *111*(D10), doi:10.1029/2005JD006617.

Park, M., et al. (2004), Seasonal variation of methane, water vapor, and nitrogen oxides near the tropopause: Satellite observations and model simulations, *Journal of Geophysical Research-Atmospheres*, *109*(D3), doi:10.1029/2003JD003706.

Parker, D. E., M. Gordon, D. P. N. Cullum, D.M. H. Sexton, C. K. Folland, and N. Rayner (1997), A new global gridded radiosonde temperature database and recalculated temperature trends, *Geophysical Research Letters 24*, 1499–1502.

Parrish, D. D., et al. (2004), Fraction and composition of NOy transported in air masses lofted from the North American continental boundary layer, *Journal of Geophysical Research-Atmospheres, 109*(D9), doi:10.1029/2003JD004226.

Peace, H., et al. (2006), Identifying the contribution of different airport related sources to local urban air quality, *Environmental Modelling & Software 21*(4), 532-538.

Petzold, A., et al. (2005), Particle emissions from aircraft engines - a survey of the European project PartEmis, *Meteorologische Zeitschrift, 14*(4), 465-476.

Pierce, R. B., et al. (2007), Chemical data assimilation estimates of continental US ozone and nitrogen budgets during the Intercontinental Chemical Transport Experiment-North America, *Journal of Geophysical Research-Atmospheres, 112*(D12), doi:10.1029/ 2006JD007722.

Pison, I. and L. Menut (2004), Quantification of the impact of aircraft traffic emissions on tropospheric ozone over Paris area, *Atmospheric Environment 38*(7), 971-983.

Pitari, G., et al. (2002), Deep convective transport in a two-dimensional model: Effects on lower stratospheric aerosols and ozone, *Meteorologische Zeitschrift, 11*(3), 187-196.

Popp, P. J., et al. (2006), The observation of nitric acid-containing particles in the tropical lower stratosphere, *Atmospheric Chemistry and Physics 6*, 601-611.

Popovicheva, O. B., et al. (2004), Water adsorption and crystallization on soot particles, *Izvestiya Atmospheric and Oceanic Physics, 40*(2), 193-201.

Popovicheva, O. B., and A. M. Starik (2007), Aircraft-generated soot aerosols: Physicochemical properties and effects of emission into the atmosphere, *Izvestiya Atmospheric and Oceanic Physics, 43*(2), 125-141.

Popovicheva, O. B., et al. (2003), Microstructure and water adsorbability of aircraft combustor soots and kerosene flame soots: Toward an aircraft-generated soot laboratory surrogate, *Journal of Physical Chemistry A 107*(47),10046-10054.

Popovitcheva, O. B., et al. (2000), Experimental characterization of aircraft combustor soot: Microstructure, surface area, porosity and water adsorption, *Physical Chemistry Chemical Physics 2*(19), 4421-4426.

Ravetta, F., et al. (2001), Experimental evidence for the importance of convected methylhydroperoxide as a source of hydrogen oxide (HOx) radicals in the tropical upper troposphere, *Journal of Geophysical Research-Atmospheres, 106*(D23), 32709- 32716.

Ramaswamy, V., et al., (2001), Stratospheric Temperature Trends: Observations and Model Simulations, *Reviews of Geophysics, 39*, 61-122.

Reiner, T., et al. (1999), Measurements of acetone, acetic acid, and formic acid in the northern midlatitude upper troposphere and lower stratosphere, *Journal of Geophysical Research-Atmospheres, 104*(D11), 13943-13952.

Richard, E. C., et al. (2006), High-resolution airborne profiles of CH4, O_3, and water vapor near tropical Central America in late January to early February 2004, *Journal of Geophysical Research-Atmospheres 111*(D13), doi:10.1029/2005JD006513.

Ridley, B. A., et al. (2005), Comments on the parameterization of lightning-produced NO in global chemistry-transport models, *Atmospheric Environment, 39*(33), 6184-6187.

Salawitch, R. J., et al. (2005), Sensitivity of ozone to bromine in the lower stratosphere, *Geophysical Research Letters, 32*(5), doi:10.1029/2004GL021504.

Sauvage, B., et al. (2007), Remote sensed and in situ constraints on processes affecting tropical tropospheric ozone, *Atmospheric Chemistry and Physics*, *7*, 815-838.

Sausen, R., et al. (2005), Aviation radiative forcing in 2000: An update on IPCC (1999), *Meteorologische Zeitschrift 14* (4), 555-561.

Schauffler, S. M., et al. (1999), Distributions of brominated organic compounds in the troposphere and lower stratosphere, *Journal of Geophysical Research-Atmospheres*, *104*(D17), 21513-21535.

Schlager, H., et al. (1999), Regional nitric oxide enhancements in the North Atlantic flight corridor observed and modeled during POLINAT 2 - a case study, *Geophysical Research Letters*, *26*(20), 3061-3064.

Schoeberl, M. R. and G. A. Morris (2000), A Lagrangian simulation of supersonic and subsonic aircraft exhaust emissions, *Journal of Geophysical Research-Atmospheres 105*(D9): 11833-11839.

Schofield, R., et al. (2004), Retrieved tropospheric and stratospheric BrO columns over Lauder, New Zealand, Journal Of Geophysical Research-Atmospheres, 109 (D14):D14304.

Schroder, F., et al. (2000), In situ studies on volatile jet exhaust particle emissions: Impact of fuel sulfur content and environmental conditions on nuclei mode aerosols, *Journal of Geophysical Research-Atmospheres*, *105*(D15), 19941-19954.

Schumann, U., et al. (2002), Influence of fuel sulfur on the composition of aircraft exhaust plumes: The experiments SULFUR 1-7, *Journal of Geophysical Research-Atmospheres*, *107*(D15), doi:10.1029/2001JD000813.

Schumann, U. and H. Huntrieser (2007), The global lightning-induced nitrogen oxides source, *Atmospheric Chemistry and Physics 7*(14), 3823-3907.

Shonija, N. K., et al. (2007), Hydration of aircraft engine soot particles under plume conditions: Effect of sulfuric and nitric acid processing, *Journal of Geophysical Research-Atmospheres 112*(D2), doi:10.1029/2006JD007217.

Simpson, I. J., et al. (2003), Airborne measurements of cirrus-activated C2Cl4 depletion in the upper troposphere with evidence against Cl reactions, *Geophysical Research Letters*, *30*(20), doi:10.1029/2003GL017598.

Singh, H., et al. (2000), Distribution and fate of selected oxygenated organic species in the troposphere and lower stratosphere over the Atlantic, *Journal of Geophysical Research-Atmospheres*, *105*(D3), 3795-3805.

Singh, H. B., et al. (2004), Analysis of the atmospheric distribution, sources, and sinks of oxygenated volatile organic chemicals based on measurements over the Pacific during TRACE-P, *Journal of Geophysical Research-Atmospheres*, *109*(D15), doi:10.1029/2003JD003883.

Singh, H. B., L. Salas, et al. (2007), Reactive nitrogen distribution and partitioning in the North American troposphere and lowermost stratosphere, *Journal of Geophysical Research-Atmospheres 112*(D12), doi:10.1029/2006JD007664.

Sinnhuber, B.M., et al. (2005), Global observations of stratospheric bromine monoxide from SCIAMACHY, *Geophysical Research Letters 32*(20): L20810.

Sinnhuber, B. M., and I. Folkins (2006), Estimating the contribution of bromoform to stratospheric bromine and its relation to dehydration in the tropical tropopause layer, *Atmospheric Chemistry and Physics*, *6*, 4755-4761.

Sioris, C. E., et al. (2006), Latitudinal and vertical distribution of bromine monoxide in the lower stratosphere from Scanning Imaging Absorption Spectrometer for Atmospheric Chartography limb scattering measurements, *Journal of Geophysical Research-Atmospheres 111*(D14), doi:10.1029/2005JD006479.

Sioris, C. E., et al. (2007), Vertical profiles of lightning-produced NO2 enhancements in the upper troposphere observed by OSIRIS, *Atmospheric Chemistry and Physics 7*(16), 4281-4294.

Smith, J. B., et al. (2001), Mechanisms for midlatitude ozone loss: Heterogeneous chemistry in the lowermost stratosphere?, *Journal of Geophysical Research- Atmospheres 106*(D1): 1297-1309.

Smyshlyaev, S. P., et al. (2003), Model study of the impact of convective processes on the gas composition of the upper troposphere and lower stratosphere, *Izvestiya Atmospheric and Oceanic Physics*, *39*(4), 432-443.

Snow, J. A., et al. (2003), Winter-spring evolution and variability of HOx reservoir species, hydrogen peroxide, and methyl hydroperoxide, in the northern middle to high latitudes, *Journal of Geophysical Research-Atmospheres*, *108*(D4), doi:10.1029/2002JD002172.

Snow, J. A., et al. (2007), Hydrogen peroxide, methyl hydroperoxide, and formaldehyde over North America and the North Atlantic, *Journal of Geophysical Research- Atmospheres*, *112*(D12), doi:10.1029/2006JD007746.

Sorokin, A., et al. (2001), On volatile particle formation in aircraft exhaust plumes, *Physics and Chemistry of the Earth Part C-Solar-Terrestial and Planetary Science*, *26*(8), 557-561.

Solomon, S., et al. (1997), Heterogeneous chlorine chemistry in the tropopause region, *Journal of Geophysical Research-Atmospheres 102*(D17): 21411-21429.

Sorokin, A., and F. Arnold (2004), Electrically charged small soot particles in the exhaust of an aircraft gas-turbine engine combustor: comparison of model and experiment, *Atmospheric Environment*, *38*(17), 2611-2618.

Sorokin, A. and F. Arnold (2006), Organic positive ions in aircraft gas-turbine engine exhaust, *Atmospheric Environment 40*(32): 6077-6087.

Sovde, O. A., et al. (2007), Aircraft pollution - a futuristic view, Atmospheric *Chemistry and Physics 7*(13), 3621-3632.

Spichtinger, P., et al. (2002), The statistical distribution law of relative humidity in the global tropopause region, *Meteorologische Zeitschrift*, *11*(2), 83-88.

Stendahl, K. (1976): Paul Among Jews and Gentiles, and Other Essays, Fortress Press, p7.

Stevenson, D. S., et al. (2004), Radiative forcing from aircraft NOx emissions: Mechanisms and seasonal dependence, *Journal of Geophysical Research-Atmospheres* 109(D17), doi:10.1029/2004JD004759.

Stevenson, D. S., et al. (2006), Multimodel ensemble simulations of present-day and near-future tropospheric ozone, *Journal of Geophysical Research-Atmospheres*, *111*(D8), doi:10.1029/2005JD006338.

Stickler, A., et al. (2006), Influence of summertime deep convection on formaldehyde in the middle and upper troposphere over Europe, *Journal of Geophysical Research-Atmospheres*, *111*(D14), doi:10.1029/2005JD007001.

Stohl, A., et al. (2002), Export of NOy from the North American boundary layer during 1996 and 1997 North Atlantic Regional Experiments, *Journal of Geophysical Research-Atmospheres*, *107*(D11), doi:10.1029/2001JD000519.

Stohl, A., et al. (2003), A new perspective of stratosphere-troposphere exchange, *Bulletin of the American Meteorological Society 84*(11), 1565.

Talukdar, R. K., et al. (2006), Uptake of HNO_3 on hexane and aviation kerosene soots, *Journal of Physical Chemistry A 110*(31), 9643-9653.

Theys, N., et al. (2007), Retrieval of stratospheric and tropospheric BrO columns from multi-axis DOAS measurements at Reunion Island (21 degrees S, 56 degrees E), *Atmospheric Chemistry and Physics 7*(18), 4733-4749.

Thornton, B. F., et al. (2003), In situ observations of ClO near the winter polar tropopause, *Journal of Geophysical Research-Atmospheres 108*(D8), doi:10.1029/2002JD002839.

Thornton, B. F., et al. (2005), Variability of active chlorine in the lowermost Arctic stratosphere, *Journal of Geophysical Research-Atmospheres 110*(D22) doi:10.1029/2004JD005580.

Thornton, B. F., et al. (2007), Chlorine activation near the midlatitude tropopause, *Journal of Geophysical Research-Atmospheres 112*(D18), doi:10.1029/2006JD007640.

Thakur, A. N., et al. (1999), Distribution of reactive nitrogen species in the remote free troposphere: data and model comparisons, *Atmospheric Environment*, *33*(9), 1403-1422.

Troller, M., A. Geiger, E. Brockmann and H.-G. Kahle, (2006), Determination of the spatial and temporal variation of tropospheric water vapour using CGPS networks, *Geophysical Journal International*, *167*, 509–520.

Tsague, L., et al. (2006), Prediction of the production of nitrogen oxide (NOx) in turbojet engines, *Atmospheric Environment*, *40*(29), 5727-5733.

Tsague, L., et al. (2007), Prediction of emissions in turbojet engines exhausts: relationship between nitrogen oxides emission index (EINOx) and the operational parameters, *Aerospace Science and Technology 11*(6), 459-463.

Tsai, F. J., et al. (2001), A composite modeling study of civil aircraft impacts on ozone and sulfate over the Taiwan area, *Terrestrial Atmospheric and Oceanic Sciences 12*(1), 109-135.

Vaughan, G., et al. (2005), Water vapour and ozone profiles in the midlatitude upper troposphere, *Atmospheric Chemistry and Physics*, *5*, 963-971.

van Noije, T. P. C., et al. (2006), Multi-model ensemble simulations of tropospheric NO2 compared with GOME retrievals for the year 2000, *Atmospheric Chemistry and Physics*, *6*, 2943-2979.

Vay, S. A., et al. (2000), Tropospheric water vapor measurements over the North Atlantic during the Subsonic Assessment Ozone and Nitrogen Oxide Experiment (SONEX), *Journal of Geophysical Research-Atmospheres*, *105*(D3), 3745-3755.

Voigt, C., et al. (2006), Nitric acid in cirrus clouds, *Geophysical Research Letters 33*(5).

Voigt, C., et al. (2007), In-situ observations and modeling of small nitric acid-containing ice crystals, *Atmospheric Chemistry and Physics 7*(12), 3373-3383.

von Kuhlmann, R. and M. G. Lawrence (2006), The impact of ice uptake of nitric acid on atmospheric chemistry, *Atmospheric Chemistry and Physics 6*, 225-235.

Wang, C., and R. G. Prinn (2000), On the roles of deep convective clouds in tropospheric chemistry, *Journal of Geophysical Research-Atmospheres*, *105*(D17), 22269-22297.

Wang, Y., et al. (2000), Evidence of convection as a major source of condensation nuclei in the northern midlatitude upper troposphere, *Geophysical Research Letters*, *27*(3), 369-372.

Wang, Y. H., et al. (2000), Influence of convection and biomass burning outflow on tropospheric chemistry over the tropical Pacific, *Journal of Geophysical Research-Atmospheres*, *105*(D7), 9321-9333.

Wang, C. X., and Z. M. Chen (2006), Effect of CH3OOH on the atmospheric concentration of OH radicals, *Progress in Natural Science*, *16*(11), 1141-1149.

Wei, C. F., et al. (2001), Modeling of ozone reactions on aircraft-related soot in the upper troposphere and lower stratosphere, *Atmospheric Environment*, *35*(35), 6167-6180.

Weisenstein, D.K., et al. (1993), Effects On Stratospheric Ozone From High-Speed Civil Transport - Sensitivity To Stratospheric Aerosol Loading, *Journal of Geophysical Research-Atmospheres, 98*(D12), 23133-23140.

Wey, C.C., et al. (2007), Overview on the aircraft particle emissions experiment, *Journal of Propulsion and Power 23*(5), 898-905.

Wey, T. and N. S. Liu (2007), Modeling jet engine aerosols in the postcombustor flow path and sampling system, *Journal of Propulsion and Power 23*(5), 930-941.

Wild, O. and M. J. Prather (2006), Global tropospheric ozone modeling: Quantifying errors due to grid resolution, *Journal of Geophysical Research-Atmospheres 111*(D11), doi:10.1029/2005JD006605.

Wilson, C. W., et al. (2004), Measurement and prediction of emissions of aerosols and gaseous precursors from gas turbine engines (PartEmis): an overview, *Aerospace Science and Technology 8(*2), 131-143.

World Meteorological Organization (2006), Scientific Assessment of Ozone Depletion 2006 (WMO/UNEP), World Meteorological Organization Global Ozone Research and Monitoring Project—Report No. 50.

Wormhoudt, J., et al., (2007), Nitrogen oxide (NO/NO2/HONO) emissions measurements in aircraft exhausts, *Journal of Propulsion and Power* 23(5), 906-911.

Wuebbles, D., et al. (2006): A report of findings and recommendations, Workshop on the Impacts of Aviation on Climate Change, June 7-9, Boston, MA.

Yang, X., et al. (2005), Tropospheric bromine chemistry and its impacts on ozone: A model study, *Journal of Geophysical Research-Atmospheres*, *110*(D23), doi:10.1029/2005JD006244.

Yelvington, P. E., et al. (2007), Chemical speciation of hydrocarbon emissions from a commercial aircraft engine, *Journal of Propulsion and Power 23*(5), 912-918.

Zhang, R. Y., et al. (2000), Enhanced NOx by lightning in the upper troposphere and lower stratosphere inferred from the UARS global NO2 measurements, *Geophysical Research Letters*, *27*(5), 685-688.

Ziereis, H., et al. (1999), In situ measurements of the NOx distribution and variability over the eastern North Atlantic, *Journal of Geophysical Research-Atmospheres*, *104*(D13), 16021-16032.

Ziereis, H., et al. (2000), Distributions of NO, NOx, and NOy in the upper troposphere and lower stratosphere between 28 degrees and 61 degrees N during POLINAT 2, *Journal of Geophysical Research-Atmospheres*, *105*(D3), 3653-3664.

In: Aviation and the Environment
Editor: Jon C. Goodman
ISBN: 978-1-60692-320-7

Chapter 2

AVIATION-CLIMATE CHANGE RESEARCH INITIATIVE (ACCRI) SUBJECT SPECIFIC WHITE PAPER (SSWP) ON UT/LS CHEMISTRY AND TRANSPORT SSWP # II, JANUARY 24, 2008

John McConnell, Wayne Evans, Jacek Kaminski, Alexandru Lupu, Lori Neary, Kirill Semeniuk, and Kenjiro Toyota
York University, Toronto Ontario, Canada

EXECUTIVE SUMMARY

The global commercial aircraft fleet currently numbers about 10,000 and flies several billion kilometres per year while burning more than 100 MT of fuel per year at high temperatures producing mostly water and CO_2. However, NOx (= $NO+NO_2$), other minor gaseous species, organic aerosols from unburnt fuel and soot and ions are also injected at cruise altitudes located in upper troposphere and lower stratosphere (UT/LS), a region particularly sensitive to atmospheric climate change.

The demand for air transportation in the US is projected to grow three fold by 2025 while similar growth is projected for the aviation industry world wide. Future climate impacts are expected to increase based on this projected aviation growth and resulting changing atmospheric conditions. These impacts relate to the impact of tripling aviation system capacity and the resulting global impact of these additional engine emissions which are estimated to be approximately twice as large as at the turn of the last century. However, if current economic projections obtain for this period, boundary layer (BL) NOx emissions may also double and hence their contribution to the UT region. In addition to global climate impacts there is also potential for even greater regional or local effects. The growth of emissions of both BL and aircraft NOx will likely lead to an increased production of ozone in the UTLS. This increase in UT/LS ozone will cause a significant increase in the radiative forcing, which in turn will contribute to global warming.

Jet traffic spends 60% of the time in the upper troposphere (UT) where current information is insufficient to make an accurate prediction of the climate impacts of increased jet traffic due to tropospheric ozone generated from aircraft NOx emissions. A review of the HOx/NOx chemistry concludes that the chemistry is fairly well known in the lower stratosphere. However, the upper troposphere HOx/NOx chemistry is uncertain as revealed by measurements of NO and OH concentrations which conflict with current model simulations from various aircraft campaigns. The aerosols from aviation emissions can also interact with the background constituents and alter the NOx and ClOx chemistry with resulting changes in regional ozone in the UT/LS.

In order to evaluate the impact of NOx from aircraft on UT ozone, the other sources of NOx in this region, such as lightning and transport from the boundary layer must be better quantified than at present. Aircraft measurements reveal "high" levels of NOx in the summer UT over North America, generated by lightning. Satellite measurements have recently been analyzed to give a more precise estimate of the global lightning source to be about 5±3 MT per year which makes it an important NOx in-situ source for the UT. Upward transport of NOx from the boundary layer is also significant. The fraction of the deep convection source of NOx from the surface sources which reaches the 10 to 13 km level is estimated to be around 10-50 %. While the uncertainty in this estimate may simply reflect the difference in the meteorology in the measurement regions, this needs to be better characterized. The parameterization for convective transport in 3D models needs to be improved. The effects of the vertical transport by the Asian monsoon and the Madden-Julian oscillation should be modelled or parameterized by models as they impact transport into the upper troposphere and stratosphere via the tropical transition layer.

In the aviation corridors, NOx is elevated above the background by aircraft emissions; the aircraft contribution may be dominant under certain meteorological conditions. The elevated NOx and ozone may persist for some time and be transported to other regions. The radiative forcing due to ozone may be much higher in some areas than on a global basis. Hence characterization of the aviation perturbations within the flight corridors needs to be improved and should be the focus of aircraft and satellite studies. This calls for additional extensive aircraft campaigns focused on the flight corridors in order to quantify the regional climate effects of aviation.

There are several recent satellites which provide new information on the NOx and nitric acid at flight levels. The data from MIPAS, ACE and AURA/MLS/HIRDLS is being applied to the NOx/HOx chemistry of the upper troposphere. Despite the excellent scientific progress now being made, future satellite instruments with enhanced capabilities are required. These enhanced capabilities should include improved vertical resolution to study the UT/LS region. Denser sampling and higher horizontal resolution are required to address the corridors issue.

There is a real concern that there will be a gap in satellite instruments suitable for UT/LS investigations in the next 5 years. Deploying ozonesondes as satellite observation gap fillers would seem to be the minimum requirement. The SHADOZ/IONS ozonesondes have proven to be highly useful for investigating ozone in the upper troposphere since they have excellent vertical resolution.

The real verification of the climate impact of increased upper troposphere ozone is the detection of changes in the ozone radiative forcing (RF) at the surface and at the top of the atmosphere. There are difficulties in measuring changes in the IPCC radiative forcing metric because of the way in which it is defined at the top of the tropopause. There are large uncertainties in the calculations of the radiative forcing metric due to a lack of knowledge of cloud effects. There need to be verifications of the radiative forcing metric by comparison against real measurements of observed surface radiative forcing and with satellite radiative trapping at the top of the atmosphere. This will need to be accomplished by concurrent simulations of surface forcing and top of the atmosphere radiative trapping with the same models used to calculate the RF metric.

There have been important advances in models since 1999. Data assimilation has proven very valuable in providing a "value-added" component to satellite data. There are new satellite instruments taking global measurements which could be used to compare with the model outputs. Multiscale models are needed to investigate the corridors aspect of the aircraft emissions and the transition to regional scale climate impacts. Parameterizations used for deep convection need to be both used consistently (with the basic dynamical model) and verified according to the scale of model.

1. Introduction

Currently, world wide, the commercial aircraft fleet numbers about 10,000 and flies several billion kilometres per year while burning more than 100 MT of fuel per year at high temperatures producing mostly water and CO_2. However, NOx (=NO+NO2), other minor gaseous species, organic aerosols from unburnt fuel and soot and ions are also injected at cruise altitudes located in upper troposphere and lower stratosphere (UT/LS) region. This region is particularly sensitive to atmospheric climate change: in the tropical tropopause layer (TTL) net heating is particularly weak and the dynamics is impacted by non-local effects such as wave breaking in the stratosphere (Holton et al., 1995; Leblanc et al, 2003). Ozone in this region is particularly important as a greenhouse gas (GHG). At mid-and high-latitudes stratospheric/tropospheric exchange (STE) occurs and delivery of ozone and other species to the troposphere is important. The state of the tropical cold point tropopause is affected by the composition and the associated thermal balance. This regulates the entry of water vapour into the middle atmosphere and hence the temperature and polar ozone chemistry in the stratosphere, which has an impact on the dynamics and thereby the UT/LS region.

Most aircraft NOx emissions are released directly into the chemically complex and radiatively sensitive UT/LS between 8-13 km. At the time of the IPCC (Penner et al., 1999) assessment, there was concern that heterogeneous chemistry following immediate conversion of sulfur to aerosols from the aircraft engines could affect the impact on ozone from the NOx emissions. Recent measurements suggest that this immediate conversion is sensitive to background conditions.

The effect of aircraft emissions on atmospheric ozone concentration depends on the altitude at which the emissions are injected. The importance of the NO catalytic production of ozone from the NOx emissions through the oxidation of methane and hydrocarbons become less effective with altitude while the catalytic ozone loss cycles become more efficient. Any uncertainties in how well we understand the atmospheric chemical and physical processes in the UT/LS affect our ability to understand the magnitude of the aviation effects on ozone and methane.

As noted above, the impact of aviation in the UT/LS was subject of an IPPC report (Penner et al., 1999) and most recently was the subject of workshop in Boston (Wuebbles, 2006). When aircraft fly in this region the NOx emitted reacts with HOx (=OH + HO2), CO and sunlight and leads to the generation of ozone. The water emitted and the ozone generated also impact the HOx which can impact methane, another GHG. Thus, from these simple examples, we see that there is an intimate and complex link between aircraft emissions and the potential for climate impact. And as the size of the fleet is expected to increase in the first part of this century it is important to consider the future impacts.

The demand for air travel in the US is projected to grow by about three fold by 2025 (see below). World wide aviation growth is also expected. Estimates of annual fuel use by 2020 annual for commercial air traffic are ~ 350 MT or 2.6 times the estimated fuel use by the global 1999 commercial fleet. This translates into global NOx emissions of ~ 1.5 MT-N from commercial air traffic or about 2.8 times the estimated 1999 NOx emissions levels. At the same time total revenue passenger kilometers are projected to increase from 3,170 billion in 1999 to 8,390 billion in 2020, or by a factor of 2.65 (Sutkus et al., 2003) These estimated increased emissions are expected to lead to global increase in ozone production with the potential for larger regional effects. The increase in UT/LS ozone will cause a significant increase in the radiative forcing, which in turn will contribute to global warming.

In this report, in addition to dealing with the topic of ozone generation and the associated direct radiative forcing and indirect forcing via its impact on methane, we were also asked to attempt an assessment of the uncertainty of dynamical influences on the impacts of aircraft emissions. We have approached this aspect of our directive by integrating it with the discussion on chemistry as it is difficult to separate the chemical impacts from transport influences.

One of the problems that arise with respect to the impact of the commercial fleet of aircraft is that we would like to be able to characterize the natural or unperturbed atmosphere so that we have a baseline for comparison. Unfortunately, we do not have that luxury since the current fleet has been flying in the UT/LS before the region has been well characterized. Thus the assessment of current impact of aircraft must be addressed via modelling studies combined with the aircraft measurement programs and models require careful evaluation.

The outline of the remainder of the report is as follows. In section two we summarize the background science focusing on the chemical and transport issues, which includes gas phase and aerosol chemistry, outlining the current state of the science including modelling, and we attempt to locate the impact of aviation within the climate change context. In section three we extract the problem areas, the areas that we see developing and areas that need substantial improvement. And we look to the other areas of the study, particularly the relationship with condensation trails and condensation cirrus. Section four sets priorities on what can be done on the short term, while section five sets recommendations and section six provides a summary.

2. Current State of Chemical and Dynamical Issues in the UT/LS

2.1. Current State of Atmospheric Chemistry in the UT/LS

In the UT/LS the chemistry is driven by the presence of ozone and water vapour: the ozone is photolysed to produce $O(^1D)$ which, with water vapour, produces HOx radicals viz.,

$$O_3 + h\upsilon \rightarrow O(^1D) + O_2$$
$$O(^1D) + H_2O \rightarrow 2OH$$

In the lower stratosphere the HOx radicals interact with NOy,[1] Cly[2] and Bry[3] and impact the ozone budget in this region. Any NO generated rapidly gets converted to a suite of NOy species in the stratosphere, including NO, NO_2, NO_3, N_2O_5, HNO_3, HNO_4, $ClNO_3$, $BrNO_3$ (and in the mesosphere N can be included).

The major source of ozone in the stratosphere is via the photolysis of O_2

$$O_2 + h\upsilon \rightarrow 2O\,.$$

As can be seen in figure 1 the net ozone source is below about 10 mb and is principally in the tropics while at high latitudes net chemical loss occurs (e.g. Cunnold et al.,1980). From this source structure the stratosphere supplies about 500 MTs of ozone annually to the troposphere, principally at high latitudes, and this represents an important component of the *net* photochemical budget of tropospheric ozone. For the most part our understanding of atmospheric chemistry outside of polar regions in the lower stratosphere appears to be reasonably well understood from satellite, balloon and aircraft measurements (e.g. Zellner, 1999).

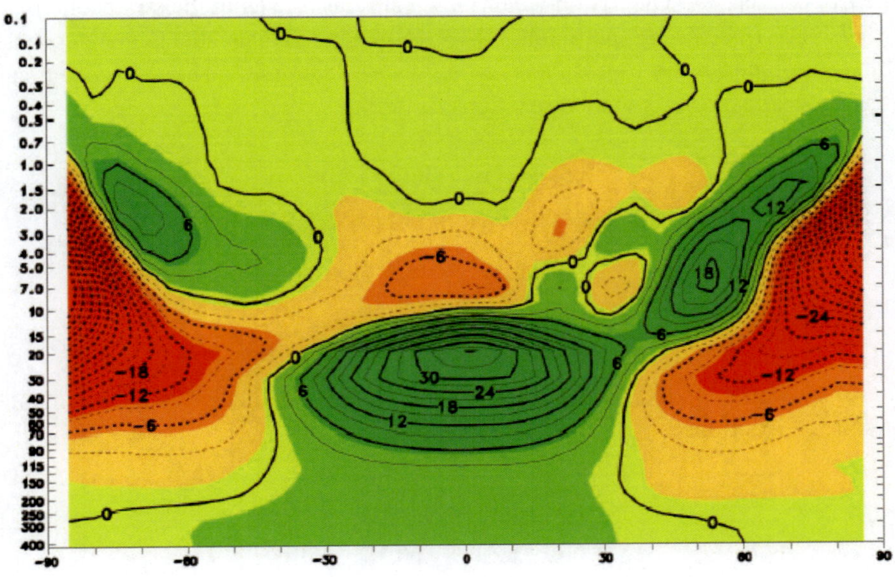

Figure 1. Net ozone production for equinox from CMAM model (10^9 ozone molecules cm^{-3}.day^{-1}) (courtesy of Stephen Beagley, 2007). [Will be cut off at 10 mb].

[1] NOy = NO + NO2 + NO3 + 2N2O5 + HNO2 + HNO3 + ClONO2 + BrONO2

[2] $Cl_y = Cl + ClO + HOCl + ClONO_2 + HCl + 2Cl_2O_2 + BrCl + 2Cl_2$

[3] $Br_y = Br + BrO + HOBr + BrONO_2 + HBr + BrCl + 2Br_2$

One of the problem areas in the stratosphere is the level of Bry. Analysis of BrO stratospheric measurements suggest that the levels of Bry in the stratosphere are about 24 pptv which is somewhat larger than that supplied by halons and CH_3Br (e.g., Salawitch et al., 2005; WMO, 2007) The discrepancy between the observations and the sources is probably due to the supply of halogenated very short lived species (WMO, 2006) which are not well mixed in the troposphere. Since Bry is important particularly for ozone loss in polar regions, knowledge of future emissions of these short lived species is important.

Another uncertainty at this point in polar regions is that about 2/3 of the loss of ozone is thought to be via the formation of the Cl_2O_2 dimer via the reaction sequence

$$ClO + ClO \xrightarrow{M} Cl_2O_2$$
$$Cl_2O_2 + h\upsilon \rightarrow Cl + ClO_2$$
$$ClO_2 + M \rightarrow Cl + O_2$$
$$2(Cl + O_3 \rightarrow ClO + O_2)$$
$$\text{Net}: 2O_3 \rightarrow 3O_2$$

However, recent laboratory measurements of Cl_2O_2 cross sections applied to the observations of the important species in the above reaction sequence, Cl_2O_2, ClO, and O_3 suggest that the reaction sequence is not rapid enough to account for the observed ozone loss (von Hobe et al., 2007; Pope et al., 2007).

In the troposphere O_3, $O(^1D)$ and water play a similar role as in the stratosphere, acting as a source of HOx radicals. In particular, the OH generated can attack many species, both organic and inorganic in the troposphere, acting as a "detergent". The main chemical source of ozone in the troposphere is via the reactions

$$NO + HO_2 \rightarrow NO_2 + OH$$
$$NO_2 + h\upsilon \xrightarrow{O_2} NO + O_3$$
$$CO + OH \xrightarrow{O_2} CO_2 + HO_2$$
$$\text{net}: CO + 2O_2 \rightarrow CO_2 + O_3$$

where the first reaction breaks the O_2 bond. Organic peroxy radicals can also generate O_3 via a similar suite of reactions

$$RH + OH \xrightarrow{O_2} RO_2 + H_2O$$
$$NO + RO_2 \rightarrow NO_2 + RO$$
$$NO_2 + h\upsilon \xrightarrow{O_2} NO + O_3$$

where RH is a gaseous organic species and further reaction of the RO can generate HOx and catalyze the formation of ozone. Thus it is clear that, under suitable conditions, NO can catalyze the generation of O_3. In addition, NO also converts HO_2 to OH.

Photochemical production of ozone in the troposphere is estimated to be about 5,000 MT/year while photochemical loss is about 4,500 MT/year leading to a net photochemical source of about 500 MT/year, i.e. of the same order to the source from the stratosphere (e.g. Stevenson et al., 2006). An important loss process for ozone is deposition to the surface over land, sea and to a less extent over ice, which amounts to a global loss of about 1,000 MT/year. In general an increase of ozone in the troposphere will lead to an increase of OH.

One of the species generated by the VOC-NOx chemistry is PAN ($CH_3CO_2NO_2$) which can be produced by the breakdown of isoprene and acetone for example. Measurements indicate that PAN is an important carrier of NOx in the UT region as it is stable at low temperatures and is relatively insoluble and slow to photolyse (e.g. Singh et al., 2006; 2007). One aspect of UT chemistry that needs to be more carefully studied is the sometimes-used assumption of photochemical steady state (PCSS) for analysis of measurements. For example, a recent study by Bertram et al. (2007) suggests that the measurement of the ratio of NOx and HNO_3 in the UT region can reveal lightning sources of NOx that occur with large scale convection. But an important corollary to their work is that quite often NOx and HNO_3 are not in PCSS due to the interaction of convection and chemistry, each with similar time scales. (See also Prather and Jacob, 1997; Lawrence and Jacob, 1998). But since convection is, in some sense, stochastic, simple photochemical calculations will be misleading and different process metrics need to be developed such as PDFs which carry the statistical information.

As noted above OH can attack organic species and one of the most important organics is methane, which is a greenhouse gas. Thus if tropospheric ozone were to increase, as a result of increased NOx, (to which aviation would contribute several percent) it is expected that OH would increase leading to a decrease in CH_4. Thus, as noted in Penner et al. (1999), there will be compensating effects in terms of radiative forcing, positive from ozone increases and negative from methane decreases. However, due to their differing lifetimes, the ozone effect will be more regional (cf. figure 2) whereas the methane impact will be global in extent (see also work by Stevenson et al., 2004).

Thus sources of NOx and hydrocarbons (HCs) are important to the ozone budget in the troposphere and it is important to have reliable estimates of their emissions.

In the cold, especially winter, free troposphere, NOx is converted and sequestered as PAN, HNO_3, HNO_4, and N_2O_5:

$$CH_3C(O)O_2 + NO_2 + M \leftrightarrow PAN + M$$
$$NO_2 + OH + M \rightarrow HNO_3 + M$$
$$NO_2 + HO_2 + M \leftrightarrow HNO_4 + M$$
$$NO_2 + O_3 \rightarrow NO_3 + O_2$$
$$NO_2 + NO_3 + M \leftrightarrow N_2O_5 + M$$

Here the $CH_3C(O)O_2$ radical is most likely produced by photochemical degradation of acetaldehyde and acetone. There are substantial contributions to the HOx radical source in the upper troposphere from acetone and other oxygenated organic compounds but with some uncertainty as regards quantitative understanding (see Section 2.2.2). N_2O_5 is also transformed to HNO_3 heterogeneously on "hygroscopic" aerosols ubiquitous in the troposphere:

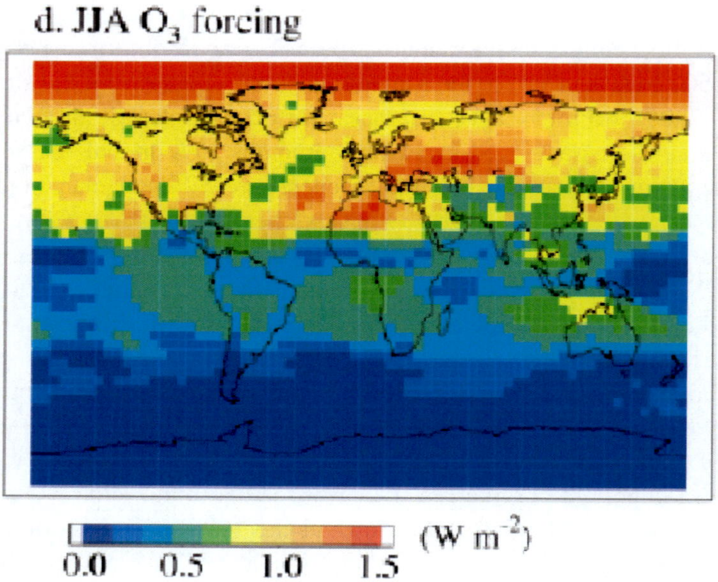

Figure 2. The instantaneous radiative forcing from tropospheric ozone since pre-industrial times of 0.49 W/m2 (Mickley et al., 2004).

$$N_2O_5 + H_2O \text{ (on aerosols)} \rightarrow 2\, HNO_3$$

According to 3-D model simulations by Dentener and Crutzen [1993], this heterogeneous reaction provides a dominant pathway for the conversion of NOx to HNO_3 in the winter troposphere and leads to significant decrease in the concentrations of ozone and OH radical (20% and 25%, respectively, on average in the northern hemisphere) particularly from winter to spring because HNO_3 is much more stable than N_2O_5 under sunlight.

Comparisons between observed and modeled NOx concentrations in the mid- to upper-troposphere from tropics to high latitudes confirmed the role of heterogeneous N_2O_5 hydrolysis, but also indicated that its reaction probability (r) should be generally smaller (by a factor of 2 or more) than assumed in Dentener and Crutzen's pioneering work (r = 0.1) [Schultz et al., 2000; Tie et al., 2003]. More stringent evidence has been obtained in the summertime lower troposphere over the United States by airborne in-situ measurements of N_2O_5 and NO_3 along with detailed aerosol measurements [Brown et al., 2006], in which the analysis of photochemical steady-state N_2O_5 concentrations inferred significant decrease in r(N_2O_5) on aerosols by more than an order of magnitude in air masses likely enriched in organic coating, nitrate content, or efflorescence of aerosols in agreement with available experimental data [Kane et al., 2001; Folkers et al., 2003; Hallquist et al., 2003; Thornton et al., 2003]. The same methodology, however, may not work well to estimate the role of the heterogeneous N_2O_5 hydrolysis in the UTLS because (as noted above) of long timescales needed to establish the photochemical steady state at cold temperatures [Brown et al., 2003]. Evans et al. [2005] compiled a new parameterization for r(N_2O_5) as a function of aerosol composition, relative humidity and temperature for the GEOS- CHEM global chemical transport model and obtained the global mean $r(N_2O_5) = 0.02$. This resulted in increases in NOx, O_3 and OH concentrations by 7%, 4%, and 8%, respectively, relative to those simulated by assuming constant $r(N_2O_5) = 0.1$. The largest changes were found in descending branches of the Hadley circulation where relative humidity is very low. Also, the new parameterization was shown to better simulate climatological values of O_3 and OH in the global troposphere for their model.

The understanding of the source and role of organic aerosols, which are very often found to be mixed internally with sulfate aerosols even in the UTLS, is still quite uncertain and requires further theoretical and experimental work for their source identification and chemical characterization and for their impacts on N_2O_5 hydrolysis [Iraci and Tolbert, 1997; Zhao et al., 2005; Heald et al., 2005; Murphy et al., 2007].

A few ppt of inorganic bromine background are likely to exist in the free troposphere (mainly via photodecomposition of bromo-carbons) as well as in the marine boundary layer (mainly via volatilization from sea-salt aerosols) and thus $BrNO_3$ hydrolysis may add to the conversion of NOx to HNO_3 contributing to the budgets of NOx and ozone:

$$BrO + NO_2 + M \rightarrow BrNO_3 + M$$

$$BrNO_3 + H_2O \text{ (on aerosols/clouds)} \rightarrow HOBr + HNO_3$$

and also leads to the conversion of water vapour to HOx since both HOBr and HNO_3 photolyze producing OH (Lary et al., 1996).

Two model studies have been performed regarding this issue in the free troposphere; von Glasow et al. [2004] found no more than a marginal impact from the heterogeneous $BrNO_3$ hydrolysis on aerosols, whereas Yang et al. [2005] showed that the surface of cloud droplets may activate this process. Since significant uncertainties exist with regard to the tropospheric source of inorganic bromine and the role of clouds for the heterogeneous reactions as well as for the scavenging of reactive gases, more work is needed for better estimates.

Aerosol emission from aircraft can be impacted by the role of organic material in the plume. However the impact appears greatest for low sulphur fuels (Yu et al., 1999).

One of the possibilities suggested for the difference between model and measured NOx/HNO_3 ratio was the uptake of HNO_3 by cirrus which also lead to denitrification if the cirrus particles are large and rapidly sedimented (e.g. Lawrence and Crutzen, 1998).

Measurements in the SOLVE (SAGE III Ozone Loss and Validation Experiment) and BIBLE (Biomass Burning and Lightning Experiment) campaigns and reported by Kondo et al. (2003) indicated that the HNO_3 uptake on cirrus clouds is very strongly temperature dependent and uptake would only be important in the UT with temperature less than ~ 215K. Ziereis et al. (2004) during the INCA campaign also found a strong temperature dependence of uptake but that, on average, only ~ 1% of NOy was found as particulate NOy. Some of the early laboratory studies suggested that ice could sequester enough HNO_3 to affect the NOx/HNO_3 photochemical ratio and this was also addressed by laboratory and model studies (cf. Hudson et al., 2002; Tabazedeh et al., 1999).

Gamblin et al. (2006a,b) have also addressed the issue of HNO_3 and NOy on cirrus using measurements from the SOLVE-I field campaign in the UT and LS over Scandinavia. They find that often NOy uptake on cirrus is important but, in the troposphere, that HNO_3 is not necessarily the main species deposited to cirrus ice and suggest that perhaps N_2O_5, HNO_4 or PAN may preferentially deposit to ice; the amount of uptake also appears sensitive to the length of time the air parcel may have been exposed to sunlight. They also consider that the apparent variability of the results in the UT may indicate that the region may not be in photochemical or physiochemical equilibrium in the mixed media situation. They also suggest (Gamblin 2007) a possible time marker or clock for cloud parcel lifetime in the UT.

2.2. Emissions

2.2.1. NO_x emissions

Emissions of NO_x into the troposphere come from a variety of sources: anthropogenic emissions, biomass burning, boreal forest fires, biogenic emissions, lightning emissions, cosmic rays and transport from the stratosphere. In the stratosphere the main source of NO_x is via the oxidation of N_2O with $O(^1D)$ with contributions from cosmic rays, auroral precipitation and sporadic solar proton events (e.g. Jackman et al., 1985; 2005).

The stratospheric source of NO to the troposphere can be estimated from the loss of N_2O in the stratosphere which is ~ 12 MT-N/year (IPCC,2001, table 4.4). About 10% of the N_2O loss occurs with reaction with $O(^1D)$ and 6% of the total yields NO (e.g. Olsen et al., 2001). There are also "small" contributions from cosmic ray precipitation ~ 0.08 MT-N/year and also auroral precipitation (Jackman et al., 1980). Most of the NO_y exits to the troposphere in the form of HNO_3 with a global flux of about 0.7 MT-N/year.

Burning of fossil fuel represents an important source of NO_x in the atmosphere with a total of about 33 MT/year and of which aviation currently contributes about 2% (see table 1). However, most of the anthropogenic emissions are into the boundary layer where the lifetime is of order of a few days whereas aviation emissions are mostly injected into the UT/LS region where the lifetime is longer, ~ months.

Biomass burning arises mostly in the tropics due to human activities (e.g. Crutzen and Andreae, 1990) such as land clearance etc and amounts to ~ 7.1T MT-N/year of which Boreal forest fires contribute about 3-8%: There is a substantial year to year variation. Boreal sources in particular may be expected to increase with changing climate (Stocks et al., 1998; Gillett et al., 2004; Flannigan et al., 2005). Emission estimates for the 1997-2006 period are available online (Randerson et al., 2007). The approach used to calculate these emissions is described in van der Werf et al. (2006), Giglio et al. (2006) and Randerson et al. (2005). The NOx emissions follow the dry seasons in the Northern and Southern hemispheres and have a year to year variability reflecting the movement of the ITCZ and the influence of El Niño (van der Werf et al., 2004, 2006). One of the uncertainties is the effective injection height of biomass burning emissions.

The amount of NOx generated by lighting is relatively uncertain and values between 2 and 12 MT-N/year are common in the literature (e.g. IPCC, 2001). Recent estimates appear to have reduced the uncertainty to about 6±2 MT- NOx/year (Martin et al, 2007) much of which is generated in the upper atmosphere. This is similar to estimate in the comprehensive review of Schumann and Huntrieser (2007) of 5±3 MT-N/year.

Table 1. Global NO_x sources (MT-N/year) for 2000 (IPCC, 2001) and 2030 Dentener et al. (2005)*

Global NOx sources	2000	Above 7 km	*2030
Fossil Fuel	33.0		40-50
Aircraft	0.7	0.6	1.5†
Biomass Burning	7.1	??	7.1
Soils	3.0**		3.0
Lightning	6*	3.6	6
Stratosphere	0.7***	0.7	0.7
Total	50.5		

* Martin et al, 2007 (The figure in brackets is the emission above 7 km.)
** Jaeglé et al., 2005.
*** Olsen et al., 2001 (and text)
† Sutkis et al. (2003).

However, we note that the analysis was a multistep process depending on chemical and dynamical modelling and observations from several different satellites and the associated error bars may be optimistic. Also reasonable estimates indicate that more than 60% of the NOx is created above 7 km at continental mid-latitudes (e.g. Pickering et al., 1998). This source of about 3.6 MT-N represents the largest in-situ source of the NOx in the upper troposphere. However, the vertical distribution of the emission source needs evaluation and towards that end the observations from the OSIRIS instrument on the Odin satellite should be useful (Sioris et al., 2007).

Cosmic ray production of NO is modulated by the Sun's magnetic field and so maximizes during solar minimum. Production maximizes ~ 12 km in polar regions due to deflection of the ions by the Earth's magnetic field. The source is ~ 0.08 MT-N/year (e.g. Jackman et al., 1980) mostly concentrated at higher latitudes.

Another "source" of NOx in the upper troposphere is that due to convection (see below). As noted above a large fraction of the NOx emissions occur at or near the surface. Thus if there is large scale convection the NOx can be lofted to the upper troposphere. Current estimates are very model sensitive and not always available from model output. Of the ~ 45 MT-N/year emitted in or near surface if 10% was lofted to the UT this would significantly impact the NOx budget in this region and this is within the range of estimates, certainly for North America as noted by Singh et al. (2007). However, during the lofting process the soluble HNO_3 would be preferentially removed in the convective tower by rainout leaving behind the NOx.

2.2.2. CO and VOC emissions

As was seen above CO plays a major role in the generation of ozone and thus knowledge of its distribution and how it might change in the future is important. Future emissions we shall address below. Current emissions, shown in table 2 are taken from the IPCC (2001) report. The sources are similar as for NOx, viz. anthropogenic, biomass burning, ocean, oxidation of methane and VOCs etc. One point to note, however, is that CO emissions from aircraft play a much more muted role than those of NOx.

The situation regarding current CO emissions seems rather uncertain, but it may also relate to model uncertainties. For example, modeling studies undertaken for the IPCC (2007) (e.g. Shindell et al., 2006) ostensibly with the similar CO emissions give very different CO fields. This is a question that needs to be addressed.

Table 2. Sources of CO (MT/year) (IPCC, 2001)

Oxidation of methane	800
Oxidation of isoprene	270
Oxidation of industrial NMHC	110
Oxidation of biomass NMHC	30
Oxidation of acetone	20
Sub-total in-situ oxidation	1230
Vegetation	150
Oceans	50
Biomass burning	700
Aircraft (Sutkis et al., 2003)	0.7
Fossil fuel	650
Sub-total direct emissions	1550
Total Sources	2780

Volatile organic compounds (VOC) include a wide variety of non-methane hydrocarbons (NMHC) and oxygenated NMHC. There are three main sources: (a) anthropogenic emissions (b) biomass burning, and (c) vegetation [IPCC, 2001] with vegetation supplying two-thirds of the global source, emitted primarily in the tropics. VOC emissions from fossil fuel usage (approximately 25% of total emissions) and biomass burning (about 5% of total emissions) have distributions similar to NO_x.

Table 3. VOC emissions *MT-C/year (IPCC, 2001)

	VOCs	Isoprene	terpene	acetone
Fossil Fuel	161			
Biomass Burning	33			
Vegetation	377	220	127	30
Total	571			

We note that the major source of emissions are natural and often are temperature sensitive and thus represent a potential changing source in a climate change scenario.

2.3. Measurements

2.3.1. Aircraft Measurements

The chemical composition of the UT/LS region has been explored by many missions using aircraft (compare Fehsenfeld et al., 2006 for a review of past work). One of the more recent experiment suites has been by the INTEX (Intercontinental Chemical Transport Experiment) mission. Singh et al. (2007), for example, found that there were occasions when the measurements of NOx and HOx and complementary species were not consistently modelled in the upper troposphere while in the lower stratosphere there was consistency between measurements and modelling. The disagreement suggests either errors in the measurements or perhaps more likely important species have not been definitively identified (such as Brx, Cly, aerosols) or error in rate constants which seems less likely.

There also has been important European research studying emissions from aircraft and their impact on atmospheric composition and climate in the UT/LS region. These studies included measurements of O_3, NOx, OH, CH_4 and water vapour (e.g. Sausen et al, 2005; Gauss et al., 2004).

2.3.2. Satellite Experiments

While aircraft campaigns reveal details of atmospheric processes satellite measurement programs with their associated validation campaigns provide global perspective. Since IPCC (1999) many satellites have been launched that have revealed the global nature of air quality in the troposphere and the status of the stratospheric ozone (e.g. Odin, Terra, Aura, SCISAT-I, ENVISAT). Most recently Calipso and CloudSat have added to the A-Train capability. MetOp, launched in 2006, has as its primary objective the gathering of meteorological data (~1 km vertical resolution of T and water vapour to about 10 km and a horizontal resolution of about 10x10 km_2) for improving weather forecasting. However, it also has an important air quality (AQ) capability via instruments such as GOME2, IASI and AVHRR (http://www.esa.int/esaEO/SEM9NO2VQUD_index_0_m.html) which can provide column or partial column information on many chemical species such as NO_2, CO, ozone, HNO_3 and aerosols of tropospheric and stratospheric interest. Global CO partial columns with ~ 20x20 km horizontal resolution have been measured by MOPITT (References for a suite of tropospheric instrument details and retrievals are given in IGAC, 2007). TES has reported partial vertical columns for CO and ozone (both tropospheric and stratospheric) while MLS/AURA operating in limb mode measures CO, as does AIRS. In spite of problems with tape on the front aperture the HIRDLS team have managed to obtain very interesting data on O_3, H_2O, CH_4, N_2O, HNO_3, CFC-11, CFC-12, temperature, cloud top pressure, and four aerosol extinctions on a standard pressure grid with ~ 1 km vertical resolution for temperature (see http://daac.gsfc.nasa.gov/Aura/HIRDLS/index.shtml, and Massie et al., 2007). Many nadir viewing surface remote sensing instruments provide estimates of aerosol optical depth (AOD), e.g. MODIS (IGAC, 2007) and aerosol properties (Hu et al., 2007). Limb viewing instruments such as OSIRIS (Murtagh et al., 2002), ACE-FTS (Bernath et al., 2005), MAESTRO (McElroy et al., 2007), the SAGE suite (e.g. Wang et al., 2006; http://www-sage3.larc.nasa.gov/) measurements provide measurements of aerosols and cirrus clouds and sub-visible cirrus in the UT/LS region.

Furthermore this information can be ingested into models via data assimilation and the information transported to other locations.

Although the prime mission of many of the satellite programs mentioned above was not to address UT/LS issues important to aviation, careful improvements in retrieval method and validation programs have lead to the generation of quantitative measurements for the UT/LS region. For example, we note that the ACE-FTS instrument on SCISAT-I simultaneously measures 20-30 species down to 5 km including many species related to AQ and biomass burning and forest fires (Bernath et al., 2005). Likewise the MLS and TES instruments, both on EOS-Aura, are being used to address AQ issues, long range transport and impacts of large scale convection (Jiang et al., 2007). However the vertical resolution for the above instruments is at least 3-4 km (e.g. ACE). Nevertheless, instruments such as SAGE, OSIRIS, MAESTRO and HIRDLS have resolution ~ 1 km suitable for UT/LS studies.

2.3.3. Aircraft Measurements Investigating Flight Corridors

The POLINAT (Pollution From Aircraft Emissions in the North Atlantic Flight Corridor (POLINAT) field campaign, using 3 aircraft and lasting from 1994 to 1998 was designed to assess the impact of commercial aircraft on the atmosphere in the vicinity of the eastern North Atlantic flight corridor by means of measurements (and modelling support) of the distribution of NO, NO_x, NO_y and other species (e.g. Schumann et al., 2000; Schlager et al., 1999; Singh et al., 1999; Ziereis et al., 2000). The measurements indicated strong latitudinal gradients in NO, NO_x, and NO_y. The research aircraft sampled plumes of various ages and saw clear signatures of air having been lofted from the lower atmosphere and also signatures of lightning NO_x. Complementary modelling studies indicated that about 50% of the NO_x in the UT/LS was from the boundary layer. This should be compared with the later work on INTEX-A over north America where the results suggested that most of the NO_x was from lightning (10% from the boundary layer) (Singh et al., 2006; 2007). This difference may reflect the different meteorological situations over the summer hot convectively unstable central USA as compared to wave cyclone systems impacting Western Europe and eastern Atlantic or simply the uncertainty in the analysis. Cooper et al. (2006) found, after allowing for recent stratospheric intrusions using FLEXPART (Stohl et al., 1998; 2005), that ~ 75% of the upper tropospheric "ozone excess" over North America is generated by lightning NO_x emissions.

The MOZAIC project, with instruments flown on commercial flights, (e.g., Cooper et al., 2006) flies in the flight corridors and represents a unique dataset for corridor studies. The MOZAIC package carries sensors to measure ozone, NO_y compounds, water vapour, temperature and relative humidity: it has been flown on over 26,600 flights from 1994 to 2006 (e.g., Marenco et al., 1998; Nedelec et al., 2003; Volz-Thomas, 2005).

2.4. Modelling Studies

Modelling studies are important for studying impacts of changing conditions either natural or anthropogenic. Current (and past) data are used for diagnosis and evaluation of models (e.g. IPCC2001, 2007) so that they may be used with confidence but with the limitations transparent.

Our understanding of the UT/LS region of the atmosphere has been advanced by chemical and dynamical modelling using different but related models spanning many scales from microphysics of cloud formation to global scale transports. Global scale models include chemical transport models (CTMs), on-line weather forecast models and climate models. Mesoscale and cloud resolving models are also being applied in an attempt to improve our understanding of processes and in particular the exchange of water between the troposphere and stratosphere.

2.4.1. Plume to Global Scale Modelling

The injection of aircraft emissions into the atmosphere occurs at the engine scale but we seek the impacts on a global scale. Current global models are not able to resolve at this scale and the process of accounting for the injection of aircraft emissions has to be parameterized and this is usually done by using models that can resolve the scales from the engine to the plume scale (loosely defined). Mesoscale modelling is then required to assess the processes of dilution from the plume scale to the grid scale that might be typical of mesoscale processes. The impact of the various assumptions of the larger scale chemistry needs to be carefully assessed.

The emitted jet engine exhaust plume is initially hot; it rapidly becomes trapped in the turbulent twin vortexes which meld in the wake and then dissipates into the background atmosphere on a longer time scale. The effect of wake dynamics on the dispersal of NO_x and HO_x was investigated by Lewellen and Lewellen (2001). It is important to consider the chemistry occurring in the aircraft plume and wake before it has been expanded to the model grid scale. Initial attempts to combine near field, far field, and global models in series (Danilin et al., 1997) were the first global impact studies to be based directly on detailed microphysics and chemical kinetics occurring in the aircraft plume and wake. From the perspective of ozone production in the UT/LS, removal of active species can occur by irreversible deposition of HO_x and NO_x source species or by conversion of more reactive nitrogen- and hydrogen-containing species into less reactive ones. Conversion of N_2O_5 into HNO_3 on sulfate or ice particles is the best established example of the latter mechanism. Soot, sulfuric acid ("sulfate"), and water-ice particles are the main condensed-phase species found in the exhaust of jet aircraft. The extent to which aircraft aerosols offset the effects of aircraft NOx emissions on atmospheric ozone depends on a variety of chemical and dynamical factors. The reactions are likely to be enhanced in the presence of increased atmospheric particulates in the plume, in contrails and in aircraft-induced cirrus clouds, which correspond to PSC type II ice from a chemical point of view. The most important heterogeneous reaction is the heterogeneous hydrolysis of N_2O_5. An important model simulation of effect of wake dynamics chemistry on NO_x and HO_x concentrations showed that the effective NO_x emissions were reduced by about 40% and the ozone perturbation by 30% (Valks and Velders, 1999).

The interaction of plume and global modelling was also investigated by Kraabøl et al. (2002) who studied the impact of aircraft emissions in an early version of the Oslo CTM including a plume model (Kraabøl et al., 2000a; b). The plume model used was a multi-cylinder model to allow for mixing and with comprehensive tropospheric chemistry. They found that the oxidation of NO_x in the plume reduces the efficiency of aircraft NO_x emissions for ozone generation by converting NO_x to HNO_3 (Kraabo et al., JGR 2002) and must be taken into account.

Meilinger et al. (2005) have developed a two-box plume model with complex gas phase and heterogeneous chemistry with a microphysical scheme. The two boxes, representing the plume and the background atmosphere, are allowed to mix using a turbulence time constant. They find that the development is quite sensitive to the conditions of the background atmosphere. They also find, similar to global scale models that above 210K hydrolysis of $BrNO_3$ and N_2O_5, enhanced by virtue of the increased aircraft induced aerosol sulphate area, is important in the ozone budget. Below 210 K growth of background particles and suppressed chlorine activation is important.

Kärcher et al. (2000) have presented an analytic parameterization for the development of nano-sized particles in the near field plume based on the amount of chemi-ions emitted, the sulphuric acid (and also the S-conversion factor) and condensable organic species which can be couple with far-plume models such as that of Danilin et al. (1997) and 3-D models to estimate the growth of aerosols in the plume and impacts in the atmosphere.

2.4.2. Global, Regional, Multiscale Models

Three-dimensional models used for studying atmospheric chemistry come in different flavours. Perhaps the least complex is the chemical transport model (CTM). In this type of model the main transport is by resolved winds supplied by another source such as weather forecast or climate model. The transport scheme adopted is critical as it must be mass conserving, stable, and not too computer intensive. Methods such as the Prather scheme (Prather, 1986) and the van Leer scheme (van Leer, 1977) seem to be among the most robust.

Many models for the IPCC(1999) report (Penner et al, 1999) were tropospheric in domain but the nature of the aircraft problem suggests that the dynamical and chemical interaction between the troposphere and stratosphere should be accurately captured. For example, in the Northern hemisphere winter more than half of aircraft emissions are directly into the lowermost stratosphere (e.g. Köhler et al., 1997). For models which do encompass the troposphere and stratosphere the resolved circulation utilized is likewise critical as it must simulate the Brewer- Dobson circulation with fidelity and stratospheric/tropospheric exchange accurately (as for example in ozone transport to the troposphere) which is important with the transport of species such as water vapour to the stratosphere. It must also represent tropical and polar quasi-horizontal transport barriers. Other features that appear to be important will be the representation of the Asian summer monsoon in transporting material to the UT/LS region (e.g. Park et al., 2007). In addition, the Madden-Julian oscillation (MJO) which is a region of intense convection in the tropics (e.g. Miura et al., 2007) is linked to the delivery of chemical species, including water to the lower stratosphere via the TTL (e.g. Wong and Dessler, 2007). Baldwin et al. (2007) also point to the influences of the stratosphere on tropospheric dynamics and also the need for more comprehensive (in terms of including the stratosphere) models.

We note that part of the ability to well represent these and other dynamical features will rest with the resolution adopted for the global CTM model as well as the metrological fields adopted. Currently the resolution for a state-of-the-art global CTM with comprehensive chemistry with both "stratospheric" and "tropospheric" chemistry is in the region of about $1°\times1°$ (and that is continuously improving with increasing computer power. Currently global weather forecast models are running with resolution ~ 25 km (e.g. ECWMF) and ~ 33 km (e.g. Canadian Meteorological Centre, Belair et al., 2007)

Other transport processes are included such as the impact of the planetary boundary layer which is an important two-way filter between the surface and the free troposphere. Large scale convection is parameterized by one of a number of schemes which can also be linked to lightning generation (e.g., Tost et al., 2007). Emissions are included but there is no standard emission suite, although several comprehensive data bases are available (e.g. GEIA, EMEFS). Comprehensive chemical schemes are now standard although what is often missing are interactive aerosol schemes and washout and rainout schemes can be quite primitive. Many species, such as ozone, are deposited on the surface with rates that depend on wind speed and surface roughness and surface type.

Regional models (or limited area models, LAMs) have a role to play in studying the dissipation of plumes of emissions perhaps injected from convection towers or aircraft. Such models enable processes to be studied at higher resolution with, one expects, higher fidelity. However, a problem of how to validate (or evaluate) these models is an issue as there is not enough data available. Nevertheless, many air quality CTM regional models have operated in the past with very limited vertical domains and physical processes are not suitable for aircraft studies. Figures 3a and 3b show NO and O_3 fields at mesoscale (15x15 km^2) resolution using a global multiscale model, GEM-AQ (Kaminski et al., 2007) and give some idea of the heterogeneity of the region.

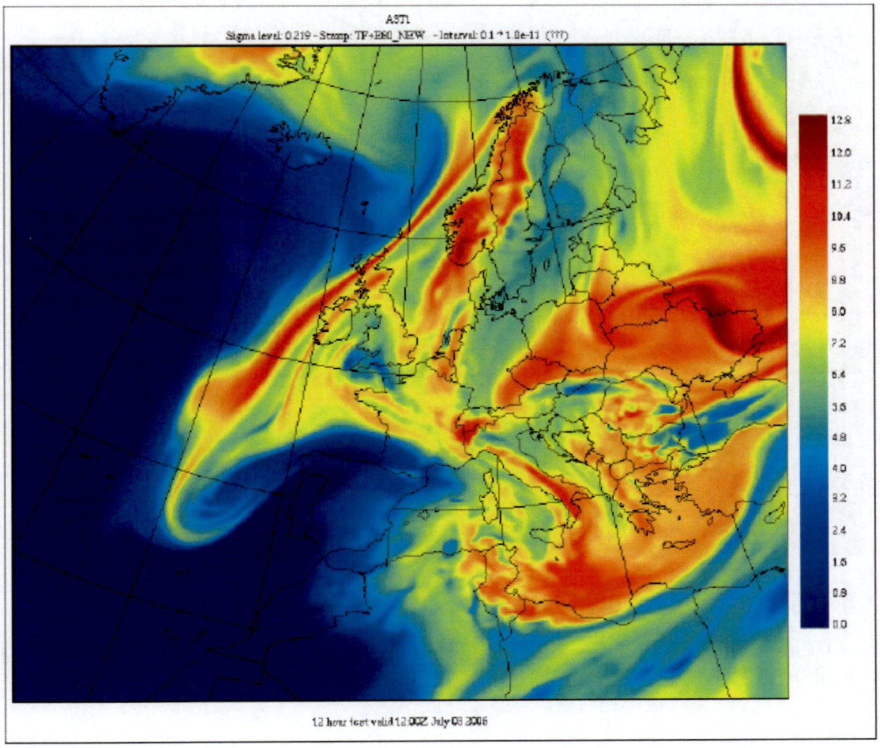

Figure 3(a). NO distribution at ~ 220 mb over eastern Atlantic and Western Europe calculated using GEM-AQ (Kaminski et al., 2007) at 15x15^2 km resolution.

One of the modelling tools that should be important in studying the different scales inherent in aircraft studies are multiscale models. These operate in (at least) two ways. One method is to run the model over the same time domain with higher and higher spatial (and temporal) resolutions and shrinking the interior grid such as has been done for air quality studies, for example, by MC2-AQ (Kaminski et al., 2002; Plummer et al., 2001; Yang et al., 2003) and also MM5 (e.g. Grell et al., 2000). A different approach is to have the interior high spatial resolution domain fixed within a lower resolution exterior domain which can be global or regional (e.g. Kaminski et al., 2007; O'Neill et al., 2007; Grell et al., 2005).

McKenna et al. (2002) have developed a new Lagrangian model for the stratosphere, CLaMS (Chemical Lagrangian Model of the Stratosphere) where the quasi-horizontal flow is driven by meteorological analysis winds with cross- isentropic flow (vertical) driven by heating rates from a radiation calculation. The model has recently been extended to the surface using a hybrid coordinate allowing a transition from potential temperature to pressure coordinates (Konopka et al., 2004) and mixing is driven by wind shear.

Another very useful analysis tool is the Lagrangian trajectory models such as FLEXPART (Stohl et al., 1998; 2005).

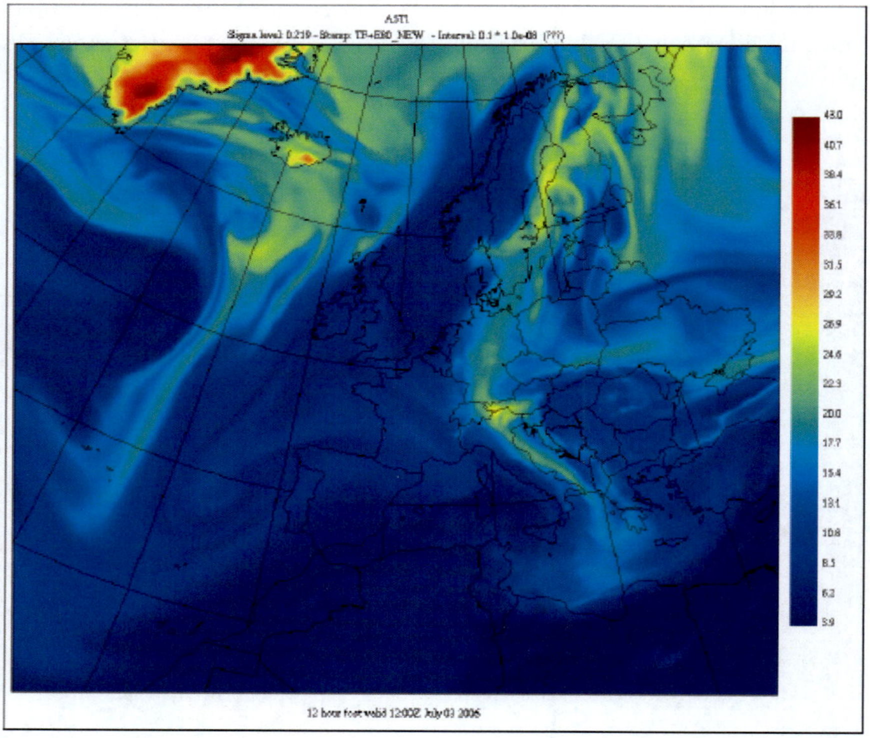

Figure 3(b). O_3 distribution at ~ 220 mb over eastern Atlantic and Western Europe calculated using GEM-AQ (Kaminski et al., 2007) at 15x152 km resolution.

2.4.3. Weather Forecast and Climate Models

CTMs are powerful tools for analysis and limited forecasting. However, for certain types of problems, such as the long terms effects of aviation, they are limited because they lack the capability to incorporate feedbacks. For tropospheric chemistry problems the main feed backs are via changing GHG fields such as methane and ozone.

Additionally, aerosols can interact with liquid and frozen water to alter cloud fields and thus impact solar and IR heating. In the stratosphere, changing water, CO_2 and ozone fields impact the heating. Modification of the heating function then alters the dynamical fields and transport characteristics. Of course, weather forecast and climate models (with and without interactive oceans) allow for some (and in some cases all) of the above feedbacks. And gaseous and aerosol fields have been added in many cases (Eyring et al., 2007b). Many models only treat the troposphere in detail (e.g., Stevenson et al., 2006; Kaminski et al., 2007) while others only have detailed chemistry with feedbacks for stratospheric species (e.g. see Eyring et al., 2006 for a suite of 13 models). But even more comprehensive on-line models with both detailed tropospheric and stratospheric chemistry with feedbacks on the meteorology are being developed (see the list in Annex I) and will be necessary for future forecasts (cf. Baldwin et al., 2007).

We note that for problems that deal with the UT/LS region that tropospheric models are not adequate. Furthermore, in order to develop an adequate Brewer-Dobson circulation the lid of the climate model must be above ~ 80 km (0.01hPa); the development of a quasi-biennial oscillation (QBO) signal requires higher vertical resolution in the stratosphere than most models have at this juncture. But that will soon change.

These models are complex and, as noted above, it is important to have confidence in the similarities while understanding the differences, the latter of which permits an estimation of our confidence in model structures. In the recent past, there have been comparison experiments of models for IPCC reports but these have not focused on the UT/LS regions (e,g, Shindell et al, 2006; Stevenson et al., 2006). In addition, the application of important metrics to evaluate the models (such as STE of ~ 500 MT-O_3/year) has only been applied in a limited fashion. This is partly because models have been in the throes of development; comparisons have been used more to deal with the issue of uncertainty in modelling rather than adopting a more systematic approach (e.g Wild, 2007). Along these lines we note that recently Eyring et al. (2006; 2007a) have published papers comparing the forecasts of chemistry climate models applied to the problem of stratospheric ozone depletion and recovery. Their study has underscored that there still are serious issues with models at the fundamental level of transport and emissions.

There are other metrics or diagnostics that can be used as a measure of how well a model is accurately representing the atmospheric situation modelled. Some of these are discussed by Pan et al. (2007) for example. CO varies rapidly across the tropopause with larger values in the troposphere. Similarly, ozone varies rapidly across the troposphere but with more in the stratosphere. Thus CO/O_3 correlations represent a very useful diagnostic to test the representativeness of model transport. Similarly, in the UT/LS regions, NO_y/N_2O correlations can be revealing of model transport. And as noted above NO_x/HNO_3 ratios in the UT should be very variable, ranging from > 1 to < 1 depending on how recent convection and lightning have occurred. Forming a suitable PDF for measurement and model ratio might be a useful diagnostic.

2.5. Future Climate Impacts

2.5.1. Climate Change

In the context of climate change it is to be expected that a future fleet will operate in a different atmosphere both in terms of meteorology (transport) and chemistry. In addition, these changes will also impact the chemistry of the UT/LS region. So that it is important to have a better understanding of changes in relative humidity in the UT/LS region (of course this also impacts contrail formation and the development of cirrus clouds.). This problem is dealt with in a different white paper. However, there are also obvious impacts on the chemistry via HO_x abundances and the production and loss of ozone and also its impact on the radiation budget in the TTL region. However, there are also chemistry-related impacts on humidity and that relates to the supply of freezing nuclei to the UT/LS region either from the lower troposphere via delivery of aerosols or from the stratosphere and mesosphere via sulphate or metallic ions generated in the mesosphere. These are issues that must be resolved in an integrated fashion rather than thought of simply as chemical or transport issues.

Future climate is likely to be different: this means that it can be anticipated that winds (transport times), cloud processes, water amounts, NO_x generation by lightning will be different. And we have noted that the relative contributions of aircraft NO_x and lightning NO_x (with a small contribution from stratospheric NO_y) are important. Nevertheless, aircraft impacts cannot be divorced from possible impacts of future surface emissions. Most climate studies using standard CO_2 increase scenarios suggest that the main impact in the future will be due to increased anthropogenic emissions (e.g. Stevenson et al., 2006) with climate impacts on biogenic emissions being of secondary import. However, with a more rapid rise in CO_2 and increased temperatures biogenic emissions grow in a non- linear fashion to become more important. In the future, anthropogenic, biogenic and biomass burning emissions, are expected to grow as population expands and the economies of India and China grow. And we note that current CO_2 emissions are growing faster than envisaged by any of the IPCC scenarios.

In future climate change scenarios, cloudiness is projected to decrease at latitudes below about 50 degrees and to increase at latitudes above 50 degrees (IPCC; 2007.p767). These changes will affect the radiative forcing from aviation ozone on a regional basis.

2.5.2. Future Aircraft Emissions

The emissions from aircraft will depend on the composition of the fuel. For example the sulphur fraction can impact the chemistry in the plume and also the generation of aerosols which are deposited in the UT/LS and may impact chemistry and cirrus formation. In the future this will depend on the development of new engines and constraints imposed by noise control for take off and landing and possibly of aircraft flying a few kilometres higher (not SSTs). For example, over the next 20 years the replacement of old aircraft with new aircraft such as the Airbus 380 and the Boeing 787 may impact future emission scenarios. Estimates of annual fuel use by 2020 annual for commercial air traffic are ~ 350 MT or 2.6 times the estimated fuel use by the global 1999 commercial fleet. This translates into global NOx emissions of ~ 1.5 MT-N from commercial air traffic or about 2.8 times the estimated 1999 NOx emissions levels. At the same time total revenue passenger kilometers are projected to increase from 3,170 billion in 1999 to 8,390 billion in 2020, or by a factor of 2.65 (Sutkus et al., 2003) These estimated increased emissions are expected to lead to global increase in ozone production with the potential for larger regional effects.

Perhaps of more importance is the advent of new air traffic control mechanisms to cut "wasted" air time and reduce fuel consumption and perhaps also the development of new routes both intra- and inter-continental. Of particular importance of the development of new routes may be the impact of increased flights over the Arctic regions, a possible impact which needs to be explored (e.g. Gauss et al., 2006).

There is also the possibility of the emission character of a future fleet to be considered. The European Commission within the fifth Framework Programme on competitive and sustainable growth supported the CRYOPLANE project to investigate aircraft fueled with liquid-hydrogen (LH_2). Since there would be no hydrocarbons (HCs) or sulphur in the fuel these type of aircraft would not emit CO_2, CO, soot or SO_2. They would emit NO_x and water. While the NO_x emissions are likely to decrease the water emissions would increase ~ 2.6. Gauss et al. (2003) investigated the impact of cryoplanes for the year 2015 using the Oslo CTM. They find that the replacement aircraft would increase water vapour near the tropopause by about 250 ppbv. Although we should note that with the current lifetime of a fleet of aircraft being ~ 30 years LH_2 planes will not be a major component of the commercial fleet for quite a few years to come.

Recent work by Søvde et al. (2007) has looked at the impact of a "mixed" fleet using an emission scenario for 2050. The "mixed" refers to a combination of subsonic and supersonic aircraft. They found impacts on the ozone due to aerosol emissions. Their analysis is limited from a climate perspective as they used meteorology for the year 2000 for both current and future scenarios.

Connectivity: Transport issues identified here are also important for water vapour and related meteors, such as cirrus, while cirrus may play a role controlling NOx distribution in the UT/LS. The aerosol, whether sulphate, organic, soot may play an important role in the chemistry of the engine vortex and plume and generation of ozone. However, perhaps more important is their putative role as FN/IN for cirrus. Again, transport is an issue.

3. Focus on Uncertainties

In this section we address the issues of uncertainties in the basic science issues concerning impacts of aircraft but with a focus more on chemistry and transport issues but include some brief comments on radiative forcing.

3.1. Chemistry and Emissions

For chemistry the issues of major uncertainty include an accurate knowledge of the tropospheric ozone budget, HOx, NOx chemistry in the UTLS which may be impacted by issues of convection lofting precursor species, possibly aerosols and affecting loss by washout. A related cloud issue is the role of HNO_3 take-up on cirrus clouds. In addition, the role of halogen chemistry is lurking on the sidelines. Certainly lower stratospheric Bry distribution (and sources) need to be resolved as does the Cl2O2 dimer problem in springtime polar regions.

The issue of the uncertainty of the background NOx emissions from lightning in the UT region and convection from the boundary layer to the UT appear to be the major issues and related corridor issues.

3.1.1. Chemistry

From section 2 it is clear that over the past decade measurements in the UT have revealed problems with our understanding of the NO_x, HO_x budget. Recent work as part of the INTEX-A campaign has been presented by Singh et al. (2006) (see also Ren et al., 2006) who, for example, note that OH measurements are substantially lower than model values for the INTEX and other campaigns (see figure 4). This uncertainties may be related to issues of steady state (used for analysis) versus a more dynamic non-steady state chemical regime driven by removal of HNO_3 within clouds and NO_x replacement by lightning and STE. Or it may be due to the lofting of precursors which may have not be measured or have escaped detection.

Issues regarding the importance of halogen chemistry are much less well defined, but could be playing a role in the HOx/NOx UT chemistry issue. For example hydrolysis of $BrNO_3$ does lead to the creation of more HOx. But at this point is more speculation on the authors' part.

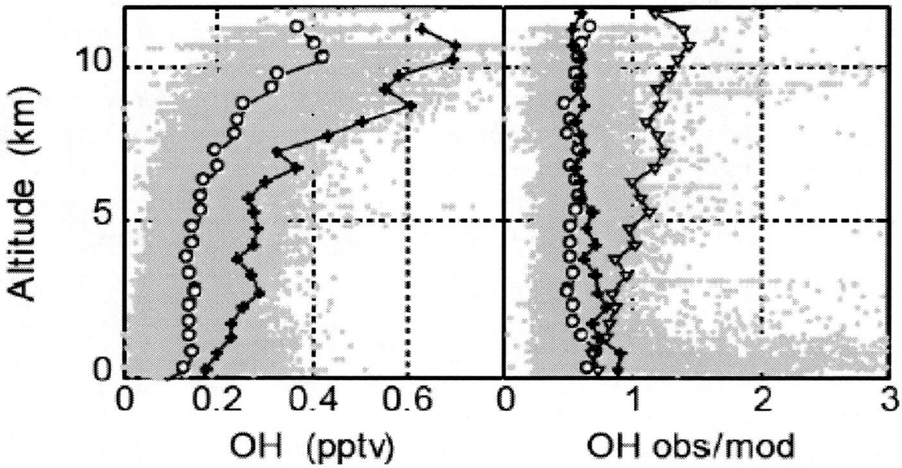

Figure 4. Comparison of the vertical profiles of (left) measured (circles) and modeled (stars) OH and (right) measured-to-modeled OH ratios during in INTEX-NA (circles), TRACE-P (stars) and PEM-TB (triangles). Individual INTEX-NA 1-minute measurements are shown (gray dots). (Ren et al., 2006).

Freezing nuclei (FN) or ice nuclei (IN) act as sources for the sublimation of water in the UT. The main sources are likely to be the lower troposphere although there is an influx of stratospheric sulphate aerosols via STE and also the possibility of a cosmic ray ionization contribution to aerosol and thus ice formation. Thus it becomes important to understand and quantify the fraction of aerosols generated that become IN and this requires more study. The issue of lower tropospheric sources of aerosols and their size distribution is likewise important although processing in clouds tends to produce aerosols in the 0.1-1.0 micron range. Thus quantification of direct sources of aerosols is quite uncertain although the situation is improving. The issue of the generation is of secondary aerosols is somewhat more uncertain, e.g. SO_2 to sulphate oxidation and, as noted above the issue of the generation of secondary organic aerosols in the UT is still an issue for research but the empirical evidence suggests that their abundance is related to tropical and Boreal fires.

The status of our knowledge of the ozone budget in the UT/LS still is more uncertain that one would like especially for the estimates of potentially small impacts of aircraft. From the above discussion we consider that the percentage contribution of air traffic emissions versus natural and other anthropogenic sources is quite uncertain in the UT.

Although we have not dealt with it, the uncertainty in the physics and microphysics of water vapour in the UT/LS is important for chemistry and ozone generation, radiative forcing and generation and sustenance of cirrus clouds.

And related to this is the issue of the impact of HNO_3 on freezing of water vapour into cirrus crystals is still uncertain.

3.1.2. Emissions

In terms of the global NO_x source the contribution from aircraft is only a few percent. However, as noted above, as viewed from the situation of the UT/LS the contribution of aircraft, ~ 0.7 MT-N/year is rather larger (~ 7%) measured against the context of ~ 4 MT-N/year from lightning in the UT, ≥ 4 MT- N/year from large scale convection (using a 10%

delivery fraction) and resolved transport and 0.8 MT-N/year from STE (N_2O and cosmic rays). However, the delivery of the aircraft NO_x is to a relatively narrow corridor which tends to enhance its impact so that locally the impact could be much larger (e.g. Schumann et al., 2000). The uncertainty in the STE of NO_y in terms of absolute amount seems rather small as it is largely constrained by the stratospheric source. However, the lightning source is still rather uncertain, although there appears to some convergence on the absolute amounts. Interestingly, the lower bounds that have appeared (~ 2 MT-N/year) are such that it increases the potential impact of aircraft. This is certainly an area that requires further study, both in terms in absolute amounts and also the vertical distribution of the source. If, as seems likely, delivery of NO_x by lightning often occurs in association with convective activity then the contributions from transport from the surface, concomitant with washout of soluble components such as HNO_3, and in-situ lightning source tends to make the unambiguous evaluation of separate lightning and convection sources difficult. It also means that the knowledge (or lack) of lower tropospheric NO_x/HNO_3 ratio takes on a greater importance. This ratio is impacted by heterogeneous chemistry and knowledge of the aerosol distribution. Furthermore the status of the amount and distribution of anthropogenic, biomass burning and biogenic tropospheric sources becomes important.

As noted in section 2, estimates of delivery of NOx from the boundary layer are uncertain and range from 10% to 50%, although this may simply reflect the different dynamical conditions over summertime north America and Europe: this may also change in a future climate state. As noted above the fraction of aircraft NOx to other sources for the year 2000 was about 0.7/(4+5+0.8) ~ 7%. As noted in the introduction, for the time frame 2020/2030 aircraft NOx emissions are expected to increase to ~ 1.5 MT-N (Sutkis et al, 2003). Anthropogenic emissions are expected to rise by a factor of 1.5 to 2 depending on the scenario followed (e.g. Dentener et al., 2005) which using 10% for the contribution to the UT give a source of ~ 7.5 to 10 MT-N/year. Even though one might anticipate more convective activity and thus more lightning in a future climate state based on energy considerations we will take the lightning contribution as fixed. In which case the ratio of aircraft to other sources in the UT is about 1.5/(4+0.8+7.5) or about 12% or less. So that the relative impact of aircraft in 2025 could be higher than at present. Moreover, if economic conditions are not conducive to such growth it is likely to impact both the air transportation sector as well as other sources of anthropogenic NOx.

3.1.3. Comprehensive Tropospheric/Stratospheric Chemistry Models

In the IPCC1999 report most of the 3D CTMs used in the study had either comprehensive tropospheric or stratospheric chemistry with background tropospheric chemistry but not both. There were few climate models with comprehensive tropospheric or stratospheric chemistry. Now both CTMs and climate models are incorporating comprehensive tropospheric and stratospheric chemistry (see annex I). Future studies of aircraft models will need to include comprehensive UT/LS chemistry.

3.2. Measurements - Species and Winds – and Analysis

It is still important to have aircraft campaigns since they can provide detailed and comprehensive in situ measurements. However, any analysis and parameterizations derived from these need to be verified for a variety of conditions. Thus the use of satellite measurements to evaluate the information on a global scale seems important. However, what is required are instruments to measure species and temperature with high vertical resolution, and in the context of contemporary experiments (see above) this means vertical resolution ~ 1 km for UT/LS measurements. These types of instruments are under study using millimetre and mid-IR in Europe and one of the experiments on ESA's list of Explorer possibilities is PREMIER (PRocess Exploration through Measurements of Infrared and millimetre-wave Emitted Radiation) (http://www.esa.int/esaCP/SEMHQH9ATME_Protecting_0.html) which could produce 3D imaging of water vapour, ozone, and other species as well as temperature, in the UT/LS and with the required vertical resolution. Interestingly, one of the goals of PREMIER is to operate synergistically with METOP (see above) together with use of models and data assimilation systems to transfer the information to the lower troposphere.

In addition, to attack the relative humidity problem in the UT, there needs to be improved instrumentation to measure air ambient temperature accurately (and precisely) to ~ 0.2°K.

Of course from a dynamical point of view, which is critical, improved wind measurements are important and ADM-aeolus (http://www.esa.int/esaLP/SEM3Y0LKKSE_LPadmaeolus_0.html) will provide improved wind measurements in the near future using Doppler wind lidar measurements with a late 2008 launch. In addition, the Canadian Space Agency is still planning to launch SWIFT (Stratospheric Wind Interferometer For Transport) (http://www.space.gc.ca/asc/eng/sciences/swift.asp) which uses ozone lines in the mid-IR and a Michelson interferometer to measure winds in the lower stratosphere.

For the past number of years it has been a goal to use chemical measurements to improve wind estimates. Recent work between ESA, Environment Canada and BIRA (Belgisch Instituut voor Ruimte-Aëronomie, Institut d'Aéronomie de Belgique: BIRA-IASB, Belgium) (e.g. de Grandpré et al., 2007) has used 4DVar and assimilation of MIPAS ozone data to improve wind estimates in the lower stratosphere using the Canadian weather forecast model, GEM (Global Environmental Multiscale) (Belair et al., 2007), and the BIRA chemical module. This is encouraging for future meteorological data. However, it is not clear in the coming years that there will be enough satellites available to provide the necessary chemical data. This is a serious problem (that also will impact the monitoring of the stratosphere and ozone recovery over the next decades).

From the above discussion it is clear that the lower atmosphere has a major impact on the UT/LS and the UT in particular on relatively short time scales. One the major impacts is the transport of species important in the ozone budget such as CO and NO_x. However, bottom up budgeting of emissions, while useful, is still quite uncertain (uncertainty depends on species). Top down assessment of tropospheric emissions (e.g. Martin et al., 2007; Jaegle et al., 2005) using satellite data, while it has limitations, is proving very useful and continued work should improve knowledge of surface emissions. Thus satellites dedicated to air quality studies such as TRAC (TRopospheric composition and Air Quality) (http://www.esa.int/esaCP/SEMHQH9ATME_Protecting_0.html) or that can be combined with other information is being done using GOME, SCHIACHY etc (Martin et al, 2007) data should be continued (see comment on PREMIER and METOP mission above).

With the rapidly changing climate over polar regions and the Arctic in particular, and with the potential for increased air traffic over the Arctic, it would seem prudent to have been observing capability for meteorology (climate) and chemical species over this region. Thus there has been recent activity to study the use of Molniya orbits (e.g. Riishojgaard, 2005). It is clear that imaging nadir viewing FTS mid-IR instruments, multi-channel near UV, Visible, NIR imagers and SCIAMACHY-type instruments can provide detailed capability to monitor this region with quasi-geostationary viewing. Several satellites with 12 hour orbits can provide continuous coverage of polar regions and extensive coverage down to mid-latitudes.

3.3. Modeling Capability

Since the IPCC 1999 report models have improved. However, actual model performance on NOx in the UT has not improved over the last decade (e.g., Singh et al, 2007). The availability of multiscale models which can be used to study plume dispersion problems has improved. Increasing computer power and storage has enabled much improved resolution for the combination of chemistry and meteorology. Gas phase and aerosol chemistry running together is now almost standard.

It is important to recall that the aviation problem is not run in isolation so that any improvement in the basic atmospheric model will lead to a better assessment of aircraft impacts. Thus we expect improved spatial resolution to continue. Currently weather forecast models are being used globally ~ 25-35 km horizontal spatial resolution (e.g. Belair at al., 2007; Jung and Leutbecher, 2007). It is likely that in the next year or two it will be possible to run global chemical weather models with similar horizontal resolution. Of course one of the problems will be the evaluation of models at this resolution which is why such as satellite with the features of PREMIER (see above) would be invaluable. In addition to improved horizontal resolution it will be important to have concomitant improved vertical resolution in order to better represent transport from the troposphere to the stratosphere and also vertical wave propagation.

There will be associated problems to be solved with the vast amounts of data generated. Some will be associated with storage, access and transfer, problems of analysis – simple comparison is limited and limiting, and, as discussed above, there will need to be much improved protocols (including correlations) for comparing chemical weather and chemistry climate models with each other and with data. Statistical analysis such as use of PDFs may become more common. The distinction between weather forecast and climate models is fast eroding as weather forecast models are run out for longer times, have improved heating codes, include ozone chemistry, have better surface schemes. However, in the short term chemistry climate models will be useful for longer simulations and thus have lower spatial resolution. Some of the issues surrounding the magnitude of a climate signal can be temporarily addressed by the using chemical weather models in a time-slice mode run with appropriate (future) SSTs from ocean-atmosphere models.

Climate chemistry models are in a state of flux at the moment. As noted above SPARC comparison studies CCMval (Chemistry-Climate Model Validation Activity for SPARC) (http://www.pa.op.dlr.de/CCMVal/) (Eyring et al., 2006; 2007a) have revealed a number of problems. One such problem is the status of the total chlorine, Cly, distribution of the models in the stratosphere. The model distributions are significantly different and the problems certainly involve transport, dynamical, age-of-air issues that need to be resolved. In addition, the input of the CFCs and related species may also be contributing to the discrepancies. The problem may be that most models use a few CFCs to represent the total input of chlorine, all with different lifetimes. Thus each model may have a distinctly different input function with an associated different time constant for flushing from the stratosphere aside from any issues regarding age of air. In addition Waugh et al. (2007) emphasize the importance of resolved transport in the lower stratosphere and the interaction between that and resolution and the representation of transport barriers.

There is a related issue with the total amount of bromine in the stratosphere. There is an important role for halogenated very short lived species and their contribution needs to be (a) resolved by measurements and (b) included in all models.

3.3.1. Resolved Transport Issues

One of the issues mentioned above is that of an apparent discrepancy in Cly distributions between models. Some of this may be attributed to the internal dynamical properties of the model (winds etc) which translate to different age-of-air between the models (e.g. Waugh et al., 2007). However, some of the discrepancy is likely due to differences in transport schemes (e.g. Wild et al., 2007). Some schemes such as semi-Lagrangian are efficient but sometimes do not have adequate conservation properties, while spectral methods, which generally have good conservation properties when strong gradients are not an issue, have problems when strong horizontal gradients create Gibbs fringes and then various types of hole-filling techniques are applied (and rarely discussed) in an attempt to maintain mass conservation which can have various disturbing pathologies. And of course, there is the associated issue of the positive part of the Gibbs fringe – how to constrain it? This is not simply a stratospheric issue. Strong species gradients arise in the troposphere (e.g. rainout of soluble species is very "spotty", in the stratosphere uptake of various species on PSCs can be very irregular, and in the mesosphere and stratosphere the formation of a strong vortex and associated descent of low mixing ratio air can create very strong horizontal gradients. Some of these issues can be ameliorated by going to higher resolution but it seems more reasonable that the community should start using more robust (if more computationally expensive) and reliable transport schemes such as the Prather scheme (Prather, 1986) or the van Leer scheme (van Leer, 1977). While some variety of transport schemes is useful, even necessary as it allows for uncertainties and variation between models some more rigorous comparisons seem necessary. A related issue is transport by sub-grid scale motions such as large scale convection. In CTMs these schemes are often/sometimes applied in an inconsistent manner as the parametrizations as the resolved winds utilized to drive the model will often have used a different parameterization and most likely at a different resolution as that used in the CTM. It would be useful to quantify the impact of such uncertainties.

3.4. Radiative Forcing Comments

The increased emissions from future air traffic will alter radiative forcing due to generation of increased ozone, increased carbon dioxide and associated decreases in methane concentrations. The major change will be due to the increased ozone in the upper troposphere, particularly in aviation corridor regions which, to some extent, is counteracted by a globally reduced methane abundance due to increased OH. And of course the relevant time constants for ozone and methane are quite different. This issue was identified at the time of the IPCC (1999) report. A recent study by Stevenson et al (2004) applied a chemistry-climate model to study the radiative forcings generated by aircraft NOx emissions through changes in ozone and methane. They injected pulses of aircraft NOx the model for a various months and investigated the impacts on ozone and methane radiative forcings. Given that the spatial distributions of the RF from aviation ozone and the RF from methane will be considerably spatially different the question remains how much compensation will occur.

The mechanism is the increase in radiative forcing at the earth's surface which causes surface and atmospheric heating. The real verification of the climate impact of increased upper troposphere ozone is the detection of changes in the ozone radiative forcing at the surface and at the top of the atmosphere. Radiative Forcing (RF) calculated at the top of the troposphere at the tropopause is used as a metric for measuring the potential for climate change. And it is often given as a simple global number (NRC, 2005). However, RF is far from uniform as is clear from the regional changes in ozone which translate into regional changes in RF (cf. figure 2), as opposed to the situation for methane where, because of its ~ 10 year lifetime, the changes in methane, and so the RF, is much more spatially uniform. Clouds can cause large changes in RF and represent a major uncertainty on the model calculation of RF.

There are difficulties in detecting changes in the IPCC radiative forcing metric because of the way in which it is defined at the top of the troposphere at the tropopause. There are large uncertainties in the calculations of the radiative forcing metric due to a lack of knowledge of cloud effects. There need to be verifications of the radiative forcing metric by comparison against real measurements of observed surface radiative forcing and with satellite radiative trapping at the top of the atmosphere. This will need to be accomplished by concurrent simulations of surface forcing and top of the atmosphere radiative trapping with the same climate models used to calculate the RF metric (Puckrin et al, 2004).

In addition, we note that RF can be readily measured at the surface where it is used as more of a climate observable than a metric. For example, the current network of pygeometers measures the long wave component of surface radiative forcing. Philipona et al. (2005) have detected the increase in total surface radiative forcing from the increase in all of the greenhouse gases. The surface forcing from tropospheric ozone itself has been measured by Evans et al. (1999) using spectral measurements of long wave radiation. The effects of clouds on surface radiative forcing need to be investigated by extensive ground measurements of surface radiative forcing by the individual greenhouse gases and particularly by tropospheric ozone. Satellite measurements of nadir outgoing long wave radiation can also be used to investigate the effects of clouds on the top of the atmosphere radiative forcing (Harries et al, 2001).

Measurements can supply RF on a regional basis. The regional nature was aptly demonstrated by the change in surface temperature range (and RF) from Sept 11-14 with no aircraft flying. For three days after September 11, the Federal Aviation Administration grounded commercial aircraft in the U.S.; there was an anomalous increase in the average diurnal temperature range of 1.2°C for the period Sept. 11-14, 2001, a change not matched in the last 30 years (Travis et al. (2002). This temporary "climate change" was due to decreased surface radiateve forcing from both contrails and tropospheric ozone.

3.5. Future Climate and Related Feedbacks

For assessments of future impacts of aviation, i.e. which might occur in a future climate, chemistry-climate models need to be used in order to permit feedbacks; CTMs will not permit feedbacks and hence the need for metrics such as radiative forcing as a proxy for change. This will apply for our subject area(s) of interest, both chemistry and transport. However, if the changes are small, several long runs will be required, with slightly different boundary conditions (such as SSTs) in order to extract statistically significant findings. This is necessary since emissions will be changing, the ozone layer will be changing, STE may alter the influx of stratospheric ozone to the troposphere, lightning production of NOx may change in a future climate and the strength of the Brewer Dobson circulation may alter, affecting water vapour amounts in the UT/LS region and this will modify the formation and impacts of contrails, contrail-cirrus and the background cirrus clouds.

3.6. Interconnectivity with Other SSWP Theme Areas

The same type of measurements and regional modelling for water vapour and NOx are required for the contrail cirrus radiative forcing theme area. Modelling of plume dispersion into the ambient atmosphere is also a common problem with the contrail cirrus theme. Another common link is the upward convection transport of smoke from biomass fires to provide CCN Cloud Condensation Nuclei) for the formation of cirrus in the UT.

4. Priorities

The following list summarizes the major uncertainties from section 3. In the list below, HP, MP and LP stand for highest priority, medium priority, and lower priority, respectively.

Chemistry

1. Our understanding of NOx/HOx chemistry in the UT is uncertain; measurements are not well reproduced by model simulations and may also be influenced by items (4) and (5) below and possibly (2) and (3). (HP)
2. There is a need to improve our understanding of the impact of background aerosols and those from aviation emissions on the background constituents. They can alter the NOx and ClOx chemistry with resulting changes in regional ozone in the UT/LS. (HP)
3. Improved estimates of the potential role of halogen chemistry are required. (MP)
4. Conversion or uptake of HNO_3 on cirrus clouds needs to be better understood. (HP)

Emissions

5. The relative contribution of aviation NOx to NOx from lightning and NOx lofted by convection from boundary layer pollution sources in the flight corridors is uncertain and needs to be better defined. (HP)

6. Chemistry processes in the UT are strongly influenced by convective processes for which species ratios can be used as measures of model performance. (HP)

Models, multiscale, global and climate

7. More plume to regional scale models for plume processing of NOx capability are required as there are very few groups at present capable of making detailed calculations. (MP)
8. Model representations of vertical transport processes from surface to tropopause need improvement (see above). (HP)
9. Ideally, models to address the climate problem should be climate models with comprehensive tropospheric and stratospheric chemistry in order to better incorporate chemical and dynamical feedbacks. If CTMs are used then they also should have comprehensive tropospheric and stratospheric chemistry. (HP)
10. Ideally, models should be able to address/simulate dynamical issues such as the Asian monsoon and the Madden-Julian oscillation in order to properly characterize upward tropical transport (MP).
11. Predictions of future climate conditions, composition and emissions are needed but should be addressed with (8) above. (HP)

Measurements

12. Measurements of vertical profiles in UT with high vertical and horizontal resolution are required and there is a role for O_3, NO2 and CO sondes. (HP)
13. There needs to be support for continued use of current satellites and analysis of concomitant data for now and new satellite instruments in the future. (HP)

Radiative Forcing Issues

14. Verification of the radiative forcing metric for ozone and methane is needed and cloud effects need to be quantified. (MP)

The understanding of aviation impacts on the climate system clearly requires a deep understanding of the *natural* atmosphere. Thus much of the above list itemizes a lack of understanding of the natural atmosphere. The following outlines ongoing work and future plans that will assist in improving our understanding of the atmosphere of the atmosphere in the UT/LS. Of course the UT/LS does not exist in isolation. Thus improved understanding of the troposphere and stratosphere in general are important.

Ongoing Measurement Programs

Clearly it is important to continue monitoring in concert with modelling and analysis. This provides continuity with current measurements and allows the time series to be extended into the future (e.g. AERONET (Aerosol Robotic Network) http://aeronet.gsfc.nasa.gov/). Maintenance of ozonesonde capability and in particularly programs such as SHADOZ (Southern Hemisphere Additional Ozonesondes) (Thompson et al., 2003a, b; http://croc.gsfc.nasa.gov/shadoz/) in concert with modelling have allowed important new understanding of tropical ozone behaviour and IONS (INTEX ozone networks study)

(Thompson et al., 2007a, b; http://croc.gsfc.nasa.gov/intexb/ions06.html) for ozone transport and the frequency of STE over North America.

In addition, there should be continuing support for analysis of current satellite experiments/instruments that have the capacity to probe the UT/LS, viz., ACE, MAESTRO, MIPAS, OSIRIS, SMR, MLS, OMI, SCHIAMACHY, TES, AIRS, IASI, HIRDLS and also instruments such as MODIS, MISR, AVHRR etc that can provide aerosol column optical depth will be important to maintain continuity into the future of basic tropospheric science. But we note that in 3-5 years there will be limited satellite availability that will be useful for UT/LS studies which is particular to aviation impacts.

Future Work

We discuss campaigns below. A critical part of the infrastructure for campaigns are aircraft, particularly aircraft that can probe the UT/LS region, such as the ER2 and Geophysica which can probe the lowermost stratosphere while other instrumented aircraft can probe the upper troposphere. It is important that these aircraft be supported. Currently the Geophysica is in need of upgrades which are ~ $1M. In addition, instrumented commercial aircraft such as the MOZAIC (Measurement of OZone by Airbus In-service aircraft) fleet also have an important role to play (http://www.cnrm.meteo.fr/dbfastex/datasets/moz.html). It would be important if other governments/aviation companies could be persuaded to participate.

The major chemistry and related issues summarized above can only be addressed with new research. For example to better characterize lightning NOx sources, and convection etc will require dedicated aircraft campaigns with possibly new instrumentation to attack problems of heterogeneous chemistry on ice and aerosols and possibly measurements of "missing" species. In addition, it will be necessary to support laboratory chemistry, particularly heterogeneous chemistry.

With respect to the development of new instrumentation an interesting possibility is the development of a NO2 sonde (e.g., Pisano et al., 1996; Sitnikov et al., 2005). This would complement the ozonesonde and would be a major asset in researching the production of ozone by aircraft since it would provide high vertical resolution profiles for NO2. Some infusion of support could lead to instruments suitable for release in concert with ozonesondes. This could be done on relatively a short time scale.

Campaigns

We consider it important continue to continue to conduct aircraft experiments and these should be coordinated where possible with satellite UT/LS measurements of gases (e.g. ACE, MIPAS, AURA (MLS, TES, HIRDLS), METOP) and aerosols. Clearly aircraft will continue to play a leading role in investigations over the next three years. The aircraft include but are not limited to the ER2, DC-8, Geophysica, WB57 and HIAPER. UAVs are a developing technology which should be exploited since they can remain aloft for several days. These will be important in addressing both summer and winter conditions and ideally there should a major campaign conducted in most continents. These would investigate the issues of (a) lofting PBL NOx sources and (b) lightning NOx (c) better characterization of corridor issues.

The impact of improving knowledge in these areas will be high as it is basic knowledge of the background atmosphere. And models simulations of same will be necessary for the simulation of aircraft impacts.

Modelling

Climate impact forecasts are dependent on models. Thus it is important to maintain and improve modelling capabilities. This would include improved horizontal and vertical resolution, improved physical processes, better emission data bases, more accurate resolved and sub-gridscale transport processes. The development of multi-scale models is needed to investigate the corridors aspect of the aircraft emissions and the transition to regional scale climate impacts. Parameterizations used for deep convection need to be both used consistently (with the basic dynamical model) and verified according to the scale of model.

It would be useful for plume-to-corridor scale models to run comparison simulations of the conversion of NOx into nitric acid in the near field plumes to better characterize the in-plume conversion of NO into HNO_3 by heterogeneous chemistry in the plume and better resolve the issue of EEIs (effective emission indices) for use in global models.

Also it will be important to develop more challenging protocols for comparing models and devise improved diagnostic schemes or metrics for comparing with measurements. This latter is particularly important as it provides useful diagnostic tools for assessing model capability. An essential component of this is improved computer resources.

Better characterization and parameterization of convection will remain an issue. One means to attack the problem is by the use of cloud resolving models (CRMs; see for example, Xie et al., 2006; Xu et al., 2006; Lopez et al., 2006) to provide a first order basis for improved parameterizations. Nevertheless, any parameterization will need to be verified against measurements.

Better models will lead to improved forecasts in terms of reliability and given current constraints the goals are eminently achievable. Elements required to address the various problems outlined above

1) Support for 3 modelling groups (a single model is useful but using different models is a more basic way of addressing uncertainty issues, as long as certain basic metrics (to be identified) are complied with.
2) Each group may require 3 PY over three years to address some of the issues.

a Campaign modelling
b Climate effects forecasting
c Access to substantial computer power.

Radiative Forcing Issues

An important goal is to reduce the errors associated with radiative forcing (RF) issues. There have only been very few and limited validation comparisons of RF with real measurements. In particular, research is necessary for the evaluation of cloud attenuation effects which represent a substantial uncertainty on model calculation of the RFM. Validation of the actual RF observable is needed to reduce GCM climate simulation errors of surface RF and surface temperatures.

In terms of impact, improving knowledge and accuracy of RF estimates would improve the uncertainties of estimates of relative contribution of aviation to global warming which are estimated using the RFM. The data could, perhaps, also lead to improvements in GCM simulations of climate changes for regions and also reduce the current error associated with the impacts of a tripled future aircraft fleet contribution to the total RFM.

The achievability is high since satellite measured databases of radiative trapping and ground surface measures of surface RF already exist. The work would consist mainly of data analysis of existing Department of Energy (DOE) Atmospheric Radiation Measurement (ARM) databases (http://www.arm.gov/) . It would be prudent to expand the current AERI (Atmosphere Emitted Radiance Interferometer) stations into a future network for surface radiative forcing.

The costs are estimated to be :

a analysis of satellite databases 5 FTEs,
b analysis of existing ARM AERI database 3 FTE,
c and network expansion $300K per station.

Timelines: analysis of AERI database 2 years, analysis of existing satellite databases: 3 years and network development 10 years.

5. Recommendations

In this section we address the "recommendations for the optimum use of current tools for modeling and data analysis" and interpret this to mean what can be done with a limited expenditure of funds (and assuming with insufficient funds for major new measurement programs initiatives). Also for the scientific domain of this SSWP we address chemistry-climate impacts of aviation. Priority one would be the impact on ozone and associated RF effects and second priority would be the effects of aerosols in the plume/contrail. In terms of uncertainty, we have noted above that the UT is a region of considerable uncertainty both in terms of background emission sources and chemistry, both gas phase and heterogeneous.

The issues raised in Sections 2 and 3 have led us to a number of recommendations for 'practical' application of the currently available information involving available measurements and current modeling capabilities. This can, perhaps, be separated into three aspects. The analysis of observations, currently collected, continuing analysis of observations taken as part of an on-going program such as a satellite experiment and the analysis using currently available models.

As noted in sections 2 and 3, recent plume and CTM modeling studies have emphasized the impacts of background aerosols (natural and non-aviation anthropogenic) and those generated by aviation combustion on the UT/LS ozone budget: plume aerosols can modify UT/LS chemistry by activating Cly to destroy ozone, while ozone oxidation reactions followed by heterogeneous reactions can result in a reduction of ozone within the plume. But these conditions are quite sensitive to background conditions. In addition, once the plume aerosols become part of the background the enhanced aerosol field can have similar impacts. These aspects could be addressed by a modelling study where the goals would be to come up with a "best" estimate of UT/LS background aerosols for current and future (~2050) conditions and also appropriate future meteorological conditions. Several models could repeat the earlier work of Meilenger et al. (2005) and Søvde et al. (2007) for these future conditions to better characterize impacts.

One issue that could also be addressed without new data would be to confirm the results of recent modelling studies on the impacts of a putative fleet of supersonic aircraft (SSA). Current estimates suggest that a fleet of SSAs would result in an increase in column ozone. But as this is impacted by aircraft-generated aerosols it would be useful to reexamine this with different fuel type.

Another recommendation is to continue extensive analysis of the current suite of satellite UT/LS measurements of gases (ACE, MIPAS, AURA/MLS, AURA/HIRDLS) in order *to better characterize the UT/LS region* so that we can be more confident of the putative impacts of aviation. But we also note that nadir viewing instruments (e.g. MOPITT, AURA/TES, MODIS etc) also yield important information on emissions and convection and also aerosol distributions. We also recommend the analysis of current data sets be accompanied by modelling analysis using 3D models and that at least a few of these should include evaluated "comprehensive" tropospheric and stratospheric chemistry in order to provide a more solid baseline. This could include both CTMs and GCMs. But there should also be an assessment of the robustness of the using appropriate diagnostics such discussed above such as STE ozone fluxes, tropospheric <OH> densities and correlations. These models could be used to assess the impact of aviation in climate mode or time-slice mode using SSTs and related surface properties from climate models. The rationale for using time-slice mode is that the aircraft climate/dynamical impacts are likely to be small and looking for climate signatures could require running many ensembles, although if sufficient models were run this might be taken as equivalent to ensembles. This would certainly supplement the assessment of climate impacts using a RF type metric.

An important recommendation concerns the radiative forcing changes due to the increased UT ozone from aircraft traffic. The uncertainty that clouds introduce to model simulations of this parameter was noted in Section 2. A request could be made to DOE ARM to process the large existing AERI database for surface radiative forcing from ozone and methane. With such a database, investigations of cloud effects on radiative forcing in corridor regions and in several climate regimes would be facilitated.

Of the issues that could be addressed is the re-evaluation of data from previous campaigns in the light of more recent understanding. And this brings up the issue of data accessibility. Is the older aircraft data accessible and could it be made available with such tools as GIOVANNI, developed by NASA for several satellite instruments? GIOVANNI is fast to use and easy to use; it permits scientists to access and work with the data easily online without importing large volumes of data.

6. Summary

The current state of knowledge on the formation of ozone and related chemistry and transport in the upper troposphere at air traffic flight levels has been surveyed. About 40% of the time aircraft fly in the lower stratosphere where the chemistry is well known and there is confidence in the projections of models. In the upper troposphere where jet traffic spends 60% of the time, the chemistry is insufficiently defined to make an accurate prediction of the climate impacts of increased jet traffic. In the upper troposphere HOx/NOx chemistry is uncertain as revealed by various aircraft campaigns. Measurements indicate that PAN is an important component of NOy in the mid to upper troposphere and is likely involved in long rage transport of NOx. Also aircraft measurements reveal "high" levels of NOy in the summer UT over NA, similar to those in the LT, but generated by lightning rather than anthropogenic emissions. The sources of NOx in the UT are not all well quantified, despite recent progress on quantifying the lightning source. The fraction of the deep convection source of NOx from the surface sources which reaches the 10 to 13 km level is estimated to be around 10 % for North American summer meteorological conditions but possibly 50% for European summer meteorological conditions. There is a need for new aircraft campaigns focused on the quantification of the NOx sources in the flight corridors.

Heterogeneous chemistry on aerosols/contrails from aircraft emissions may alter ozone loss chemistry in the regional background atmosphere and modify ozone production from NOx emissions. Plume to corridor studies would be useful to better characterize the conversion of NOx into HNO_3 in the near field plumes. These effects need to be characterized/confirmed for different fuel types.

The evidence of the climate impact of increased upper tropospheric ozone due to air traffic is the detection of changes in the ozone radiative forcing at the surface and at the top of the atmosphere. The radiative forcing due to ozone may be higher in some regional areas than on a global basis. There are uncertainties in the calculations of the radiative forcing metric mainly due to a lack of knowledge of cloud effects. There have been few verifications of the RF metric with real measurements of observed surface radiative forcing or with satellite radiative trapping at the top of the atmosphere.

The characterization of aviation impacts in details within the corridor is limiting progress and should be the focus of aircraft and satellite studies. However, it is rendered difficult by the relatively small effect on synoptic scales. There are several recent satellites which provide new information on the NOx and nitric acid at flight levels and this information could perhaps be applied to the problem.

In concert with aircraft campaigns, satellites experiments with improved vertical resolution are needed to study the UT/LS region.

There is a concern that there will be a gap in satellite instruments suitable for UT/LS investigations. The SHADOZ/IONS ozonesondes have proven to be highly useful for investigating ozone in the upper troposphere and could be used to partially fill such a gap. There have been too few aircraft campaigns focused on the flight corridors and which are coordinated with satellite overpass measurements.

There have been large advances in models since 1999. Data assimilation has proven very valuable in providing "value added" information from satellite data. Multiscale models are needed to investigate the corridors aspect of the aircraft emissions and the transition to regional scale climate impacts. Parameterizations used for deep convection need to be both used consistently (with the basic dynamical model) and verified according to the scale of model. Finally it is recommended that climate assessments of the impacts of a future fleet should use chemistry-climate models with comprehensive tropospheric and stratospheric chemistry at as high a resolution as feasible.

We caution that although current measurements can yield improved results by the application of more sophisticated models, it is unlikely that accurate simulations of aircraft emissions impacts on UT/LS ozone and resulting radiative forcing of climate will be possible without information from new satellite and aircraft missions and expanded sounding systems such as ozonesondes and lidars.

References

Baldwin, M. P., M. Dameris, T. G. Shepherd, How will the stratosphere affect climate change, *Science,* 316, 1576-1577, 2007.

Belair, S. and 34 authors, Operational implementation of a 33-km version of GEM for global medium-range predication at CMC, *CMOS bulletin SCMO,* 35, 77-84, 2007.

Bernath, P. F., and 45 authors, Atmospheric Chemistry Experiment (ACE): Mission Overview, *Geophys. Res. Lett.,* 32, L15S01, doi:10.1029/2005GL022386, (2005).

Bertram, T.H., and 23 authors, Direct Measurements of the Convective Recycling of the Upper Troposphere, *Science,* 315, 816-820, 2007.

Brown, S. S., T. B. Ryerson, A. G. Wollny, C. A. Brock, R. Peltier, A. P. Sullivan, R. J. Weber, W. P. Dube, M. Trainer, J. F. Meagher, F. C. Fehsenfeld, and A. R. Ravishankara, Variability in nocturnal nitrogen oxide processing and its role in regional air quality, *Science,* 311, 67-70, 2006.

Brown, S. S., H. Stark, and A. R. Ravishankara, Applicability of the steady state approximation to the interpretation of atmospheric observations of NO_3 and N2O5, *J. Geophys. Res.,* 108(D17), 4539, doi:10.1029/2003JD003407, 2003.

Cooper, O. R., A. Stohl, M. Trainer, A. M. Thompson, J. C. Witte, S. J. Oltmans, G. Morris, K. E. Pickering, J. H. Crawford, G. Chen, R. C. Cohen,T. H. Bertram, P. Wooldridge, A. Perring, W. H. Brune, J. Merrill, J. L. Moody, D. Tarasick, P. Nédélec, G. Forbes, M. J. Newchurch, F. J. Schmidlin, B. J. Johnson, S. Turquety, S. L. Baughcum, X. Ren, F. C. Fehsenfeld, J. F. Meagher, N. Spichtinger, C. C. Brown, S. A. McKeen, I. S. McDermid, and T. Leblanc, Large upper tropospheric ozone enhancements above midlatitude North America during summer: In situ evidence from the IONS and MOZAIC ozone measurement network, *J. Geophys. Res.,* 111, D24S05, doi:10.1029/2006JD007306, 2006.

Crutzen, P., and Andreae, M. O., Biomass burning in the tropics: Impact on atmospheric chemistry and biogeochemical cycles, *Science, 250*, 1669–1678, 1990.

Danilin, M.Y., J.M. Rodriguez, M.K.W. Ko, D.K. Weisenstein, R.C. Brown, R.C. Miake-Lye, and M.R. Anderson, Aerosol particle evolution in an aircraft wake: implications for the high-speed civil transport fleet impact on ozone, *J. Geophys. Res.,*102, 21453-21463, 1997:

de Grandpré, J., C. Charette, P. Gauthier, S.Chabrillat, A. Robichaud, Y. Rochon, Y.Yang and R. Ménard, Development and Validation of a Chemical Weather System at Environment Canada, 13th July, IUGG, Perugia, 2007.

Dentener, F. J. and P. J. Crutzen, Reaction of N2O5 on tropospheric aerosols: Impact on the global distributions of NOx, O_3, and OH, *J. Geophys. Res.*, 98, 7149-7163, 1993.

Dentener, F., D. Stevenson, J. Cofala, R. Mechler, M. Amann, P. Bergamaschi, F. Raes, and R. Derwent,The impact of air pollutant and methane emission controls on tropospheric ozone and radiative forcing: CTM calculations for the period 1990–2030, *Atmos. Chem. Phys.*, 5, 1731–1755, 2005.

Evans, M. J. and D. J. Jacob, Impact of new laboratory studies of N2O5 hydrolysis on global model budgets of tropospheric nitrogen oxides, ozone, and OH, *Geophys. Res. Lett.*, 32, L09813, doi: 10.1029/2005GL022469, 2005.

Evans, W.F.J. and E. Puckrin, Remote Sensing Measurements of Tropospheric Ozone by Ground-Based Thermal Emission Spectroscopy, *J. Atmos. Sci.*, 56, pp 311-318,1999.

Eyring,V.,and 39 others, Assessment of temperature, trace species, and ozone in chemistry-climate model simulations of the recent past, *J. Geophys. Res,* 111, D22308, doi:10.1029/2006JD007327, 2006.

Eyring,V.,and 38 others, Multi-model projections of stratospheric ozone in the 21st century, submitted to *J. Geophys. Res,* Spring 2007a.

Eyring, V., Stevenson, D. S., Lauer, A., Dentener, F. J., Butler, T., Collins, W. J., Ellingsen, K., Gauss, M., Hauglustaine, D. A., Isaksen, I. S. A., Lawrence, M. G., Richter, A., Rodriguez, J. M., Sanderson, M., Strahan, S. E., Sudo, K., Szopa, S., van Noije, T. P. C., and Wild, O.: Multi-model simulations of the impact of international shipping on Atmospheric Chemistry and Climate in 2000 and 2030, *Atmos. Chem. Phys.*, 7, 757-780, 2007.

Fehsenfeld, F. C., G. Ancellet, T. S. Bates, A. H. Goldstein, R. M. Hardesty, R. Honrath, K. S. Law, A. C. Lewis, R. Leaitch, S. McKeen, J. Meagher, D. D. Parrish, A. A. P. Pszenny, P. B. Russell, H. Schlager, J. Seinfeld, R. Talbot, and R. Zbinden, International Consortium for Atmospheric Research on Transport and Transformation (ICARTT): North America to Europe—Overview of the 2004 summer field study, *J. Geophys. Res.*, 111, D23S01, doi:10.1029/2006JD007829, 2006.

Flannigan, M. D., B. D. Amiro, K. A. Logan, B. J. Stocks, and B. M. Wotton, Forest fires and climate change in the 21st century, Mitigation and Adaptation Strategies for Global Change, DOI: 10.1007/s11027-005-9020-7, 2005.

Folkers, M., Th. F. Mentel, and A. Wahner, Influence of an organic coating on the reactivity of aqueous aerosols probed by the heterogeneous hydrolysis of N2O5, *Geophys. Res. Lett.*, 30(12), 1644, doi:10.1029/2003GL017168, 2003.

Gamblin B., O. B. Toon, M. A. Tolbert, Y. Kondo, N. Takegawa, H. Irie, M. Koike, J. O. Ballenthin, D. E. Hunton, T. M. Miller, A. A. Viggiano, B. E. Anderson, M. Avery, G. W. Sachse, J. R. Podolske, K. Guenther, C. Sorenson, and M. J. Mahoney, Nitric acid condensation on ice: 1. Non-HNO_3 constituent of NOY condensing cirrus particles on upper tropospheric, *J. Geophys. Res.*, 111, D21203, doi:10.1029/2005JD006048, 2006a.

Gamblin B., et al., Correction to "Nitric acid condensation on ice: 1. Non-HNO 3 constituent of NO Y condensing on upper tropospheric cirrus particles", *J. Geophys. Res.*, 111, D24210, doi:10.1029/2006JD008266, 2006b.

Gamblin B., et al., Nitric acid condensation on ice: 2. Kinetic limitations, a possible "cloud clock" for determining cloud parcel lifetime, *J. Geophys. Res.,* 112, D12209, doi:10.1029/2005JD006049, 2007.

Gauss, M., I. Isaksen, V. Grewe, M. Köhler, D. Hauglustaine, D. Lee, Impact of Aircraft NOx Emissions: Effects of Changing the Flight Altitude. In R. Sausen, C. Fichter and G. Amanatidis (eds.): European Conference on Aviation, Atmosphere and Climate (AAC). Proceedings of an International Conference, Friedrichshafen, Germany, 30 June to 3 July 2003, EUR 21051, 122-127, 2004.

Gauss, M., I. S. A. Isaksen, D. S. Lee, and O. A. Søvde, Impact of aircraft NOx emissions on the atmosphere – tradeoffs to reduce the impact, *Atmos. Chem. Phys.,* 6, 1529–1548, 2006.

Gauss, M., I.S.A. Isaksen, S. Wong and W.-C. Wang, Impact of H_2O emissions from cryoplanes and kerosene aircraft on the atmosphere, *J. Geophys. Res.,* 108, D10, 4304, doi.1029/2002JD002623, 2003.

Giglio, L., G.R. van der Werf, J.T. Randerson, G.J. Collatz, and P.S. Kasibhatla, Global Estimation of Burned Area using MODIS Active Fire Observations, *Atmos. Chem. Phys.,* 6, 957-974. SRef-ID: 1680-7324/acp/2006-6-957, 2006.

Gillett, N. P., A. J. Weaver, F. W. Zwiers, M. D. Flannigan, Detecting the effect of climate change on Canadian forest fires, *Geophys. Res. Lett.,* 31, L18211, doi:10.1029/2004GL020876, 2004.

Grell, G. A., S. Emeis, W. R. Stockwell, T. Schoenemeyer, R. Forkel, J. Michalakes, R. Knoche and W. Seidl, Application of a multiscale, coupled M5/chemistry model to the complex terrain of the VOTALP valley campaign, *Atmos. Environ.,* 34, 1435-1453, 2000.

Grell, G. A., S. E. Peckham, R. Schmitz, S.A. McKeen, G. Frost, W. C. Skamarock, and B. Eder, Fully coupled online chemistry within the WRF model, *Atmos. Environ.,* 39, 6957-6975, 2005.

Hallquist, M., D. J. Stewart, S. K. Stephenson, R. A. Cox, Hydrolysis of N2O5 on sub-micron sulfate aerosols, *Phys. Chem. Chem. Phys.,* 5, 3453-3463, 2003.

Harries, J. E., H. E. Brindley, P. J. Sagoo and R.J. Bantges, Increases in greenhouse forcing inferred from the outgoing longwave radiation spectra of the Earth in 1970 and 1997, *Nature.* 410, 355-357, 2001.

Heald, C. L., D.J. Jacob, R.J. Park, L.M. Russell, B.J. Huebert, J.H. Seinfeld, H. Liao, and R. J. Weber, A large organic aerosol source in the free troposphere missing from current models, *Geophys. Res. Lett., 32,* L18809, doi:10.1029/2005GL023831, 2005

Holton, J. R., P. H. Haynes, M. E. McIntyre, A. R. Douglass, R. B. Rood, L. Pfister, Stratosphere-Troposphere Exchange, *Rev. Geophys,* 33, 403-439, 1995.

Hu, R.-M., R. V. Martin, and T. D. Fairlie, Global retrieval of columnar aerosol single scattering albedo from space-based observations *J. Geophys. Res.,* 112, D02204, doi:10.1029/2005JD006832, 2007.

Hudson, P. K., j. E. Shilling, M. A. Tolbert, and O. B. Toon, Uptake of nitric acid on ice at tropospheric temperatures: implications for cirrus clouds, *J. Phys. Chem., A,* 106, 9874-9882, 2002.

IGAC: IGACtivities Newsletter of of the International Global Atmospheric Chemistry Project, Satellite Observations of Tropospheric Trace Gases and Aerosols: An Emerging and Compelling Component of the Science of Earth Observation and Global Change, Vol 35, March, 2007. (http://www.igac.noaa.gov/newsletter/igac35/Mar_2007_IGAC_35.pdf)

IPCC, Climate Change 2001: The Scientific Basis. Contribution of Working Group I to the Third Assessment Report of the Intergovernmental Panel on Climate Change [Houghton, J.T.,Y. Ding, D.J. Griggs, M. Noguer, P.J. van der Linden, X. Dai, K. Maskell, and C.A. Johnson (eds.)]. Cambridge University Press, Cambridge, United Kingdom and New York, NY, USA, 881pp. 2001 IPCC, Climate Change 2007: The Physical Science Basis. Contribution of Working Group I to the Fourth Assessment Report of the Intergovernmental Panel on Climate Change [Solomon, S., D. Qin, M. Manning, Z. Chen, M. Marquis, K.B. Averyt, M. Tignor and H.L. Miller (eds.)]. Cambridge University Press, Cambridge, United Kingdom and New York, NY, USA, 996 pp, 2007.

Iraci, L. T. and M. A. Tolbert, Heterogeneous interaction of formaldehyde with cold sulfuric acid: Implications for the upper troposphere and lower stratosphere, *J. Geophys. Res.,* 102, 16,099-16,107, 1997.

Jackman, C. H., J. E. Frederick, and R. S. Stolarski, Production of odd nitrogen in the stratosphere and mesosphere: An intercomparison of sources and strengths, *J. Geophys. Res.,* 85, 7495-7505, 1980.

Jackman, C. H., and R. D. McPeters, The response of ozone to solar proton events during solar cycle 21: A theoretical interpretation, *J. Geophys. Res.,* 90, 7955–7966. 1985.

Jackman, C. H., M. T. DeLand, G. J. Labow, E. L. Fleming, D. K. Weisenstein, M. K. W. Ko, M. Sinnhuber, and J. M. Russell, Neutral atmospheric influences of the solar proton events in October–November 2003, *J. Geophys. Res.,* 110, A09S27, doi:10.1029/2004JA 010888. 2005.

Jaeglé, L, L. Steinberger, R. V. Martin, and K. Chance, Global partitioning of NOx sources using satellite observations: Relative roles of fossil fuel combustion, biomass burning and soil emissions, *Farad. Discuss.,*130, 407–423, DOI: 10.1039/b502128f, 2005.

Jiang, J., N. J. Livesey, H. Su. L. Neary, and J. C. McConnell, Upper tropospheric Asian Outflow to North America: Connecting surface emission, convective uplift and long-range transport of carbon monoxide, accepted by *Geophys. Res. Lett.,* July, 2007.

Jung, T. and M. Leutbecher, Performance of the ECMWF forecasting system in the Arctic during winter, Q. J. Roy. *Met. Soc.,* 133, 1327-1340, 2007

Kane, S. M., F. Caloz, and M.-T. Leu, Heterogeneous uptake of gaseous N2O5 by (NH4)2SO4, NH4HSO4, and H2SO4 aerosols, *J. Phys. Chem.,* 105, 6465-6470, 2001.

Kaminski, J. W., D. A. Plummer, L. Neary, J. C. McConnell, J. Struzewska, and L Lobocki, First application of MC2-AQ to multiscale air quality modelling over Europe, *J. Phys. Chem. Earth,* 24 August, 8 pages, 2002.

Kaminski, J. W., L. Neary, J. Struzewska, J. C. McConnell, A. Lupu, J. Jarosz, K. Toyota, S. L. Gong, X. Liu, K. Chance and A. Richter, A Multiscale Air Quality Model – GEM-AQ: I – Evaluation of global gas phase results, *ACPD.,* 7, 14,895–14,937, 2007.

Kärcher, B., R. P. Turco, F. Yu, M. Y. Danilin, D. K. Weisenstein, R. C. Miake- Lye, R. Busen, A unified model for ultrafine aircraft particle emissions, *J. Geophys. Res.,* 105, 29379-29386, 10.1029/2000JD900531, 2000.

Köhler, I., R. Sausen and R. Reinberger, Contributions of aircraft emissions to the atmospheric NOx content, *Atmos. Envir.* ,31, 1801-1818, 1997.

Kondo, Y. et al., Uptake of reactive nitrogen on cirrus cloud particles in the upper troposphere and lowermost stratosphere, *Geophys Res. Lett.,* 30, 1154, doi:10.1029/2002GL016539, 2003.

Konopka, P., G. Günther, R. Müller, F. H. S. dos Santos, C. Schiller, F. Ravegnani, A. Ulanovsky, H. Schlager, C. M. Volk, S. Viciani, L. L. Pan, D.-S. McKenna, and M. Riese, Contribution of mixing to upward transport across the tropicaltropopause layer (TTL), *Atmos. Chem. Phys.*, 7, 3285–3308, 2007.

Kraabøl, A. G., P. Konopka, F. Stordal, H. Schlager, Modelling chemistry in aircraft plumes 1: comparison with observations and evaluation of a layered approach, *Atmos. Envir.*, 34, 3939-3950, 2000.

Kraabøl, A. G. and F. Stordal, Modelling chemistry in aircraft plumes 2: the chemical conversion of NOx to reservoir species under different conditions, *Atmos. Envir.*, 34, 3951-3962, 2000.

Kraabøl, A. G.,T.K. Berntsen, J.K. Sundet, and F. Stordal, Impacts of NOx emissions from subsonic aircraft in a global threedimensional chemistry transport model including plume processes, *J. Geophys. Res.*, 107, doi:10.1029/2001JD001019, 2002.

Lary, D. J., M. P. Chipperfield, R. Toumi, T. Lenton, Heterogeneous atmospheric bromine chemistry, *J. Geophys. Res.*, 101, 1489-1504, 10.1029/95JD02839, 1996.

Lawrence, M. G. and P. J. Crutzen, The impact of cloud particle gravitational settling on soluble trace gas distributions, *Tellus*, 50B, 263-289, 1998.

Leblanc, T., I. S. McDermid and A. Hauchecorne , Lidar observations of ozone in the subtropical eastern Pacific UT/LS and its relationship to meridional transport: Effect of Rossby Waves in the tropopause region, *Geophys. Res. Abs.*, Vol. 5, 14783, European Geophysical Society 2003.

Lewellen, D. C. and W. S. Lewellen, The Effects of Aircraft Wake Dynamics on Contrail Development, *J. Atmos. Sci.*, 58, 390-406 , 2001.

Lewellen, D. C., W. S. Lewellen, Effects of aircraft wake dynamics on measured and simulated NOx and HOx wake chemistry, *J. Geophys. Res.*, 106, 27661- 27672, 10.1029/2001JD000531, 2001.

Lopez, J.P., A.M. Fridlind, H.-J. Jost, M. Loewenstein, A.S. Ackerman, T.L. Campos, E.M. Weinstock, D.S. Sayres, J.B. Smith, J.V. Pittman, A.G. Hallar, L.M. Avallone, S.M. Davis, and R.L. Herman, CO signatures in subtropical convective clouds and anvils during CRYSTAL-FACE: An analysis of convective transport and entrainment using observations and a cloud-resolving model. *J. Geophys. Res.*, 111, D09305, doi:10.1029/2005JD006104. 2006.

Marenco, A.. V. Thouret, P. Nédélec, H. Smit, M. Helten, D. Kley, F. Karcher,P. Simon, K. Law, J. Pyle, G. Poschmann, R. Von Wrede, C. Hume, T. Cook, Measurement of ozone and water vapor by Airbus in-service aircraft: The MOZAIC airborne program, An overview, *J. Geophys. Res.*, 103, 25,631–25,642, 1998.

Martin,R. V., B. Sauvage,I. Folkins, C. E. Sioris, C. Boone, P. Bernath, and J. Ziemke, Space-based constraints on the production of nitric oxide by lightning, *J. Geophys. Res.*, 112, D09309, doi:10.1029/2006JD007831, 2007.

Massie, S. T. and 22 others, Validation of HIRDLS Observations of PSCs and Subvisible Cirrus, *J. Geophys. Res.*, doi:10.1029/2007JD008788, in press, 2007.

McElroy, C. T., C. R. Nowlan, J. R. Drummond, P. F. Bernath, D. V. Barton, D. G. Dufour, C. Midwinter, R. B. Hall, A. Ogyu, A. Ullberg, D. I. Wardle, J. Kar, J. Zou, F. Nichitiu, C. D. Boone, K. A. Walker, and N. Rowlands, The ACE- MAESTRO instrument on SCISAT: description, performance, and preliminary results, *Applied Optics*, 46, 4341-4356, 2007.

McKenna, D. S., Konopka, P., Grooß, J.-U., Günther, G., Müller, R., Spang, R., Offermann, D., and Orsolini, Y.: A new Chemical Lagrangian Model of the Stratosphere (CLaMS): Part I Formulation of advection and mixing, *J. Geophys. Res.,* 107, 4309, doi:10.1029/2000JD000114, 2002.

Meilinger, S. K., B. Kärcher, and Th. Peter, Suppression of chlorine activation on aviation-produced volatile particles, *Atmos. Chem. Phys.,* 2, 307–312, 2002.

Meilinger, S. K., B. Kärcher, and Th. Peter, Microphysics and heterogeneous chemistry in aircraft plumes – high sensitivity on local meteorology and atmospheric composition, *Atmos. Chem. Phys.,* 5, 533–545, 2005.

Mickley, L. J., D. J. Jacob, B. D. Field, and D. Rind, Climate response to the increase in tropospheric ozone since preindustrial times: A comparison between ozone and equivalent CO_2 forcings, 109, D05106, doi:10.1029/2003JD003653, 2004

Miura, H. et al., A Madden-Julian Oscillation event realistically simulated by a Global Cloud-Resolving Model, *Science,* 318, 1763-1765, 2007.

Murphy, D. M., D. J. Cziczo, P. K. Hudson, and D. S. Thomson, Carbonaceous material in aerosol particles in the lower stratosphere and tropopause region, *J. Geophys. Res.,* 112, D04203, doi:10.1029/2006JD007297, 2007.

Murtagh, D., U. Frisk, F. Merino, M. Ridal, A. Jonnson, J. Stegman, G. Witt, P. Eriksson, C. Jimenez, G. Megie, J. de Noe, P. Ricaud, P. Baron, J. R. Pardon, A. Hauchcorne, E. J. Llellwelyn, D.A. Degenstein, R.L. Gattinger, N.D. Lloyd, W.F.J. Evans, I.C. McDade, C.S. Haley, C. Sioris, C. von Savigny, B.H. Solheim, J.C. McConnell, K. Strong, E.H. Richardson, G.W. Leppelmeier, E. Kyrola, H.Auvinen and L. Oikarinen, An overview of the ODIN atmospheric mission, *Can. J. Phys.,* 80, 309-319, 2002

Nedelec, P., Cammas, J.-P., Thouret, V., Athier, G., Cousin, J.-M., Legrand, C., Abonnel, C., Lecoeur, F., Cayez, G., and Marizy, C.: An improved infrared carbon monoxide analyser for routine measurements aboard commercial Airbus aircraft: technical validation and first scientific results of the MOZAIC III programme, *Atmos. Chem. Phys.,* 3, 1551-1564, 2003.

NRC, Radiative forcing of climate change: expanding the concept and addressing uncertainties, National academies Press, 2005

Olsen, S.C., C. A. McLinden, and M.J. Prather, Stratospheric N2O-NOy system: testing uncertainties in a three-dimensional framework, *J. Geophys. Res.,* 106, 28,771-28,784, 2001.

O'Neill, N. T., M. Campanelli, A. Lupu, S. Thulasiraman, J. S. Reid, M. Aube, L. Neary, J.W. Kaminski. J.C. McConnell, Optical evaluation of the GEM-AQ air quality model during the Quebec smoke event of 2002: performance criteria for extensive and intensive optical variables, *Atmos. Envir.,* 40, 3737-3749, 2006.

Pan, L. L., J. C. Wei, D. E. Kinnison, R. R. Garcia, D. J. Wuebbles, and G. P. Brasseur, A set of diagnostics for evaluating chemistry-climate models in the extratropical tropopause region, D09316, doi:10.1029/2006JD007792, 2007

Park, M., W. J. Randel, A. Gettelman, S. T. Massie, and J. H. Jiang, Transport above the Asian summer monsoon anticyclone inferred from Aura Microwave Limb Sounder tracers, *J. Geophys. Res.,* 112, D16309, doi:10.1029/2006JD008294, 2007.

Penner, J.E., D.H. Lister, D.J. Griggs, D.J. Dokken, M. McFarland (Eds).1999: Aviation and the global atmosphere.- A special report of IPCC working groups I and III. IPCC.- Cambridge University Press. Cambridge UK and New York, NY, USA, 365 pp.(1999).

Philipona, R., B. Dürr, A. Ohmura, and C. Ruckstuhl, Anthropogenic greenhouse forcing and strong water vapor feedback increase temperature in Europe, *Geophys. Res. Lett.,* 32, L19809, doi:10.1029/2005GL023624, 2005.

Pickering, K. E., Y. Wang, W.-K. Tao, C. Price, and J.-F Muller, Vertical distribution of lightning NOx for use in regional and global transport models, *J. Geophys. Res.,* 103, 31,203-21,216, 1998.

Pisano, J.T., J.W. Drummond and D. R. Hastie, A Lightweight NO2 Instrument for Vertical Height Profiles, *J. Atm. Ocean Tech.,* 13, 400-406, 1996.

Plummer, D. A., J. C. McConnell, J. Drummond, J. Narayan, V. Young, D. R. Hastie, Assessment of emissions data for the Toronto region using aircraft- based measurements and an air quality model, *Atmos. Environ*., 35, 6543- 6463, 2001.

Pope,F.D., J.C. Hansen, K.D. Bayes, R.R. Friedl, and S.P. Sander, 2007, Ultraviolet absorption spectrum of chlorine peroxide, ClOOCl, *J. Phys Chem. A,* 111 (20): 4322-4332. Prather, M. J., Numerical advection by conservation of second-order moments, *J. Geophys. Res.,* 91, 6671–6681, 1986.

Prather, M. J., and D. J. Jacob, A persistent imbalance in HOx and NOx photochemistry of the upper troposphere driven by deep tropical convection, *Geophys. Res. Lett.,* 24, 3189–3192, 1997.

Puckrin, E. , W.F.J. Evans, J. Li and H. Lavoie, Comparison of Clear-Sky Greenhouse Fluxes Simulated With Radiative Transfer Models, *Can. J Remote Sensing,* 30, 903-912, 2004.

Randerson, J. T., G. R. van der Werf, G. J. Collatz, L. Giglio, C. J. Still, P. Kasibhatla, J. B. Miller, J. W. C. White, R. S. DeFries, and E. S. Kasischke, Fire emissions from C3 and C4 vegetation and their influence on interannual variability of atmospheric CO_2 and d13 CO_2, *Global Biogeochemical. Cycles,* 19, GB2019, doi:10.1029/2004GB002366, 2005.

Randerson, J. T., G. R. van der Werf, L. Giglio, G. J. Collatz, and P.S. Kasibhatla. Global Fire Emissions Database, Version 2 (GFEDv2.1). Data set. Available on-line [http://daac.ornl.gov/] from Oak Ridge National Laboratory Distributed Active Archive Center, Oak Ridge, Tennessee, U.S.A, 2007.

Ren, X., W. H. Brune, J. Mao, R. B. Long, R. L. Lesher, J. H. Crawford, J. R. Olson, G. Chen, M. A. Avery, G. W. Sachse, G. S. Diskin, J. D. Barrick, L. G. Huey, A. Fried, R. C. Cohen, B. Heikes, P. Wennberg, H. B. Singh, D. R. Blake, R. E. Shetter, HOx Observation and Model Comparison during INTEX- NA 2004, http://www.espo.nasa.gov/docs/intex-na/HOx.pdf, 2006.

Riishojgaard, L. P, An update on the Molniya orbit imager, (http://www.eumetsat.int/groups/cps/documents/document/pdf_conf_p47_s4_13_ riishojga_v.pdf), circa 2005

Salawitch, R. J., D. K. Weisenstein, L. J. Kovalenko, C. E. Sioris, P. O. Wennberg, K. Chance, M. K. W. Ko, and C. A. McLinden, Sensitivity of ozone to bromine in the lower stratosphere, *Geophys. Res. Lett.,* 32, L05811, doi:10.1029/2004GL021504, 2005.

Sausen, R., I. Isaksen, V. Grewe, D. Hauglustaine, D.S. Lee, G. Myhre, M.O. Köhler, G. Pitari, U. Schumann, F. Stordal and C. Zerefos,: Aviation radiative forcing in 2000: An update on IPCC (1999). *Meteorol. Z. 14,* 555-561, 2005 Schlager, H., P. Schulte, F. Flatøy, F. Slemr, P. Van Velthoven, H. Ziereis and U. Schumann, Regional nitric oxides ehancements in the North Atlantic flight corridor observed and modeled during POLINAT 2, a case study, *Geophys. Res. Lett,* 26, 3061-3064, 1999.

Schultz, M. G., D. J. Jacob, J. D. Bradshaw, S. T. Sandholm, J. E. Dibb, R. W. Talbot, and H. B. Singh, Chemical NOx budget in the upper troposphere over the tropical South Pacific, *J. Geophys. Res.*, 105(D5), 6669–6680, 2000.

Schumann, U., H. Schlager, F. Arnold, J. Ovarlez, H. Kelder, Ø. Hov, G. Hayman, I. S. A. Isaksen, J. Staehelin, P. D. Whitefield Pollution from aircraft emissions in the North Atlantic flight corridor: Overview on the POLINAT projects. *J. Geophys. Res.,* 105, 3605–3632, 2000.

Schumann, U. and H. Huntrieser, The global lightning-induced nitrogen oxides source, *Atmos. Chem. Phys.,* 7, 3823–3907, 2007.

Shindell, D. T., and 42 authors, Multimodel simulations of carbon monoxide: Comparison with observations and projected near-future changes, 111, D19306, doi:10.1029/2006JD007100, 2006.

Singh, H. B., A. M. Thompson, and H. Schlager, SONEX airborne mission and coordinated POLINAT 2 activity: Overview and accomplishments, *Geophys. Res. Lett,* 25, 3053-3056, 1999.

Singh H. B., W. H. Brune, J. H. Crawford, D. J. Jacob, P. B. Russell, Overview of the summer 2004 Intercontinental Chemical Transport Experiment–North America (INTEX-A), *J. Geophys. Res.,* 111, D24S01, doi:10.1029/2006JD007905, 2006.

Singh, H.B., L. Salas, D. Herlth, R. Kolyer, E. Czech, M. Avery, J.H. Crawford, R.B. Pierce, G.W. Sachse, D.R. Blake, R.C. Cohen, T.H. Bertram, A. Perring, P.J. Wooldridge, J. Dibb, G. Huey, R.C. Hudman, S. Turquety, L.K. Emmons, F. Flocke, Y. Tang, G.R. Carmichael, and L.W. Horowitz, Reactive nitrogen distribution and partitioning in the North American troposphere and lowermost stratosphere, *J. Geophys. Res.,* 112, D12S04, doi:10.1029/2006JD007664, 2007.

Sioris, C. E., C. A. McLinden, R. V. Martin, B. Sauvage, C. S. Haley, N. D. Lloyd, E. J. Llewellyn, P. F. Bernath, C. D. Boone, S. Brohede, and C. T. McElroy, Vertical profiles of lightning-produced NO2 enhancements in the upper troposphere observed by OSIRIS, *Atmos. Chem. Phys.,* 7, 4281–4294, 2007.

Sitnikov, N. M. , A. O. Sokolov, F. Ravegnani, V. A. Yushkov and A. E. Ulanovskiy , A Chemiluminescent Balloon-Type Nitrogen Dioxide Meter for Tropospheric and Stratospheric Investigations (NaDA), *Instruments and Experimental Techniques,* 48 (3), 400-405, doi:10.1007/s10786-005-0070-6, 2005.

Søvde, O. A., M. Gauss, I. S. A. Isaksen, G. Pitari, and C. Marizy, Aircraft pollution – a futuristic view, *Atmos. Chem. Phys.,* 7, 3621–3632, 2007.

Stevenson, D. S., R. M. Doherty, M. G. Sanderson, W. J. Collins, C. E. Johnson and R. G. Derwent (2004), Radiative forcing from aircraft NOx emissions: Mechanisms and seasonal dependence, *J. Geophys. Res.,* 109, D17307, doi:10.1029/2004JD004759.

Stevenson, D.S., F.J. Dentener, M.G. Schultz, K. Ellingsen, T.P.C. van Noije, O. Wild, G. Zeng, M. Amann, C.S. Atherton, N. Bell, D.J. Bergmann, I. Bey, T. Butler, J. Cofala, W.J. Collins, R.G. Derwent, R.M. Doherty, J. Drevet, H.J. Eskes, A.M. Fiore, M. Gauss, D.A. Hauglustaine, L.W. Horowitz, I.S.A. Isaksen, M.C. Krol, J.-F. Lamarque, M.G. Lawrence, V. Montanaro, J.-F. Mueller, G. Pitari, M.J. Prather, J.A. Pyle, S. Rast, J.M. Rodriguez, M.G. Sanderson, N.H. Savage, D.T. Shindell, S.E. Strahan, K. Sudo, S. Szopa, Multi-model ensemble simulations of present-day and near-future tropospheric ozone, *J. Geophys. Res.,* 111, D08301, doi:10.1029/2005JD006338, 2006.

Stocks, B. J., M. A. Fosberg, T. J. Lynham, L. Mearns, B. M. Wotton, Q. Yang, J-Z. Jin, K. Lawrence, G. R. Hartley, J. A. Mason and D. W. McKenny, Climate Change and Forest Fire Potential in Russian and Canadian Boreal Forests, *Climatic Change,* 38, 1-13, 1998.

Stohl, A., M. Hittenberger, and G. Wotawa, Validation of the Lagrangian particle dispersion model FLEXPART against large scale tracer experiment data, *Atmos. Environ.,* 32, 4245– 4264, 1998.

Stohl, A., C. Forster, A. Frank, P. Seibert, and G. Wotowa, Technical note: The Lagrangian particle dispersion model FLEXPART version 6.2, *Atmos. Chem. Phys.,* 5, 2461–2474, 2005.

Sutkus Jr., D. J., S. L. Baugham, and D. Dubois, Commercial Aircraft Emission Scenario for 2020: Database and development and Analysis, NASA/CR-2003- 212331, pp59, 2003.

Tabazadeh, A., O.B. Toon and E. J. Jensen, A surface chemistry model for non-reactive trace gas adsorption on ice: implications for nitric acid scavenging by cirrus, *Geophys. Res. Lett,* 26, 2211-2214, 1999.

Thompson, A.M., J.C. Witte, R.D. McPeters, S.J. Oltmans, F.J. Schmidlin, J.A. Logan, M.Fujiwara, V.W.J.H. Kirchhoff, F. Posny, G.J.R. Coetzee, B. Hoegger, S. Kawakami, T. Ogawa, B.J. Johnson, H. Vömel and G. Labow, Southern Hemisphere Additional Ozonesondes (SHADOZ) 1998-2000 tropical ozone climatology 1. Comparison with Total Ozone Mapping Spectrometer (TOMS) and ground-based measurements, *J. Geophys. Res.,* Vol. 108 No. D2, 8238, doi: 10.1029/2001JD000967, 30 January 2003a.

Thompson, A.M., J.C. Witte, S.J. Oltmans, F.J. Schmidlin, J.A. Logan, M. Fujiwara, V.W.J.H. Kirchhoff, F. Posny, G.J.R. Coetzee, B. Hoegger, S. Kawakami, T. Ogawa, J.P.F. Fortuin, and H.M. Kelder, Southern Hemisphere Additional Ozonesondes (SHADOZ) 1998-2000 tropical ozone climatology 2. Tropospheric variability and the zonal wave-one, *J. Geophys. Res.,* Vol. 108 No. D2,8241, doi: 10.1029/2002JD002241, 31 January 2003b.

Thompson, A.M., J.B. Stone, J. C. Witte, S. K. Miller, R. B. Pierce, R. B. Chatfield, S.J. Oltmans, O. R. Cooper, A.L. Loucks, B. F. Taubman, B.J. Johnson, E. Joseph, T.L. Kucsera, J. T. Merrill, G. A. Morris, S. Hersey, G. Forbes, M. J. Newchurch, F. J. Schmidlin, D. W. Tarasick, V. Thouret and J.- P.Cammas, IONS-04 (INTEX Ozonesonde Network Study, 2004): 1.

Summertime Upper Tropospheric/Lower Stratosphere Ozone over Northeastern North America, *J. Geophys. Res.,* doi: 10.1029/2006JD007441, 112, D12S12, 2007a.

Thompson, A. M., J. B. Stone, J. C. Witte, S. K. Miller, S. J. Oltmans, T. L. Kucsera, K. L. Ross, K. E. Pickering, J.T. Merrill, G. Forbes, D. W. Tarasick, E. Joseph, F. J. Schmidlin, W. W. McMillan, J. Warner, E. J. Hintsa and J. E. Johnson, IONS-04 (INTEX Ozonesonde Network Study, 2004): 2.

Tropospheric Ozone Budgets and Variability over Northeastern North America, *J. Geophys. Res.,* doi: 10.1029/2006JD007670, 112, D12S13, 2007b.

Thornton, J. A., C. F. Braban, and J. P. D. Abbatt, N2O5 hydrolysis on sub-micron organic aerosols: the effect of relative humidity, particle phase, and particle size, *Phys. Chem. Chem. Phys.,* 5, 4593–4603, 2003.

Tie, X., et al., Effect of sulfate aerosol on tropospheric NOx and ozone budgets: Model simulations and TOPSE evidence, *J. Geophys. Res.,* 108(D4), 8364, doi:10.1029/2001JD001508, 2003.

Tost, H., P. Jöckel, and J. Lelieveld, Lightning and convection parameterisations – uncertainties in global modelling, *Atmos. Chem. Phys.*, 7, 4553–4568, 2007.

Travis, D. J., A. M. Carleton, and R. G. Lauritsen, Contrails reduce daily Temperature range. *Nature*. 418, 601, 2002.

Valks, P.J.M. and G.J.M. Velders, The present-day and future impact of NOx emissions from subsonic aircraft on the atmosphere in relation to the impact of NOx surface source, *Ann. Geophys.*, 17,1064-1079, DOI:10.1007/s00585-999- 1064-7,1999.

van der Werf, G. R., Randerson, J. T., Collatz, G. J., Giglio, L., Kasibhatla, P. S., Arellano, A. F., Olsen, S. C., and Kasischke, E. S., Continental-scale partitioning of fire emissions during the 1997 to 2001 El Nino/La Nina period, *Science*, 303, 73–76, 2004.

van der Werf, G.R., J.T. Randerson, L.Giglio, G.J. Collatz, and P.S. Kasibhatla, Interannual variability in global biomass burning emission from 1997 to 2004, *Atmos. Chem. Phys.*, 6, 3423-3441, 2006.

van Leer, B., Towards the Ultimate Conservative Difference Scheme. IV. A New Approach to Numerical Convection, *J. Comput. Phys.*, 23, 276-299, 1977. 51 of 61

Volz-Thomas, A., Berg, M., Heil, T., Houben, N., Lerner, A., Petrick, W., Raak, D., and Pätz, H.-W.: Measurements of total odd nitrogen (NOy) aboard MOZAIC in-service aircraft: instrument design, operation and performance, *Atmos. Chem. Phys.*, 5, 583-595, 2005.

von Glasow, R., R. von Kuhlmann, M. G. Lawrence, U. Platt, and P. J. Crutzen, Impact of reactive bromine chemistry in the troposphere, *Atmos. Chem. Phys.*, 4, 2481–2497, 2004.

von Hobe, M., R. J. Salawitch, T. Canty, H. Keller-Rudek, G. K. Moortgat, J.-U. Grooß, R. Müller, and F. Stroh, Understanding the kinetics of the ClO dimmer cycle, *Atmos. Chem. Phys.*, 7, 3055 – 3069, 2007.

Wang, P.-H., J. Fishman, V. L. Harvey, and M. H. Hitchman, Southern tropical upper tropospheric zonal ozone wave-1 from SAGE II observations (1985- 2002), *J. Geophys. Res.*,111, 2006, Doi: 10.1029/2005JD006221, 2006.

Waugh, D. W., S. E. Strahan, and P. A. Newman, Transport and odelling of stratospheric inorganic chlorine, *Atmos. Chem. Discuss*, 7, 8597-8616, 2007.

Wuebbles, D., A Report of Findings and Recommendations of the" Workshop on the Impacts of Aviation on Climate Change, June 7-9,2006 , Boston, MA, August 31, 2006.

Wild, O., Modelling the global tropospheric ozone budget: exploring the variability in current models, *Atmos. Chem. Phys.*, 7, 2643-2660, 2007

WMO, Scientific Assessment of Ozone Depletion: 2006, Global Ozone Research and Monitoring Project—Report No. 50, 572 pp., Geneva, Switzerland, 2007.. Wong, S. and A. E. Dessler, Regulation of H_2O and CO in tropical tropopause layer by the Madden-Julian oscillation *J. Geophys. Res.*, 112, D14305, doi:10.1029/2006JD007940, 2007

Xie, S., Simulations of midlatitude frontal clouds by single-column and cloud- resolving models during the Atmospheric Radiation Measurement March 2000 cloud intensive operational period. *J. Geophys. Res.*,110,: D15S03, 2005.

Xu , K.-M., R. T. Cederwall , L. J. Donner , W. W. Grabowski, F. Guichard, D. E. Johnson , M. Khairoutdinov, S. K. Krueger, J. C. Petch, D. A. Randall , C. J. Seman , W.-K. Tao , D. Wang, S. C. Xie , J. J. Yio , M.-H. Zhang, An intercomparison of cloud-resolving models with the atmospheric radiation measurement summer 1997 intensive observation period data, Quart. *J. Royal Met. Soc.*, 128, 593 – 624, 2006.

Yang, R, J., A. G. Xia, D. V. Michelangeli, D. A. Plummer, L. Neary, J. W. Kaminski, and J. C. McConnell, Evaluating a Canadian regional air quality model using ground-based observations in North-Eastern Canada and United States, *J. Environ. Monit.,* 5, 40-46, 2003.

Yang, X., R. A. Cox, N. J. Warwick, J. A. Pyle, G. D. Carver, F. M. O'Connor, and N. H. Savage, Tropospheric bromine chemistry and its impacts on ozone: A model study, *J. Geophys. Res.,* 110, D23311, doi:10.1029/2005JD006244, 2005

Yu, F., R. P. Turco, and B. Karcher, The possible role of organics in the formation and evolution of ultrafine aircraft particles, *J. Geophys. Res.,* 104, 4079-4087, 1999

Zhao, J., N. P. Levitt, and R. Zhang, Heterogeneous chemistry of octanal and 2, 4- hexadienal with sulfuric acid, *Geophys. Res. Lett.*, 32, L09802, doi:10.1029/2004GL022200, 2005.

Zellner R, Global Aspects of Atmospheric Chemistry, p182, *Topics in Physical. Chemistry,* Vol. 6), Springer, New York, 1999.

Ziereis, H., H. Schlager, P. Schulte, P. F. J. van Velthoven, F. Slemr, Distributions of NO, NOx, and NOy in the upper troposphere and lower stratosphere between 28° and 61°N during POLINAT 2, *J. Geophys. Res.,* 105, 3653-3664, 2000.

Ziereis, H., A. Minikin, H. Schlager, J. F. Gayet, F. Auriol, P. Stock, J. Baehr, A. Petzold, U. Schumann, A. Weinheimer, B. Ridley, and J. Ström, Uptake of reactive nitrogen on cirrus cloud particles during INCA, Geophys. Res. Lett., 31, L05115, doi:10.1029/ 2003GL018794. 2004.

ANNEX I: MODELS WITH COMPREHENSIVE TROPOSPHERIC AND STRATOSPHERIC CHEMISTRY

CTMs

Bousserez, N. et al. Evaluation of the MOCAGE chemistry transport model during the ICARTT/ITOP experiment. J. Geophys. Res., 112, D10S42, doi:10.1029/2006JD007595 (2007).

Kinnison, D. E., G. P. Brasseur, S. Walters, R. R. Garcia, D. R. Marsh, F. Sassi, V. L. Harvey, C. E. Randall, L. Emmons, J. F. Lamarque, P. Hess, J. J. Orlando, X. X. Tie, W. Randel, L. L. Pan, A. Gettelman, C. Granier, T. Diehl, U. Niemeier, and A. J. Simmons, Sensitivity of chemical tracers to meteorological parameters in the MOZART-3 chemical transport model, J. Geophys. Res., 112, D20302, doi:10.1029/2006JD007879, 2007.

Rotman, D. A., C. S. Atherton, D. J. Bergmann, P. J. Cameron-Smith, C. C. Chuang, P. S. Connell, J. E. Dignon, A. Franz, K. E. Grant, D. E. Kinnison, C. R. Molenkamp, D. D. Proctor, and J. R. Tannahill, IMPACT, the LLNL 3-D global atmospheric chemical transport model for the combined troposphere and stratosphere: Model description and analysis of ozone and other trace gases, 109, D04303, doi:10.1029/2002JD003155, 2004.

Climate models

Shindell, D. T, G. Faluvegi, N. Unger, E. Aguilar, G. A. Schmidt, D. M. Koch, S. E. Bauer, and R. L. Miller, Simulations of preindustrial, present-day, and 2100 conditions in the

NASA GISS composition and climate model G-PUCCINI, Atmos. Chem. Phys., 6, 4427–4459, 2006. [Prather Transport]

Marsh, D.R., R.R. Garcia, D.E. Kinnison, B.A. Boville, F. Sassi, and S.C. Solomon, Modeling the whole atmosphere response to solar cycle changes in radiative and geomagnetic forcing, J. Geophys. Res., in press, 2007.

Jöckel, P. et al. The atmospheric chemistry general circulation model ECHAM5/MESSy1: consistent simulation of ozone from the surface to the mesosphere. Atmos. Chem. Phys. 6, 5067-5104 (2006).

Annex II. Suggested Structure for Each SSWP

1. Introduction and Background specific to your theme area
2. Review of specific theme

a. Current state of science
b. Critical role of the specific theme
c. Advancements since the IPCC 1999 report
d. Present state of measurements and data analysis
e. Present state of modeling capability/best approaches
f. Current estimates of climate impacts and uncertainties
g. Interconnectivity with other SSWP theme areas

3. Outstanding limitations, gaps and issues that need improvement

a. Science
b. Measurements and analysis
c. Modeling capability
d. Interconnectivity with other SSWP theme areas

4. Prioritization for tackling outstanding issues based on their

a. Impact
b. Ability to improve the climate impacts estimates with reduced uncertainties
c. Practical use (e.g. model improvement, sensitivity analysis, metric development etc.)
d. Achievability
e. Estimated cost
f. Timeline

5. Recommendations for best use of current tools for modeling and data analysis

a. Options
b. Supporting rationale
c. How to best integrate best available options?
6. Summary

ANNEX III: QUESTIONS TO BE CONSIDERED IN REPORT

Please note that each of the SSWP is meant to provide recommendations on:

* Improvement needed to advance the state of science and modelling capability; and
* best use of the present state of science and modeling capability to
+ better quantify magnitude of climate impacts of aviation and associated uncertainties for present and future condition
+ develop metrics to measure these impacts on all relevant scales

It is expected that each SSWP will also address the following common questions to the best extent possible:

* What are the key science questions/issues specific to aviation induced climate change impacts for the present and future conditions?
* What are the state of science, present modeling capability, and observation databases available to answer these key questions?
* What are the controlling factors: scientific knowledge vs. modelling capability vs. computational resources given that aviation-induced atmospheric perturbations range from plume to global scales?
* What are the presently available best options among existing models and their individual modules to isolate and estimate atmospheric and radiative perturbations due to aviation emissions? How well these models perform to simulate the state of the background atmosphere due to all non-aviation sources?
* How to best integrate available modeling options to simulate atmospheric perturbations due to aviation and how to evaluate the model performance to characterize the aviation-induced perturbations?
* What are the gaps and uncertainties in science? What are the limitations in observations and modeling tools to answer the key questions?
* With no further scientific knowledge, how and with what level of uncertainties, can the key questions be answered today using the best available modeling tools?
* If the gaps were to be addressed, would the ability to answer the key questions get any better? If so, to what possible extent and within what possible timeframe?

The review panel has made some suggestions that are general and applicable to all selected proposals:

* Within the sphere of your own SSWP, reach beyond the issues that were listed in the solicitation. Our stated questions/issues were merely the sample to provide some guidance as well as relevance and they were not intended to limit your scope of work.
* Be comprehensive in preparing the SSWP by including the review on the state of science, gaps and uncertainties, modeling capabilities and current state of relevant measurements and analysis of the existing data with a focus on present and future climate impacts due to aviation.
* Do not limit the scope of SSWP to your own activities. Outreach the latest efforts of the entire scientific community as a whole.

* Wherever possible, maintain the interconnectivity among themes of all other SSWPs. Study of climate impacts needs to be carried out within the one-atmosphere framework through interrelated processes irrespective of how they are distinctly classified as dynamical, transport, chemical, microphysical, optical and/or radiation.

* Identify the best modeling and analysis options presently available and provide the supporting rationale.

* Address the gaps and uncertainties (in the state of science, modelling capabilities and their practical applications), identify the key areas of improvements and prioritize them based on their practical achievability as well as associated timelines.

In: Aviation and the Environment
Editor: Jon C. Goodman

ISBN: 978-1-60692-320-7

Chapter 3

CLIMATE IMPACT OF CONTRAILS AND CONTRAIL CIRRUS SSWP # IV, JANUARY 25, 2008

U. Burkhardt, B. Kärcher, H. Mannstein and U. Schumann
DLR Institute for Atmospheric Physics, Oberpfaffenhofen, Germany

EXECUTIVE SUMMARY

Generally, the climatic impact of air traffic (of which a substantial part may be due to contrails and contrail cirrus) today (year 2000) amounts to 2-8% of the global radiative forcing associated with climate change. Due to the projected increase in air traffic [ICAO, 2007] the relative importance of air traffic is going to increase drastically. In the long term it may well be, that the most serious threat to the continued growth of air travel is its impact on climate [Green, 2005]. In view of the societal relevance and economic importance of sustainable growth of global aviation, it would be appropriate that the climate science community received sufficient funding, allowing significant progress estimating climate impacts, in order to ensure that political decisions are based on increasingly sound scientific knowledge. Aircraft-induced cloudiness, which comprises contrail cirrus and modification of cirrus by aircraft exhaust soot emissions are the most uncertain component in aviation climate impact assessments [IPCC, 2007]. Since they may be the largest component in aviation radiative forcing aircraft-induced cloudiness and contrail cirrus in particular requ ire a largeresearch effort.

Contrails develop at lower relative humidity than natural cirrus and therefore can increase high cloudiness and change the radiation budget significantly in or near regions with high air traffic density. Several studies have inferred coverage due to line-shaped contrails in limited areas using satellite data. In situ measurements have been made analyzing young contrails regarding their ice water content, particle sizes and particle habit, some of which have not been fully mined. Radiative transfer models estimate the radiative forcing due to individual contrails. Current estimates of global contrail radiative forcing are based on climate model simulations using a simple parameterization for line-shaped contrail coverage and their ice water content.

From those studies we know that line-shaped contrail coverage in areas of high traffic density may be as large as a few percent. The optical properties of the probed contrails are distinct from natural cirrus with line-shaped contrails consisting of a larger number of smaller particles. The optical properties of isolated line-shaped contrails change the radiative balance in a way that they cause in the majority of situations warming of the atmosphere. Line-shaped contrails are estimated to cause a global radiative forcing of about 10 mW/m2.

Contrary to the IPCC, we judge the state of science regarding contrail radiative forcing to be poor. Regional line-shaped contrail coverages have not been inferred from satellite data in a way that warrants intercomparability. Detection thresholds have not been properly estimated and discussed so that a comparison with model estimates is hampered. Available airborne measurements of contrails suffer from the poor detectability and characterization of size and habit of small ice crystals typical for contrails. Estimates of global line-shaped contrail coverage are all based on one single contrail parameterization approach using a climate model. Most of those studies use the same global model, the same tuning data set and the same assumptions about contrail ice microphysics and optical depth so that similar results are not surprising. There exists a pending dissent about contrail optical properties from Lidar measurements and satellite retrievals estimating larger mean optical depth than suggested by global models. Due to the problems using observational data for model validation the extent of the disagreement is not known. Radiative transfer simulations find only modest changes of contrail radiative forcing due to three-dimensional effects, the inclusion of the diurnal cycle of air traffic and other factors. However, even the most advanced radiative transfer studies have not yet incorporated best knowledge of key parameters such as contrail ice microphysics in order to place conclusive bounds on associated uncertainties.

Therefore, in a first step, we recommend that more independent studies and sensitivity experiments should be performed estimating the climate impact of line-shaped contrails so that proper error bars of radiative forcing can be inferred. Better observational data together with the associated detection thresholds and efficiencies must be obtained that can be used for constraining contrail model parameterizations and model validation. A uniform data set estimating line-shaped contrail coverage from satellites globally is needed.

Without further progress in contrail modeling we will not be able to answer questions about the climate impact of future air traffic scenarios. Radiative forcing estimates due to contrails cannot be simply scaled with an increased air traffic since future air traffic is forecasted to increase mainly in the more humid subtropics of southeast and east Asia. Model estimates of radiative forcing are mainly describing the effect of contrails in the areas of strongest current air travel, the extratropics. Observational studies as well have been focusing on the mid latitudes. In the subtropics there is little observational evidence of the optical properties and radiative effects of contrails and it is not known how well current contrail parameterizations will perform in the tropics.

Until now not even an accepted (IPCC-level) best estimate of radiative forcing due to aircraftinduced cirrus changes exists and the state of knowledge is generally judged to be very low. Persistent contrails spreading into long-lived contrails cirrus decks covering s ubstantialregional areas is observed at times, hence significant atmospheric effects can be expected. However, there is no robust estimate for the additional coverage due to contrail cirrus. The small- and synoptic-scale meteorological conditions supporting contrail cirrus development (including relative humidity and wind shear) appear to be highly variable. Therefore, no simple relationship exists between contrail age and linearity and detectability.

The optical properties of contrail cirrus are not known and radiative forcing due to contrail cirrus has not been estimated. Therefore, current estimates of total aviation-induced radiative forcing likely lack an important contribution. Contrail cirrus may not only change the radiative balance due to an increase in cloud coverage or optical thickness of existing cirrus but also by modifying the upper tropospheric moisture budget and by replacing or changing natural cirrus, should the optical properties of contrail cirrus remain distinct from natural cirrus. Cirrus changes due to the emission of soot particles from aircraft jet engines are much less certain. The icenucleation behavior of fresh soot emissions is probably poor according to in situ data, but regional or large-scale effects on cirrus properties and coverage cannot be ruled out. Progress in this area requires a targeted field study demonstrating the ability of aging aircraft soot particles to form ice at lower relative humidities than ice nuclei from other sources or the ability to change particle size spectra in cirrus.

We propose introducing contrail cirrus as a new, purely anthropogenic ice cloud type and recommend studying the whole life cycle of contrails. On the one hand in situ and remote sensed observations of aged contrail cirrus are needed. On the other hand contrails should be treated in global models as an independent cloud class together with their associated ice water content. The formation of contrail cirrus from individual young contrails over a wide range of spatial scales requires a special model study. Also identification of aviation induced cloudiness in observations needs further studies.

Avoiding persistent contrail formation due to suitable operational (real time) changes in air traffic management may provide a clue for efficiently reducing the aviation climate impact due to persistent contrails on a short time scale. Weather forecast models may be used to predict areas in which contrails form and persist with similar limitations as climate models and would therefore benefit from the climate model research. This predictive capability is a prerequisite for the development of mitigation strategies.

INTRODUCTION AND BACKGROUND

Changes in cirrus cloudiness caused by contrails, contrail cirrus and soot particles together are denoted as aircraft-induced cloudiness (AIC) [Forster *et al.*, 2007]. Persistent contrails spread considerably during their life time and transform from line-shaped (or linear) into more irregularly formed contrail cirrus. Contrail cirrus is composed of irregularly-shaped ice crystals that, just like natural cirrus, reflect solar radiation and trap outgoing longwave radiation [Platt, 1981; Stephens and Webster, 1981]. Radiative effects of cirrus and contrails have been addressed in several review or overview articles [Liou, 1986; Graßl, 1990; Parungo, 1995; Fabian and Kärcher, 1997; Fahey *et al.*, 1999; Lee *et al.*, 2000; Schumann and Ström, 2001; Minnis, 2003; Schumann, 2002, 2005]. Observations reveal that young contrail ice crystals have smaller effective diameters than natural cirrus [Sassen, 1979; Betancor Gothe and Graßl, 1993; Gayet *et al.*, 1996; Petzold *et al.*, 1997; Lawson *et al.*, 1998; Heymsfield *et al.*, 1998; Poellot *et al.*, 1999; Schröder *et al.*, 2000; Febvre *et al.*, 2008]. Such comparatively small particle sizes render the radiative impact of contrails different from that of most natural cirrus clouds during at least part of the contrail life cycle. The contrail radiative effect is thought to be a net warming as longwave heating dominates over shortwave cooling owing to the relatively small visible optical thickness (< 0.5) of most contrails probed in field measurements [Duda and Spinhirne, 1996; Jäger *et al.*, 1998; Meyer *et al.*, 2002; Duda *et al.*, 2004; Minnis *et al.*, 2005; Palikonda *et al.*, 2005; Atlas *et al.*, 2006]. Additional cloudiness may also be induced by aviation due to the possible influence of aviation aerosol (mainly soot emissions) on cirrus clouds [Ström and Ohlsson, 1998; Hendricks *et al.*, 2005]. In the literature this effect was termed soot cirrus [Schumann, 2006].

Generally, the climatic impact of air traffic (of which a substantial part may be due to contrails and contrail cirrus) today (year 2000) amounts to 2-8% of the global radiative forcing associated with climate change. Since air traffic has been increasing on average by 5% per year since the 1990 (twice as fast as the global economy), emissions have been increasing, even though fuel consumption per passenger kilometer has been reduced significantly. Due to the projected increase in air traffic [ICAO, 2007] the relative importance of air traffic is going to increase drastically. Additionally, air traffic is concentrated in certain regions that experience much larger climate impact due to air traffic than the global mean. In the long term it may well be, that the most serious threat to the continued growth of air travel is its impact on climate [Green, 2005]. The planned introduction of emission trading schemes must be based on a solid scientific basis which is currently still lacking especially for non-CO_2 emissions [IPCC, 2007; Wuebbles and Ko, 2007]. The necessary scientific research would support the strategic planning of the Joint Planning and Development Office (JPDO) to develop the Next Generation Air Transportation System (NextGen) and the European vision 2020 of the Advisory Council for Aeronautics Research in Europe (ACARE), as well as informing the International Civil Aviation Organization (ICAO) through its Committee on Aviation Environmental Protection (CAEP) on how scientific knowledge may be used to improve assessments of environmental health and welfare impacts of aviation environmental policy. In view of the societal relevance and economic importance of sustainable growth of global aviation, it would be appropriate that the climate science community received sufficient funding, for supporting not only applied but also basic research allowing real progress estimating climate impacts, in order to ensure that political decisions are based on increasingly sound scientific knowledge.

Contrails are short-lived when forming in dry air. They are persistent and grow in terms of their horizontal coverage and ice water content whenever the air masses, in which they reside, stay saturated or supersaturated with respect to the ice phase [Brewer, 1946]. Icesupersaturated regions and cirrus occurrences are closely tied to synoptic weather patterns [Detwiler and Pratt, 1984; Schumann, 1996; Kästner *et al.*, 1999; Spichtinger *et al.*, 2003a, 2005; Haag and Kärcher, 2004; Gettelman *et al.*, 2006; Carleton *et al.*, 2007] and mesoscale vertical air motion variability [Kärcher and Ström, 2003]. Typical thicknesses of icesupersaturated layers are 500 m at middle and high latitudes [Spichtinger *et al.*, 2003b; Treffeisen *et al.*, 2007; Rädel and Shine, 2007a] limiting the vertical extent of contrail cirrus. Occasionally much deeper layers have been observed indicated by contrail fall streaks [Knollenberg, 1972; Konrad and Howard, 1974; Schumann, 1994; Atlas *et al.*, 2006]. Contrail outbreaks, which describe clusters of persistent contrails that spread in suitable weather conditions, indicate that ice supersaturated layers can be horizontally extended (at least up to 35,000 km2) [Minnis *et al.*, 1998] and can last for many hours [Detwiler and Pratt, 1984; Mannstein *et al.*, 1999; DeGrand *et al.* 2000; Duda *et al.*, 2001, 2004, 2005]. Since the beginning of jet air traffic, it is known that contrail cirrus can appear without natural cirrus when atmospheric conditions do not support natural cloud formation, enhancing natural cloud coverage [e.g., Kuhn, 1970; Detwiler and Pratt, 1984; Schumann and Wendling, 1990].

Contrails or contrail clusters are also observed in conjunction with cirrus clouds depending on the synoptic situation [Sassen, 1997; Immler *et al.*, 2007]. Atmospheric feedbacks presumably exist between persistent contrails and natural cirrus, because they share the same condensable water vapor reservoir.

Until now only young or linear contrails have been subject to observational and theoretical analyses. Jet exhaust contrails form by condensation of the emitted water vapor mainly on coemitted aerosol particles [Busen and Schumann, 1995; Gierens and Schumann, 1996; Schumann, 1996; Kärcher, 1996; Schröder *et al.*, 1998; Kärcher *et al.*, 1996, 1998; Schumann *et al.*, 1996; 2002]. Dynamical processes related to the decay of aircraft vortices determine the number and mass of contrail ice crystals that survive in ice-supersaturated air [Lewellen and Lewellen, 2001; Unterstrasser *et al.*, 2008]. The effective diameters of observed ice crystals in young contrails are initially ~1 μm and increase with contrail age [Sassen, 1979; Gayet *et al.*, 1996; Freudenthaler *et al.*, 1996; Strauss *et al.* 1997; Petzold *et al.*, 1997; Goodman *et al.*, 1998; Lawson *et al.*, 1998; Heymsfield *et al.*, 1998; Sassen and Hsueh, 1998; Poellot *et al.*, 1999; Schröder *et al.*, 2000; Del Guasta and Niranjan, 2001; Febvre *et al.*, 2008]. Individual contrails can persist for many hours with radiative processes affecting contrail longevity and growth [Kuhn, 1970; Knollenberg, 1972; Gierens, 1994]. Contrail cirrus are frequently observed to spread, inducing additional cirrus cloud coverage. This contrail cirrus can only to some extent be distinguished from natural cirrus using satellites by tracking. Microphysical properties of this aged and hence nonlinear contrail cirrus depends on the amount of water vapor available in the ambient air and less on the moisture input from the aircraft [Schumann, 2002]. In a sheared environment, the increase in horizontal coverage is dependent on the vertical extent of the contrail, which is in turn controlled by ice crystal sedimentation and hence vertical layering of supersaturation. The size of the ice crystals in contrail cirrus and their sedimentation properties may depend on the initial number of ice crystals formed in the young contrail [Schumann, 1996] and on the processing of the ice crystals in the wake vortices [Lewellen and Lewellen, 2001]. Otherwise, the temporal evolution of initially linear contrails into spreaded contrail cirrus [Reinking, 1968; Gierens, 1998; Minnis *et al.*, 1998; Schröder *et al.*, 2000; Atlas *et al.*, 2006] is controlled by atmospheric state variables and dynamical processes (e.g., relative humidity, temperature, vertical shear of the horizontal wind field perpendicular to the contrail axis, horizontal advection and diffusion, vertical air motion). Coverage due to nonlinear contrail cirrus has not been simulated yet. Attempts to estimate contrail cirrus coverage and optical depth from remote sensed data are considered very uncertain [Fahey *et al.*, 1999; Sausen *et al.*, 2005]. The number and size distribution of ice crystals in nonlinear contrail cirrus is not known. Remote sensing observations may miss linear contrails with a width lower than the pixel size. Aviation-induced cloudiness components, nonlinear contrail cirrus and soot cirrus are indistinguishable from background cirrus. So far, the IPCC has assigned a best estimate of radiative forcing to linear contrails only [Fahey *et al.*, 1999; Forster *et al.*, 2007].

Observational tools include Lidar and Radar instruments, satellite sensors and standard cloud physics instrumentation onboard high flying aircraft, but available measurements do not cover the full contrail cirrus life cycle [SSWP Key Theme 4]. Lidar and Radar have been used to conduct case studies of contrails or to develop local contrail statistics [Konrad and Howard, 1974; Kästner *et al.*, 1993; Freudenthaler *et al.*, 1996; Sassen, 1997; Jäger *et al.*, 1998; Uthe *et al.*, 1998; Sassen and Hsueh, 1998; Del Guasta and Niranjan, 2001; Sussmann and Gierens, 2001; Immler *et al.*, 2007]. Observational studies of regional contrail coverage have been reported, including visual inspection of satellite images and automated algorithms to identify linear objects in satellite scenes [Joseph *et al.*, 1975; Carleton and Lamb, 1986; Lee *et al.*, 1989; Schumann and Wendling, 1990; Bakan *et al.*, 1994, Mannstein *et al.*, 1999; Chen *et al.*, 2001; Meyer *et al.*, 2002, 2007; Minnis *et al.*, 2003, 2005; Palikonda *et al.*, 2005; Duda *et al.*, 2005; Stuefer *et al.*, 2005; Mannstein and Schumann, 2005]. For uniform detection of contrails during day and night, most studies used infrared satellite images which may detect preferably contrails effective in the infrared and may underestimate the fraction of contrails with high solar albedo. Time series composed of studies using different remote sensing instruments suffer from different false alarm rates and detection efficiencies. A global, homogeneous analysis of coverage and optical properties by linear contrails is still missing. Number and size of ice crystals and optical depth of non line-shaped contrail cirrus cannot be observed since they can generally not be distinguished from natural clouds. Therefore optical properties need to be modeled.

The restriction to polar orbiting satellites results in a temporal sampling of contrails that interferes with the daily pattern of air traffic. Any attempt to relate observed contrail cirrus coverage to air traffic has to rely on a precise knowledge of real air traffic movements. Such information is available only regionally. Available trend analyses are considered uncertain, because the aviation signal is difficult to isolate and the trends of natural cirrus cloud amounts may have many causes [Chagnon, 1981; Liou et al., 1990; Boucher, 1999; Zerefos *et al.*, 2003; Minnis *et al.*, 2004; Stubenrauch and Schumann, 2005; Stordal *et al.*, 2005; Travis *et al.*, 2007; Eleftheratos *et al.*, 2007]. Attempts to attribute observed cirrus trends to aviation cannot discriminate among contrail and soot effects and natural trends. Contrail and soot effects on cirrus therefore need to be analyzed separately using improved correlation analysis of observations or modeling tools. Observation-based studies have discussed the contrail effect on surface temperature and diurnal temperature range [Travis *et al.*, 2005; Ponater *et al.*, 2005; Hansen *et al.,* 2005].

Modeling approaches comprise microphysical process models [Kärcher *et al.*, 1995; Brown *et al.*, 1997; Kärcher, 1998; Yu and Turco, 1998], Large-Eddy simulations (LES) [Boin and Levkov, 1994; Gierens, 1996; Chlond, 1998; Jensen *et al.*, 1998a; Gierens and Jensen, 1999; Khvorostyanov and Sassen, 1998; Sussmann and Gierens, 1999, 2001; Chen and Lin, 2001; Lewellen and Lewellen, 2001; Ström and Gierens, 2002; Paoli *et al.*, 2004; Unterstrasser *et al.*,, 2008], radiative transfer calculations and radiative forcing estimates [Fortuin *et al.*, 1995; Strauss *et al.,* 1997; Schulz, 1998; Liou *et al.*, 1998; Minnis *et al.*, 1999; Meerkötter *et al.*, 1999; Myhre and Stordal, 2001; Chen *et al.*, 2001; Stuber *et al.*, 2006; Gounou and Hogan, 2007; Stuber and Forster, 2007; Rädel and Shine, 2007b] and global or regional modelling [Liou *et al.*, 1990; Rind *et al.*, 1996; Sausen *et al.*, 1998; Wang *et al.*, 2001; Duda *et al.*, 2005; Ponater *et al.*, 1996, 2002, 2005; Marquart *et al.*, 2003; Fichter *et al.*, 2005; Hansen *et al.*, 2005]. Process-based and LES models covered only contrail formation or early stages of the transformation into cirrus. Most radiation models use optical depth and ice water content in a parametric manner instead of representing realistic values and respective variability. However, the latter is important, as climate forcing is known to be strongly influenced by regional and seasonal forcing patterns. Only few attempts have been undertaken to investigate the global impact of linear contrails with climate models. Contrail modeling relies on the successful simulation of the large scale climate and is hampered by the limited information of observed [SSWP Key Theme 3] and simulated [Kärcher *et al.*, 2006;

Tompkins *et al.*, 2007; Liu *et al.*, 2007; Gettelman and Kinnison, 2007] upper tropospheric supersaturation.

We propose to introduce contrail cirrus as a new, purely anthropogenic ice cloud type in global models for the following reasons. Contrail cirrus have distinct optical properties and interact with the moisture field and natural cirrus. Individual contrails have been tracked for long periods of time (as long as 17 hours) in satellite imagery [Minnis *et al.*, 1998], and this does not seem an upper limit of possible life times. Therefore they can be advected, spread and condense water causing considerable cloud coverage away from the source areas. Contrail cirrus can significantly change cirrus coverage in the vicinity of air traffic routes and alter radiative fluxes. The prospect of climate change and rapidly increasing demands for air transportation emphasizes the need to study contrail cirrus. To enable an environmentally sustainable development of air traffic in the future is a major motivation for research in this area, with strong links to research efforts aiming at understanding dynamical, microphysical and chemical processes in the upper troposphere and lower stratosphere region. Section 2 reviews the current state of science, focusing on advancements since the 1999 IPCC report. Section 3 discusses limits of available methods and identifies research issues that urgently need improvement to enable scientific progress. Section 4 prioritizes outstanding issues. Section 5 provides recommendations to maximize science output. The SSWP closes with a summary in section 6 and a comprehensive list of references.

2. REVIEW

A. Current State of Science

Thermodynamic Conditions for Contrail Formation

Contrails were first observed in 1915, but it took more than 25 years to provide proper explanations. Early theories of contrail formation before 1940 (reviewed by Schumann [1996]) considered various details of mixing of the engine heat, moisture and particle emissions in the exhaust jet behind the engine with ambient air and various microphysical details of particles, and liquid or ice particle formation. It was therefore a major progress when Schmidt [1941], and later Appleman [1953], explained the formation of contrails purely thermodynamically without the need to consider details of jet mixing and particle microphysics. The only assumption needed is about whether contrail particles form at liquid water or ice saturation. Schmidt [1941] assumed contrail formation at ice saturation. Several other studies at the same time, as reviewed in Schumann [1996], see e.g. Brewer [1946], provided clear evidence that contrail formation requires liquid saturation. This has been confirmed in many follow-on studies [Schumann *et al.*, 1996; Jensen *et al.*, 1998b; Kärcher *et al.*, 1998; Schumann, 2000].

The thermodynamic theory assumes isobaric mixing of both specific heat (enthalpy) and water vapor concentration in the exhaust at equal rates after complete combustion with ambient air without other sources and losses (such as radiative heating). This approach ignores details of the initial split of exhaust energy in internal energy and kinetic energy in the exhaust jet [Schumann, 1996, 2000] and initial variations of the heat/moisture ratio between the core and bypass parts of the engine jets [Schumann *et al.*, 1997]. The thermodynamic theory also ignores details of initial visibility [Appleman, 1953]. All these issues impact the predicted threshold temperature below which contrails form by up to ~1 K only. However, it was found important to note that only part of the chemical fuel energy is converted to heat in the engine exhaust. A fraction з, corresponding to the overall propulsion efficiency is used to propel the aircraft against its drag. This fact was known in principal already to Schmidt [1941], but the explicit inclusion of the overall propulsion efficiency з = F V/(mF Q) as a function of aircraft speed V, fuel mass flow rate mF and engine thrust F (or specific fuel consumption per thrust, SFC = mF/F) was first identified in Busen and Schumann [1995] and explained in detail by Schumann [1996], and later confirmed by various studies [Schumann, 2000; Schumann *et al.*, 2000; 2002; Detwiler and Jackson, 2002]. Still details of mixing and microphysics [Kärcher *et al.*, 1996, Paoli *et al.*, 2004; Vatazhin *et al.*, 2007] matter for formation of particles in the young contrail, their visibility, radiative effects, and possibly also for their later fate [Schumann, 1996].

According to basic thermodynamics, the maximum temperature and minimum relative humidity at which contrails form (i.e., threshold conditions) are determined by ambient temperature, pressure and relative humidity, specific heat of fuel combustion, emission index of water vapor, and the overall aircraft propulsion efficiency. The amount of fuel consumption is as such unimportant for contrail formation, but often used as a proxy for flown kilometers or water vapor emissions.

Microphysics of Contrail Formation

The sole empirical constraint consists of assuming that at least plume water saturation is required to nucleate contrail particles. Observations of contrail formation in threshold conditions at very low and normal fuel sulfur content showed small visible differences in both contrail onset and appearance [Busen and Schumann, 1995]. Numerical simulations [Kärcher *et al.*, 1995] consistent with observations [Schumann *et al.*, 1996] suggested that emitted soot particles must be involved as nucleation centers for contrail ice particles, as close to the formation threshold, liquid plume aerosols (consisting of water, sulfuric acid and organics) forming at subsaturations relative to ice do not freeze rapidly. Visible contrail formation in threshold conditions within one wingspan behind jet engines at very low fuel sulfur content was rather explained by the rapid formation and subsequent freezing of a (partial) water coating on ~104 cm-3 exhaust soot particles [Kärcher *et al.*, 1996]. The coating is enhanced by condensation of sulfuric acid created by oxidation fuel sulfur precursor gases [Schumann *et al.*, 1996]. In-situ measurements provided quantitative indication that a significant part of soot emissions contributed to contrail ice formation [Schröder *et al.*, 1998; Schumann *et al.*, 2002]. Threshold conditions, the water saturation criterion and the impact of fuel sulfur content, have been confirmed by in-situ measurements within the measurement uncertainties [Schumann *et al.*, 1996; Jensen *et al.*, 1998b; Kärcher *et al.*, 1998; Schumann, 2000].

A sufficient number of ice crystals are needed to make the contrail visible very quickly [Schumann, 1996]. Those are provided by exhaust soot particles acting as nucleation centers. Emitted metal particles, that have been found as residual in contrail ice particles [Twohy and Gandrud, 1998; Petzold *et al.*, 1998], or entrained ambient particles are not abundant enough. If ambient temperatures decrease below the formation threshold, plume supersaturations increase, leading to activation of the large reservoir of liquid plume particles (exceeding that of soot particles by orders of magnitude) in addition to soot and increasing the number of nascent contrail ice crystals up to ten times [Kärcher *et al.*, 1998]. The initial contrail ice particle number is limited to ~105 cm-3, because they remove the excess supersaturation within fractions of a second.

Large plume cooling rates (~1 K/ms) exert a strong dynamical control on contrail formation. This causes properties of nascent contrails to be rather insensitive to details of the ice nucleation process. In fact, contrail formation can be explained by homogeneous freezing of the water droplets either containing soot cores or sulfuric acid traces as passive inclusions. The assumption of perfect ice nucleation behavior of the majority of freshly emitted soot particles would contradict observational evidence as contrails then would become visible under threshold conditions significantly closer to the jet engine exit as soon as ice saturation is reached. It should be noted that contrails become visible within meters from the engine exit if the ambient air temperature is more than an order 10 K cooler than the threshold temperature. However, it cannot be excluded that a small fraction of the coated soot particles nucleate ice without passing a water activation stage [Kärcher *et al.*, 1996, Schumann *et al.*, 1996]. Some contrail observations would be consistent with such soot particles forming ice from about 140% relative humidity over ice up to water saturation [Kärcher *et al.*, 1998]. If that happens, contrails would also form in a small temperature range above the threshold temperature if the plume did not reach water saturation, but the contrails would stay invisible.

Wake Processing of Contrails

During jet mixing the gas and particle mixing ratios decrease until the jet plumes become captured in a pair of trailing vortices after several seconds of plume age. When the ambient air is ice-supersaturated and secondary ice nucleation occurs on ambient aerosols, contrail regions that formed at the plume edges or in upwelling limbs of the vortices may contain a few much larger crystals [Heymsfield *et al.*, 1998]. At this point, the majority of ice particles are still very small (mean diameters 0.5-1 μm) and their total concentrations are reduced considerably (by up to a factor 500) [Schröder *et al.*, 2000]. The capturing virtually suppresses further mixing until the vortices become unstable and break-up after 1-3 min [Lewellen and Lewellen, 1996]. The aircraft influence on wake dynamics ceases after several Brunt-Väisälä periods (several 10 min). Hereafter, plume dispersion is under the control of atmospheric turbulence, gravity waves and wind shear (dispersion regime) [Schumann *et al.*, 1995, 1998; Gerz *et al.*, 1996]. Contrails persist and further accumulate ice mass only at ambient ice supersaturation.

At low ambient shear, vortex dynamics is the primary determinant of the vertical extent of young contrails [Sussmann and Gierens, 1999] and can have dramatic impact on ice crystal properties. Ice crystal number densities can be significantly reduced during adiabatic compression that results from the downward motion of the vortex system (typically ~300 m at a few m/s) [Lewellen and Lewellen, 2001]. The sinking induces baroclinic instability at the top of the vortex pair from which a few ice particles can escape (secondary vortex).

Systematic analyses of the wake effects on young contrail properties are hampered by the large number of influencing factors. Contrail properties depend on ambient stability, turbulence conditions and aircraft type, as well as on ambient temperature and supersaturation. Almost all ice crystals survive in the sinking primary vortices at ambient supersaturations exceeding 30%, which are rare [Spichtinger *et al.*, 2003a; Gettelman *et al.*, 2006]. The surviving ice particle fraction decreases with decreasing supersaturation and is smallest (factor 100 reduction) for the highest temperatures still allowing contrail formation, because sublimation rates are fastest [Unterstrasser *et al.*, 2008]. The secondary vortex is favored in wakes of heavy aircraft at slight ambient supersaturations, resulting in faint contrails at the original cruising altitude [Sussmann and Gierens, 2001]. The exact loss of ice crystal number is difficult to quantify accurately because this depends on the spread of ice particle sizes which is only poorly known. In-situ measurements reveal a range of total ice crystal concentrations between 10-1000 cm-3 after a few minutes of plume age [Gayet *et al.*, 1996; Heymsfield *et al.*, 1998; Schröder *et al.*, 1998, 2000; Febvre *et al.*, 2008], a spread consistent with the variability in the wake processes discussed above.

Ice mass is typically concentrated within ~200 m deep vertical layers that extend ~100 m horizontally after vortex breakup, determined by wake dynamics. Despite wake-induced variations, the total contrail ice mass after the early dispersion regime is roughly given by the saturation vapor excess and is therefore strongly temperature-dependent [Lewellen and Lewellen, 2001]. Ice crystal number densities generally remain high enough (exceeding ~1 cm-3) to ensure depletion of saturation vapor excess and therefore thermodynamic equilibrium in young contrails. This view is consistent with observations which additionally point to a large variability range in ice water content [Schumann, 2002]. Presumably, in the dispersion regime the respective ice water path will be largely determined by the vertical extent of the supersaturated layer in which the contrail particles sediment given sufficiently high supersaturations.

The effective ice crystal size is affecting the optical depth and radiative forcing of ice clouds. It varies in proportion to the inverse of the cubic root of ice crystal number and is therefore expected to exhibit a certain variability range (~1001/3 ЎЦ 5) [Meerkötter *et al.*, 1999]. The high number of small particles [Petzold *et al.*, 1997] implies high optical extinctions due to contrails a few minutes old, as confirmed by in-situ measurements [Febvre *et al.*, 2008]. Another factor affecting radiative effects is the shape of ice crystals. Replica images reveal that the majority of ice particles in young (< 30-60 min) contrails bear a quasi-spherical shape (droxtals) [Gayet *et al.*, 1996; Schröder *et al.*, 2000], but other crystal habits have been detected as well sometimes even in young (< 15 min) contrails [Strauss *et al.*, 1997; Goodman *et al.*, 1998; Lawson *et al.*, 1998; Febvre *et al.*, 2008]. The factors determining the ice particle shapes in aging contrails remain unclear but may include factors such as pressure, temperature, relative humidity and vertical velocity.

Development of Contrail Cirrus

Aircraft measurements of contrail ice particle size distributions only exist for line-shaped contrails because nonlinear contrails are very difficult to identify for pilots without additional support. Most of these probed contrails were less than 1 h old [Gayet *et al.*, 1996; Schröder *et al.*, 2000; Febvre *et al.*, 2008; for new evaluations we refer to one SSWP from Key Theme 4]. The data indicate smaller mean ice particles sizes in contrails than found in cirrus clouds developing in similar conditions. Typical effective diameters and total ice water contents in contrails at least 3 min old range from 2.5-10 μm and 2-5.5 mg/m3, respectively, at temperatures near 218 K [Schröder *et al.*, 2000]. These values are systematically smaller than those measured with the same instrumentation in nearby cirrus at similar temperatures. At higher temperatures sizes and ice water content can be larger [Heymsfield *et al.*, 1998]. When sorted according to their age, contrail ice particle concentrations have been shown to decrease (due to plume mixing) and effective diameters to increase (due to condensation), approaching typical values characteristic for the small particle mode found in midlatitude cirrus clouds (0.3-30 cm-3 and 20-30 μm, respectively). Despite significant differences in ice particle number and size, the scattering phase function, asymmetry parameter and optical extinction may not always differ substantially between natural cirrus and young (15-20 min) contrails [Febvre *et al.*, 2008].

Cirrus clouds including subvisible cirrus exhibit a wide range of morphologies and microphysical properties, depending on formation mechanisms and ambient conditions [Dowling and Radke, 1990]. Not much is known about the properties of older contrails and contrail cirrus because of the lack of in-situ observations and the difficulty to simulate those clouds with process models owing to the increased spatial and extended time scales. Contrail cirrus particle sizes and concentrations may approximate those of natural cirrus with time but there may remain differences in geometry and vertical distribution of cloud ice and particle shapes. One observed contrail with unknown age contained near spherical particles with an effective diameter of 30-36 μm and an ice water content of 18 mg/m3 [Gayet *et al.*, 1996]. The cirrus cloud probed nearby was characterized by values of 48-60 μm and 15-50 mg/m3, respectively. The larger effective diameter in the cirrus was brought about by a second, large particle mode centered at a maximum particle dimension of 300-400 μm and containing only few irregular crystals. Causes for the generation of a large particle mode include aggregation and sedimentation, as well as early nucleation of few efficient heterogeneous ice nuclei. A large particle mode has not been detected in contrails. Whether such a mode can develop during the contrail life cycle remains open, and its potential impact on radiative forcing remains to be studied. Sedimentation is likely to be more important in cirrus than in young contrails. The few existing data [Schumann, 2002] do not allow drawing general conclusions on difference between contrail cirrus and natural cirrus at similar temperatures.

At a given layer depth, large supersaturation leads to rapid ice particle growth and sedimentation, limiting the contrail life time. Contrail fall streaks may extend temporarily into subsaturated air thus causing the contrail to have a larger vertical extent than the depth of the supersaturated layer. Three studies report heavily precipitating contrails with unusually deep fallstreaks, large ice water content and very large maximum crystal dimensions (> 1 mm) [Knollenberg, 1972; Schumann and Wendling, 1990; Atlas *et al.*, 2006]. It is conceivable that such geometrically (and presumably optically thick) contrails develop only at large layer thicknesses (> 1-2 km) and high persistent supersaturations (> 20-30%), both of which are rare events [Gierens *et al.*, 1999a; Spichtinger *et al.*, 2003b]. More commonly contrails experience lower supersaturations (< 15%) and evolve in supersaturated layers ~ 500 m deep.

One numerical study revealed that interactions between radiation and dynamics can affect the early development of contrails [Jensen *et al.*, 1998a]. The numerous ice crystals in a young contrail in a sheared environment subject to ice supersaturation absorb upwelling longwave radiation. The resulting strong diabatic heating drives turbulence-induced updrafts (updraft speeds 5-8 cm/s, exceeding synoptic values), enhancing the vertical depth and changing the contrail microstructure. Radiative cooling in the top layers opens the possibility of secondary ice formation by homogeneous freezing there by generating high supersaturations. These processes are most effective in a neutrally or unstably stratified atmosphere. Such interactions are known to occur in cirrus clouds as well, potentially prolonging their life time [Dobbie and Jonas, 2001]. Under less humid and more stably stratified ambient conditions, radiative effects have been shown to be less important [Gierens, 1996; Chlond, 1998].

Contrail cirrus have been shown to survive for many hours and hence synoptic processes become significant. Contrail cirrus can be advected over long distances during their life time.

Given upper tropospheric wind speed of 30 m/s, they move at ~100 km/h possibly into regions with little or no air traffic. The vertical shear of the horizontal wind spreads contrails into tilted layers. The vertical wind shear perpendicular to the contrail axis pulls the contrail apart at a rate proportional to the contrail height. Spreading rates observed at a midlatitude site range from 18-140 m/min [Freundenthaler *et al.*, 1995], causing line-shaped contrails to grow quickly to widths of several km. Turbulence that is connected with strong shear causes contrails to loose their line shape. When acting in isolation, wind shear reduces the ice water path and optical depth in each vertical contrail column, but at the same time increases the horizontal coverage. The overall effect on radiative forcing is not clear, as this is determined by the product of coverage and optical depth. Given the variability in upper tropospheric shear rates [Dürbeck and Gerz, 1996] and a typical spread of ice particle growth rates (0.3-2 μm/min) there is no unique relationship between contrail age and linearity, nor between age and optical depth (visibility).Virtually no information is available about the coverage due to older and nonlinear contrails mainly since in satellite images, contrail cirrus cannot be identified when they loose their line shape and/or cease to be visibly brighter than cirrus due to the high concentration of small ice particles.

Supersaturation with respect to the ice phase is a prerequisite for contrail cirrus to persist and to accumulate ice mass [Brewer, 1946]. Synoptical processes determine the regional areas of ice supersaturation [Spichtinger *et al.*, 2005] and therefore control to a large extent the life time of contrail cirrus. Contrails develop often in special synoptic situations like ahead of a warm front and are connected with cirrus clouds [Detwiler and Pratt, 1984; Kästner *et al.* 1999, Sassen 1997, Immler *et al.,* 2007]. Hence contrails appear before natural cirrus form. According to satellite images, line-shaped contrails already showing a significant degree of spreading often appear in clusters (outbreaks) in heavily traveled areas [Schumann and Wendling, 1990; Mannstein *et al.*, 1999; DeGrand *et al.*, 2000]. In supersaturated areas mean relative humidity as well as its variability is high [Gettelman *et al.*, 2006]. This is obvious when comparing areas with a large frequency of supersaturation with the spatial distribution of high clouds, both exhibiting similar patterns. During the contrail life time, the synoptic and mesocale variability of the atmosphere influence the contrails in the same way as natural cirrus. This variability stems from fluctuations of temperature and moisture which have a variety of sources, including gravity and orographic waves, convection, and wind shear induced turbulence, among others. It leads to local cooling and heating. Whether contrail cirrus properties are sensitive to such forcings depend on the relative magnitude of these dynamical and microphysical time scales (e.g., for ice mass growth). Given identical dynamical forcings, microphysical changes may be different at different stages of the contrail cycle because the associated time scales in turn are determined by particle number and size.

Contrail cirrus competes with natural cirrus for condensable water and therefore has the potential of delaying cirrus onset and replacing natural cirrus. Sedimentation of contrail ice crystals may lead to additional drying of upper tropospheric air masses but it is not known whether this transport is enhanced by contrail cirrus due to the additional cloudiness or reduced due to the smaller mean particle size. On the one hand, inferred statistical connections between changes in cirrus cloudiness and air traffic using remote sensed data are uncertain. On the other hand, the contrail cirrus life cycle has not been represented in global models yet.

Trends of Cirrus Cloudiness

Remote sensing methods cannot distinguish between aged contrail cirrus and natural cirrus. Further insight into trends of cirrus cloud coverage that could at least in part be forced by air traffic is gained by monitoring cirrus coverage and relating it to air traffic. Ground-based observations of monthly mean high cloud coverage show a step-like increase around 1965, possibly correlated with the onset of jet air traffic. Coverage increases more r apidly during1965-1982 than before the jet era 1948-1964 [Liou *et al.*, 1990] possibly due to the introduction of jet aircraft in air travel. Contrails may be responsible for degradation in the observability of the solar corona and photosphere in the period 1961-1978 (Schumann, 2002). Based on ship- and ground-based observations, a change in the occurrence frequency of cirrus was found to be correlated with aviation fuel consumption and was largest in the main flight corridors over the north east of the U.S.A. and the northern Atlantic [Boucher, 1999]. A similar study based on satellite data reported consistency in trends of cirrus and linear contrail amounts over the USA [Minnis *et al.*, 2004]. They used the 300 hPa moisture fields from the NCEP (National Center for Environmental Prediction) reanalysis data as a proxy for natural cirrus coverage and found a 1% increase of contrail cirrus per decade over the continental US. Removing ENSO, NAO and QBO trends from time series of cirrus occurrence and eliminating the effects of convection and changing tropopause temperature revealed increases in cirrus trends in regions with high air traffic density [Zerefos *et al.*, 2003]. Contrary to these works, another satellite study suggests extra cirrus coverage over in regions with high air traffic density over Europe but remains inconclusive because other factors impacting high cloudiness have not been removed from the data set [Stordal *et al.*, 2005]. Satellite data for trends in high cloud amount and retrieved upper tropospheric humidtity showed a clear positive trend in the high cloud occurrence over the North Atlantic flight corridor when the humidity was insufficient for cirrus formation but allowed persistent contrail formation [Stubenrauch and Schumann, 2005]. Two months of cirrus cover deduced from METEOSAT data and actual air traffic data from EUROCONTROL suggested a strong linear growth of cloud coverage and air traffic density which eventually becomes saturated when approaching the fractional coverage of ice-supersaturation [Mannstein and Schumann, 2005]. Later the correlation was shown to be inconclusive because of natural spatial variations of cirrus coverage in the domain investigated [Mannstein and Schumann, 2007].

These few attempts to infer relationships between cirrus amount and aviation suffer from the poor knowledge of trends in natural cirrus and their dependence on a plethora of dynamical factors acting from the mesoscale up to planetary scales and by aerosol-related processes affecting upper tropospheric ice initiation. Further, these approaches are unable to discriminate between contrail cirrus effects and effects caused by aircraft soot emissions. The latter could modify the cirrus properties and indirectly the background moisture f ields inwhich contrails grow. Hence, current trend analyses invoking a contrail impact are noteworthy but not conclusive.

Contrail Effects on the Radiation Budget

The radiative impact of clouds depends strongly on cloud optical depth and their inhomogeneity [Fu *et al.*, 2000]. A few satellite studies inferred probability distributions of linear contrail optical depths at visible wavelengths, which are a useful measure of this inhomogeneity [Meyer *et al.*, 2002; Minnis *et al.*, 2005; Palikonda *et al.*, 2005]. These distributions exhibit maxima in the range 0.1-0.4, consistent with optical depth values derived in several case studies. It is conceivable that contrail cirrus developing from the secondary vortex, contrail cirrus that is subject to large wind shear, or evaporating contrails become subvisible. While Lidar data point to the existence of subvisible contrails [Sassen, 1997; Immler *et al.*, 2007], quantitative evidence on larger spatial scales is lacking as satellite sensors are not capable of detecting contrails with low (perhaps < 0.05-0.2) visible optical depths. Optical depth distributions of thin cirrus clouds detected by Lidar [Immler and Schrems, 2002] exhibit a shape that is skewed towards small values, comprising a significant fraction of subvisible clouds (optical depth < 0.01-0.03) even at midlatitudes. Subvisible cirrus cause a small radiative forcing per area but if occurring frequently may have a large effect.

Global radiative forcing estimates obtained using general circulation models (GCMs) depend crucially on the assumptions made about the optical depth of the contrails. When simulating contrails offline, optical properties of contrails are assumed constant and radiative forcing estimates are simply scaled linearly with optical depth. Only one climate modeling approach [Ponater *et al.*, 2002; Marquart *et al.*, 2003] attempts the simulation of the regional variability of the optical properties of contrails.

Another crucial factor affecting radiative forcing of contrail cirrus is their coverage. Analyses that consider at least regional scales (useful for global model validation) must rely on satellite remote sensing techniques. Only few studies investigated sufficiently long time series of contrails to provide average regional linear contrail coverage over western Europe, USA and the greater Thailand region [Bakan *et al.*, 1994; Meyer *et al.*, 2002, 2007; Palikonda *et al.*, 2005]. Some of these observations reach back to the 1980s, requiring scaling with average fuel consumption to obtain estimates for more recent air traffic which introduces an unspecified uncertainty. Specifications of what has been observed in terms of false alarm rates and other technical issues of the detection algorithm, optical depth detection limits and detection efficiency, average optical depth and width and associated variability and ice crystal effective sizes or other optical properties are vague or missing in most cases. Therefore, the inferred coverage is difficult to compare among each other and are of limited use for model validation. Current estimates of the global distribution of linear contrail coverage diagnosed with a climate model rely on sorting out contrails with minimum optical depths < 0.02 to compare with observed coverages [Ponater *et al.*, 2002; Marquart *et al.*, 2003, Fichter *et al.*, 2005]. Choosing different lower detection limits would result in global mean and especially in regional changes of simulated coverage. Global mean coverage due to line-shaped contrails are estimated to range between 0.04% and 0.09%.

The effect of contrail cirrus on the radiation budget depends on the size, habit, number and vertical distribution of crystals, surface albedo, solar zenith angle, height and thickness of contrail, spatial inhomogeneity and presence of clouds and water vapor column below the contrails (effecting brightness temperature). During night radiative forcing is always positive. Contrails with optical depths of 0.2-1 have been shown to exert a net warming in the chosen combinations of controlling parameters [Meerkötter *et al.*, 1999] even though parameter combinations could be specified that could cause net cooling [Myhre and Stordal, 2001; Mannstein and Schumann, 2005; Schumann, 2005; Sausen *et al.*, 2005]. The effect of contrails replacing cirrus has not yet been studied. If contrails should replace natural cirrus on a larger scale, and if aged contrails retain different optical properties (many small ice particles) then it is conceivable that even though those contrails are warming the net forcing of contrails replacing natural cirrus is a cooling. Moreover, contrails may increase cirrus optical thickness beyond the point where this increase causes a cooling.

Mitigation options such as fuel additives or cryoplane technology are not expected to decrease contrail radiative forcing significantly [Marquart *et al.*, 2001, 2005; Gierens, 2007], whereas changes in flight levels can change contrail coverage significantly [Sausen *et al.*, 1998; Fichter *et al.*, 2005] making contrail avoidance due to flight rerouting a viable option.

B. Critical Role of Contrails and Contrail Cirrus

Contrail cirrus are the most obvious effect of air traffic but are presently the most uncertain component in aviation climate impact assessments. Since they may be the largest component in aviation radiative forcing they require a large research effort.

Contrails develop at lower relative humidity than natural cirrus and therefore increase high cloudiness. This increase can be significant in or near regions with high air traffic density. Contrails just as natural clouds are a major part of the climate system changing the radiation budget. Due to differences in the ice particle size distributions and in horizontal and vertical cloud structure the optical properties of and radiative forcing by contrails are different to those of natural clouds. Furthermore, the multitude of possible parameter combinations (e.g., solar zenith angle, surface albedo and overlap with natural cloudiness) makes contrail radiative impact extremely space- and time-dependent. In any case, all studies currently available have indicated a time mean global net warming effect on the atmosphere [Sausen *et al.*, 2005]. Contrails may also change the radiation budget by changing the optical properties of natural cirrus or even preventing natural cirrus from forming.

The additional coverage caused by aviation is predicted to grow strongly due to a forecasted increase in air traffic increasing the radiative effect of contrails. Radiative forcing estimates due to contrails cannot be simply scaled with an increased air traffic since future air traffic is forecasted to increase mainly in the more humid subtropics of southeast and east Asia. Model estimates of radiative forcing are mainly describing the effect of contrails in the areas of strongest current air travel, the extratropics. Observational studies as well have been focusing on the mid latitudes. In the subtropics there is little observational evidence of the optical properties and radiative effects of contrails. Additionally, air traffic in the future will take place in an already changed climate, that is itself subject of research.

Contrary to CO_2, contrails and the possible indirect soot effect have a short life time, probably not much longer than days to weeks. On short term, contrails have a far larger climate impact than CO_2 emissions. Contrail avoidance therefore reduces the climate effects of aviation on the short term. This may be achieved by flight rerouting, which is discussed also for future minimization of NOx-induced ozone changes due to aviation. More careful flight routing with most accurate meteorological data may also help to reduce fuel consumption. Advanced air traffic management operations have the potential to reduce contrail formation by avoiding flights through supersaturated regions. Because route optimizations need to take also the effects of NOx and CO_2 emissions into account it is not clear whether flight rerouting reduces contrail induced radiative forcing. The inclusion of such climate aspects in aircraft design or air traffic management tools have been proposed but not yet fully analyzed.

A greater portion of the upper troposphere will support contrail formation if future aircraft should have greater overall propulsion efficiency. Reductions of soot emissions due to improved engine technology may only change contrail properties if the reductions are very large but would not avoid contrails, since ambient aerosol particles would replace them as nucleation centers. However, reductions of soot emissions would diminish possible sootinduced changes of cirrus clouds.

C. Advancements Since the IPCC 1999 Report

Remote Sensing

An automated satellite-based detection algorithm for line-shaped contrails has been published [Mannstein *et al.*, 1999] which was applied by several groups using AVHRR data over Europe [Meyer *et al.* 2002], the continental USA [Duda *et al.*, 2004; Palikonda *et al.*, 2005], eastern north Pacific [Minnis *et al.*, 2005] and southeast and east Asia [Meyer *et al.*, 2007]. First estimates of the amount of older contrail cirrus which cannot be identified by their line shape have been given by Minnis [2004] and Mannstein and Schumann [2005]. Some detailed Lidar and in-situ case studies have added knowledge on structure and optical parameters of individual contrails [Freudenthaler *et al.*, 1995, 1996; Atlas *et al.*, 2006; Febvre *et al.*, 2008]. Several cirrus trend analyses have been carried out after 1999 [Zerefos *et al.*, 2003; Minnis *et al.*, 2004; Stordal *et al.*, 2005; Stubenrauch and Schumann, 2005] (section 2.a), but detected cirrus changes could not be unambiguously ascribed to aviation.

Upper Tropospheric Humidity and Supersaturation

In the last years increased effort has been put into obtaining reliable statistics of supersaturation. MLS and more recently AIRS retrievals have been used to infer global supersaturation statistics [Spichtinger *et al.*, 2003a; Gettelman *et al.*, 2006] showing extended areas with large frequencies of supersaturation in the upper troposphere, reaching in the midlatitudes maxima of up to 30%. The overall frequency of supersaturation in those studies is relatively uncertain but agrees relatively well with estimates from measurements along commercial flight routes (MOZAIC) in the upper troposphere of ~13% [Gierens *et al.*, 1999a]. Minimum frequencies are found in equatorial areas. The large scale structures of supersaturation resemble those of humidity. The vertical extent of supersaturated areas has been estimated using humidity-corrected radiosonde data [Spichtinger *et al.*, 2003b; Rädel and Shine, 2007a]. Parameterizations of cloud microphysics describing the formation of ice crystals at substantial supersaturation have been devised [see Kärcher *et al.*, 2006 for most recent developments including the impact of heterogeneous ice nuclei] and implemented in climate models [Lohmann and Kärcher, 2002]. Nevertheless cloud coverage has remained to be uniquely dependent on humidity making microphysics and cloud coverage inconsistent. Other modeling approaches consist of simply changing the humidity threshold of cloud coverage to a higher supersaturated value, neglecting the fact that cirrus form and evaporate at different humidities. As long as the life cycle of cirrus is not consistently modeled, there will be a need to parameterize contrail formation.

Wake Processes

To describe contrail formation and the early interaction of contrails with wing tip vortices a highly sophisticated two phase flow model has been developed [Paoli *et al.*, 2004]. LES methods have been used for the carrier phase, solving the fully compressible 3D Navier Stokes equations and water vapor, while a Lagrangian particle tracking approach has been adapted to ice formation from exhaust soot particles. Simulated mixing histories of air parcels and probability distributions for ice particle size and water vapor reveal much of the complex structure of nascent contrails as a result of strong interactions between particle microphysics and turbulence. In general terms, the overall findings of much simpler approaches using classical mixing assumptions [Kärcher *et al.*, 1996, 1998] have been confirmed. Threedimensional LES studies with a simpler treatment of ice microphysics have also been presented shedding light on the evolution of contrails during the vortex phase [Lewellen and Lewellen, 2001]. A similar 2D-approach has been developed recently aiming at a more systematic survey of atmospheric parameters influencing contrails up to the dispersion phase [Unterstrasser *et al.,* 2008]. Contrary to the prior approaches, the latter model can straightforwardly be extended to study the contrail-to-cirrus transition on larger scales using LES methods. After 1999, a 2D cloud-resolving model has been employed to simulate contrails up to 30 min of age [Chen and Lin, 2001], yielding information on ice crystal size distributions similar to the model of Jensen *et al.* [1998a]. Both works agree upon the importance for radiative processes in simulating young contrails. A regional climate model has been fed with results from the cloud-resolving simulations (contrail coverage, effective sizes and short-/longwave optical depths) to estimate the climate impact of contrail layers in an area surrounding Taiwan using ensemble simulations [Wang *et al.*, 2001]. The regional study concluded that contrail radiative forcing is dominated by contrail coverage and radiative properties play a smaller role owing to the spatial inhomogeneity of the coverage.

Contrail Coverage

First global estimates of the radiative forcing due to contrails were derived in 1998 based on the calculation of potential contrail coverage from offline calculations using temperature, humidity and pressure from ECMWF reanalyses and folding this potential contrail coverage with some measure of flight density [Sausen *et al.*, 1998]. Since then this approach has been upgraded resulting in a climate model parameterization of contrails, calculating online contrail occurrence and the contrail ice water content from the condensable water at the time step [Ponater *et al.*, 2002; Marquart *et al.*, 2003]. This allows for capturing the dependence of contrail occurrence on the weather regime and the regional and temporal variability of contrail ice content. Still this method consists of calculating the potential contrail coverage and tuning the simulated contrail coverage to the observed contrail coverage over a selected region assuming that the tuning coefficient is temporally and locally universal. Most if not all studies scaled the contrail coverage to that derived by Bakan *et al.* [1994] for Europe (30°W-30°E, 35°N -75°N). Other approaches still calculate contrail coverage by folding offline the potential contrail coverage with data of flight density or use estimates of contrail coverage from older studies and assume globally and temporally fixed optical depth. They concentrate on using more sophisticated radiative transfer models analyzing the variability of radiative forcing due to different background parameters [Meerkötter *et al.*, 1999], due to the daily cycle of air traffic [Myhre and Stordal, 2001; Stuber *et al.*, 2006; Stuber and Forster, 2007] and 3D effects on radiative transfer [Chen *et al.*, 2001; Gounou and Hogan, 2007]. It was generally found that global contrail radiative forcing does not vary strongly depending on the radiation code used; it may depend, however, strongly on the method to treat cloud overlap in GCMs [*Marquart and Mayer*, 2002]. However, most of these studies apply the same contrail ice crystal size distribution [Strauss *et al.*, 1997] and ignore vertical variability in ice water content and effective radii, so that large deviations between these estimates may not be expected in the first place. Chen *et al.* [2001] rely on simulated ice crystal spectra showing a persistent small particle mode at 2-3 μm within supersaturated air leading to net cooling by contrails. This persistent small particle mode disagrees with earlier findings by Schröder *et al.* [2000]. According to these observations the small particles grow substantially within the first 30 minutes. Differences may also be caused by a more advanced treatment of horizontal inhomogeneities in radiative transfer and the different atmospheric background in the subtropics.

Contrail Optical Depth

Lower estimates of radiative forcing due to line-shaped contrails since 1999 are based on lower estimates of mean contrail optical depth. Usually a global mean optical depth of 0.1 is assumed instead of 0.3 used by IPCC [1999]. It is unclear whether this lower optical depth of contrails is more realistic. Satellite observations estimate mean optical depth of contrails to range between 0.2-0.4 over the U.S.A. [Minnis *et al.*, 2005; Palikonda *et al.*, 2005] and 0.05-0.2 over Europe [Meyer *et al.*, 2002]. For comparison, Ponater *et al.* [2002] compute mean visible optical depths of 0.1-0.13 and 0.06-0.09, respectively, over these regions, with individual values covering several orders of magnitude. The models compute mean values for ensembles of contrails within rather large grid boxes (e.g. ~300 x 300 km2 for T30 resolution), while the observations provide optical depth for individual contrails or contrail clusters at the cloud scale or the scale of satellite resolution. Presently, one cannot decide how accurate the model results are.

A reason for the low contrail optical depth simulated by the ECHAM4-GCM may be a general low bias in the ice water content of natural clouds. Comparison with observational data hints at an underestimation of ice water content and effective radii of cirrus by ECHAM4 [Lohmann *et al.*, 2007]. Furthermore, it is assumed that the condensable water at one time step (of 30 min) is a good proxy for the ice water content of the contrail while ice water content of natural cirrus is accumulated using a prognostic variable. It is likely that the spatial and temporal variability of the ice water content and therefore of optical depth as simulated by Ponater *et al.* [2002] is more realistic than the overall amount. On the other hand, it is also unclear whether observations of optical depth are representative since the optical depth detection threshold of satellite sensors is usually not specified and detectability may be biased towards optically thicker contrails. The decrease in radiative forcing due to the studies performed after 1999 has been compensated by the use of an air traffic inventory for 2002 which includes an increase in air traffic from the 1992 values used earlier [Sausen *et al.*, 2005; Forster *et al.*, 2007].

Future Scenarios

Recently mitigation studies have been performed, analyzing the use of fuel additives [Gierens, 2007], cryoplane propulsion [Marquart *et al.,* 2001, 2005] and flight level changes [Fichter *et al.*, 2005] indicating that flight rerouting may be the most successful mitigation option (section 2.e).

D. Present State of Measurements and Data Analysis

Relative Humidity

Relative humidity measurements in the upper troposphere and lower stratosphere with sufficient vertical resolution are needed in order to validate the humidity fields simulated by global models which is the basis for modeling the occurrence and optical properties of contrails and natural cirrus. Relative humidity is difficult to measure in the upper troposphere and lower stratosphere.

Satellite observations are employed to infer supersaturation [Spichtinger *et al.,* 2003a; Gettelman *et al.*, 2006] without having been designed to measure relative humidities. The influence cirrus clouds have on the inferred values is uncertain. While spatial patterns of relative humidity are reliable, the inferred magnitudes of supersaturations are highly uncertain. Most satellite instruments suffer from relatively coarse horizontal and/or vertical resolution, greatly limiting their use for detailed model validation. The applicability of more recent improved instrumentation such as CALIPSO/CloudSat or Odin-SMR [Ekström *et al.*, 2007] in contrail research remains to be shown. A promising, still developing technique for the retrieval of moisture is the GPS tomography [Troller *et al.*, 2006], but for the time being the resolution and the sensitivity to the low water vapor contents in the tropopause region is not sufficient.

Older operational radiosondes are known to have dry biases and cannot be employed to measure relative humidity reliably at altitude without suitable corrections [Miloshevich et al., 2001]. However, carefully calibrated and corrected RS80A radiosondes [Nagel *et al.*, 2001] and follow up instruments can be used to detect supersaturated vertical layers [Spichtinger *et al.*, 2003b; Rädel and Shine, 2007a]. In-situ (airborne and balloon-borne) research instruments measure clear-sky relative humidity relatively well in the extratropical regions (with an uncertainty of about ±10 % relative humidity) and are therefore the best option for contrail studies. To date, the MOZAIC program [Marenco *et al.*, 1998] provides 9 years of insitu measurements onboard commercial aircraft along major flight routes and is a powerful data source not only for relative humidity [Gierens *et al.*, 1999a] but also for temperature and water fluctuations [Gierens *et al.*, 2007]. Aircraft measurements of water vapor with forwardfacing inlets are more difficult inside clouds because care has to be taken to avoid the additional detection of small particles. However, there are known differences in the water vapor measurement between different in-situ instruments and satellite retrievals [Kley *et al.*, 2000] and between different satellite retrievals [Read *et al.*, 2007]. These pending discrepancies between different in situ instruments are most relevant in the cold atmosphere mostly found in the tropical tropopause region [Peter *et al.*, 2007] too high to be affected by commercial subsonic air traffic. More information on this subject will be provided by SSWPs of Key Theme 3.

Contrail Measurements

Aircraft measurements use a suite of techniques to describe the number size distribution, ice water content and scattering phase function of contrail particles and are the best tool to provide a detailed optical and microphysical characterization of contrails. Fresh contrails that can clearly be related to their source aircraft are straightforward to probe (near field measurements), although only very experienced pilots can handle the perils of close encounters between contrail-producing and contrail-detecting aircraft. Due to the highly turbulent microstructure of fresh contrails, many crossings at similar plume ages are needed to obtain statistically robust data. It has been recognized for a long time that measurements of particle concentrations with optical particle spectrometers that have inlets could overestimate the number concentrations of small ice crystals due to shattering of large ice crystals on the inlets or aircraft bodies. This is highly relevant to young contrails, as those are expected to consist of high numbers of small crystals. As laid out in an SSWP in Key Theme 4, existing contrail measurements appear to be reliable mainly because large crystals (> 100 μm) are largely absent in fresh contrails, so that available near-field measurements are in agreement with theoretical expectations [Kärcher *et al.*, 1998].

Much less is known about the contrail life cycle and their decay, which is essential for the assessment of the radiative forcing of contrails and contrail cirrus. Only few measurements, mostly from Lidar or airborne measurements exist for the first 60 minutes. Some case studies using satellite imagery discuss the evolution of radiative parameters and forcing of contrail clusters for up to about 6 hours [Duda *et al.*, 2001]. These measurements do not allow representative statistics. Tracking of contrails and contrail outbreaks from satellite data has been performed in some case studies, but a general description of the life cycle and the resulting radiative forcing of contrails and contrail cirrus is available only for some specific cases [Minnis *et al.*, 1998; Schumann, 2002].

The climate impact of contrails and contrail cirrus cannot be measured directly, it has to be derived from model studies. Measurements and data analysis is necessary to validate their optical properties over the life cycle.

Visual or photographic detection of contrail occurrence [Bakan *et al.*, 1994] covers only linear contrails similar to those detectable in satellite measurements. The frequency of occurrence at a given site indicates how often the thermodynamic formation conditions are met. The occurrence of linear contrails observed from airports in the contiguous United States has been analyzed by Minnis *et al.* [2003]. It shows a strong seasonal cycle and regional differences, but almost no daily cycle, pointing to the very dense air traffic during the daylight observing period. About 80-90% of the contrails occurred along with cirrus clouds. In Fairbanks, Alaska, a region with relatively low air traffic, Stuefer *et al.* [2005] validated mesoscale model forecasts of contrail formation conditions by visual observation of single aircraft. Sassen [1997] published the results from 10 years of observations of high clouds and contrails at the FARS site in Utah using a combined data set of all-sky camera images, polarization Lidar and radiometers. These ground based Lidar observations of contrails rely on the drift of contrails directly over the fixed location. Lidar observations resolve contrails and cirrus clouds with a high spatial resolution and give information on altitude, optical depth and, in combination with near-IR spectroscopy [Langford *et al.*, 2005] also on ice particle size.

Ground-based contrail observations have been performed using a scanning Lidar with tracking capability operated near Garmisch-Partenkirchen, Germany [Jäger *et al.*, 1998]. The tracking of many single contrails in meteorological conditions favorable for their formation and persistence yielded surveys of the evolution of cross section, height and optical parameters (e.g., depolarization) up to 60 min of estimated contrail age [Freudenthaler et al., 1995, 1996]. In combination with an air traffic data base [Garber *et al.*, 2005], MODIS satellite images and ground based photographs, Atlas *et al.* [2006] have assessed the age of 18 contrails probed by the NASA/GSFC micropulse Lidar. This Lidar provided information on particle fall speeds and estimated sizes, optical extinction coefficients, optical, and ice water path for contrails and their fall streaks with ages up to 2 hours. Lidar observations of thin cirrus and contrails have been carried out in Lindenberg, Germany [Immler *et al.*, 2007]. The classification of cirrus clouds was aided by a CCD camera, and a high resolution radiosonde corrected for the dry bias was also operated at this site. In 90% of the cases where ice supersaturation was indicated by the radiosonde, cirrus clouds have been detected. As the Lidar is capable of detecting very thin cirrus (visible optical depths of 10-4 or less), a high detection frequency may not come as a surprise. Cirrus including a large fraction of subvisible cirrus have been observed in 55% of all observations. Visual inspection of the camera images showed that 5% of the observed cirrus could be identified as aging contrails, but only ~10% of the identified contrails were line-shaped. The rest consisted of significantly spreaded contrail cirrus or was connected to pre-existing cirrus.

Few measurements of contrails with airborne Lidar systems have been reported. During the ICE89 campaign Kästner *et al.* [1993] derived optical depths of three contrails and surrounding cirrus clouds and compared them to optical depth values derived from AVHRR data. They found an agreement within 10% between both methods and optical depths of the contrails were by 0.1-0.25 higher than in the cirrus. A scanning Lidar system was also flown on the NASA DC-8 aircraft during the SUCCESS campaign [Uthe *et al.*, 1998]. This system was primarily designed to help locate and direct the DC-8 into thin cirrus and contrail layers, but also provided high resolution data on the vertical cloud structure. Although the recorded Lidar data have not been fully analyzed, they have been claimed to be useful for the interpretation of data collected in-situ and from radiometric sensors and for inferring optical and radiative cloud properties.

Cirrus Detection Using Satellite Imagery

New passive instruments of unprecedented quality like the Spinning Enhanced Visible and Infra-Red Imager (SEVIRI) aboard the geostationary Meteosat Second Generation (MSG) allow for the first time the quantification of cloud properties during the life cycle of clouds from space. The new "MSG cirrus detection algorithm" (MeCiDA) has been developed using the seven infrared channels of SEVIRI [Krebs *et al.*, 2007] thus providing a consistent scheme for cirrus detection at day and night. MeCiDA combines morphological and multi-spectral threshold tests and detects optically thick and thin ice clouds. The thresholds were determined by a comprehensive theoretical study using radiative transfer simulations as well as manually evaluated satellite observations. The results have been validated by comparison with the Moderate Resolution Imaging Spectroradiometer (MODIS) Cirrus Reflection Flag: An extensive comparison showed that 81% of the pixels were classified identically by both algorithms. On average, MeCiDA detected 60% of the MODIS cirrus. The lower detection efficiency of MeCiDA was caused by the lower spatial resolution of MSG/SEVIRI, and the fact that the MODIS algorithm uses infrared and visible radiances for cirrus classification. The advantage of MeCiDA compared to retrievals for polar orbiting instruments like MODIS or previous geostationary satellites is that it allows the derivation of quantitative data every 15 min, 24 h a day. This high temporal resolution allows the study of diurnal variations and life cycle aspects. MeCiDA has been used to derive cirrus coverage over Europe and the North Atlantic for a complete year in the frame of the ESA project CONTRAILS (Validation of the Eurocontrol contrail detection model with satellite data).

Contrail Detection Using Satellite Imagery

Linear contrails with widths of the order of the horizontal resolution of satellite sensors (~1 km for NOAA AVHRR type sensors) or larger are an obvious feature in satellite imagery, but automated or manual identification of contrail cirrus is not possible. Besides geometrical matters, contrails detected by Lidar or by ground-based observers may not be detectable by satellites owing to their moderately high optical depth detection thresholds (mostly ~0.1 in the visible wavelength range).

The main criteria to identify a linear contrail in satellite imagery are its shape and its contrast to the background. The best contrast for young contrails composed of small ice particles is usually found in the difference of temperatures measured in the thermal infrared split window channels (at ~11μm and 12μm) originally designed to retrieve the sea surface temperature. Over a homogeneous surface and a dry atmosphere even fresh contrails with a width smaller than the image resolution can be identified, but classical retrieval methods for the optical properties fail in this case, because the true width remains unknown.

Purely visual interpretation of satellite images is influenced by human factors but is usually more efficient in detecting linear contrails than automated methods. The first published visual data analysis was performed by Bakan *et al.* [1994]. An automated contrail detection algorithm [Mannstein *et al.*, 1999] has been applied to AVHRR over Europe [Meyer *et al.,* 2002], the continental USA [Duda *et al.*, 2004; Minnis *et al.*, 2005; Palikonda *et al.*, 2005] and southeast and east Asia [Meyer *et al.*, 2007].

The contrail detection algorithm indicates pixels in satellite infrared data which are covered by linear contrails. Contrail width (wider than pixel size) and length are easily derived from the resulting contrail mask. Integration over larger areas and/or times enables derivation of contrail coverage. In case studies with additional information on air traffic and wind conditions it is also possible to derive spreading rates [Duda *et al.*, 2004]. Based on the automated identification of linear contrails, their optical properties and radiative forcing has been estimated from the brightness temperature difference between the contrails and their surrounding assuming that the contrail temperature equals the atmospheric temperature at the same altitude [Meyer *et al.*, 2002, Minnis *et al.*, 2005; Palikonda *et al.*, 2005]. For all derived parameters it has to be kept in mind, that they are related to the spatial resolution of the sensor, which is in the order of at least 1.3 x 1.3 km2 in the nadir of the satellite for the AVHRR, which was used for all of these studies. The sub-pixel variation of contrails is not considered in the algorithms.

False alarms in the contrail detection algorithm are usually [90% according to Meyer *et al.*, 2002] caused by natural cirrus clouds with a shape similar to contrails, the detection efficiency decreases with increasing the background inhomogeneity, which might also be caused by other contrails. Tuning the algorithm to a low false alarm rate reduces also the detection efficiency, enhancing the detection efficiency results in a higher false alarm rate. The false alarm rate can be reliably determined statistically from observations in regions without air traffic, but the detection efficiency has to be assessed by comparison to visual inspection, as no other truth is available.

A major problem with this algorithm is its sensitivity to minor differences in the spectral and spatial performance of the sensor. Both, detection efficiency and false alarm rate have to be determined for each instrument independently by visual inspection, which introduces a high level of uncertainty. A direct cross calibration between different instruments is not possible because the satellites are on different orbits. Therefore, time series using contrail analyses of different satellites suffer from large uncertainties.

Another more general problem of the interpretation of data from satellites in a sunsynchronous orbit is the sampling at slowly drifting local times. The sampling interferes in this case with the daily pattern of air traffic, resulting in aliasing effects. For case studies, the algorithm has also been applied to MODIS, A(A)TSR, MSG and GOES data, the latter in geostationary orbits allowing for nearly continuous observation at the expense of the high resolution of the polar orbiters.

The majority of contrail studies using satellite data are based on the thermal infrared channels, but contrails are also detectable in the visible and near infrared part of the spectrum [Minnis, 2003]. A systematic derivation of optical properties like optical depth, effective particle size, ice water content, or particle number using these channels has not been reported.

E. Present State of Modeling Capability

Contrails and contrail cirrus form at lower relative humidities than natural clouds and therefore change the overall cloud coverage. The additional cloud coverage due to contrail cirrus together with the specific optical properties of contrail cirrus, that are different from those of natural cirrus, are the two main factors influencing the radiative forcing due to contrail cirrus. Therefore a realistic radiative forcing, and thus a realistic climatic impact of contrail cirrus, can only be obtained if the estimates of contrail cloud coverage and optical properties of contrails are themselves realistic. Detailed modeling of contrails in concert with field observations help to parameterize the processes as a function of large-scale meteorology. So far a more process-oriented treatment of contrails in large scale models is missing. These are until now only based on the criterion for contrail formation and inventories of air traffic and constrained by contrail statistics obtained from satellite observations. In a climate model, only the effect due to linear contrails has been simulated so far and contrail coverage has been limited to source areas.

Large-Eddy Simulations

The interplay between near-field observations and models of contrail formation have unraveled many features of the contrail formation process. It can be explained sufficiently well with existing knowledge and does not introduce significant uncertainty in models describing the subsequent evolution [Kärcher, 1999]. Contrail evolution in the vortex regime is mostly described by 2D or 3D LES, few of which have been coupled to a simplified description of the ice phase using bulk microphysical methods [Lewellen and Lewellen, 2001; Unterstrasser *et al.*, 2008]. Those simulations are complex and suggest a range of factors influencing contrail development up to the dispersion regime. Comparisons to Lidar measurements in case studies [Sussmann and Gierens, 1999] showed that these models are sufficiently well developed to study the impact of wake dynamics and atmospheric parameters on the contrail ice mass, but must invoke assumptions about underlying ice crystal size distributions to make inferences about particle number densities. The latter point is an important constraint when employing such information in global model contrail parameterization schemes. Besides contrail ice mass, information on ice particle number is required to estimate effective crystal sizes for use in the radiation schemes. This information must currently be drawn from in-situ measurements.

A study of the mesoscale evolution of contrail cirrus has been performed and parameters such as initial crystal number, shear and supersaturation have been varied [Jensen *et al.*, 1998a]. Radiation was found to be important for contrail development. Large-Eddy simulations could also be used to study the contrail to cirrus transition on successively increasing spatial and temporal scales, but such efforts have not yet been reported. In such an approach, it is not clear at which point large-scale processes take over a dominant role in determining contrail cirrus evolution and their interaction with cirrus. Contrail properties simulated by LES models can be prescribed in regional models in order to estimate the regional climate effect of contrails [Wang *et al.*, 2001].

Global Modeling

The overall synoptic situation connected with supersaturated regions can probably be well represented in weather forecast and climate models, although low resolution models only allow for a statistical subgrid scale description. Humidity is a critical variable in atmospheric models due to the strong influence of subgrid scale processes and the presence of strong spatial humidity gradients. While areas of supersaturation can be identified in such models, the prediction of its magnitude and small-scale variability is much more demanding. Most global models do not allow supersaturation on the grid scale but rely on assumptions about its subgrid variability in their cloud schemes. Only the ECMWF integrated forecast system currently allows for explicit ice supersaturated states that are consistently simulated with cirrus cloud fraction although cirrus microphysics is still highly simplified [Tompkins *et al.*, 2007]. Few climate models prognose explicit supersaturation [Wilson and Ballard, 1999; Lohmann and Kärcher, 2002; Liu *et al.*, 2007] but simulate cloud coverage inconsistent with ice microphysics.

Contrails cannot be treated as a mere source term to the cloud parameterization in a climate model because of differences in the number and particle size distribution of contrail cirrus and natural cirrus. Instead contrails need to be parameterized in global models. Most global modeling approaches rely on a variation of a single contrail cover parameterization proposed by Sausen *et al.* [1998]. Contrail cloud coverage is introduced into the model which must be parameterized consistently with the model's cloud physics. In the absence of explicit supersaturation in the model a potential contrail coverage is defined. The potential contrail coverage is the area which would support contrail formation. This critical relative humidity for contrail formation is then made consistent with the cloud parameterization. Since contrails can form at a lower relative humidity than natural clouds, the critical humidity for contrail formation in the GCM grid box is defined as a combination of the two critical humidities. Potential contrail coverage is then limited by natural cirrus coverage. The dependency between natural cirrus coverage and relative humidity is unchanged. As a result the contrail parameterization can only simulate an additional coverage due to contrails and cannot replace natural cirrus. Potential contrail coverage is usually interpreted as the maximally attainable additional coverage due to contrails. This is not correct since only the formation conditions of contrails are modeled and not the persistence conditions. Once contrails are formed they can persist whenever the air is supersaturated, or in the modeling framework moister than a specified critical humidity. Potential contrail coverage was calculated using ECMWF reanalysis data or ECMWF operational data or simulations of the global climate model ECHAM4, which originated from an older version of the ECMWF model.

In order to arrive at a global contrail coverage, potential contrail coverage is then folded with an air traffic inventory [Gierens *et al.*, 1999b; Marquart *et al.*, 2003; Ponater *et al.*, 2002; Sausen *et al.*, 1998]. In Sausen *et al.* [1998] folding was done linearly and with the square root of air traffic, the latter to account for saturation effects such as contrail merging and consumption of condensable water. Gierens [1998] argues that in the presence of advection, saturation effects are not likely to happen. Duda *et al.* [2005] determines that the folding should be done with the fourth root of air traffic. In most studies the global DLR and the newer AERO2K data set have been used. As a measure of air traffic density mostly fuel usage or flown kilometers were used. Using flown kilometers or fuel usage was shown to lead to different results especially in the long distance flight corridors [Gierens *et al.*, 1999b] but since flown kilometers are not available in some data sets, fuel usage is often used. The folding of the air traffic data with the potential contrail coverage is either done online or in most cases offline. The resulting field describes the frequency of contrail formation.

The computed frequency of contrail formation is then related to the observed coverage of line-shaped contrails by a tuning coefficient. Within the parameterization contrails exist only for one time step. This results in a contrail coverage that is limited to the areas of air traffic. Advection, spreading and persistence of contrails is not covered. Contrail coverage is always zero in areas of no air traffic while in reality strong winds in the upper troposphere can advect contrails over hundreds of kilometers. The tuning coefficient is set so that the calculated lineshaped contrail coverage agrees with the observed contrail coverage over a particular area without taking into account physical mechanisms. This tuning coefficient is assumed to be temporally and spatially constant. Until now the mean European contrail coverage of Bakan *et al.* [1994] was always used for tuning the parameterization. Rädel and Shine [2007b] estimate that contrail coverage may be significantly changed when using a model that simulates explicitly supersaturation instead of the potential contrail coverage as defined by Sausen *et al.* [1998].

Duda *et al.* [2005] use in a very similar approach to Sausen *et al.* [1998] but employ a short term regional forecasting model, different definition of potential contrail coverage and a different tuning data set. The regional air traffic inventory of Garber *et al.* [2005], which describes air traffic over the contiguous U.S.A. is used. Potential contrail coverage is calculated from a regional forecasting model that contains explicit supersaturation, as the frequency at which persistent contrails can form. Comparison of the patterns of simulated contrail coverage with the satellite inferred contrail coverage quantified the influence of parameters such as the relative humidity threshold and order of relationship between air traffic and contrail coverage.

All global studies restrict themselves to studying only line-shaped contrails. Contrail cirrus coverage cannot be estimated in global climate models using the parameterization used for line-shaped contrails because no estimates of contrail cirrus coverage exist since older contrail cirrus is difficult to distinguish from natural cirrus. Instead a process based approach must be chosen in order to simulate contrail cirrus coverage and ice water content.

Optical Properties and Radiative Forcing

Most parameterizations make only crude assumptions about the optical depth of contrails. In most offline studies, optical thickness has been set to a temporally and spatially constant value. The choice of this value has a large impact on the resulting radiative forcing. Since observations and climate model simulations point at optical thickness being very variable in time and space, any kind of constant optical depth value introduces an error in the radiative forcing calculations. Even if an average value of contrail optical depth is chosen and only global radiative forcing is of interest, the forecasted increase of air traffic in the subtropics is likely to result in a change in the mean contrail optical depth. Only one global contrail parameterization simulates optical properties of contrails as a function of ambient conditions online in the climate model [Ponater *et al.*, 2002; Marquart *et al.*, 2003; Ponater *et al.*, 2005]. In contrails water condenses just like in natural cirrus a fraction of the moisture excess but since contrails exist only for one time step the contrail ice water content equals the condensed water at the time step. Contrary to cirrus they do not accumulate ice water in the model.

It is not clear how contrails overlap but the overlap assumption influence the simulated contrail coverage and the resulting radiative forcing very strongly. Usually it is assumed that contrails overlap randomly because they are far from filling the potentially contrailsupporting area and the flights are assumed to not overlap. This assumption will be especially justified if contrails are allowed to advect away from the source area [Gierens, 1998]. In areas where a substantial fraction of the potential contrail-supporting area is filled up, contrail overlap depends on the overlap of those areas. The finding that the vertical depth of supersaturated areas is small hints at random overlap being a reasonable choice even when the contrail-supporting area is filled up. Nevertheless, for the radiative calculations, maximum random overlap of contrails and natural cirrus is assumed.

Feedbacks of contrails on the simulated climate have not been studied yet. It is not clear whether contrails dry the atmosphere more strongly than air traffic moistens it. Furthermore it is not clear if contrails can replace natural clouds by a significant amount and if they replace natural clouds how large the net radiative forcing might be due to the different optical properties of contrails and natural clouds.

Offline Radiative Transfer Models

As an alternative to global modeling, radiative transfer models have been used calculating the effect of prescribed contrails. Some of these models have underlying ice water paths, effective ice crystal radii and ice crystal shapes as a basis to parameterize their microphysics. Those parameterizations have been optimized for cirrus clouds based on the state of knowledge in the mid 1990s and include high values for the ice water path and effective radius that are not representative for contrail cirrus [Plass *et al.*, 1973; Fu and Liou, 1993; Fortuin *et al.*, 1995]. Up to date, most studies actually prescribing contrail microphysical properties base their results on a single ice particle size distribution [Strauss *et al.*, 1997], although in-situ data show a marked temporal evolution of radiatively relevant contrail properties and effective sizes are smaller initially [Schröder *et al.*, 2000]. The plane parallel assumption for contrails adopted by virtually all studies causes contrail radiative forcing to grow strictly in proportion to the fractional coverage, disregarding inhomogeneity effects [Schulz, 1998; Chen *et al.*, 2001; Gounou and Hogan, 2007]. Improved optical data sets for ice crystal radiative properties that have become available [Yang *et al.*, 2000, 2005] have not yet propagated into radiative transfer models used to study contrails. Attempts to overcome the assumption of vertical homogeneity of contrail optical properties have not been reported.

Future Air Traffic and Mitigation Scenarios

Projected air traffic rise causes an increase of linear contrail coverage [Gierens *et al.*, 1999b] assuming that the atmosphere stays the same and when allowing for climate change [Marquart *et al.*, 2003]. The expected rise depends strongly on the used air traffic scenario [Gierens *et al.*, 1999b]. However, the simulated development of upper tropospheric relative humidity and cirrus clouds in a future climate, and hence their impact on contrails, must be considered uncertain due to known difficulties of representing these variables in climate models. Changes in propulsion efficiency cause contrail formation at higher temperatures and therefore at lower altitudes [Schumann, 1996]. Areas in which potentially contrails can form are increased by more than 10% when changing the propulsion efficiency by 0.1 [Sausen *et al.*, 1998]. Actual contrail coverage, on the other hand, is expected to increase by only 0.1% since air travel usually takes place in areas colder than the temperature formation threshold. Flight level changes have a major impact on contrail coverage [Sausen *et al.*, 1998; Fichter *et al.*, 2005]. Air traffic at about 10 km has the strongest impact on radiative forcing [Rädel and Shine, 2007b] and should therefore be avoided if the contrail impact is to be minimized. If the cryoplane technology is used aerosol output is decreased and moisture output increased lowering the relative humidity threshold for contrail formation and causing an increase in ice crystal sizes. The former leads to an increase in contrail coverage and the latter possibly to a reduction in optical thickness. Both effects are estimated to cancel when calculating radiative forcing [Marquart *et al.*, 2003]. Fuel additives designed to change the ice nucleation behaviour of exhaust soot particles are not likely to have a significant impact on contrail radiative forcing because the formation process is not sensitive to details of the ice formation process [Gierens, 2007].

F. Current Estimates of Climate Impacts and Uncertainties

Contrails increase the planetary albedo and hence cause a negative radiative forcing in the shortwave (SW) range. Contrail temperature is usually lower than the brightness temperature of the atmosphere without the contrails. Therefore, contrails induce a positive radiative forcing in the longwave (LW) range. The net radiative forcing is the difference between the SW and LW values. In most cases, the net radiative forcing is positive at the top of the atmosphere. Radiative forcing is mostly negative at the Earth's surface, in particular during daytime. The radiative forcing increases with ice water path, or optical depth, and with contrail coverage [Meerkötter *et al.*, 1999]. For a given ice water path, the SW dominates over the LW effect for sufficiently small effective ice crystal radii. The cross-over point depends also on assumed ice crystal habit [Zhang *et al.*, 1999]. We emphasize that the net radiative forcing is generally the difference between two large values: negative SW forcing and positive LW forcing. Hence, any small error in either of them has a large impact on the computed net effect.

Global mean radiative forcing estimates for persistent line-shaped contrails have been reported that differ by a factor five. This results mainly from the use of different values for contrail coverage and optical depth. For 1992 air traffic, Marquart *et al.* [2003], Myhre and Stordal [2001], and Minnis *et al.* [1999] have yielded values of 3.5 mW/m2, 9 mW/m^2, and 17 mW/m^2, respectively. Minnis *et al.* [1999] assume global contrail coverage of 0.1% for 1992, an optical thickness of 0.3, contrails at 200 hPa altitude, and hexagonal ice particles;

they also included a simplified diurnal cycle with a globally uniform 2:1 day-to-night ratio. Myhre and Stordal [2001] use the same optical depth and coverage but find smaller radiative forcing values because of different approaches for the daily traffic cycle, scattering properties of ice particles, and contrail altitude. Marquart *et al.* [2003] normalize the contrail cove rage computed with a climate model by reference to more recent satellite observations [Meyer *et al.*, 2002] implying a smaller global contrail coverage (0.05-0.07% for 1992); they compute smaller optical thickness values [Ponater *et al.*, 2002], including the daily cycle, improved altitude distributions of the contrails, and an update of the LW radiation scheme of the global model [Marquart and Mayer, 2002]. Their radiative forcing result for 1992 is five times smaller than the value used in the IPCC [1999] assessment.

IPCC [2007] adopted the result of Sausen *et al.* [2005] to conclude that the best estimate for the radiative forcing of persistent line-shaped contrails for aircraft operations in 2000 is 10 mW/m^2. The value is based on independent estimates derived from Myhre and Stordal [2001] (15 $mW\ m^{-2}$) and Marquart *et al.* [2003] (6 mW/m^2). The two values were used by the IPCC [2007] to set the uncertainty range of a factor of two. This best estimate is significantly lower than the IPCC [1999] value of 34 mW/m^2, linearly scaled from 1992 to 2000 air traffic. The change results from reassessments of persistent linear contrail coverage and lower optical depth estimates, as detailed above. The new estimates include diurnal changes in the shortwave solar forcing, which decreases net forcing for a given contrail cover by about 20%.

Regional cirrus trends were used as a basis to compute a global mean radiative forcing value for AIC (aircraft-induced cloudiness) in 2000 of 30 mW/m^2 with a range of 10-80 mW/m^2 [Stordal *et al.*, 2005; Sausen *et al.*, 2005]. This value is not considered a best estimate because of the uncertainty in the optical properties of AIC and in the assumptions used to derive AIC coverage. However, this value is in agreement with the upper limit estimate for AIC radiative forcing in 1992 of 26 mW/m^2 derived from surface and satellite cloudiness observations [Minnis *et al.*, 2004]. A value 30 mW/m^2 is close to the upper limit estimate 40 mW/m^2 derived for AIC without line-shaped contrails in IPCC [1999].

A by far larger climate impact has been deduced by Minnis *et al.* [2004], who have analyzed a cirrus trend of ~1%/decade over the continental USA between 1971 and 1995, which was attributed almost exclusively due to air traffic increase during the period. Assuming an optical depth of 0.25 this increase of high clouds was calculated to induce a global mean radiative forcing of up to 25 mW/m^2 and a surface temperature response of 0.2-0.3 K/decade in the region of the forcing, which would explain practically all observed warming over the respective area between 1973 and 1994. In response to the Minnis *et al.* [2004] conclusion, contrail forcing was examined by Shine [2005] and in two global climate modeling studies [Hansen *et al.*, 2005; Ponater *et al.*, 2005]. These studies stressed that it is not possible to derive a regional climate response from a regional climate forcing and concluded that the surface temperature response calculated by Minnis *et al.* [2004] is too large by about one order of magnitude. For the Minnis *et al.* [2004] result to be correct, the climate efficacy of contrail forcing would need to be much greater than that of other forcing terms (e.g., CO_2). Instead, model simulations hint at a smaller efficacy of contrail forcing than equivalent CO_2 forcing [Hansen *et al.*, 2005; Ponater *et al.*, 2005].

For contrail cirrus, no reliable estimate of the optical properties and of the radiative forcing exists. The IPCC estimate of an upper bound of radiative forcing of 40 mW/m^2 by contrail- and soot-induced cirrus changes is based on the assumptions of 0.2% global additional cirrus coverage with an optical thickness of 0.3 (same as for line-shaped persistent contrails) [IPCC, 1999] . Both assumptions are very uncertain. The optical properties of the contrail cirrus are likely different from that of line-shaped contrails. The radiative forcing depends nonlinearly on the optical depth. It increases approximately linearly for small optical depth values, reaches a maximum in between 2 and 5 and may be negative for optical depth values larger than 10 [Meerkötter *et al.*, 1999]. Contrails within cirrus may enhance the optical depth of the cirrus beyond the limit where an increase in optical depth causes a reduction of the radiative forcing. Hence, a reliable estimate of the radiative forcing by contrail cirrus cannot be given. For 1% additional cirrus cloud coverage regionally (optical depth 0.28), a regional surface temperature increase of the order 0.1 K was expected from a study by Strauss *et al.* [1997]. With a 2D radiative convective model, a 1 K increase was found in surface temperature over most of the Northern Hemisphere for additional cirrus coverage of 5% [Liou *et al.*, 1990].

For 1% additional cirrus cloud coverage globally (optical depth 0.33) a general circulation model coupled to a mixed layer ocean model computed 0.43 K global warming [Rind *et al.*, 2000]. Ponater *et al.* [2005] find a smaller specific climate response from contrails than for CO_2 increases in their climate model: the equilibrium response of surface temperature to radiative forcing from contrails is 0.43 K/(W m^{-2}) while 0.73 K/(W $m\text{-}^2$) for CO_2. For a global contrail coverage of 0.06 % and 0.15 %, with mean radiative forcing of 3.5 mW/m^2 and 9.8 mW/m^2 in 1992 and 2015, respectively (optical depth 0.05-0.2 depending on region and season, Meyer *et al.* [2007]), the computed transient global mean surface temperature increase until 2000 amounts to ~0.0005 K in this model [Ponater *et al.*, 2005].

Contrails cool the surface during the day and heat the surface during the night, and hence reduce the daily temperature amplitude. The net effect depends strongly on the daily variation of contrail coverage. A reduction of solar flux by an order 50 W/m^2, as measured by Sassen [1997], is to be expected locally in the shadow of optically thick (optical depth > 1) contrails. The surface LW forcing is small because of the shielding of terrestrial radiation by water vapor in the atmosphere above the surface. Hence, the Earth's surface locally receives less solar energy in the shadow of contrails [Sassen, 1997]. This does not exclude a warming of the atmosphere-surface system driven by the net flux change at the top of the atmosphere [Meerkötter *et al.*, 1999]. As shown by a 1D radiation-convection model, vertical heat exchange in the atmosphere may cause a warming of the surface even when it receives less energy by radiation [Strauss *et al.*, 1997].

Travis *et al.* [2002] claimed observable increases in the daily temperature range due to reduced contrails in the three days period of September 11-14, 2001, when air traffic over parts of the USA was reduced. They report that the daily temperature range was 1 K above the 30-year average for the three days grounding period, which was interpreted as evidence that jet aircraft do have an impact on the radiation budget over the USA. Several studies discussed these findings and pointed out that the statistical significance is weak and does not allow for strong conclusions [Schumann, 2005; Forster *et al.*, 2007; Dietmüller *et al.*, 2007]. Moreover, unusually clear weather in that region could also explain the observed daily temperature range [Kalkstein and Balling, 2004; Travis *et al.*, 2007].

Radiative forcing due to contrails is expected to increase in future due the projected increase in air traffic. Marquart *et al.* [2003] simulated the radiative forcing due to the increase of air traffic and due to climate warming. By 2015 the radiative forcing of line-shaped persistent contrails is simulated to be 9.4 mW/m^2 and by 2050 14.8 mW/m^2, compared to 3.5 mW/m^2 in 1992. Neglecting climate change the radiative forcing would be larger since the simulated temperatures in the tropical upper troposphere would be colder and the frequency of contrail formation larger than when allowing for climate change.

The majority of global contrail studies rely on a single modeling approach to simulate lineshaped contrail coverage, relying on assumptions such as constant tuning factor, representativeness of the coverages reported by Bakan *et al.* [1994] and constant optical depth. Furthermore studies are not independent since they are carried out with only few different models and always tuned to the same observed contrail coverage over Europe. Newer and lower estimates of radiative forcing are partly based on the assumption of a lower constant optical depth than in the 1990s. One-dimensional radiation schemes seem to agree on RF due to linear contrails and therefore do not add to the range of forcing estimates. However, most of these studies apply the same contrail ice crystal size distribution [Strauss *et al.*, 1997] so that the uncertainty in radiative forcing may be underestimated. Three-dimensional effects on radiative transfer are not insignificant but are not considered in global models yet.

Rädel and Shine [2007b] estimate the combined error due to the assumption of constant optical depth and due to the use of scaling factors for tuning the contrail coverage to be about 60%. Additionally smaller errors due to assumptions of ice crystal parameters, neglect of 3D radiative transport, assumption of constant engine parameters, diurnal cycle of contrail coverage, errors due to the cancellation of between long wave and short wave forcings. All errors together are estimated to account for a factor of two in net radiative forcing.

G. Interconnectivity with Other SSWP Theme Areas

Our theme is closely connected to theme area 3 in terms of observations of ice supersaturation in the upper troposphere and lower stratosphere, both globally from satellites and locally from aircraft or balloons. We have emphasized that such measurements are difficult to perform and, in the case of remotely sensed data, highly uncertain. While we principally understand the causes of supersaturation, prediction in global models is at its infancy. A physically consistent representation of supersaturation, ice microphysics and coverage of contrail cirrus and natural cirrus including their subgrid-scale features requires new modeling approaches.

Our theme is also connected to theme area 4 regarding measurements of contrail cirrus. A global homogeneous data set of relevant contrail cirrus properties (primarily optical depth and coverage) is not available. We have emphasized that available measurements (comprising Lidar and Radar instruments, satellite sensors and standard cloud physics instrumentation onboard high flying aircraft) do not cover the full contrail cirrus life cycle. Virtually all quantitative in-situ information available covers only contrail ages up to ~30-60 min or perhaps up to ~2-3 h when tracking individual contrails in remotely sensed data.

3. Outstanding Limitations, Gaps and Issues Requiring Improvement

A. Science

Representing Contrail Life Cycle in Global Models

It is currently not possible to simulate the complete life cycle of contrail cirrus (i.e. fractional coverage, microphysical properties, radiative forcing) from formation to decay. The radiative effect of short lived (up to ~30 min) and non line-shaped contrails has not been properly discussed yet. Further, physical mechanisms that remain unconsidered by current approaches include advective transport of contrail cirrus out of the major contrail-forming areas. Interactions of contrails with the moisture field and cirrus clouds cannot be treated well in current models. Contrail cirrus taps condensable water and might remove the moisture by sedimentation, therefore changing the relative humidity. This may cause the atmosphere not to reach or to reach later the moisture thresholds for formation of natural cirrus therefore delaying cirrus onset.

More emphasis has to be put in estimating the climate effect of contrail cirrus. New process based methods have to be developed since results cannot be tuned to observations. These efforts would benefit from a better knowledge of the temporal development of contrail properties from in-situ and remote sensing measurements. It remains unclear if and, if at all, when contrails acquire similar properties as natural cirrus. The apparent lack of aged contrail cirrus measurements hinders progress in this area.

Contrail Cirrus Optical Depth and Coverage

Any confidence in estimated global radiative forcing of contrail cirrus will remain low unless the underlying optical depth mean and variability of contrail cirrus has been fully explored and the radiation schemes in global models have been adapted to contrail-specific optical properties. Clouds produce different flux changes depending on the environmental circumstances (cloud, surface or atmospheric properties). As a class, thin cirrus cool the surface and exert a net warming within and at the top of the atmosphere [Chen *et al.*, 2000]; optically thicker cirrostratus and anvil cirrus still warm the atmosphere on the whole but cool the surface and top of the atmosphere. This annual and global mean picture derived from ISCCP-D2 data has largely confirmed earlier studies regarding cirrus radiative forcing [Hartmann *et al.*, 1992], but still contains substantial simplifications in treating the vertical layering of cloud, the radiative transfer in cirrus, and in assumptions about the nighttime radiative fluxes, so that these findings cannot be viewed as a final conclusion. Contrail cirrus belonging to the class of thin cirrus are therefore also expected to warm the atmosphere on average. However, a host of underlying factors controlling radiative forcing by contrail cirrus, including their radiative impact when coexisting with cirrus, need to be explored further to build more confidence in predictions of their net radiative effect.

The actual radiative relevance of clouds is also controlled by the product of their typical spatial coverage and their frequency of occurrence (cloud amount). In the case of contrails the latter is determined by the formation probability along aircraft flight paths while the former is more closely tied to the factors controlling atmospheric supersaturation, transport and contrail dissipation. The total coverage is the sum of coverage due to line-shaped contrails and contrail cirrus. A separate estimate of the latter contribution has not yet been reported. IPCC [2007] estimates the ratio of total (contrail cirrus plus soot cirrus) coverage due to aircraftinduced cloudiness to that of persistent linear contrails in the range 1.8-10 [Minnis *et al.*, 2004; Mannstein and Schumann, 2005]. The upper bound is currently not supported by Mannstein and Schumann [2007]. The study by Stubenrauch and Schumann [2005] would imply even smaller lower bounds [Schumann, 2005]. Locally, this ratio is ill-defined if considering regions into which contrails have been advected but where air traffic is low or absent.

A further open question is the radiative effect of producing contrails inside existing cirrus (or other high level clouds). Such contrails may increase the optical depth of the combined cirrus/contrail systems compared to the cirrus, or high level cloud, alone. If the optical depth is thick already (~3-6), then an increase in optical depth may cause a cooling. If t he opticaldepth is small, the increase in optical depth will still cause a warming. The imp ortance of thiseffect depends on at least three factors. (i) The relative frequency of occurrence of contrails inside thick cirrus (high level clouds) compared to contrails outside cirrus or in thin cirrus; (ii) the change in optical depth for solar and terrestrial radiation caused by the contrail forming inside the existing high level cloud; (iii) the gradient of the radiative forcing with optical depth. To our knowledge this problem has not been studied yet, but without solving it, one cannot exclude that contrails cool.

Soot Effects

Whereas it is feasible to use as a first step a proxy for supersaturation when simulating contrails, the simulation of the soot effect relies on the explicit simulation of supersaturation. The lack of consistency between ice supersaturation, cirrus microphysics and cirrus cloud coverage in most global models currently does not allow the simulation of the indirect effect on climate induced by soot emissions with confidence. Satellites cannot discriminate between pure contrail effects and soot effects on cirrus, therefore hampering a sound model validation of contrail impact on climate.

To tackle the soot effect, an in-situ experiment should be designed to demonstrate the iceforming capability of aircraft soot emissions (experimentum crucis). Such a measurement should be performed first in relatively unpolluted air because the background cirrus in flight corridors could already be affected by aviation soot. The soot should be emitted along with tracers marking the air mass. Difficulties in interpretation may arise from dynamical effects that can easily mask aerosol-induced cirrus changes and the impact of ice nuclei from other sources such as mineral dust.

Metrics

The climate impact of contrails is usually reported in terms of global mean or regional mean contrail-cirrus cover [Sausen *et al.*, 1998], and forcing in terms of shortwave (SW), longwave (LW) and net (LW+SW) radiative forcing values [Minnis *et al.*, 1999]. In order to assess the climate impact one needs to know the equilibrium global mean surface temperature change ΔT per net radiative forcing (RF), $\Delta T = \lambda_{contrail}$ RF, or the efficacy, i.e. the value $\lambda_{contrail} / \lambda_{CO_2}$ relative to that for RF from CO_2 concentration changes [Hansen *et al.*, 2005; Ponater *et al.*, 2005]. Since contrails are strongly correlated with air traffic density, even when accounting for drift of contrails during their life-time, the contrail-induced climate impact occurs mainly at northern midlatitudes [Minnis *et al.*, 2004]. Moreover, contrails cirrus is special in respect to its potential impact on the hydrological cycle, with many still unexplored mechanisms.

Generally, our gap analysis is in agreement with the findings of the 2006 Boston Workshop on climate impacts of aviation summarized by Wuebbles and Ko (2007) (http://web.mit.edu/aeroastro/partner/reports/climatewrksp-rpt-0806.pdf).

B. Measurements and Analysis

Upper Tropospheric Relative Humidity

The global distribution of humidity in the upper troposphere is not well determined since satellites have a low resolution in the area of the tropopause. Relative humidity is even more uncertain since it relies on consistent temperature and humidity measurements. In satellite data, only large-scale features such as geographical or seasonal patterns are robust features, the magnitude of inferred supersaturation is uncertain [Gettelman *et al.*, 2006]. Such observations need to be refined and continued to infer reliable statistics and better quantitative information on magnitude, frequency of occurrence, and variability of supersaturation.

Despite pending issues in measuring relative humidity in-situ, ice supersaturation can be measured with aircraft with sufficient accuracy in the extratropics. Those measurements are particularly useful for aviation-related research (and for the general understanding of upper tropospheric/lower stratospheric processes as well) when performed on a regular basis on commercial aircraft. In-flight measurements using the already existing Tropospheric Aircraft Meteorological Data Relay (TAMDAR) system should routinely include reliable humidity measurements at flight level, thus providing a climatology of relative humidity and cirrus coverage. TAMDAR and NOAA's Water Vapor Sensing System 2 (WVSS2) might improve the situation in the future when adapted to and used at cruising altitudes.

Whereas the formation of ice supersaturation in the extratropics is understood (Section 2.a), a reliable statistic of the magnitude, vertical layering and horizontal extent of supersaturation are not available. Field measurements support the predominance of homogeneous freezing as a major source of cloud ice mass [Jensen *et al.*, 2001]. This inference has been made based on frequently measured maximum supersaturations being consistent with the homogeneous freezing process and the frequent occurrence of a large number of small ice crystals [Kärcher and Ström, 2003; Gayet *et al.*, 2004; Hoyle *et al.*, 2005]. However, it is not clear why high ice supersaturations can persist in the presence of cold thin cirrus and why some values are exceptionally high (above the homogeneous freezing level) outside of clouds at very low (< 200 K) temperatures [Jensen *et al.*, 2005; Peter *et al.*, 2007].

Remote Sensing of Contrails

Global coverage by linear contrails, their optical properties and also the related radiative cloud forcing are in principle deducible from satellite measurements. Instruments like MODIS on Aqua and Terra or the A(A)TSR(2) series on ERS1, ERS2 and ENVISAT offer this possibility, as their data is available in a resolution of ~1 km for nearly the whole globe. A systematic study of contrail cover from AVHRR on METOP, AATSR and MODIS (~1 km resolution), in connection with air traffic data may provide very useful results for model validation.

A fine tuning of the automated contrail detection algorithm to these instruments followed by a thorough characterization of the performance in terms of detection limits, false alarm rates and detection efficiencies are necessary prerequisites. Contrails may be detected as soon as they show a significant contrast from background in terms of measurable radiation, but quantification of what can be detected and what not is difficult. False alarm rates and error bounds limit accuracy of contrail cover deduced from NOAA AVHRR channels to ~0.1% cover [Meyer *et al.*, 2002]. The optical depth values may be uncertain to an order 0.05 or more. Error bounds on detection limits, effective radius, life time, spreading rates, etc. have still to be determined.

The transition of linear contrails into contrail cirrus, which cannot be identified from shape, will remain poorly defined. Polar orbiting satellites observe clouds only once in long periods (typically a day) and can therefore not be used to follow the life cycles of individual contrails. Tracking of contrails and contrail cirrus in data from geostationary satellites with a high temporal resolution (5 min in MSG 'rapid scan', 1 min in GOES) can be used to retrieve a portion of the life cycle and the radiative forcing, ice water path, optical depth and effective particle size as function of contrail age and in relation to ambient conditions. Because of the lower spatial resolution of sensors in geostationary orbit this approach can detect only thicker and wider contrails. Systematic studies of such kind have still to be performed. For the interpretation of all measurements it is an advantage to know precisely the actual air traffic, which might have caused the observed contrails. Such data sets are usually not available to the research community. Moreover, knowledge of actual wind fields, temperature and humidity fields is needed at high spatial and temporal resolution to check for contrail formation threshold conditions and to identify the lateral displacement of contrails for given meteorology. Such data can be made available from meteorological analyses from numerical weather prediction centers.

Because of the sensor dependence of the observed cloud properties, satellite observation results (such as cloud cover) cannot be compared with model results directly. For proper comparison of satellite data and model results, one should apply a sensor simulator to the model results which simulates what the sensor would see for the given model state. Such an approach is essential for model validation.

One should note that the value of contrail cirrus coverage may depend strongly on its definition, namely whether it includes only the coverage observable to a specific sensor, or whether is limited to contrail cirrus above a certain optical depth threshold. Hence, optical depth and contrail coverage should always be reported together with the implied thresholds.

Identification and Characterization of Aged Contrail Cirrus

A major limitation in studies of older contrail cirrus is the difficulty to track single contrails with time, or to detect a contrail once it has lost its line shape. Ground-based Lidar can follow linear contrail evolution for a certain time, limited by the wind speed advecting the contrails away from the site. Research aircraft pilots quickly lose track of contrails without additional guidance, for instance from concomitant satellite observations. These are the major reasons for the lack of in-situ or Lidar measurements of contrail cirrus. Poor airborne sampling statistics for evolving large ice crystals and the difficulty in determining the exact sampling position (and hence, infer contrail age from measurements of NO) remain serious problems in any aircraft-based measurement. In-flight measurements of radiative fluxes of aging contrails should be easier to perform, but this requires two aircraft for a proper characterization of up- and downwelling flux densities. As very limited information is available from both in-situ and remote sensing measurements, and measurement uncertainties are often not clearly quantified, the optical properties of even line-shaped contrails and their subsequent time evolution remain a matter of debate. Remote sensing of the optical parameters of ice clouds relies on assumptions about the shape and size distribution of ice crystals. The lack of precise information from direct measurements leads to uncertainties regarding their radiative impact. Even though observational case studies would provide useful information for validation of process models, these measurements do not allow representative statistics. A general description of the contrail cirrus life cycle and the resulting radiative forcing of contrails and contrail cirrus is therefore hardly achievable without the help of models.

Correlations between Cirrus Coverage and Air Traffic

The statistical analysis of the correlation between cirrus properties and air traffic data may be the only method allowing the determination of AIC from observations [Mannstein and Schumann, 2005]. The method may be used to study cirrus not only in terms of cover but also directly in terms of radiation signals measurable from satellites. The method is attractive in principle, because it offers chances to detect the mean life time of contrail-cirrus.

For proper interpretation of such correlation results one has to know any cross-correlation of the observables with other parameters, such as geographical latitude and longitude because of land-ocean contrasts. Model results are useful to identify such cross-correlations [Mannstein and Schumann, 2007]. Even for nonzero cross-correlations, the method may be useful to determine upper bounds on the amount of AIC changes. Moreover, the same kind of statistical analysis may also be applied to model results, which helps not only to identify cause-effect relationships but also supports validation.

From ongoing work, we see chances that such methods provide useful correlation analyses for regions over the globe where the natural variability of cirrus statistics is small.

This method requires input in terms of temporally highly resolving geostationary satellite data over long periods and large regions (continents, oceans, hemispheres), together with information on air traffic movements at high spatial (~50 km) and temporal (~1 h) resolution, and corresponding meteorological analysis data for the same regions and time periods.

C. Modeling Capability

Scale Problem

One of the key problems in cloud and, even more so, in contrail modeling is the large range of spatial scales involved. The scale of young contrails (width ~50 m, comparable to the aircraft wing span) up to the scale of ice supersaturated regions (~500 km). Contrails can either be simulated using a Eulerian or a Lagrangian approach.

In a Lagrangian approach one would follow a finite set of typical individual contrail segments over their life-time and derive estimates of the properties of the ensemble of all contrails from the Lagrangian contrail segments. This approach could be an extension of a Gaussian plume model used to simulate the highly inhomogeneous concentration field of emitted trace species in a flight corridor [Schumann and Konopka, 1994; Schumann *et al.*, 1995; Schlager *et al.*, 1997]. The idea of this approach has been demonstrated in an idealized simulation of contrail coverage [Gierens, 1998]. Such a model needs input in terms of spatially and temporally highly resolved air traffic movement data. Moreover, the contrail model needs meteorological data input (temperature, humidity, horizontal and vertical wind) from a numerical weather prediction model, preferably one which simulates ice supersaturation [Tompkins *et al.*, 2007]. The change in contrail properties of a Lagrangian contrail segment with time for given ambient conditions can be parameterized based on detailed contrail simulations [Unterstrasser *et al.*, 2008]. Offline simulations would be useful for applications such as route optimization and for direct comparison with observations. For climate simulations it is necessary to simulate climate feedbacks which require an online scheme. This would be computationally extremely expensive.

Using the Eulerian approach (parameterization) contrail cirrus properties are described by a suitable set of variables at each grid point of a global circulation model. The variables have to characterize fractional coverages and mean properties of contrails of various ages within a grid cell of the Eulerian model. The model simulates the variation of contrails by integrating budget equations including contrail cirrus sources and sinks, in time and space. Such a parameterization is an extension of a GCM cloud scheme. It allows accounting for the feedback of contrail cirrus on the ambient (cloudy) atmosphere. Such a model is suitable for analysis of contrail cirrus both in the present and in a future climate.

Uncertainties in Global Modeling

Climate models need to be improved in two main aspects. In order to reduce the uncertainty regarding contrail radiative forcing the simulation of upper tropospheric fields needs to be improved and validated. A realistic simulation of the upper tropospheric relative humidity field is crucial since the frequency of contrail occurrence and the optical properties of contrails are strongly dependent on the relative humidity field. The coverage due to line-shaped contrails has been shown to agree reasonably well with observations in specified areas. Nevertheless, the method is unsatisfactory relying on the assumption of a constant scaling between contrail formation frequency and coverage. This assumption is likely to introduce errors especially calculating coverage for future scenarios in which air traffic increases in areas the parameterization was not tuned to. The optical properties of contrails are still under debate, with the modeling community usually assuming or simulating a mean optical depth of ~0.1. Some remote sensing observations suggest similar values and other remote sensing observations, including Lidar and high resolution remote sensing, deriving optical depths of 0.3 or 0.4.

Improvements and Validation Necessary for Relative Humidity and Cloud Coverage

It has become apparent that many climate models have problems simulating the humidity field in the upper troposphere. Models often have problems representing moisture in the area of the tropopause [John and Soden, 2007]. Errors in the upper tropospheric humidity field and associated errors in the temperature field, that manifest themselves often dramatically in the model's cold bias, have an impact on the simulated contrail statistics. Only recently more observations of the upper tropospheric humidity field (MOZAIC, MLS, AIRS) have become available enabling the validation of climate models in the upper troposphere. When evaluating the water budget in climate models, the emphasis is usually put into warm clouds. The microphysics of ice clouds has not been systematically evaluated and may be even used for tuning the model [DelGenio, 2002; Jakob, 2002]. Consequently there are indications that the optical properties of natural clouds may not be represented well at least in some climate models. Specifically it has been noted that the ice water content and effective ice crystal radii are too small in the ECHAM4 climate model [Lohmann *et al.*, 2007]. Size spectra that have a large impact on the microphysics and on the optical properties of clouds have not yet been updated according to the newest measurement results.

Climate models do not explicitly capture the formation of cirrus clouds. Nearly all climate models diagnose cirrus coverage in the same way as coverage due to water clouds, purely from the surrounding humidity, and apply saturation adjustment. They do not allow for explicit supersaturation relative to ice. Some modules have been developed to represent icesupersaturation in global models [Kärcher *et al.*, 2006; Tompkins *et al.*, 2007] that might in the long term lead to a sufficiently accurate physically-based parameterization of contrail development. Future weather forecast and climate models must increase their vertical resolution to enable the simulation of stacked thin layers of supersaturation. They must include proper parameterizations for subgrid-scale dynamical processes that drive ice nucleation, and adapt their cloud schemes to cirrus clouds consistent with observations. The introduction of supersaturation at the grid scale of such models, however, requires current cloud fraction parameterization to be fundamentally modified to be consistent with known cirrus microphysics and supersaturation [Kärcher and Burkhardt, 2008]. A consistent cirrus coverage defining the formation and the evaporation of cirrus at different relative humidity levels and allowing for non-equilibrium states has not been implemented yet. Once this is implemented and validated for natural cirrus, parameterizations of contrails can be based on the improved physics.

Improvements Necessary for Contrail Cirrus

Meanwhile, contrails can be parameterized requiring a proxy for supersaturation instead of the explicit representation of supersaturation, as applied for natural clouds. This approach has been used successfully for simulating line-shaped contrails. Line-shaped contrail coverage has been simulated by tuning an area-averaged coverage to observational data and assuming a globally and temporally constant tuning coefficient. This approach precludes the simulation of the contrail life cycle and assumes that the ice water content can be estimated from the condensable water at a single time step. Because of the former the estimation of global mean radiative forcing due to aircraft-induced cloud changes has until now been limited to the forcing due to line-shaped contrails. Contrail cirrus cannot be modeled globally with existing methods so that a best estimate of radiative forcing due to contrail cirrus does not exist.

One possibility that may lead to substantial progress in global modeling is a process-based treatment of contrail cirrus as an individual cloud type with specific sources and sinks. Such an approach will allow uncertainties to be systematically reduced by properly representing and evaluating the processes that determine the entire contrail cirrus life cycle. Instead of constraining contrail coverage, the processes influencing contrail cirrus coverage must be identified, described and adequately constrained. The contrail cirrus parameterization should have a similar amount of subgrid scale information as the natural cloud scheme. Microphysical process rates have to be adjusted to contrails. In this way an independent estimate of line-shaped contrail coverage may be obtained that does not suffer from assuming a constant tuning coefficient and estimating contrail ice water content from the model state at a single time step. Furthermore the coverage due to contrail cirrus and the associated ice water content could be simulated. Nevertheless, as long as natural cirrus coverage is only diagnosed natural cirrus limits contrail cirrus coverage. Therefore contrail cirrus can replace natural cirrus and compete for condensable water with natural cirrus only in a limited way.

A realistic simulation of the interaction between contrail cirrus and natural cirrus may be achieved by calculating both coverages prognostically. A prognostic treatment of natural cirrus as suggested by Kärcher and Burkhardt [2008] enables the use of different formation and evaporation humidity levels for natural cirrus and therefore the simulation of supersaturation.

Radiation

Radiation codes in GCMs have a number of deficiencies that make the estimation of contrail radiative forcing uncertain. Three-dimensional effects in radiative transfer are thought to be non-negligible but are not covered routinely even in sophisticated radiative transfer calculations [Gounou and Hogan, 2007]. The microphysical basis for the application of radiative transfer simulations should be improved using real contrail size spectra and realistic vertical layering. This may eventually lead to improved radiation schemes for GCMs for contrail cirrus. Removing uncertainties in contrail radiative forcing must face the general difficulty that the net radiative effect of contrails and cirrus is difficult to evaluate accurately because it results from counteracting effects of large shortwave and longwave forcing terms.

Validation

Besides model development and improvement it is indispensable to also focus on validation. A GCM should be validated using statistical and climatological data. Generally the humidity fields and cloud coverages and optical properties simulated by climate models need to be validated. Suitable data to do this are just becoming accessible. Furthermore fields and frequency of supersaturation simulated by GCMs need to be validated. Available in-situ data for young contrails (up to 30-60 min age corresponding to one GCM time step, section 2.a) could be used to check whether contrail ice water contents are properly initialized in processbased contrail parameterization schemes. Although LES models are available to simulate individual contrails and their evolution within a few hours, those approaches are computationally demanding and are not straightforward to use for GCM validation. Available in-situ measurements provide only snapshots of possible contrail realizations. There still is a marked gap of climatological data describing contrail and contrail cirrus coverages and optical properties that are needed for the validation of simulated contrail and contrail cirrus coverages. Often the conditions under which contrails could be detected are not specified in detail and different observation-based statistics may have different detection thresholds. When developing process-based parameterizations of contrail cirrus coverage, data describing those processes, such as spreading, ice particle sizes and initial conditions after formation, are needed to constrain the parameterization. This calls for novel and innovative theoretical methods to infer contrail cirrus microphysical and optical properties on a statistical basis.

Another problem for GCM validation using remote sensing is the difficulty to discriminate between contrail cirrus and possible effects caused by aircraft soot emissions in such data. As a first step, it would be necessary to demonstrate experimentally whether soot modifies cirrus cloudiness (section 3.a). Even if aircraft aerosols should not lead to a significant change in cirrus cloudiness, properties of aerosols from other sources would still be required to predict the formation of natural cirrus by homogeneous and heterogeneous ice nucleation.

D. Interconnectivity with Other SSWP Theme Areas

Limitations, gaps and issues requiring improvement connect to SSWP theme areas 3 and 4 covering upper tropospheric relative humidity and contrail-specific microphysics. We recall our statements in section 2.g. Concerning uncertainties in developing appropriate metrics to describe aviation-induced climate change, we refer to the SSWPs from theme area 7.

4. Prioritization for Tackling Outstanding Issues

Modeling and validation (A)

In the last section a number of open issues were identified that preclude progress in estimating the global climatic impact of contrails. Some of those issues are known shortcomings in climate models. Eliminating those shortcomings, which relate to the moisture budget and cloud representation in the models, may require several years of attention but would be required to reduce uncertainty of the estimates of the global climate effect of contrails and contrail cirrus. The improvements would increase confidence in our ability to simulate contrails only on the long time scale and therefore would not reduce uncertainty of the climate forcing of contrails for quite a few years to come.

On the other hand existing contrail parameterizations should be tested regarding the tuning and validated with more observational data and contrail resolving models as they become available. The high degree of interdependency of current results on global persistent linear contrail radiative forcing that arises from the use of identical data sets for tuning and validation should be reduced. Furthermore parameterizations should be based o n processes so that only those processes would need to be constrained. Schemes should be extended covering not only contrails but also contrail cirrus. Radiative parameterizations and overlap calculations should be expanded to cover not only natural cirrus but also contrails. In the future improved parameterizations could then be implemented in models that have an improved representation of the moisture budget and cirrus representation.

1. Improving and validating the representation of the moisture and clouds in the upper troposphere in atmospheric models.
2. Including microphysical parameterizations leading to supersaturation in atmospheric models and representing processes of cirrus formation.
3. Testing the sensitivity of current contrail parameterization to tuning and assumptions influencing optical properties of contrails.
4. Development of a process based contrail / contrail cirrus parameterization for use in climate models.
5. Improving representation of radiative response to contrail cirrus in atmospheric models (including optical properties and cloud overlap).

6. Analyzing under which conditions contrails cool when forming inside high level clouds.
7. Development of a plume-based contrail model to simulate the scale transition for fresh contrails to extended contrail cirrus decks.
8. Inclusion of the plume-based contrail model in a weather forecasting model for shortterm prediction and validation purposes (long term goal may be inclusion in climate models).
9. Models need to be validated with a number of observational data sets. Critical observations include absolute and relative humidity, ice water content, ice particle size distributions and habit, optical depth, vertical motion, wind shear, turbulence, etc. Further measurement campaigns are needed. CALIPSO and CLOUDSAT data should be analyzed regarding the detection of contrail cirrus.

Remote Sensing and in-Situ Experiments (B)

An important issue is the quantification of aviation induced cloud changes AICC (including contrail cirrus, soot cirrus, changes to existing cloud systems; and changes in terms of coverage, microphysical and optical properties, radiative forcing etc.) from observations. Besides the modeling approach described in (A), we suggest a strategy to determine AICC directly by remote sensing. A second important issue is the homogeneous analysis of the precise coverage and properties of line-shaped contrails over a large region of the Earth with specified accuracy. Finally, one needs specific in situ soot experiments with aircraft soot sources in remote regions and measurements of the soot impact on cirrus that may form or may be changed due to the presence of soot [Kärcher *et al.*, 2007].

1. Improving and validating the representation of contrail and cirrus remote sensing analysis schemes providing cloud coverage, optical thickness, brightness temperature, reflectance, microphysical properties, contrail age, etc.
2. Provision of simple aircraft impact prediction tools such as contrail cover as a function of air traffic with prescribed spreading and life time [Gierens, 1998; Mannstein and Schumann, 2005].
3. Testing of correlations between observed cloud properties (from B1) and predicted aircraft impact (from B2) and investigation of any cause-and-impact relationship.
4. Improved determinations of the line-shaped contrail coverage and properties of lineshaped contrails over many regions of the Earth.
5. Soot experiments investigating the impact of soot on cirrus in the atmosphere.

B1-B3 would be similar to the approach tried by Mannstein and Schumann [2005]. Instead of comparing cirrus coverage and air traffic data over Europe, areas need to be selected where air traffic induces an observable change. The observations (B1) would be based on METEOSAT (MSG) cirrus observations; in a first step cirrus cover is used as observable; in a second step radiances can be employed additionally [Krebs *et al.*, 2007; Mannstein and Schumann, 2005]. The simple model (B2) simulates contrail coverage along aircraft flight paths as a function of contrail age with a few free parameters (e.g., contrail lifetime) [Mannstein and Schumann, 2005]. From correlating these results (B3), the amount and radiance contributions of aviation-induced cirrus changes are determined including best fitting model parameters. In a next step, one might also correlate a more advanced contrail prediction scheme (driven with meteorological analysis data) to observations to determine further AIC parameters.

B4 would make use of a generalized (i.e. used with various sensors) version of the automated satellite-based detection algorithm for line-shaped contrails [Mannstein *et al.*, 1999], which was applied by several groups using AVHRR data over Europe [Meyer *et al.* 2002], the continental USA [Duda *et al.*, 2004; Palikonda *et al.*, 2005], eastern north Pacific [Minnis *et al.*, 2005] and southeast and east Asia [Meyer *et al.*, 2007]. The method should be applied to AVHRR, MODIS, A(A)TSR, MSG and GOES data, the latter in geostationary orbits allowing for nearly continuous observation at the expense of the high resolution of the polar orbiters.

B5 would be similar to the SUCCESS [Toon and Miake-Lye, 1998] and the SULFUR experiments [Schumann *et al.*, 1996, 2002]. The experiment should allow tackling the sootcirrus issue. The in-situ experiment should be designed to demonstrate the ice-forming capability of aircraft soot emissions. Soot source can be either a dedicated soot generator, or a strongly sooting engine or a modern normal engine with typical soot properties but low soot emission amounts, depending on the measurement methods used to detect the soot source. Such a measurement should be performed first in relatively unpolluted air (perhaps in the Southern Hemisphere, Punta Arenas) because the background cirrus in flight corridors could already be affected by aviation soot. The soot should be emitted along with tracers marking the air mass. It might be advisable to investigate in addition the cirrus properties in regions with high soot loading from other sources (biomass burning, surface traffic sources, etc.) injected into the upper troposphere by convection or large-scale cyclonic events. However, this will require a far larger experimental set-up then the initial idea to follow the fate of soot emissions from a well defined source.

A. Impact

Modeling and Validation (A)

A1 and A2 would have a large impact on the reliability of contrail simulations. Until now contrails are simulated by models that have known biases and that have not been rigorously tested regarding the moisture and clouds in the upper troposphere. More model development and improvement is needed so that we can be more confident about contrail simulations.

A3 would give us an improved estimate of the uncertainty of existing estimates of contrail radiative forcing that may still be underestimated due to the fact that most estimates use only slight variations of the same method and largely identical data sources.

A4 would be a completely new approach and has therefore the ability to give us an independent estimate of contrail radiative forcing. Furthermore this approach would for the first time enable the estimation of the effect of contrail cirrus.

A5. Cloud overlap assumptions and radiative response are not yet adapted to contrails or/and the coexistence of contrails and clouds. But different cloud overlap assumptions and assumptions about particle size and habit have a strong impact on the radiative forcing estimates.

A6. This approach requires (i) a statistical model of cloud properties (frequency distribution of high level clouds of various optical thickness); (ii) a model study to understand the microphysical differences between contrails forming in cloud free air from contrails forming inside clouds; and (iii) radiative transfer calculations to determine the change in SW and LW radiative forcing values due to inserting a contrail into the high level cloud.

A7. Simulating scales from ~50 m (width of young contrails) to ~500 km (grid scale of global models) as a function of aircraft movements, aircraft emissions, altitude, ambient temperature (including stratification), humidity, vertical and horizontal wind (including rising motion and wind shear), turbulence, ambient aerosols, ambient clouds, requires special model development.

A8. Including the plume-based contrail model in a NWP model (such as the ECMWF-IFS) would allow comparison with individual (past and new, in situ and remote sensing) observations. Moreover the plume-based contrail model in the NWP can be used to predict contrail coverage at time scales needed for air traffic management to minimize the effect of contrails. Model results obtained with a GCM in climate mode could on the other hand be compared only to observations in a statistical sense except when nudging the GCM with observational fields.

A9 would support the development of contrail cirrus parameterizations or simulations.

Remote Sensing and in Situ Measurements (B)

The activity could provide upper and lower bounds on aviation impact on cloud changes. Furthermore activities B1 and B4 are critical for validating global model simulations of contrail cirrus A8, see outline below.

B. Ability to Improve Climate Impacts with Reduced Uncertainties

Uncertainty of radiative forcing due to contrails has not yet been properly estimated. Therefore research should not aim at reducing error bars but at developing independent approaches and using those approaches to estimate sensitivities to assumptions.

Modeling and Validation (A)

A1 might not reduce uncertainties of contrail radiative forcing unless the representation of supersaturation has been validated itself. Application using several different host GCMs is likely to increase uncertainties since contrail forcing estimates have until now been mainly calculated using a single model (ECHAM) or using the related (ECMWF model).

A2 might actually first lead to an increase of the uncertainty since more processes need to be represented or parameterized in models. Those processes need to be constrained and validated with observational data which are scarce.

A3 would not reduce the uncertainty but yield more reliable estimates of uncertainty.

A4 would give an independent estimate of linear contrail radiative forcing and therefore may increase the estimate of uncertainty. In the case of contrail cirrus this approach would give a first estimate and at the same time could be used to provide an estimate of uncertainty.

A5 might be able to decrease uncertainty due to providing a mean and variability of cloud optical properties since different assumptions in cloud optical properties were the main reason for different radiative forcing estimates.

A6 needs to be solved to exclude or confirm the potential of contrails-in-cirrus inducing cooling.

A7 and A8. The plume-based contrail model can be used to test simulations of supersaturation in NWP models by comparing predicted and observed contrail cirrus. Hence, the activity also contributes to improving global models and their ability to simulate climate impacts with reduced uncertainty and to determine strategies to reduce this climate impact.

A9 is crucial for reducing uncertainties in models.

Remote Sensing and in Situ Measurements (B)

Activity (B) would provide an observational basis for an assessment of future climate change due to aviation impact on cloud changes.

Improved and validated contrail and cirrus remote sensing analysis schemes are required to obtain data on cloud coverage, optical thickness, brightness temperature, reflectance, microphysical properties, contrail age, etc.

By correlating results from aircraft impact prediction tools with observed cirrus properties, insight on cause-and-impact relationships between air traffic and cirrus changes and constraints on important model parameters can be obtained.

A uniform approach to determine the line-shaped contrail coverage and properties of lineshaped contrails over many regions of the Earth would provide data from which the global amount of line-shaped contrail cover could be determined experimentally; moreover these results would be essential for model constraining and validation.

By measuring the properties of soot, cirrus and other aerosol behind a soot source, one learns about the change of soot with time and about the soot impact on cirrus. We believe that such an experiment is essential to tackle the soot-cirrus issue.

C. Practical Use

Results from (A) and (B) would contribute to the next available IPCC assessment of global climate change and for related ICAO activities.

A1, A2 and A5 are prerequisites for a microphysically consistent simulation of ice clouds and their optical properties in general. After development they contribute to a better estimation of the climatic impact of contrail cirrus only in conjunction with A3 and A4. A3 and A4 are based on existing methods and need validation A9. This activity will lead immediately to more realistic estimates of the radiative forcing of contrails and contrail cirrus and the associated uncertainty. A6 would reduce the uncertainty on the lower bound of the radiative forcing by contrails (relevant also for contrail-cirrus). A7 and A8 combined enable the validation of contrail simulations and therefore are not of immediate practical use. Using A7 and A8 for air traffic management on the other hand would be immediately useful.

B1 is crucial for validating contrail models. B2 and B3 would provide model-independent data on AIC. B3 provides one basis for validating global contrail models. B5 is crucial to understanding soot ageing and soot-cirrus interaction and for demonstrating a measurable impact of soot on cirrus.

D. Achievability

Many of the suggested subjects require cooperation of researchers across several fields including basic research. This underlines the need for cooperation beyond several institutes. In most of the subjects DLR (internally and with external partners) is already active. Those areas have additionally been indicated below in order to facilitate cooperation.

Modeling and Validation (A)

A1. Due to the availability of new satellite based data sets in the upper troposphere (e.g., AIRS) validation should now be possible. However, it must be recognized that remote sensing of humidity and clouds itself is fraught with significant uncertainties. A number of transport schemes for climate models have been developed that need to be implemented (if they aren't already) and validated in the upper troposphere. Cooperation in the field of remote sensing is necessary.

A2. Only recently supersaturation has been included in a few models [Tompkins *et al.*, 2007] but not always consistently with microphysics or cloud coverage parameterizations. This work should be continued [Kärcher and Burkhardt, 2008] and extended to include recent advances in ice nucleation microphysics [Hendricks *et al.*, 2005].

A3 is straightforward.

A4 requires expertise in both atmospheric dynamics and cloud microphysics. A process based parameterization needs to be consistent with the existing model cloud scheme. Therefore such a parameterization will vary depending on the host cloud scheme. The development, introduction and validation of such a scheme into the ECHAM GCM is currently followed by U. Burkhardt and B. Kärcher at DLR.

In A5 radiative transfer models and LES models can be used studying properties and radiative effects of individual contrails. LES modeling of contrail development and the transition into cirrus is performed by S. Unterstrasser and K. Gierens.

A6. The cloud properties may be derived from CALIPSO data. The study to understand the differences between contrails forming in cloud free air from contrails forming inside clouds can be performed with an LES-model [Unterstrasser *et al.*, 2008]. The radiative transfer calculations can be performed with existing tools [Meerkötter *et al.*, 1999]. A preliminary study has been started by U. Schumann and R. Meerkötter.

A7 and A8. The following ingredients for the plume-based contrail model are available: Gaussian plume models, meteorology from a NWP model, validation data (including MODIS, MSG observations, CALIPSO, Cloudsat, in situ data, LES model results), aircraft movement data base for periods for which MSG-data are available. Corresponding work has been started within the European Integrated Project QUANTIFY by K. Gierens, U. Schumann and QUANTIFY-partners.

A9. A large community is required to tackle validation issues, including validation of cloud and moisture variables retrieved by remote sensing via in-situ measurements and advanced cirrus modeling. In support of the latter, I. Sölch and B. Kärcher currently couple a multiscale LES model with a sophisticated aerosol-ice-radiation package to simulate cirrus by means of Lagrangian tracking, an approach opening up new ways of analyzing cirrus clouds in conjunction with field measurements.

Remote Sensing and in Situ Measurements (B)

B1: A good basis is the method MeCiDA developed at DLR because of its suitability for geostationary satellites and all day and night times. So far MeCiDA has been used to derive cirrus coverage over Europe and the North Atlantic for a complete year.

B2: This requires input in terms of actual aircraft movements. A dataset is needed including 3D position vectors as a function of time along the flight paths for each aircraft. The type of aircraft and engine has to be known for emission estimates. The data should be available for the region covered by geostationary satellites (i.e., Europe, North Atlantic, Eastern North America) and should be available for the time periods for which satellite observations are being performed. Unless better data get available, the use the global data set from AERO2K for the year 2002 is recommended, or special data sets provided for smaller regions e.g., by EUROCONTROL (Europe and Atlantic, year 2004) and DFS (Germany, Sept. 2002).

B3: Limited experience exists in correlating observed and predicted contrail cover. Since results of correlation analyses are easily misinterpreted regions have to be selected where only the aircraft impact is relevant. Alternatively modeling is needed to discriminate between aircraft impact and other reasons for cloud changes. Presently, K. Graf, H. Mannstein, B. Mayer and U. Schumann at DLR are working on this topic.

B4 requires the application of the algorithm of Mannstein *et al.* [1999] to as many remote sensing data sets as possible covering a large part of the globe, with quantifiable and comparable accuracy.

B5 would make use of a suitable soot source (the source could be a normal aircraft engine, but the plume soot particles should be easily traceable for at least hours and hundreds of kilometers) and at least one research aircraft measuring aerosol and cirrus properties. The measurement should be performed in relatively unpolluted air because the background cirrus in flight corridors could already be affected by aviation soot. The soot should be emitted along with tracers marking the air mass. To overcome possible difficulties in interpretation, the project needs to be supported by proper model activities, addressing, e.g., dynamical effects that can mask aerosol-induced cirrus changes and the impact of IN from other sources such as mineral dust. The experiment could be performed with or including the new High Altitude and Long Range Research Aircraft (HALO) research aircraft, which should become operational in summer 2009. A first demonstration mission CIRRUS-ML is being prepared under the coordination of DLR. HALO will be equipped with a powerful set of aerosol and cirrus instruments. HALO will also be available for emission and identification of a passive tracer gas (H. Schlager and others). A laboratory-style aircraft engine soot generator has been developed at DLR Stuttgart; its use for airborne applications could be studied. The experiment can also be performed with US-aircraft (DC-8, HIAPER) or Russian aircraft. Cooperation with the Atmospheric Soot Network (http://www.asn.u-bordeaux.fr) on this topic would also be possible.

E. Estimated Cost

Modeling and Validation (A)

Costs are determined by individual salaries of experienced research scientists (timelines are suggested in section 4.f) and the use of observational tools needed for validation purposes. Computing costs should also be considered.

Remote Sensing and in Situ Measurements (B)

Cost besides salaries include those for obtaining and evaluating satellite data and aircraft movement data bases as well as designing and carrying out a large-scale field campaign including personnel preferably in the southern hemisphere including the development of a proper soot source.

F. Timeline

The necessary research can be performed within the time frame associated with projected doubling of air traffic, as estimated below.

Modeling and Validation (A)

A1 and A9. Model development and validation is an ongoing process and makes progress when new data sources become available. It is usually required that both modelers and data teams work closely together. It is difficult to associate a timeline because the amount of dedicated work depends on the type of model improvement and validation parameter and progress in verifying retrievals.

A2. Development of a theoretical ice nucleation scheme that describes physically based ice particle formation in cirrus requires at least one year of work of an experienced research scientist (1 PY). Its implementation in a climate model and thorough testing requires ~2 PY. Achieving consistency between microphysics and cloud coverage in the model is even more time-consuming. We estimate 1 PY to develop a consistent cirrus cloud scheme and ~3 PY for implementation and validation depending on the original model's cloud scheme. Adapting to an improved radiation scheme would be a significant additional effort (2-3 PY).

A3 would require few months work testing the impact of one parameter change.

A4 and A5 would require ~2 PY each, covering the design and development of the parameterization (A4) and performing detailed contrail studies as a basis for upgrading radiation parameterizations (A5).

A6. A preliminary study can be performed within a few months time.

A7 and A8. the initial model development until demonstration of the feasibility and first validation results requires ~2 PY for 3 years, plus support by the NWP team, and the team providing input in terms of observation data and aircraft movement data base.

Remote Sensing and in Situ Measurements (B)

B1, B2 and B3 would require funding of at least ~3 PY for 3 years.

B4 requires cooperation of teams working in the field of contrail detection, and access to all relevant satellite data around the globe. The initial phase would be devoted to a careful comparison and adjustment of the detection algorithm. Thereafter, a large set of data would be processed. This expensive task may require ~10 PY within a 3 year period.

B5 may require 2 PY to develop an appropriate soot source and 5 PY for experiment and analyses. Parts of this work can be done in parallel.

5. RECOMMENDATIONS

Pure literature research or compilations of existing knowledge is not going to advance science any further. There are definite gaps of understanding (see section 3) that need to be addressed before any more definite conclusions about climate forcing of contrails can be drawn. Methods exist that could be applied gaining e.g. a homogeneous data base of contrail properties from remote sensing. Progress in simulating climate forcing due to contrails requires considerable effort developing new concepts.

A. Options

Options depend strongly on the amount of funding and support available. Activities (A) and (B) can be carried out simultaneously. They both offer large advances in understanding and potentially lead to significant progress within 3-5 years. However, problems are highly complex so that final conclusions cannot be drawn in such a short period. With the proper timing, this research may contribute considerably to the upcoming (fifth) IPCC report or a second dedicated IPCC aviation assessment.

Modeling and Validation (A)

To make headway in evaluating the climatic impact of contrails and contrail cirrus, we recommend concentrating efforts on both, climate and radiative transfer modeling and on improving the data basis needed for validating those models. On the one hand a combination of remote sensing, along the lines explained in section 4, and in situ measurements would be useful and on the other hand LES and simple modeling in order to provide validation data sets or enhance process understanding.

Without new concepts in global modeling, no true progress estimating the climate impact of contrails will be made. Physically-based parameterizations describing microphysical and optical properties of contrail cirrus need to get developed and realized in different global models to ensure independent estimates.

On the long term, treating supersaturation, contrail cirrus, soot cirrus and natural cirrus consistently, global models will be able to provide more robust predictions of radiative forcing with reduced uncertainty.

A large effort needs to be put into obtaining validation data sets in order to constrain global model parameterizations. The data sets must be exactly characterized by the thresholds of the observational tools in order to enable a direct comparison with global model output.

Observations (see below) and global modeling should be accompanied by modeling of contrail cirrus on the cloud scale and by radiative transfer simulations. Together they benefit model development and remote sensing alike by providing or depending our understanding of processes.

With a range of matured climate models, we finally recommend to carry out IPCC-type assessment simulations focusing on the contrail climate impact. To this end, emission scenarios need to be employed that capture the most recent estimates of future air traffic and climate change parameters.

Remote Sensing and in Situ Measurements (B)

Remote sensing provides regional statistics of alterations of coverage and contrail optical properties. Analysis tools (such as those developed at DLR) should be applied to global observations yielding homogeneous data sets. Those could also be used in conjunction with improved methods in order to investigate possible correlations between air traffic and high cloudiness changes quantifying the aircraft-induced component. Activity B4 requires the application of an automated detection algorithm (such as that one developed at DLR) to different satellite sensors.

In parallel, we recommend to carry out in-situ measurements, preferably of old contrail cirrus. Such measurements must be carefully designed and supported by on-line meteorological analyses to enable probing of contrails in later stages of their life cycle. Again, remote sensing including Lidar can be employed in support of this goal by locating and tracking individual contrails and guiding the aircraft experimenters. In-situ measurements should cover both, microphysics and radiation, ideally using a number of research aircraft at the same time. Those measurements should also address the soot impact on cirrus.

The activity B5 requires the characterization of the soot source, the knowledge of the exact position of the aircraft and measurements of the undisturbed meteorology.

B. Supporting Rationale

The rationale behind our recommendation is that one approach alone or several approaches in isolation are insufficient to improve the current state of knowledge. Only when all options noted above are tied together can significant progress be made and uncertainties reduced.

C. How to Best Integrate Best Available Options

A 10 year research plan, organized in two steps, should suffice to address the most pressing issues raised in this SSWP. Research must be closely coordinated with the scientific community interested in upper tropospheric / lower stratospheric transport, chemistry and aerosol and cloud physics. Moreover, the research should be embedded in general climate and climate mitigation research activities. The design and performance of a large-scale measurement campaign must involve experimentalists, modelers and theoreticians alike. Coordinated model assessments of aviation-induced climate change could take place in an early stage after about 3 years and at the end of the research project. Funding must be large enough to integrate the international science community and to enable several independent approaches.

We recommend an intense cooperation between the US-agencies (FAA, NASA, NSF) with European agencies (DLR, EU, DFG). We also recommend an intense cooperation between research-oriented teams and agencies or companies having access to details on air traffic (e.g. EUROCONTROL, FAA, ICAO), and engine emissions. For direct access to meteorological fields inclusion of teams from the leading weather services may be helpful.

For the purpose of maximum acceptance and maximum use of existing knowledge, we recommend performing these projects in an environment of open information exchange and open participation. The classical "Virginia Beach" meetings as in 1992-1997 should be revived.

6. Summary

A number of issues were identified indicating pressing research need regarding better validation data sets and climate model improvements. Long term efforts are required both in observations and modeling, developing new process parameterizations for ice clouds and their radiative effects, since model improvements are interdependent. Nevertheless, improvements building on the current state of cloud parameterizations in climate models could also lead to significant progress in understanding the aviation impact on climate at a shorter time scale.

References

Appleman, H., 1953: The formation of exhaust condensation trails by jet aircraft. *Bull. Amer. Meteorol. Soc. 34*, 14-20.

Atlas, D., Z. Wang and D.P. Duda, 2006: Contrails to cirrus – Morphology, microphysics, and radiative properties. *J. Appl. Meteorol. Clim. 45*, 5-19.

Bakan, S., M. Betancor, V. Gayler and H. Graßl, 1994: Contrail frequency over Europe from NOAA satellite images. *Ann. Geophys. 12,* 962-968.

Betancor-Gothe, M. and H. Graßl, 1993: Satellite remote sensing of the optical depth and mean crystal size of thin cirrus and contrails. *Theor. Appl. Clim. 48*, 101-113.

Boin, M. and L. Levkov, 1994: A numerical study of contrail development. *Ann. Geophys. 12*, 969-978.

Boucher, O., 1999: Air traffic may increase cirrus cloudiness. *Nature 397,* 30-31.

Brewer, A.W., 1946: Condensation trails. *Weather 1*, 34-40.

Brown, R.C., R.C. Miake-Lye, M.R. Anderson and C.E. Kolb, 1997: Aircraft sulphur emissions and the formation of visible contrails. *Geophys. Res. Lett. 24,* 385-388.

Busen, R. and U. Schumann, 1995: Visible contrail formation from fuels with different sulphur contents. *Geophys. Res. Lett. 22*, 1357-1360.

Carleton, A.M. and P.J. Lamb, 1986: Jet contrails and cirrus clouds: A feasibility study employing high resolution satellite imagery. *Bull. Amer. Meteorol. Soc. 67*, 301-309.

Carleton, A.M., D. J. Travis, K. Master and S. Vezhapparambu, 2007: Composite atmospheric environments of jet contrail outbreaks for the United States. *J. Appl. Meteorol. Climatol.,* accepted.

Changnon, S.A. Jr., 1981: Midwestern cloud, sunshine and temperature trends since 1901 – Possible evidence of jet contrail effects. *J. Appl. Meteorol. 20*, 496-508.

Chen, T., W.B. Rossow and Y. Zhang, 2000: Radiative effects of cloud-type variations. *J. Clim. 13*, 264-286.

Chen, J.-P. and R.-F. Lin, 2001: Numerical simulation of contrail microphysical and radiative properties. *TAO 12,* 137-154.

Chen, J.-P., W.-H. Lin and R.-F. Lin, 2001: Estimation of contrail frequency and radiative effects over the Taiwan area. *TAO 12*, 155-178.

Chlond, A., 1998: Large eddy simulations of contrails. *J. Atmos. Sci. 55,* 796-819. DeGrand, J.Q., A.M. Carleton, D.J. Travis and P.J. Lamb, 2000: A satellite-based climatic description of jet aircraft contrails and associations with atmospheric conditions 1977-79. *J. Appl. Meteorol. 39*, 1434-1459.

DelGenio, A.D, 2002: GCM Simulations of cirrus for climate studies; In: Cirrus. D.K. Lynch, K. Sassen, D. O'C. Starr and G. Stephens (Eds.), pp. 310–326. New York, Oxford University Press.

Del Guasta, M. and K. Niranjan, 2001: Observation of low depolarization contrails at Florence (Italy) using a 532-1064 nm polarization Lidar. *Geophys. Res. Lett. 28,* 4067-4070.

Detwiler, A. and R. Pratt, 1984: Clear-air seeding: Opportunities and strategies. *J. Wea. Mod. 16*, 46-60.

Detwiler, A. and A. Jackson, 2002: Contrail formation and propulsion efficiency. *J. Aircr. 39*, 638-644.

Dietmüller, S., M. Ponater, R. Sausen, K.P. Hoinka and S. Pechtl, 2007: Contrails, natural cirrus, and diurnal temperature range. *J. Clim.,* submitted.

Dobbie, S. and P.R. Jonas, 2001: Radiative influences on the structure and lifetime of cirrus clouds. *Q. J. R. Meteorol. Soc. 127*, 2663-2682.

Dowling, D.R. and L.F. Radke, 1990: A summary of the physical properties of cirrus clouds. *J. Appl. Met. 29*, 970-978.

Dürbeck, T. and T. Gerz, 1996: Dispersion of aircraft exhausts in the free atmosphere, *J. Geophys. Res. 101*, 26,007-26,016.

Duda, D.P. and J.D. Spinhirne, 1996: Split-window retrieval of particle size and optical depth in contrails located above horizontally inhomogeneous clouds. *Geophys. Res. Lett. 23*, 3711-3714.

Duda, D.P., P. Minnis and L. Nguyen, 2001: Estimates of cloud radiative forcing in contrail clusters using GOES imagery. *J. Geophys. Res. 106,* 4927-4937.

Duda, D.P., P. Minnis, L. Nguyen and R. Palikonda, 2004: A case study of the development of contrail clusters over the Great Lakes. *J. Atmos. Sci. 61,* 1132-1146.

Duda, D.P., P. Minnis and R. Palikonda, 2005: Estimated contrail frequency and coverage over the contiguous United States from numerical weather prediction analyses and flight track data. *Meteorol. Z. 14*, 537-548.

Ekström, M., P. Eriksson, B. Rydberg and D.P. Murtagh, 2007: First Odin sub-mm retrievals in the tropical upper troposphere: humidity and cloud ice signals. *Atmos. Chem. Phys.* 7, 459-469.

Eleftheratos, K., C.S. Zerefos, P. Zanis, D.S. Balis, G. Tselioudis, K. Gierens and R. Sausen, 2007: A study on natural and manmade global interannual fluctuations of cirrus cloud cover for the period 1984-2004. *Atmos. Chem. Phys. 7*, 2631-2642.

Fabian, P. and B. Kärcher, 1997: The impact of aviation upon the atmosphere. *Phys. Chem. Earth 22*, 503-598.

Fahey, D.W., U. Schumann, S. Ackerman, P. Artaxo, O. Boucher, M.Y. Danilin, B. Kärcher, P. Minnis, T. Nakajima and O.B. Toon, 1999: Aviation-produced aerosols and cloudiness. In: Aviation and the Global Atmosphere. A Special Report of IPCC Working Groups I and III [Penner, J.E., D.H. Lister, D.J. Griggs, D.J. Dokken and M. McFarland (eds.)]. Cambridge University Press, New York, USA.

Febvre, G., J.-F. Gayet, B. Kärcher, A. Minikin, V. Shcherbakov, O. Jourdan, U. Schumann R. Busen, H. Schlager and M. Fiebig, 2008: On optical and microphysical characteristics of contrails and cirrus. *J. Geophys. Res.,* in preparation.

Fichter, C., S. Marquart, R. Sausen and D.S. Lee, 2005: The impact of cruise altitude on contrails and related radiative forcing. *Meteorol. Z. 14*, 563-572.

Forster, P., V. Ramaswamy, P. Artaxo, T. Berntsen, R. Betts, D.W. Fahey, J. Haywood, J. Lean, D.C. Lowe, G. Myhre, J. Nganga, R. Prinn, G. Raga, M. Schulz and R. Van Dorland, 2007: Changes in atmospheric constituents and in radiative forcing. *In: Climate Change 2007: The physical science basis. Contribution of Working Group 1 to the 4th Assessment Report of the Intergovernmental Panel on Climate Change* [Solomon, S., D. Qin, M. Manning, Z. Chen, M. Marquis, K.B. Averyt, M. Tignor and H.L. Miller (eds.)]. Cambridge University Press, Cambridge, UK, and New York, USA.

Fortuin, J.P.F., R. Van Dorland, W.M.F. Wauben and H. Kelder, 1995 : Greenhouse effects of aircraft emissions as calculated by a radiative transfer model. *Ann. Geophys. 13*, 413-418.

Freudenthaler, V., F. Homburg and H. Jäger, 1995: Contrail observations by ground-based scanning Lidar: Cross-sectional growth. *Geophys. Res. Lett. 22*, 3501-3504.

Freudenthaler, V., F. Homburg and H. Jäger, 1996: Optical parameters of contrails from Lidar measurements: Linear depolarization. *Geophys. Res. Lett. 23,* 3715-3718.

Fu, Q. and K.N. Liou, 1993: Parameterization of the radiative properties of cirrus clouds. *J. Atmos. Sci. 50*, 2008-2025.

Fu, Q., B. Carlin and G. Mace, 2000: Cirrus horizontal inhomogeneity and OLR bias. *Geophys. Res. Lett, 27*, 3341-3344.

Garber, D.P., P. Minnis and K.P. Costulis, 2005: A commercial flight track database for upper tropospheric aircraft emission studies over the USA and southern Canada. *Meteorol. Z. 14*, 445-452.

Gayet, J.-F., G. Febvre, G. Brogniez and H. Chepfer, W. Renger and P. Wendling, 1996: Microphysical and optical properties of cirrus and contrails. *J. Atmos. Sci. 53,* 126-138.

Gerz, T., T. Dürbeck and P. Konopka, 1998: Transport and effective diffusion of aircraft emissions. *J. Geophys. Res. 103*, 25,905-25,913.

Gettelman, A. and D.E. Kinnison, 2007: The global impact of supersaturation in a coupled chemistry-climate model. *Atmos. Chem. Phys. 7,* 1629-1643.

Gettelman, A., E.J. Fetzer, A. Eldering and F.W. Irion, 2006: The global distribution of supersaturation in the upper troposphere from the Atmospheric Infrared Sounder, *J. Clim. 19,* 6089-6103.

Gierens, K., 1994: The influence of radiation on the diffusional growth of ice crystals. *Beitr. Phys. Atmos. 67*, 181-193.

Gierens, K., 1996: Numerical simulations of persistent contrails. *J. Atmos. Sci. 53,* 3333-3348.

Gierens, K., 1998: How the sky gets covered with condensation trails. *Meteorol. Z. 7*, 181-187.

Gierens, K. and U. Schumann, 1996: Colors of contrails from fuels with different sulphur contents. *J. Geophys. Res. 101*, 16,731-16,736.

Gierens, K. and E. Jensen, 1998: A numerical study of the contrail-to-cirrus transition. *Geophys. Res. Lett., 25*, 4341-4344.

Gierens, K., U. Schumann, M. Helten, H. Smit and A. Marenco, 1999a: A distribution law for relative humidity in the upper troposphere and lower stratosphere derived from three years of MOZAIC measurements. *Ann. Geophys. 17*, 1218-1226.

Gierens, K., R. Sausen and U. Schumann, 1999b: A diagnostic study of the global distribution of contrails, Part II: Future air traffic scenarios. *Theor. Appl. Climatol. 63*, 1-9.

Gierens, K., R. Kohlhepp, N. Dotzek and H.G.J. Smit, 2007: Instantaneous fluctuations of temperature and moisture in the upper troposphere and tropopause region. Part 1: Probability densities and their variability. *Meteorol Z. 16,* 221-231.

Goodman, J., R.F. Pueschel, E.J. Jensen, S. Verna, G.V. Ferry, S.D. Howard, S.A. Kinne and D. Baumgardner, 1998: Shape and size of contrail ice particles. *Geophys. Res. Lett. 25,* 1327-1330.

Gounou, A. and R.J. Hogan, 2007: A sensitivity study of the effect of horizontal photon transport on the radiative forcing of contrails. *J. Atmos. Sci. 64,* 1706-1716.

Graßl, H., 1990: Possible climatic effects of contrails and additional water vapor. In: Air Traffic and the Environment – Background, Tendencies and Potential Global Atmospheric Effects. U. Schumann (ed.), Lecture Notes in Engineering, *Springer Berlin,* 124-137.

Green, J.E., 2005: Future aircraft – greener by design? *Meteorol. Z. 14,* 583-590.

Haag, W. and B. Kärcher, 2004: The impact of aerosols and gravity waves on cirrus clouds at midlatitudes. *J. Geophys. Res. 109,* D12202, doi:10.1029/2004JD004579.

Hansen, J., M. Sato, R. Ruedy, L. Nazarenko, A. Lacis, G.A. Schmidt, G. Russell, I. Aleinov, M. Bauer, S. Bauer, N. Bell, B. Cairns, V. Canuto, M. Chandler, Y. Cheng, A. DelGenio, G. Faluvegi, E. Fleming, A. Friend, T. Hall, C. Jackman, M. Kelley, N. Kiang, D. Koch, J. Lean, J. Lerner, K. Lo, S. Menon, R. Miller, P. Minnis, T. Novakov, V. Oinas, J. Perlwitz, J. Perlwitz, D. Rind, A. Romanou, D. Shindell, P. Stone, S. Sun, N. Tausnev, D. Thresher, B. Wielicki, T. Wong, M. Yao and S. Zhang, 2005: Efficacy of climate forcings, *J. Geophys. Res. 110,* D18104, doi:10.1029/2005JD005776.

Hartmann, D.L., M.E. Ockert-Bell and M.L. Michelsen, 1992: The effect of cloud type on the Earth's energy balance: Global analysis. *J. Clim. 5*, 1281-1304.

Hendricks, J., B. Kärcher, U. Lohmann and M. Ponater, 2005: Do aircraft black carbon emissions affect cirrus clouds on the global scale? *Geophys. Res. Lett. 32*, L12814, doi:10.1029/2005GL022740.

Heymsfield, A.J., R.P. Lawson and G.W. Sachse, 1998: Growth of ice crystals in a precipitating contrail. *Geophys. Res. Lett. 25,* 1335-1338.

ICAO, 2007: Environmental Report 2007. Environmental Unit of the International Civil Aviation Organization, 231pp.

Immler, F. and O. Schrems, 2002: LIDAR measurements of cirrus clouds in the northern and southern midlatitudes during INCA (55oN, 53oS): A comparative study. *Geophys. Res. Lett. 29,* 1809, doi:10.1029/2002GL015077.

Immler, F., R. Treffeisen, D. Engelbart, K. Krüger and O. Schrems, 2007: Cirrus, contrails, and ice supersaturated regions in high pressure systems at northern mid latitudes. *Atmos. Chem. Phys. Discuss. 7,* 13,175-13,201.

Jäger, H., V. Freudenthaler and F. Homburg, 1998: Remote sensing of optical depth of aerosols and clouds related to air traffic. *Atmos. Environ. 32,* 3123-3127.

Jakob, C., 2002: Ice clouds in numerical weather prediction models–progress, problems and prospects. Cirrus; In: Cirrus. D.K. Lynch, K. Sassen, D. O'C. Starr and G. Stephens (Eds.), *Oxford University Press*, New York, 327-345.

Jensen, E.J., A.S. Ackerman, D.E. Stevens, O.B. Toon and P. Minnis, 1998a: Spreading and growth of contrails in a sheared environment. *J. Geophys. Res., 103*, 13,557-13,567.

Jensen, E. J., O.B. Toon, S. Kinne, G.W. Sachse, B.E. Anderson, K.R. Chan, C.H. Twohy, B. Gandrud, A. Heymsfield and R.C. Miake-Lye, 1998b: Environmental conditions required for contrail formation and persistence. *J. Geophys. Res.*, *103*, 3929-3936.

Jensen, E.J., O.B. Toon, S.A. Vay, J. Ovarlez, R. May, P. Bui, C.H. Twohy, B. Gandrud, R.F. Pueschel and U. Schumann, 2001: Prevalence of ice-supersaturated regions in the upper troposphere: Implications for optically thin ice cloud formation. *J. Geophys. Res. 106*, 17,253-17,266.

Jensen, E.J., J.B. Smith, L. Pfister, J.V. Pittman, E.M. Weinstock, D.S. Sayres, R.L. Herman, R.F. Troy, K. Rosenlof, T.L. Thompson, A.M. Fridlind, P.K. Hudson, D.J. Cziczo, A.J. Heymsfield, C. Schmitt and J.C. Wilson, 2005: Ice supersaturations exceeding 100% at the cold tropical tropopause: Implications for cirrus formation and dehydration. *Atmos. Chem. Phys. 5*, 851-862.

John, V.O. and B.J. Soden, 2007: Temperature and humidity biases in global climate models and their impact on climate feedbacks. *Geophys. Res. Lett. 34*, L18704, doi:10.1029/2007GL030429.

Joseph, J.H., Z. Levin, Y. Mekler, G. Ohring and J. Otterman, 1975: Study of contrails observed from ERTS I satellite imagery. *J. Geophys. Res. 80*, 366-372.

Kärcher, B., 1996: Aircraft-generated aerosols and visible contrails. *Geophys. Res. Lett. 23*, 1933-1936.

Kärcher, B., 1998: Physicochemistry of aircraft-generated liquid aerosols, soot, and ice particles: 1. Model description. *J. Geophys. Res., 103*, 17,111-17,128.

Kärcher, B., Th. Peter and R. Ottmann, 1995: Contrail formation: Homogeneous nucleation of H2SO4 / H_2O droplets. *Geophys. Res. Lett. 22,* 1501-1504.

Kärcher, B., Th. Peter, U.M. Biermann and U. Schumann, 1996: The initial composition of jet condensation trails. *J. Atmos. Sci. 53,* 3066-3083.

Kärcher, B., R. Busen, A. Petzold, F.P. Schröder, U. Schumann and E.J. Jensen, 1998: Physicochemistry of aircraft generated liquid aerosols, soot, and ice particles, 2. Comparison with observations and sensitivity studies. *J. Geophys. Res. 103*, 17,129-17,148.

Kärcher, B. and J. Ström, 2003: The roles of dynamical variability and aerosols in cirrus cloud formation. *Atmos. Chem. Phys. 3,* 823-838.

Kärcher, B., J. Hendricks and U. Lohmann, 2006: Physically-based parameterization of cirrus cloud formation for use in global atmospheric models. *J. Geophys. Res. 111,* D01205, doi:10.1029/2005JD006219.

Kärcher, B., O. Möhler, P.J. DeMott, S. Pechtl and F. Yu, 2007: Insights into the role of soot aerosols in cirrus cloud formation. *Atmos. Chem. Phys. 7,* 4203-4227.

Kärcher, B. and U. Burkhardt, 2008: Prediction of cirrus clouds in general circulation models. *Q. J. R. Meteorol. Soc.*, submitted.

Kästner, M., K.T. Kriebel, W. Renger, G.H. Ruppersberg and P. Wendling, 1993: Comparison of cirrus height and optical depth derived from satellite and aircraft measurements. *Mon. Wea. Rev. 121*, 2708-2717.

Kästner, M., R. Meyer and P.Wendling, 1999: Influence of weather conditions on the distribution of persistent contrails. *Meteorol. Appl. 6*, 261-271.

Kalkstein, A.J. and R.C. Balling Jr., 2004: Impact of unusually clear weather on United States daily temperature range following 9/11/2001. *Climate Res. 26*, 1-4.

Khvorostyanov, V.I. and K. Sassen, 1998: Cloud model simulation of a contrail case study: Surface cooling against upper tropospheric warming. *Geophys. Res. Lett. 25,* 2145- 2148.

Kley, D., J.M. Russell III and C. Phillips, 2000: SPARC Assessment of upper tropospheric and stratospheric water vapor. WCRP-113, WMO/TD-No.1043, SPARC Report No.2, 312pp.

Knollenberg, R.G., 1972: Measurements of the growth of the ice budget in a persisting contrail. *J. Atmos. Sci. 29,* 1367-1374.

Konrad, T.G. and J.C. Howard, 1974: Multiple contrail streamers observed by Radar. *J. Appl. Meteorol. 13*, 563-572.

Krebs, W., H. Mannstein, L. Bugliaro and B. Mayer, 2007: Technical note: A new day and night-time Meteosat Second Generation Cirrus Detection Algorithm MeCiDA. *Atmos. Chem. Phys. 7*, 6145-6159.

Kuhn, P.M., 1970: Airborne observations of contrail effects on the thermal radiation budget. *J. Atmos. Sci. 27,* 937-943.

Langford, A.O., R.W. Portmann, J.S. Daniel, H.L. Miller, C.S. Eubank, S. Solomon and E.G. Dutton, 2005: Retrieval of ice crystal effective diameters from ground-based nearinfrared spectra of optically thin cirrus. *J. Geophys. Res.* 110, D22201, doi:10.1029/2005JD005761.

Lawson, R.P., A.J. Heymsfield, S.M. Aulenbach and T.L. Jensen, 1998: Shapes, sizes and light scattering properties of ice crystals in cirrus and a persistent contrail during SUCCESS. *Geophys. Res. Lett. 25,* 1331-1334.

Lee, D.F., 1989: Jet contrail identification using the AVHRR split window. *J. Appl. Meteorol. 28*, 993-995.

Lee, D.S., P.E. Clare, J. Haywood, B. Kärcher, R.W. Lunnon, I. Pilling, A. Slingo, and J.R. Tilston, 2000: Identifying the uncertainties in radiative forcing of climate from aviation contrails and aviation-induced cirrus. Report DERA/AS/PTD/CR000103, 69pp.

Lewellen, D.C. and W.S. Lewellen, 1996: Large-eddy simulations of the vortex-pair breakup in aircraft wakes. *AIAA J. 34*, 2337-2345.

Lewellen, D.C. and W.S. Lewellen, 2001: The effects of aircraft wake dynamics on contrail development. *J. Atmos. Sci. 58,* 390-406.

Liou, K.N., 1986: Influence of cirrus clouds on weather and climate processes: A global perspective. *Mon. Wea. Rev. 114,* 1167-1199.

Liou, K.N., S.C. Ou and G. Koenig, 1990: An investigation of the climatic effect of contrail cirrus. In: Air Traffic and the Environment – Background, Tendencies and Potential Global Atmospheric Effects. U. Schumann (Ed.), Lecture Notes in Engineering, *Springer Berlin*, 154-169.

Liou, K.N., P. Yang, Y. Takano, K. Sassen, T. Charlock and W. Arnott, 1998: On the radiative properties of contrail cirrus. *Geophys. Res. Lett. 25,* 1161-1164.

Liu X., J.E. Penner, S.J. Ghan and M. Wang, 2007: Inclusion of ice microphysics in the NCAR community atmospheric model version 3 (CAM3). *J. Clim. 20,* 4526-4547.

Lohmann, U., P. Stier, C. Hoose, S. Ferrachat, S. Kloster, E. Roeckner and J. Zhang, 2007: Cloud microphysics and aerosol indirect effects in the global climate model ECHAM5-HAM. *Atmos. Chem. Phys. 7*, 3425-3446.

Lohmann, U. and B. Kärcher, 2002: First interactive simulations of cirrus clouds formed by homogeneous freezing in the ECHAM GCM. *J. Geophys. Res. 107*, 4105, doi:10.1029/2001JD000767.

Mannstein, H. and U. Schumann, 2005: Aircraft induced contrail cirrus over Europe. *Meteorol. Z. 14*, 549-554.

Mannstein, H. and U. Schumann, 2007: Corrigendum. *Meteorol. Z. 16*, 131-132.

Mannstein, H., R. Meyer and P. Wendling, 1999: Operational detection of contrails from NOAA AVHRR-data. *Int. J. Remote Sens. 20,* 1641-1660.

Marquart, S., R. Sausen, M. Ponater and V. Grewe, 2001: Estimate of the climate impact of cryoplanes. *Aerospace Sci. Technol. 5,* 73-84.

Marquart, S. and B. Mayer, 2002: Towards a reliable GCM estimation of contrail forcing. *Geophys. Res. Lett. 29*, 1179-1182.

Marquart, S., M. Ponater, F. Mager and R. Sausen, 2003: Future development of contrail cover, optical depth, and radiative forcing: Impacts of increasing air traffic and climate change. *J. Clim. 16*, 2890-2904.

Marquart, S., M. Ponater, L. Ström and K. Gierens, 2005: An upgraded estimate of the radiative forcing of cryoplane contrails. *Meteorol. Z. 14,* 573-582.

Meerkötter, R., U. Schumann, D.R. Doelling, P. Minnis, T. Nakajima and Y. Tsushima, 1999: Radiative forcing by contrails. *Ann. Geophys. 17,* 1080-1094.

Meyer, R., H. Mannstein, R. Meerkötter, U. Schumann and P. Wendling, 2002: Regional radiative forcing by line-shaped contrails derived from satellite data. *J. Geophys. Res. 107*, 4104, doi:10.1029/2001JD000426.

Meyer, R., R. Buell, C. Leiter, H. Mannstein, S. Pechtl, T. Oki and P. Wendling, 2007: Contrail observations over Southern and Eastern Asia in NOAA/AVHRR data and comparisons to contrail simulations in a GCM. *Int. J. Remote Sens. 28*, 2049-2069.

Miloshevich, L.M., H. Vömel, A. Paukkunen, A.J. Heymsfield and S.J. Oltmans, 2001: Characterization and correction of relative humidity measurements from Väisälä RS80- A radiosondes at cold temperatures. *J. Atmos. Oceanic. Technol. 18*, 135-156.

Minnis, P., 2003: Contrails. In: Encyclopedia of Atmospheric Sciences, J. Holton, J. Pyle, and J. Curry (Eds.), *Academic Press London,* 509-520.

Minnis, P., 2005: Response to comment on "Contrails, cirrus trends, and climate". *J. Clim. 18*, 2783-2784.

Minnis, P., D.F. Young, D.P. Garber, L. Nguyen, W.L. Smith Jr. and R. Palikonda, 1998: Transformation of contrails into cirrus during SUCCESS. *Geophys. Res. Lett. 25,* 1157-1160.

Minnis, P., U. Schumann, D.R. Doelling, K. Gierens and D.W. Fahey, 1999: Global distribution of contrail radiative forcing. *Geophys. Res. Lett. 26,* 1853-1856.

Minnis, P., J.K. Ayers, M.L. Nordeen and S.P. Weaver, 2003: Contrail frequency over the United States from surface observations. *J. Clim. 16*, 3447-3462.

Minnis, P., J.K. Ayers, R. Palikonda and D. Phan, 2004: Contrails, cirrus trends, and climate. *J. Clim. 17,* 1671-1685.

Minnis, P., R. Palikonda, B.J. Walter, J.K. Ayers and H. Mannstein, 2005: Contrail properties over the eastern North Pacific from AVHRR data. *Meteorol. Z. 14,* 515-523.

Myhre, G. and F. Stordal, 2001: On the tradeoff of the solar and thermal infrared impact of contrails. *Geophys. Res. Lett. 28,* 3119-3122.

Nagel, D., U. Leiterer, H. Dier, A. Kats, J. Reichard and A. Behrend, 2001: High accuracy humidity measurements using the standardized frequency method with a research upper-air sounding system. *Meteorol. Z. 10*, 395-405.

Palikonda, R., P. Minnis, D.P. Duda and H. Mannstein, 2005: Contrail coverage derived from 2001 AVHRR data over the continental United States of America and surrounding areas. *Meteorol. Z. 14,* 525-536.

Paoli, R., J. Hélie and T. Poinsot, 2004: Contrail formation in aircraft wakes. *J. Fluid Mech. 502*, 361-373.

Parungo, F., 1995: Ice crystals in high clouds and contrails. *Atmos. Res. 38,* 249-262.

Penner, J.E., D.H Lister, D.J. Griggs, D.J. Dokken and M. McFarland, 1999: Aviation and the global atmosphere – A special report of IPCC working groups I and III. Intergovernmental Panel on Climate Change. *Cambridge University Press,* 365pp.

Peter, Th., C. Marcolli, P. Spichtinger, T. Corti, M.B. Baker and T. Koop, 2006: When dry air is too humid. *Science 314*, 1399-1402.

Petzold, A., R. Busen, F.P. Schröder, R. Baumann, M. Kuhn, J. Ström, D.E. Hagen, P.D. Whitefield, D. Baumgardner, F. Arnold, S. Borrmann and U. Schumann, 1997: Nearfield measurements on contrail properties from fuels with different sulfur content. *J. Geophys. Res. 102*, 29,867-29,880.

Petzold, A., J. Ström, S. Ohlsson and F.P. Schröder, 1998: Elemental composition and morphology of ice-crystal residual particles in cirrus clouds and contrails. *Atmos. Res. 49*, 21-34.

Plass, G.N., G.W. Kattawar and F.E. Catchings, 1973: Matrix operator theory of radiative transfer. *Appl. Opt. 12*, 314-329.

Platt, C.M.R. 1981: The effect of cirrus of varying optical depth on the extraterrestrial net radiative flux. *Q. J. R. Meteorol. Soc. 107*, 671-678.

Poellot, M.R., W.P. Arnott and J. Hallett, 1999: In-situ observations of contrail microphysics and implications for their radiative impact. *J. Geophys. Res. 104*, 12,077-12,084.

Ponater, M., S. Brinkop, R. Sausen and U. Schumann, 1996: Simulating the global atmospheric response to aircraft water vapor emissions and contrails: A first approach using a GCM. *Ann. Geophys. 14*, 941-960.

Ponater, M., S. Marquart and R. Sausen, 2002: Contrails in a comprehensive global climate model: Parameterization and radiative forcing results. *J. Geophys. Res., 107,* 4164, doi:10.1029/2001JD000429.

Ponater, M., R. Sausen, S. Marquart and U. Schumann, 2005: On contrail climate sensitivity. *Geophys. Res. Lett. 32,* L10706, doi:10.1029/2005GL022580.

Rädel, G. and K. Shine, 2007a: Evaluation of the use of radiosonde humidity data to predict the occurrence of persistent contrails. *Q. J. R. Meteorol. Soc. 133,* 1413-1424.

Rädel, G. and K. Shine, 2007b: Influence of aircraft cruise altitudes on radiative forcing by persistent contrails. *J. Geophys. Res. 112,* submitted.

Read, W.G., et al., 2007: Aura Microwave Limb Sounder upper tropospheric and lower stratospheric H_2O and relative humidity with respect to ice validation. *J. Geophys. Res.*, 112, D24S35, doi:10.1029/2007JD008752.

Reinking, R., 1968: Insolation reduction by contrails. *Weather 23*, 171-173.

Rind, D., P. Lonergan, and K. Shah, 1996: Climatic effect of water vapor release in the upper troposphere. *J. Geophys. Res.*, 101, 29395-29405, doi:10.1029/96JD02747.

Rind, D., P. Lonergan and K. Shah, 2000: Modeled impact of cirrus cloud increases along aircraft flight paths. *J. Geophys. Res. 105*, 19,927-19,940.

Sassen, K., 1979: Iridescence in an aircraft contrail. *J. Opt. Soc. Am. 69,* 1080-1083.

Sassen, K., 1997: Contrail-cirrus and their potential for regional climate change. *Bull. Amer. Meteorol. Soc. 78,* 1885-1903.

Sassen, K. and C.-Y. Hsueh, 1998: Contrail properties derived from high-resolution polarization Lidar studies during SUCCESS. *Geophys. Res. Lett. 25,* 1165-1168.

Sausen, R., K. Gierens, M. Ponater and U. Schumann, 1998: A diagnostic study of the global distribution of contrails, Part I: Present day climate. *Theor. Appl. Clim. 61*, 127-141.

Sausen, R., I. Isaksen, V. Grewe, D. Hauglustaine, D.S. Lee, G. Myhre, M.O. Köhler, G. Pitari, U. Schumann, F. Stordal, and C. Zerefos, 2005: Aviation radiative forcing in 2000: An update on IPCC (1999). *Meteorol. Z. 14,* 555-561.

Schlager, H., P. Konopka, P. Schulte, U. Schumann, H. Ziereis, F. Arnold, M. Klemm, D.E. Hagen, P.D. Whitefield, and J. Ovarlez, 1997: In situ observations of air traffic emission signatures in the North Atlantic flight corridor. *J. Geophys. Res. 102*, 10,739-10,750.

Schmidt, E., 1941: Die Entstehung von Eisnebel aus den Auspuffgasen von Flugmotoren. In: Schriften der Deutschen Akademie der Luftfahrtforschung, Vol. 44. Verlag R. Oldenbourg, München, Berlin, 1-15.

Schröder, F.P., B. Kärcher, A. Petzold, R. Baumann, R. Busen, C. Hoell and U. Schumann, 1998: Ultrafine aerosol particles in aircraft plumes: In-situ observations. *Geophys. Res. Lett. 25,* 2789-2792.

Schröder, F.P., B. Kärcher, C. Duroure, J. Ström, A. Petzold, J.-F. Gayet, B. Strauss, P. Wendling and S. Borrmann, 2000: The transition of contrails into cirrus clouds. *J. Atmos. Sci. 57,* 464-480.

Schulz, J., 1998: On the effect of cloud inhomogeneity and area averaged radiative properties of contrails. *Geophys. Res. Lett. 25,* 1427-1431.

Schumann, U., 1994: On the effect of emissions from aircraft engines on the state of the atmosphere. *Ann. Geophys. 12*, 365-384.

Schumann, U., 1996: On conditions for contrail formation from aircraft exhausts. *Meteorol. Z. 5*, 4-23.

Schumann, U., A. Dörnbrack, T. Dürbeck and T. Gerz, 1997: Large-eddy simulation of turbulence in the free atmosphere and behind aircraft. *Fluid Dyn. Res. 20*, 1-10.

Schumann, U., H. Schlager, F. Arnold, R. Baumann, P. Haschberger and O. Klemm, 1998: Dilution of aircraft exhaust plumes at cruise altitudes. *Atmos. Environ. 32*, 3097-3103.

Schumann, U., R. Busen and M. Plohr, 2000: Experimental test of the influence of propulsion efficiency on contrail formation. *J. Aircr. 37*, 1083-1087.

Schumann, U., 2000: Influence of propulsion efficiency on contrail formation. *Aerospace Sci. Technol. 4*, 391-401.

Schumann, U., 2002: Contrail Cirrus. In: Cirrus. D.K. Lynch, K. Sassen, D. O'C. Starr and G. Stevens (Eds.), *Oxford University Press*, New York, 231-255.

Schumann, U., 2005: Formation, properties and climate effects of contrails. *C. R. Physique 6,* 549-565.

Schumann, U., 2006: Climate change impact of air traffic, paper presented at 25th International Congress of the Aeronautical Sciences, DGLR, Hamburg, proceedings available from DGLR, http://www.icas2006.org/index2.php, paper number 199, pp. 7.

Schumann, U. and P. Wendling, 1990: Determination of contrails from satellite data and observational results. In: Air Traffic and the Environment – Background, Tendencies and Potential Global Atmospheric Effects. U. Schumann (Ed.), Lecture Notes in Engineering, *Springer Berlin*, 138-153.

Schumann, U. and P. Konopka, 1994: A simple estimate of the concentration field in a flight corridor. In: U. Schumann and D. Wurzel (eds.): Impact of Emissions from Aircraft and Spacecraft upon the Atmosphere. Proceed. Intern. Sci. Colloquium, Köln (Cologne), Germany, April 18-20, 1994, DLR-Mitt. 94-06, 354-359.

Schumann, U., P. Konopka, R. Baumann, R. Busen, T. Gerz, H. Schlager, P. Schulte and H. Volkert, 1995: Estimate of diffusion parameters of aircraft exhaust plumes near the tropopause from nitric oxide and turbulence measurements. *J. Geophys. Res. 100,* 14,147-14,162.

Schumann, U., J. Ström, R. Busen, R. Baumann, K. Gierens, M. Krautstrunk, F.P. Schröder, and J. Stingl, 1996: In-situ observations of particles in jet aircraft exhausts and contrails for different sulfur containing fuels. *J. Geophys. Res. 101,* 6853-6869.

Schumann, U. and J. Ström, 2001: Aviation impact on atmospheric composition and climate. In: European research in the atmosphere 1996-2000: Advances in our understanding of the ozone layer during THESEO. European Commission, Brussels, 257-307.

Schumann, U., F. Arnold, R. Busen, J. Curtius, B. Kärcher, A. Petzold, H. Schlager, F. Schröder, K.-H. Wohlfrom, 2002: Influence of fuel sulfur on the composition of aircraft exhaust plumes: The experiments SULFUR 1-7. *J. Geophys. Res. 107,* doi:10.1029/2001JD000813.

Shine, K.P., 2005: Comments on "Contrails, cirrus clouds, and climate". *J. Clim. 18,* 2781-2782.

Spichtinger, P., K. Gierens and W. Read, 2003a: The global distribution of ice-supersaturated regions as seen by the microwave limb sounder. *Q. J. R. Meteorol. Soc. 129*, 3391-3410.

Spichtinger, P., K. Gierens, U. Leiterer and H. Dier, 2003b: Ice supersaturation in the tropopause region over Lindenberg, Germany. *Meteorol. Z. 12*, 143-156.

Spichtinger, P., K. Gierens and H. Wernli, 2005: A case study on the formation and evolution of ice supersaturation in the vicinity of a warm conveyor belt's outflow region. *Atmos. Chem. Phys. 5*, 973-987.

Stephens, G.L. and P.J. Webster, 1981: Clouds and climate: Sensitivity of simple systems. *J. Atmos. Sci. 38*, 236-247.

Stordal, F., G. Myhre, E.J.G. Stordal, W.B. Rossow, D.S. Lee, D.W. Arlander and T. Svendby, 2005: Is there a trend in cirrus cloud cover due to aircraft traffic? *Atmos. Chem. Phys. 5*, 2155-2162.

Strauss, B., R. Meerkötter, B. Wissinger, P. Wendling and M. Hess, 1997: On the regional climatic impact of contrails: microphysical and radiative properties of contrails and natural cirrus clouds. *Ann. Geophys. 15,* 1457-1467.

Ström, J. and S. Ohlsson, 1998: In-situ measurements of enhanced crystal number densities in cirrus clouds caused by aircraft exhaust. *J. Geophys. Res. 103*, 11,355-11,361.

Ström, L. and K. Gierens, 2002: First simulations of cryoplane contrails. *J. Geophys. Res. 107*, doi:10.1029/2001JD000838.

Stubenrauch, C. and U. Schumann, 2005: Impact of air traffic on cirrus coverage. *Geophys. Res. Lett. 32,* L14813, doi:10.1029/2005GL022707.

Stuber, N., P. Forster, G. Rädel and K. Shine, 2006: The importance of the diurnal and annual cycle of air traffic for contrail radiative forcing. *Nature 441,* 864-867.

Stuber, N. and P. Forster, 2007: The impact of diurnal variations of air traffic on contrail radiative forcing. *Atmos. Chem. Phys. 7,* 3153-3162.

Stuefer, M., X. Meng and G. Wendler, 2005: MM5 contrail forecasting in Alaska. *Mon. Wea. Rev. 133*, 3517-3526.

Sussmann, R. and K. Gierens, 1999: Lidar and numerical studies on the different evolution of vortex pair and secondary wake in young contrails. *J. Geophys. Res. 104*, 2131-2142.

Sussmann, R. and K. Gierens, 2001: Differences in early contrail evolution of two-engine versus four-engine aircraft: Lidar measurements and numerical simulations. *J. Geophys. Res. 106*, 4899-4911.

Toon, O.B. and R.C. Miake-Lye, 1998: Subsonic aircraft: Contrail and cloud effects special study (SUCCESS). *Geophys. Res. Lett. 25*, 1109-1112.

Tompkins, A., K. Gierens and G. Rädel, 2007: Ice supersaturation in the ECMWF Integrated Forecast System, *Q. J. R. Meteorol. Soc. 133,* 53-63.

Travis, D.J., A.M. Carleton and R.G. Lauritsen, 2002: Contrails reduce daily temperature range. *Nature 418*, 601.

Travis, D.J., A.M. Carleton, J.S. Johnson and J.Q. DeGrand, 2007: US jet contrail frequency changes: influences of jet aircraft flight activity and atmospheric conditions. *Int. J. Climatol. 27*, 621-632.

Treffeisen, R., R. Krejci, J. Ström, A.C. Engvall, A. Herber and L. Thomason, 2007: Humidity observations in the Arctic troposphere over Ny-Ålesund, Svalbard, based on 15 years of radiosonde data. *Atmos. Chem. Phys. 7*, 2721-2732.

Troller, M., A. Geiger, E. Brockmann, J.-M. Bettems, B. Bürki and H.-G. Kahle, 2006: Tomographic determination of the spatial distribution of water vapor using GPS observations. *Adv. Space Res. 37*, 2211-2217.

Twohy, C.H. and B.W. Gandrud, 1998: Electron microscope analysis of residual particles from aircraft contrails. *Geophys. Res. Lett. 25*, 1359-1362.

Unterstrasser, S., K. Gierens and P. Spichtinger, 2008: The evolution of contrail microphysics in the vortex phase. *Meteorol. Z.*, accepted.

Uthe, E.E., N.B. Nielsen and T.E. Osberg, 1998: Airborne scanning Lidar observations of aircraft contrails and cirrus clouds during SUCCESS. *Geophys. Res. Lett. 25*, 1339-1342.

Vatazhin, A.B., V.E. Kozlov, A.M. Starik and E.K. Kholshchevnikova, 2007: Numerical modeling of the formation of aerosol particles in jet engine plumes. *Fluid Dynamics 42*, 33-43.

Wang, W.-C., W. Gong and J.-P. Chen, 2001: SUNYA regional model simulation of radiative forcing and climate impact due to contrails over regions around Taiwan. *TAO 12,* 179-194.

Wuebbles, D.W. and M. Ko, 2007: Evaluating the impacts of aviation on climate change. *EOS 88*, 157-160. See also: Workshop on the Impacts of Aviation on Climate Change –A Report of Findings and Recommendations, June 7-9, 2006, Boston, MA, NASA/FAA

Joint Planning and Development Office, Environmental Integrated Project Team, 58 pp, August 2006.

Yang, P., K.N. Liou, K. Wyser and D. Mitchell, 2000: Parameterization of the scattering and absorption properties of individual ice crystals. *J. Geophys. Res. 105*, 4699-4718.

Yang, P., H. Wei, H.L. Huang, B.A. Baum, Y.X. Hu, G.W. Kattawar, M.I. Mishchenko and Q. Fu, 2005: Scattering and absorption property database for nonspherical ice particles in the near- through far-infrared spectral region. *Appl. Opt. 44*, 5512-5523.

Yu, F. and R.P. Turco, 1998: Contrail formation and impacts on aerosol properties in aircraft plumes: Effects of fuel sulfur content. *Geophys. Res. Lett. 25,* 313-316.

Zerefos, C.S., K. Eleftheratos, D.S. Balis, P. Zanis, G. Tselioudis and C. Meleti, 2003: Evidence of effect of aviation on cirrus cloud formation. *Atmos. Chem. Phys. 3*, 1633-1644.

Zhang, Y., A. Macke and F. Albers, 1999: Effect of crystal size spectrum and crystal shape on stratiform cirrus radiative forcing. *Atmos. Res. 52*, 59-75.

In: Aviation and the Environment
Editor: Jon C. Goodman

ISBN: 978-1-60692-320-7

Chapter 4

ACCRI THEME 4: CONTRAILS AND CONTRAIL-SPECIFIC MICROPHYSICS

***Andrew Heymsfield*[*1], *Darrel Baumgardner*[*2], *Paul DeMott*[*3], *Piers Forster*[*4], *Klaus Gierens*[*5], *Bernd Kärcher*[*6] and *Andreas Macke*[*7]**

[1] National Center for Atmospheric Research; Boulder, Colorado, USA
[2] Universidad Nacional Autónoma de Mexico; Mexico City, Mexico
[3] Colorado State University; Ft. Collins, Colorado USA
[4] School of Earth and Environment; University of Leeds, Leeds, LS2 9JT, UK
[5] DLR-Institut fir Physik der Atmosphäre; Oberpfaffenhofen, D-82234 Wessling, Germany
[6] DLR-Institut fir Physik der Atmosphäre Oberpfaffenhofen, 82234 Wessling, Germany
[7] Leibniz-Institut fir Meereswissenschaften; IFM-GEOMAR; D-24105 Kiel, Germany

EXECUTIVE SUMMARY

Theme 4 of the ACCRI, "Contrails and Contrail-Specific Microphysics", reviews the current state of understanding of the science of contrails: 1) how they are formed, 2) their microphysical properties as they evolve, 3) how they develop into contrail cirrus and if their microphysical properties can be distinguished from natural cirrus, 4) their radiative properties

---

* heyms1@ucar.edu
* darrel@servidor.unam.mx
* pdemott@lamar.colostate.edu
* P.M.Forster@leeds.ac.uk
* klaus.gierens@dlr.de
* bernd.kaercher@dlr.de

and how they are treated in global models and 5) the ice nucleating properties of soot aerosols and whether these aerosols can nucleate cirrus crystals.. Key gaps and underlying uncertainties in our understanding of contrails and their effect on local, regional and global climate are identified and recommendations are provided for research activities that will remove or decrease these uncertainties.

Contrail formation is described by a simple equation that is a function of atmospheric temperature and pressure, specific fuel energy content, specific emission of water vapor and the overall propulsion efficiency. Thermodynamics is the controlling factor for contrail *formation* whereas the physico-chemistry of the emitted particles acts in a secondary role. The criteria for contrail formation determine whether a contrail will form but does not predict whether the contrail will persist or spread into an extensive cirrus-like cloud.

Contrail ice crystals are captured within the downward-travelling vortex pair generated by the aircraft, descending with an average speed of about 2 m/s, which induces adiabatic compression, heating, and sublimation. This phase reduces the ice particle concentration and the contrail will persist and spread only possible if the ambient air is supersaturated with respect to ice.

At formation, the ice number concentration (~1 0^4 -10^5 cm^{-3}) and size (several tenths of tm) are mainly determined by the plume cooling rates. During the early stages of a contrail's development, the total ice crystal concentratio ns are of order 10^3-10^4 cm^{-3} and mean diameters of 5 tm. In the vortex (descending) phase, the concentrations diminish to an order of 10 -100 cmand mean sizes increase up to 10 tm diameter. Continued evolution of the size distributions depends on the ambient relative humidity. Observations of the ice crystal shapes during the early phase of contrail formation and beyond are sparse yet important for estimating the radiative properties of young contrails.

When a contrail is first formed the aircraft's contribution of water vapor to the contrail is appreciable. Soon thereafter the ice water content (IWC) increases and is modulated based on the ambient (environmental) water vapor density. The IWC can be quantified using a simple model that converts the ambient (environmental) water vapor supersaturation into condensate.

The fate of a contrail depends on the environmental relative humidity with respect to ice (RHi). In an ice supersaturated environment, contrail microphysical properties are similar to natural cirrus, i.e., concentrations of ice crystals larger than 100 tm in diameter are of order 10-100 l^{-1} and the habits are bullet rosettes; however, many previous measurements of size distributions in the presence of ice crystals several hundred microns and above have been dominated by artifacts resulting from the shattering of crystals on the microphysical probes' inlets. It is therefore difficult to differentiate the particle size distributions (PSD) in cases where large crystals exist in contrails, and that the time and transition toward natural cirrus properties is not well defined based on current data.

Soot particles emitted by aircraft jet engines may perturb cirrus properties and alter cirrus coverage without contrail formation being involved. Aircraft engine exhaust that does not form contrails, and contrails that evaporate, provide sources of enhanced aerosol particle number concentrations di- rectly in regions where cirrus may be forming. Consideration also needs to be given to the role of prior contrail and cloud processing on "preactivating" exhaust ice nuclei.

* amacke@ifm-geomar.de

The nucleation process(es) involved in producing cirrus ice crystals from aircraft exhaust soot aerosols when contrails do not form or from residual soot following contrail evaporation are highly uncertain. Ice formation by black carbon particles in general remains poorly understood. Most studies have been conducted at temperatures warmer than 235K and only a few have carefully quantified the freezing fraction of soot particles on a single particle basis. Furthermore, the studies most relevant to cirrus formation have used idealized soot particles of unknown relevance to aircraft exhaust soot. Thus, in contrast to the level of knowledge of the composition of aircraft exhaust emissions reflected in the literature, relatively little is known about their role in ice formation, motivating a strong need for further systematic laboratory and in-situ studies.

Aviation soot emissions may change the number of ice crystals in cirrus by several tens of p ercent according to global model studies; however, due to the limited and inconclusive results on ice nucleating behavior of soot particles, global models that address soot-induced cirrus can only provide preliminary parametric studies exploring possible uncertainties of changes in cirrus properties.

Radiative forcing from contrails depends on many factors: contrail coverage, ice water path, optical properties, geometry, time of day, size and location, age and persistence, background cloud iness and surface albedo. Contrails reflect solar radiation leading to a negative forcing and absorb/trap longwave radiation causing a positive forcing. The net forcing of a contrail is expected to be a positive forcing; however, the cancellation of shortwave and longwave terms of roughly equal magnitude means that the radiative impact is very sensitive to any error in either term. After coverage, which is poorly known, ice water path and optical properties are the largest sources of uncertainty.

Most contrail-related subscale processes are not represented in current large-scale models, with the notable exception of the thermodynamic conditions for contrail formation. Until recently GCMs have not carried ice supersatuation, so estimates of available ice have had to be obtained from parameterizations that assume that ice exists above a relative humidity threshold less than 100%. These schemes are also diagnostic as one time step does not know about contrails at any previous timesteps; therefore, assumptions are also needed about contrail lifetime. The lack in climate models of physical and radiative interaction between contrails and their moist environment renders impossible a meaningful determination of global contrail effects on the water budget in the upper troposphere. Constraining the radiatively important ice crystal size and the IWC or the optical depth of contrails is needed before an accurate estimate of global contrail radiative forcing can be made. While several additional factors including ice crystal shape add to the uncertainty in the radiative forcing, it is the contrail coverage, the IWC and the crystal size that are key to providing an accurate forcing estimate.

The IPCC fourth assessment estimated the linear contrail radiative forcing for 2005 to be 0.01 Wm^{-2} but with a low level of scientific understanding and a factor of three uncertainty in its magn itude. Clearly, significant refinements to the estimates will require improvements in the knowledge and representations of (a) contrail and cirrus ice microphysics and radiative properties, (b) the global distribution of upper-tropospheric ice supersaturation, and (c) improved treatment of contrail and cirrus microphysics, radiation, aerosols, vertical motions driving ice supersaturation, and inte r- actions of contrails with their environment, in global models. Acquiring the information that is needed to make these improvements will require targeted field studies and new instruments that overcome the uncertainties and limitations that impeded previous studies.

1. Introduction and Background

The continuing growth of airborne transportation is accompanied by increased emissions of gases and particles whose impact on climate, regionally and globally, remains highly uncertain due to limited information on the emission properties and the complex physical processes that govern how these emissions interact with the environment over multiple spatial and temporal scales (Wuebbles, 2006). Given the projection that the demand for air transportation services could grow by a factor of three by 2025 (Next Generation Air Transportation System, 2004), it is imperative that the environmental impact of the current fleet of aircraft is evaluated so that the impact of further growth can be accurately assessed.

Linear contrails, products of aircraft emissions in the upper atmosphere, are arguably the most visible human influence on the Earth's climate. They are high level ice clouds formed under specific atmospheric conditions that usually have more of a climate warming influence by trapping longwave radiation rather than a cooling effect by reflecting incoming solar radiation.. The environmental conditions determine whether they appear at all and if they do appear, whether they persist from minutes to hours and if they spread to form wide decks of cirrus-like clouds or even act to seed or enhance clouds that have already formed. The studies that have attempted to quantify their warming effect suggest that the climate forcing could be comparable to that from aviation's CO_2 emissions, but the magnitude is uncertain.

This section of the SSWP, "contrails and contrail-specific microphysics" assesses the current understanding of the science of contrail formation, what we know about their microphysical and radiative properties as a function of time following their formation, and the limitations and uncertainties that are obstacles to accurately predicting how contrails from current and future aircraft fleets impact regional and global climate change. The discussions herein reflect those that were outlined for contrail formation and climate impact during the 2006 workshop on aviation and its impact on climate change (Wuebbles, 2006). They are presented here in greater scientific detail and additional information is presented on the current state of information from field projects that measured contrail properties and much more attention is given to in-situ instrumentation for measuring contrail properties and the limitations associated with these sensors.

2. Review of the Specific Theme

2.1. Current State of Science

2.1.1. Range of Conditions for Formation of Contrails, Their Persistence and Evolution into Cirrus

Contrail formation is determined almost exclusively by basic thermodynamics and the atmospheric conditions in which engine emissions are released. Contrail formation is described by a simple equation containing atmospheric temperature and pressure, specific fuel energy content, specific emission of water vapour and the so-called overall propulsion efficiency. This equation is known as the Schmidt-Appleman-Criterion (SAC), which has been formulated in a convenient format by Schumann (1996) based on earlier work by Schmidt (1941) and Appleman (1953). The SAC states that a contrail forms during the plume expansion process if the mixture of exhaust gases and ambient air transiently reaches or surpasses saturation with respect to liquid water. The fact that the mixture must reach water saturation (and not only ice saturation or any other relative humidity) is the only empirical component of the thermodynamic approach and it is the only part that is related to the ice - forming properties of emitted particles. It simply means that the emitted particles act primarily as cloud condensation nuclei (CCN) and are poor ice forming nuclei (IN), i.e. they first activate into liquid droplets that freeze afterwards. The validity of the SAC description has been demonstrated and confirmed from various research flights (Busen and Schumann, 1995; Kärcher et al., 1998; Jensen et al., 1998; IPCC 1999). These validations also demonstrate that the thermodynamics is the controlling factor for contrail *formation*, not the physico-chemistry of the emitted particles; however, as we note later, the latter does exert a weak influence on contrail *properties*.

The mixing process is assumed to take place isobarically, so that on a T (absolute temperature) – e (partial pressure of water vapour in the mixture) diagram the mixing (phase) trajectory appears as a straight line. The slope of the mean phase trajectory in the turbulent exhaust field, G (units Pa/K), is characteristic for the respective atmospheric situation and aircraft/engine/fuel combination and given by

$$G = \frac{EI_{H2O}\, p c_p}{\varepsilon Q(1-\eta)}$$

where is the ratio of molar masses of water and dry air (0.622), c_p is the isobaric heat capacity of air (1004 J/kg K) and p is the ambient air pressure. G depends on the emission index of water vapour, EI_{H2O} (1.25 kg per kg kerosene burnt), the chemical heat content of the fuel, Q (43 MJ per kg of kerosene), and on the overall propulsion efficiency,, of the aircraft engine. Modern airliners have a propulsion efficiency of approximately 0.35, and therefore produce contrails in less cold air than older aircraft (Schumann, 2000). Improved jet engine technology may therefore enhance contrail formation.

One can formulate the SAC condition as a criterion for the maximum temperature, T_c, that would allow contrail formation for given conditions of the ambient relative humidity and pressure:

$$T_c(G,e) = T_{LM}(G) - \frac{e^*[T_{LM}(G)] - e}{G} \tag{1}$$

where e is the ambient water vapor partial pressure, e^* is the saturation vapor pressure (wrt liquid water) and T_{LM} the maximum temperature that allows contrail formation when the ambient relative humidity (wrt water) is 100% (i.e. e=e^*). There is a large degree of uncertainty in predicting T_c because of the uncertainties in determining the parameters that control the magnitude and behaviour of G. One particularly difficult quantity to measure is. It varies from aircraft to aircraft, cannot be accurately determined at ground conditions and is more difficult to evaluate at cruise altitudes. As shown in figure 1 for errors in T_c due to uncertainties in, a typical uncertainty inof 0.02 leads to an uncertainty in T_c of about ±0.5 K. This error propagates into the uncertainty in the relative humidity calculated from water vapor pressure measurements, i.e. at upper tropospheric temperatures, such an error in T_c of ±0.5 K corresponds to an error in the threshold relative humidity of about ±5% (in RH units).

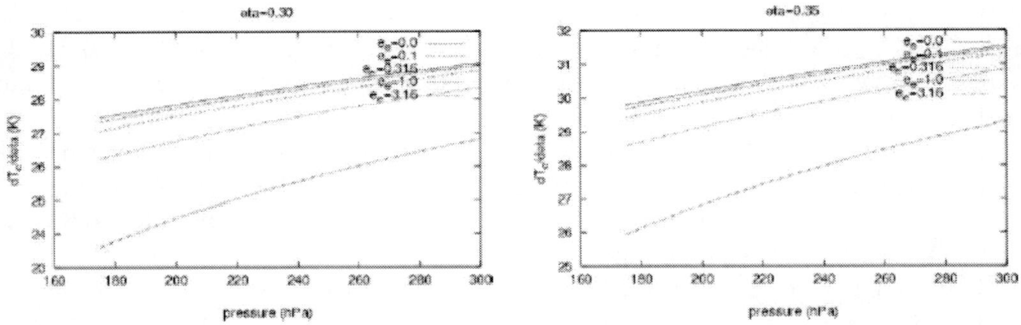

Figure 1. Error in the determination of the contrail formation threshold temperature *Tc* due to uncertainties in the determination of the overall propulsion efficiency, |, for older aircraft with | =0.3 (left panel) and more modern aircraft with | =0.35 (right panel).

The SAC only determines contrail formation and not what follows, i.e. its persistence or evolution into extensive, cirrus-like cloud. Persistence is possible only if the ambient air is supersaturated with respect to ice such that once formed, ice crystals in the plume can grow until the air becomes subsaturated.

2.1.2. Chemical and Microphysical Mechanisms that Determine the Evolution of Emissions from the Engine Exit to Plume Dispersion

The initial composition of jet contrails are determined by processes occurring within approximately one wingspan behind the aircraft: chemical and water activation of combustion particles, i.e. soot aerosols, and the subsequent formation of ice on some of these particles (Kärcher et al., 1996). The contribution of soot particles to contrail formation at temperatures near T_c was inferred from theoretical studies in the cooling plume of the homogeneous freezing potential of fully liquid, volatile acidic plume particles that start forming before the threshold conditions for ice formation (Kärcher et al., 1995). It was found that volatile particles do not freeze homogeneously in plumes that are barely supersaturated with respect to water as a result of their very small sizes (a few nm) relative to soot particles (> 10 nm). Soot is formed from sulphurous and carbonaceous compounds during combustion. The sulphurous component, in the form of sulphuric acid, H2 SO4, increases together with water vapour by condensation after emission (Kärcher, 1998), potentially altering the ice nucleation process involving soot particles; however, if there is sufficient moisture in the plume as it rapidly cools, the soot particles can acquire a liquid water coating that instantaneously freezes into ice once the plume achieves water supersaturation. This can occur even if hygroscopic $_{H2SO4}$ particles are absent and the soot surfaces have poor water adsorption properties, similar to graphite (Kärcher et al., 1996). In the latter case, the degree of required water supersaturation may depend on the hydrophilic/hydrophobic character of the soot particles (Popovicheva et al., 2007). It was also concluded, based on an observation of contrail formation very close to T(Busen and Schumann, 1995), that soot particles with $H_{2SO4/H2O}$ coa tings may nucleate ice heterogeneously slightly below water saturation (Kärcher et al., 1996); however, due to the extremely rapid increases in relative humidity, such a small difference in ice nucleation behavior does not significantly affect the SAC criteria within the uncertainties of measurements that determine T_c. In contrast, an assumption of near perfect ice nucleation of exhaust soot, i.e. contrail formation at or substantially above plume ice saturation, would clearly contradict observational evidence (Kärcher et al., 1998).

The number concentration (~10^4-10^5 cm^{-3}) and size (fractions of a micrometer) of contrail ice particles at formation is mainly determined by the very high plume cooling rates of order 1000 K/s (Kärcher et al., 1998). Soot particle concentrations in aircraft plumes are typically of this order (Petzold et al., 1 998b, 1999) and, once frozen, are sufficiently high to shut off further ice nucleation by depleting the excess water vapor. This predominant dynamical control renders the ice nucleation properties of the particles in the contrail plume relatively unimportant. Increasing the fuel sulphur content leads to more rapid growth of soot particle coatings and potentially activates the entire soot particle reservoir (Schumann et al., 1996; Gierens and Schumann, 1996). Cold ambient temperatures and increased fuel sulphur content lead to slightly more and smaller contrail particles, also because homogeneous freezing of the more numerous water-activated liquid exhaust droplets that take dominance over soot-induced ice formation (Kärcher et al., 1998). Some further turbulent mixing with ambient air and depositional growth of contrail ice particles occurs until the capture of individual plumes in the vortices suppresses the mixing. At this point, the majority of the ice crystals are still very small (diameters < 1 tm) but their concentration has decreased to ~10^3-10^4 cm^{-3}. The processes that occur during the downward displacement and break-up of wake vortices are thought to be primarily responsible for the observed variability in the number concentrations of young contrail ice particles. Numerical studies (Sussmann and Gierens, 1999) show that a variable fraction of the initial ice crystals sublimate during the vortex phase, depending on the ambient humidity, temperature, stability and turbulence.

Contrail ice crystals are captured within the downward-travelling vortex pair. They descend with an average speed of about 2 m/s that implies adiabatic compression and heating. The heating can be computed assuming a dry adiabatic lapse rate (which can be safely assumed since the evaporating ice mass and latent heating is small) such that a vertical displacement of 300 m heats the air by 3 K and decreases the relative humidity (wrt ice) by 30% within the plume. Hence, a fraction of the contrail ice sublimates during the vortex phase unless the ambient RH exceeds 130%. Surviving fractions of contrail ice (by number and mass) of the order 10^{-3} at ice saturation increase in a power law fashion with increasing supersaturation (Unterstrasser et al., 2007). The power law exponent increases strongly with ambient temperature such that the relationship between the su rviving ice fraction and supersaturation becomes more sensitive as the temperature increases.

The adiabatic sinking of the vortex pair leads to a baroclinic instability around the upper stagnation point of the pair. Ice crystals can escape from the vortex system and remain behind. The ice in this so-called secondary wake (which is merely a small fraction of the initially produced ice) is not subject to adiabatic compression and survives when the ambient air is supersaturated, resulting in a faint but persistent contrail. This contrail consists either of the ice in the secondary wake plus the ice that survives the adiabatic heating in the primary vortex or of the secondary wake alone. The numerical studies indicate a broad range of initial conditions (ice mass and number concentration) for the later evolution of contrails.

2.1.3. Role of Emission Characteristics and Plume Processes on the Large-Scale Aviation Impact

As detailed above, nascent contrail properties are not very sensitive to the emission characteristics of kerosene-fuelled jet engines as the initial number of contrail ice particles is limited by the very high cooling rates in the plume rather than by soot emission indices. The slight alteration of nascent contrail properties by high sulphur emissions is of little practical relevance given that the fuel sulphur content is expected to decrease rather than increase in the future (Wuebbles, 2006). These conclusions might be tempered if future alterations to fuels and combustion parameters would enhance the ice nucleating properties of soot particles (see section 2.1.4); however, the e xpected impact of changes in things like fuel additives is negligible (Gierens, 2007). The most influential factors that modify contrail properties are those that occur during and after vortex break-up as previously described in 2.1.2. The contrail -to-cirrus transition is further controlled by the moisture fields (the topic of key theme 3) and the distribution of vertical shear in the horizontal wind. The generation of individual contrail-cirrus may be sensitive to vertical gradients of thermodynamic parameters and turbulence levels. In heavily travelled regions within or near flight corridors persistent contrails do not appear as single objects whereby the spreading of multiple contrails leads to contrail decks in which the evolution of one contrail cannot be considered separately from the others.

2.1.4. Potential for Cirrus Formation/Modification due to Aviation Soot Emissions

Soot particles emitted by aircraft jet engines, in the absence of contrail formation, may also perturb cirrus properties and alter cirrus coverage. Aircraft engine exhaust that does not form contrails and contrails that evaporate provide sources of enhanced concentrations of aerosol particles in regions where cirrus may eventually form. The potential perturbation likely occurs on regional scales because the residence time of aerosols in the upper tropopause is of the order of days to several weeks, depending on the location of the emissions, the season and the latitude. This aerosol indirect effect is mentioned within the Intergovernmental Panel on Climate Change Fourth Assessment Report (Denman et al. 2007) As such it is an atmospheric component that has medium potential impact on overall aerosol forcing of climate but with very low understanding. The magnitude of the cloud perturbation (e.g. changes in ice cloud particle effective radius) depends on the ice nucleating ability of the rele ased soot particles, on the efficiency of existing aerosol particles to nucleate ice, on temporal interactions of the soot particles with ambient gases and aerosols, on the abundance of water vapor (H_2O), and on dynamic processes setting the stage for the generation of clouds in ice supersaturated regions (Haag and Kärcher, 2004). In fact, an understanding of cirrus formation itself is tantamount to estimating any aircraft soot impact on cirrus.

A good level of understanding of the key, microphysical factors in cirrus formation has evolved over the last 20 years. An important process for ice formation in cirrus is homogeneous freezing of liquid-containing aerosol particles. This process for sulfuric acid and other liquid aerosols appears to be at least quantitatively well understood (Sassen and

Dodd, 1988; Heymsfield and Sabin, 1989; Jensen and Toon, 1994; Koop et al., 2000; DeMott, 2002; Lin et al. 2002; Möhler et al., 2003; Haag et al., 2003; Koop, 2004). Pure liquid droplets (or highly diluted liquid aero sol particles) freeze with predictable nucleation rates at water saturation and at approximately 235K Haze particles freeze in a similar manner at progressively more subsaturated conditions as the temperature decreases. Presumably, this process sets an upper bound (figure 2) on the ice supersaturation conditions needed for cirrus cloud formation in the upper troposphere, without other factors that might inhibit ice formation by this process.

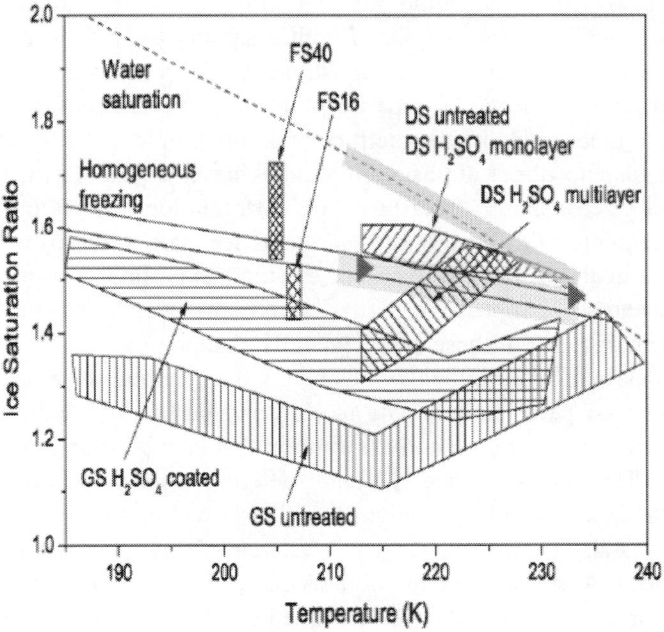

Figure 2. Range of ice formation conditions of some different types of soot particles in the cirrus cloud regime, adapted from Kärcher et al. (2007). All ice formation conditions represent small fractions (<1%) of particles activating for different types. GS represent graphite spark generator soot tested in cloud expansion chamber studies by Möhler et al. (2005a). FSxx represents flame generator soot of xx% organic content in similar chamber studies by Möhler et al. (2005b). DS represent commercial soot particle samples tested in a continuous flow diffusion chamber (CFDC) by DeMott et al. (1998). Yellow shading indicates the range of conditions of low temperature ice formation by fresh biomass combustion particles in a CFDC (DeMott et al. 2008). Red arrow points represent CFDC ice formation conditions for real combustor soot selected at a mobility size of 250 nm (Koehler et al. 2008). Light green shading indicates CFDC ice formation conditions for 50 nm fresh exhaust particles from burning jet fuel in a laboratory burner (DeMott et al. 2002). The lower and upper threshold curves for homogeneous freezing are calculated for 1% of solution particles with radii 1 μm and 0.1 μm freezing within 100 s and 1 s, respectively.

During cirrus formation there is a competition between homogeneous freezing of soluble aerosol particles and potentially more efficient heterogeneous ice nucleation by some fraction of insoluble aerosol particles, such as dust (e. .g., Zuberi et al. 2002; DeMott et al. 2003a,b; Richardson et al. 2007), crystallized organic and inorganic phases of soluble aerosols (e.g., Zuberi et al., 2001; Abbatt et al., 2006; Zobrist et al., 2006; Shilling et al., 2006; Beaver et al.,

2006) or soot (e.g., Kärcher et al. 1996; Jensen and Toon, 1997; DeMott et al. 1997; Kärcher et al. 2007). This competition is constrained by the available water vapor and therefore only a fraction of all aerosols can influence the microphysical composition of cirrus. Thus the dynamics that drives supersaturation and the aerosols that drive ice nucleation are intertwined in cirrus much as in any cloud (Kärcher and Ström, 2003; Haag and Kärcher, 2004).

On the premise that aircraft soot particles are effective ice nuclei (IN) under conditions that do not already favor homogeneous freezing of liquid aerosol particles and that cirrus formation is dynamically triggered by slow synoptic uplift, cloud resolving simulations have shown that the resulting cirrus are more stable and have different areal coverage and optical properties than cirrus formed on liquid particles in the absence of aircraft soot (Jensen and Toon, 1997). The main problem associated with assessing the role of aircraft soot in cirrus cloud formation is to unambiguously demonstrate that ice nucleates mainly on the primary exhaust soot particles or secondary particles formed by condensation and coagulation. Designing airborne measurements that help resolve this problem is very challenging. In fact, it must be noted that no direct atmospheric studies have yet validated the impacts of both homogeneous and heterogeneous freezing on cirrus formation directly in the atmosphere, i.e. matching measurements with predictions of cloud ice particle distributions directly from aerosol and dynamical properties. Such a "closure" experiment requires high resolution measurements of aerosol composition, IN, relative humidity, vertical motion, and cloud ice particle size distribution. All of these are technically challenging.

Atmospheric variability in vertical winds below the synoptic scale often seems to control the cooling rates of air parcels containing ice-forming particles (Kärcher and Strom, 2003), challenging the validity of the common assumption that cirrus formation is triggered by slow synoptic uplift. The cooling rate history determines the relative contributions of heterogeneous and homogeneous ice nucleation and how many aerosol particles nucleate to form cirrus ice crystals (DeMott et al., 1997; Gierens, 2003; Kärcher and Lohmann, 2003; Kärcher et al., 2006). Atmospheric cooling rates are difficult to determine experimentally and to describe and predict by models. Taken together, this renders a separate experimental evaluation of dynamical and aerosol effects on cirrus formation very difficult.

Measurements have indicated that regions of the upper troposphere are sometimes highly supersaturated with respect to ice, exceeding values typically required for, at temperatures warmer than needed for homogeneous freezing. There are possible physical mechanisms for such high supersaturations (e.g., Jensen et al., 2005; Peter et al., 2006); however, measuring humidity at low temperatures remains challenging and global prediction of ice supersaturation is still in its infancy. Fortunately, in the extratropics at altitudes where most current air traffic occurs, observed levels of supersaturation are consistent with homogeneous ice nucleation as outlined by Koop et al. (2000).

The following sections review the present state of knowledge of the ice nucleation potential of aircraft exhaust particles based on laboratory and field studies and scenarios for the potential for cirrus modification due to aircraft emissions.

2.1.4.1. Laboratory Evidence Regarding Ice Formation by Black Carbon Particles and Aircraft Exhaust Soot

Ice formation by black carbon particles in general remains poorly understood. It is further necessary in this regard to distinguish what is known about ice formation by soot particles in general and what is known about those particularly relevant to aircraft emissions. In the former case, it is apparent that some bla ck carbon containing particles act as ice nuclei, as carbonaceous particles have been found as one of the major types of apparent nuclei of ice crystals formed on aerosols sampled from the background upper troposphere and cold regions of the lower troposphere (Chen et al. 1998; Rogers et al. 2001). The efficiency of black carbon particles acting as IN depends very sensitively, but in unclear ways, on temperature and supersaturation, the soot size and surface oxidation characteristics, and as for any ice nucleus on the mechanism of ice formation (DeMott 1990; Gorbunov et al. 2001; Diehl and Mitra, 1998; MOhler et al. 2005a,b; Dymarska et al. 2006). Most studies have been focused at temperatures warmer than 235K and only a few have carefully quantified the freezing fraction of soot particles on a single particle basis. Only a few other laboratory studies have addressed the ice nucleation properties of soot particles at lower temperatures (DeMott et al. 1999; MOhler et al. 2005a,b). Furthermore, the studies most relevant to cirrus formation have used idealized soot particles of unknown relevance to aircraft exhaust soot. Thus, in contrast to the level of knowledge of the composition of aircraft exhaust emissions reflected in the literature, relatively little is known about their role in ice formation.

Figure 2, adapted from Kärcher et al. (2007) summarizes studies of soot ice formation in the cirrus regime. Results from DeMott et al. (1999) for commercial black carbon particles (240 nm mode diameter "lamp black" from Degussa Corporation, Frankfurt/Main, Germany) frozen in a continuous flow diffusion chamber are indicated by DS in the figure. The aerosol particles were treated in some cases by exposure to H_2SO_4 molecules to simulate contrail and atmospheric processing. Untreated particles and those with lower H_2SO_4 coverage ("monolayer") activated ice only at a relative humidity close to water saturation. With greater H2 SO4 coverage ("multilayer") ice nucleation by 1% of the particles occurred for ice saturation ratios (Si) less than required for homogeneous freezing of soluble particles (Koop et al., 2000) at T<220 K. Although the mechanism that apparently renders multilayer-treated DS as more efficient IN is unclear at the moment, it does not appear to be present for untreated or thinly coated particles.

Möhler et al. (2005a) used fractal-like, agglomerated soot particles (100 nm mode diameter) from a graphite spark generator (GS in figure 2) for ice nucleation studies in the AIDA cloud chamber of Forschungszentrum Karlsruhe. Untreated GS particles showed S (0.1 freezing) far below the homogeneous freezing threshold conditions. Coating with H_2 SO4 (volume fractions of 20-80%) drove threshold ice formation conditions significantly higher, toward the homogeneous freezing condition compared to the untreated GS particles.

In contrast to these results favoring a possible role of soot particles as low temperature deposition and immersion freezing nuclei are results for actual collected and re -dispersed combustor soot containing particles (Koehler et al. 2008), smaller soot-containing particles from high temperature stove combustion of jet fuel (DeMott et al. 2002) and soot -containing biomass combustion particles (DeMott et al. 2008). In all of these cases the soot particles were present within internally mixed particles that included soluble organic and inorganic matter. These results, for arguably more realistic atmospheric particles, indicate no special ice nucleating properties to initiate freezing for S lowe r than required for homogeneous freezing in the cirrus temperature regime. Ice formation conditions are indistinguishable or inhibited with respect to homogeneous freezing.

Hydrophilic and hydrophobic organic carbon (OC) is also present in aircraft-emitted aerosols (Demerdjian et al. 2007) or is contained in ambient aerosols that ultimately mix with and contribute to the chemical composition of the plume particle mixtures. Möhler et al.

(2005b) employed the AIDA chamber to investigate the ice nucleation ability of soot particles with different OC content generated in a propane burner with different fuel-to-air ratios. Flame soot particles with 16% OC mass content (FS16 in figure 2) nucleated ice at Si only modestly below and indistinguishable from homogeneous freezing thresholds whereas ice was initiated from flame soot with 40% OC content (FS40 in figure 2) at Si values exceeding those required for homogeneous freezing. Although these experiments indicate suppressed IN activity for certain thicker OC coatings, the general impact of OC may depend on the nature of the combustion process and fuel, e.g. IN activity may differ for jet fuel combustion versus biomass burning.

The noted differences in the IN activity of soot particles appear related to different physical or chemical surface properties of the various particles. Most aircraft emitted soot particles are smaller than 100 nm in diameter and have highly complex chemistry. The apparent lack of IN activity of particles representative of those found in jet exhaust is informative; however the lack of understanding of the physical and chemical mechanisms of this activity suggests strong caution on generalizing behaviors in the ambient upper troposphere. The role of prior contrail and cloud processing of exhaust ice nuclei, i.e. "preactivation" (Roberts and Hallett, 1968), also needs to be better understood and taken into account. Clearly there is a strong need for further systematic laboratory and in-situ studies.

2.1.4.2. Atmospheric Evidence for Aircraft Soot Impact on Cirrus

Evidence exists that cirrus ice crystal residue particles sampled from within aircraft corridors contain enhanced numbers of soot particles (Ström and Ohlsson, 1998). Nevertheless, this observation alone is insufficient to demonstrate that soot particles have actually played a role in the process of ice formation. Instead, soot could be included as a passive tracer within freezing aerosol particles or could be scavenged by ice crystals after the cloud has formed. More recent studies inferred strong differences between the S formation conditions of cirrus between the Northern and Southern hemispheres (Ström et al. 2003; Haag et al. 2003; Kärcher and Ström, 2003), but this could not be clearly attributed to either aircraft or other anthropogenic aerosols. Nevertheless, the number concentrations of cirrus ice crystals and the insoluble residuals of these crystals were found to increase proportionally (Seifert et al. 2004).

Only a single study of the concentrations and chemical composition of ice nuclei in and out of aircraft exhaust trails has been made directly (Chen et al. 1998). The aircraft IN instrument could only measure to temperatures at the warm limit of most cirrus clouds (~235K). In this temperature regime no difference was seen in the carbonaceous component of ice nuclei in and out of plumes. As in the background free troposphere, crustal, carbonaceous and metallic particles dominated the compositions of heterogeneous ice nuclei. The only difference in composition within exhaust trail regions was in the numbers of metallic particles. This could indicate a source from aircraft due to the presence of metal in combustion particles from jet engines (Demerdjian et al. 2007). Nevertheless, the concentrations of ice nuclei within dry contrails were indi stinguishable from those in the ambient free troposphere. This result is emphasized in figure 3, where the correlation coefficient of IN, condensation nuclei (CN) and NO are shown for the data set taken during the NASA Subsonic Cloud and Contrail Effects Special Study (SUCCESS) experiment (Rogers et al. 1998; Rogers et al. 2008). While CN and NO concentrations show some positive correlation, IN at T>235K correlate with neither. It is not known if these results can be extrapolated to lower temperatures.

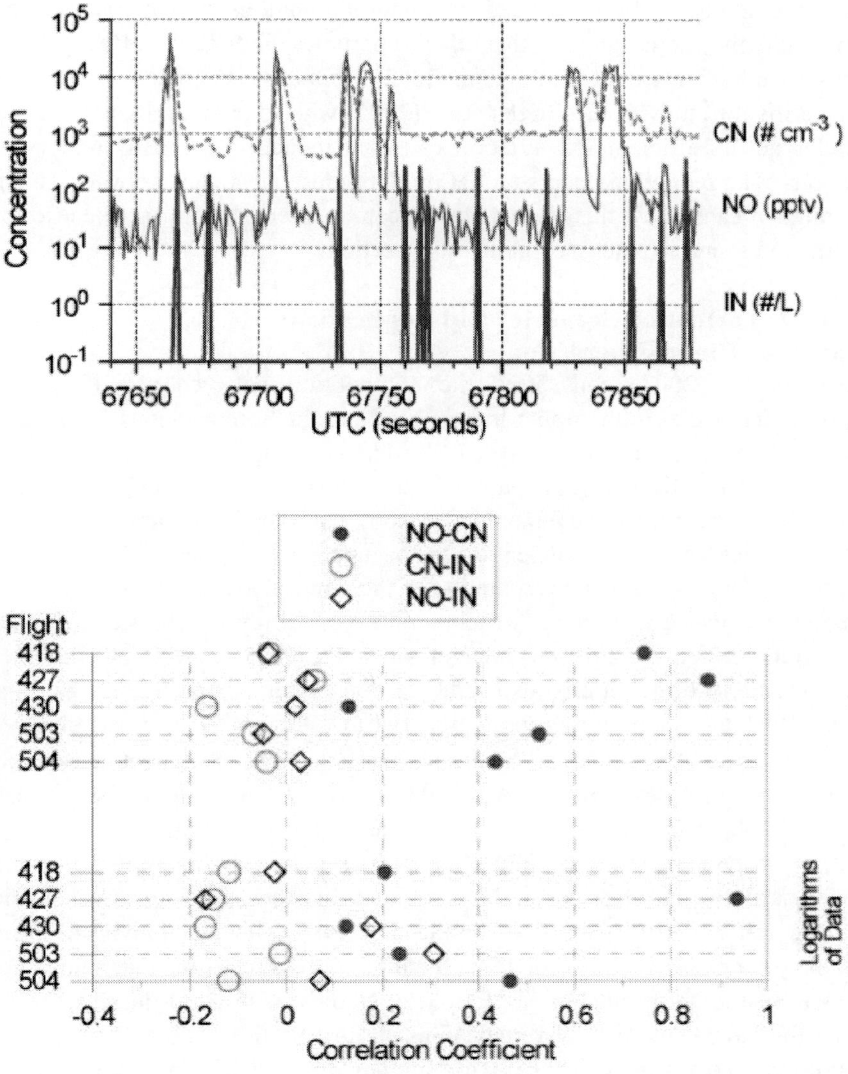

Figure 3. Top panel: Concentrations at 1 Hz of CN (dash line), NO (thin) and IN (thick) for 240 s during several DC-8 aircraft penetrations of a T-39 aircraft exhaust plume on one day during the SUCCESS experiment. IN were processed at -30°C in the immersion freezing regime (supersaturated with respect to water) in the top panel results. Ambient temperature was -43°C. Bottom panel: Summary of correlation coefficients between NO, IN (T range of 235 to 250K) and CN for five flights, distinguishing (top) all values and (bottom) logarithms of values which excludes seconds when ice nuclei counts were zero.

In-situ measurements of the ice nucleation efficacy of exhaust have never been obtained for the temperature conditions representative of contrail or cirrus formation. There is a glaring scientific need for improved real-time measurement capabilities of ice nucleation involving ambient aerosols at cirrus levels. It is also notable that ground based studies of aircraft emissions, such as the PARTEMIS program (Wilson et al. 2004; Petzold et al. 2005), have studied the CCN activity of exhaust particles, but never the IN activities.

Despite the limited and inconclusive results from the afore mentioned laboratory studies, soot particles continue to be prescribed in model simulations as a major source of IN ice formation with the consequence that these particles indirectly affect climate model simulations to a major extent (Lohmann, 2002; Lohmann and Diehl, 2006). Specific to aircraft effects is the study of Hendricks et al. (2005) who found that, by assuming all aircraft soot particles act as efficient IN, aviation causes a reduction of cirrus ice crystal number concentrations. The reduction occurs at NH midlatitudes and ranges between 10 and 60%, in terms of annual means. How this impact ensues can be inferred from the specifics of the soot-cirrus interaction scenario discussed in the next section.

2.1.4.3. Cirrus Formation Scenarios and Implications of Aircraft Soot-Cirrus Interactions

Kärcher et al. (2007) considered the competition between aircraft generated soot particles, other ice nuclei in the upper troposphere and the homogeneous freezing process for liquid aerosol particles using a phys ically-based parameterization scheme for cirrus formation in adiabatically rising air parcels. This paper recognized that in order to have an important role as heterogeneous ice nuclei soot particles from aircraft emissions must compete with other known ice nucleators in the upper troposphere. Mineral dust particles, including those expected to be representative of the atmosphere, are known to act as efficient heterogeneous IN over a wide range of temperature conditions (Zuberi et al., 2002; Hung et al., 2003; Archeluta et al., 2005; Möhler et al., 2006; Knopf and Koop, 2006; Kanji and Abbatt, 2006). Field observations also indicate that mineral particles, fly ash and metallic particles far from their source regions may serve at times as IN in cirrus clouds - in some cases without being associated with significant acidic or other condensed components (Heintzenberg et al., 1996; Chen et al., 1998; DeMott et al., 2003a; Sassen et al., 2003; Cziczo et al., 2004; Twohy and Poellot, 2005; Richardson et al., 2007).

Figure 4, from Kärcher et al. (2007), provides a description of the expected role of aircraft soot particles on cirrus formation provided that the ice nucleation behavior of such soot particles can be fully quantified. In the absence of such quantitative information an idealized, single threshold values for IN activation was assumed for that study. As a general conclusion from the abovementioned laboratory studies of mineral dust particles, onset ice saturation values for heterogeneous nucleation on dust particles are as low as Si =1-1.25 in cirrus conditions; however, Si is1.3.1.4 for smaller dust particles and 1.35.1.5 for kaolinite and montmorillonite immersed in aqueous (NH4)2SO4 droplets. These values can be used to constrain the competition between ice formation by soot particles and other atmospheric ice nuclei. The total number densities of ice crystals, ni, as a function of soot particle number density, ns is shown in figure 4. Besides ns, other crucial parameters that were varied in the calculations were: (i) the vertical wind speed, w, ranging from purely synoptic uplift (top panel) via mean values typically observed on the mesoscale (middle panel) to lee wave or convective forcing (bottom panel); (ii) the abundance of mineral dust particles as indicated by the legends, with concentrations nd increasing with w; and (iii) the threshold ice saturation ratios Scr for ice nucleation by the soot particles as given at the top of the figure, whereby the highest value is below but close to the homogeneous freezing threshold. The results are insensitive to variations of the liquid particle properties. Other assumptions about IN sizes are based on observational evidence, as described by Kärcher et al. (2007).

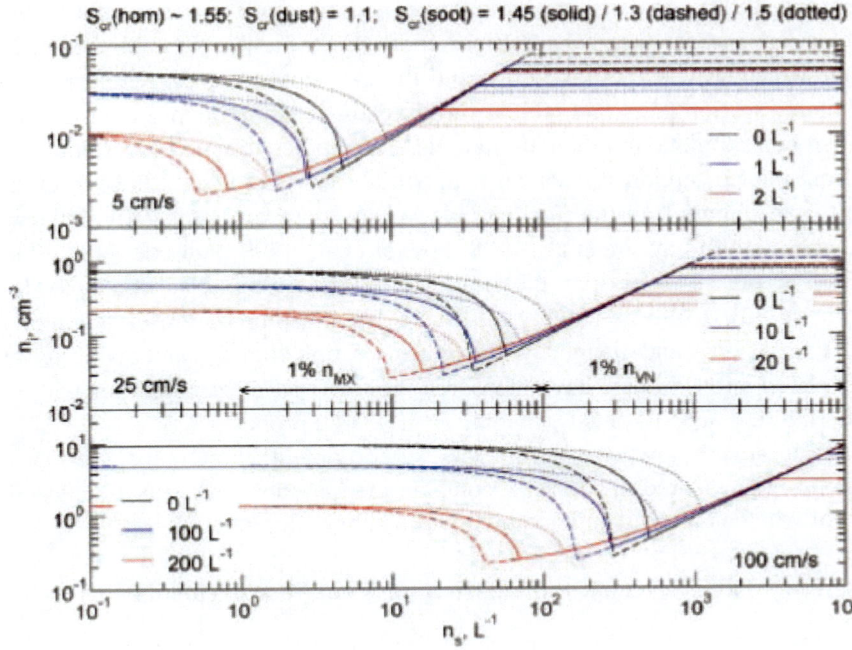

Figure 4. The calculated total number density of nucleated ice crystals as a function of assumed soot particle number density, from Kärcher et al. (2007). The curves result from competition of three particle types during ice formation in adiabatically rising cirrus air parcels. Results are shown for updraft speeds of 5 cm/s (typical synoptic-scale vertical wind), 25 cm/s (corresponding to typically observed background mesoscale temperature fluctuations), and 100 cm/s (strong updrafts in orographic waves or near convection). The air parcels start rising at 250 hPa and 220K at ice saturation and contain natural mineral dust particles with concentrations noted in the legends, a wide range of soot particle concentrations ns as indicated, and 500 cm.3 sulfuric acid particles ("hom" nuclei). The ice nucleation thresholds of these particle types are given at the figure top. The two arrows in the middle panel indicate the range of number concentrations of black carbon-containing type MX (aircraft exhaust coagulated with ambient particles) and VN (volatile/non-volatile mixed exhaust particles)particles typical for far-field plume ages up to 2 days, assuming that 1% of those are active IN.

In all panels in figure 4 the black curves denote the dependence of ni versus ns without interference by dust particles. In the limit of small ns, ni arises exclusively from homogeneous freezing of liquid particles and soot does not play any role in ice formation. The number of ice crystals formed by homogeneous freezing increases from 0.05 cm.3 to 10 cm.3 when the updraft speed increases from 5 cm s-1 to 100 cm s-1, highlighting the sensitivity of ni on the cooling rate. Increasing nd in this limit (from black via blue to red curves) decreases ni roughly by a factor of 2 (blue curves) and 5 (red curves) in each panel. These ice crystals, forming early (at lower S) during cooling, enhance the water vapor losses due to depositional growth and thereby reduce the rate of increase of supersaturation leading to fewer liquid particles freezing homogeneously.

Given that soot-induced changes in cirrus are not readily observable, if they indeed exist, the next section and those that follow return to a discussion of contrail cirrus.

2.1.5. Contrail and Cirrus Cloud Observations

Contrail microphysical properties—the ice water content (IWC), the total ice particle surface area (A) or extinction (σ), total ice particle number concentration (Nt), ice particle size distributions (PSD), ice particle mean number and mass or volume-weighted diameters (*D*, *D*m, v) and ice particle shapes have been reported in a number of published articles over the past three decades. These observations and the instruments used to collect the data, are shown in table 1. Section 2.2.4 and table 4 further elaborates on the measurement techniques used to collect contrail data and their limitations. The studies to date have sampled contrails properties from their inception (Goodman et al., 1998; Schröder et al., 2000) through to their development into contrail cirrus for periods of an hour or later (Knollenberg, 1972; Heymsfield et al., 1998; Lawson et al., 1998; Poellot et al., 1999; Schröder et al., 2000, Atlas et al., 2006). The observations span the temperature range -78C (Gao et al., 2004) to -30C (Gayet et al., 1996). Contrail ice crystals have been captured, preserved and analyzed (Goodman et al., 1998) and imaged with high-resolution digital cameras (Lawson et al., 1998). Most observations to date have measured the microphysical properties with optical 1-D light scattering (e.g., FSSP) or 2-D imaging (e.g., 2D-C) probes.

In the discussion below, we report ice water content (IWC) and ice particle size distributions (PSD) measured in contrails, compare them to those in cirrus clouds and address the question of whether contrail and natural cirrus can be differentiated.

Table 1. Key Contrail Microphysical Measurements

Study	Contrail Type/numbers	Key Findings	Particle Probe1
Knollenberg (1972)	Contrail>cirrus uncinus	Unexpectedly large IWC --Contrail evolved into natural cirrus --Some crystals developed to >0.5 mm	1D-C (75 μm-2.175 mm)
Gayet et al. (1996)	Contrail cirrus	Nt (>50 μm) up to 0.175 cm-3, larger than natural cirrus	2D-C (25-800 μm)
Goodman (1996)	Contrail	1 min. after generation --Nt~5-10 cm^{-3} -- *D*v 4-5 μm --Habits predom. plates --Shapes formed when crystals D>0.5 μm	Wire Impactor (>0.5 μm)
Heymsfield et al (1998)	Contrail>cirrus	Nt=10-100 cm- *D*=1-10 μm Contrail visible for >6 hours	MASP (0.3-20 μm) VIPS (20-200 μm) PI (50-1000 μm) 2D-C (50-1600 μm)
Lawson et al. (1998)	Contrail>cirrus	Crystal habits: columns and bullet rosettes When time>40 minutes, 1-20 micron crystals in contrail core with Nt~1 cm^{-3}	MASP (0.3-20 μm) PI (50-1000 μm)

Table 3. (Continued).

Study	Contrail Type/numbers	Key Findings	Particle Probe1
Poellot et al. (1999)	21 contrail clouds	Nt > 10 cm^{-3} *D*~10 μm	FSSP-100 (2-47 μm) 1D-C (20-600 μm) 2D-C (33-1056 μm)
Schröder et al. (2000)	12 Contrail Flights --Sampled up to 30 min. from generation	Nt > 100 cm^{-3} *D*=1-10 μm Ice Particles Spherical	FSSP-300 (0.3-20 μm) FSSP-100 (6–98 μm) 2D-C (20-650 μm) Replicator (>2 μm)
Schumann (2002)	Contrail>cirrus	Compilation of contrail IWC estimates	Many instruments
Gao et al. (2004)	Cold contrail	Presence of new class of HNO_3 containing ice crystals at T<202K	MASP [TDL]
Atlas et al. (2006)	Contrail cirrus	Tracked contrails with satellite and lidar	Ground-based Lidar MODIS

1D-C: One-dimensional Optical Array Probe
2D-C Two-dimensional Optical Array Probe
FSSP: Forward Scattering Spectrometer Probe
MASP: Multi-angle Aerosol Spectrometer Probe
MODIS: Moderate-Resolution Imaging Spectroradiometer
PI: Particle Imaging Nephelometer
Replicator: Continuous impactor-type probe producing ice crystal crystal replicas
TDL: Tunable Laser Diode Hygrometer
VIPS: Video Ice Particle Sampler- continuous impactor-type probe producing videos of ice crystals
Wire Impactor: Impactor-type ice crystal replication technique.

2.1.5.1. Ice Water Content

The ice water content of contrails has largely been estimated from particle size distributions measured by optical spectrometer probes (see Section 2.2.4). This calculation requires assumptions about the ice crystal shapes and their masses and is subject to a factor of two or more uncertainties[4] that extend beyond the issues of measuring contrail PSD (Section 2.2.4). Direct measurements of the IWC can now be made by such probes as a counterflow virtual impactor (CVI, Twohy et al., 1997) and the Fast In situ Stratospheric Hygrometer (FISH). Early on in the lifetime of a contrail, however, the ice crystals are often smaller five microns, the lower size limit that can be collected by CVI and related bulk samplers, or fall below the IWC (~0.001 g m-3) that can be detected by them. Details of their operating principles are given in Section 2.2.4.

Schumann (2002) summarized most IWC estimates in contrails and fit the temperature (T)-dependent relationship to these observations:

[4] Early on in the lifetime of a contrail the PSD are dominated by small ice crystals, which, for the calculations of the IWC, are considered to be solid ice spheres for lack of information on crystal masses (e. g., Schröder et al., 2000). This can lead to obvious uncertainties and usually overestimates IWC by up to a factor of two to three (see crystals collected in contrails ~40 to 80 seconds after the contrail was generated, Goodman et al., 1998).

$$IWC(g/m^3) = \exp(6.97 + 0.103T[^\circ C]) \times 10^3. \tag{2}$$

This curve, plotted in figure 5, is derived from data that spans the temperature range -67 to -30C and IWC from 0.001 to 0.07 g/m3, encompassing the majority of published contrail data acquired to date.

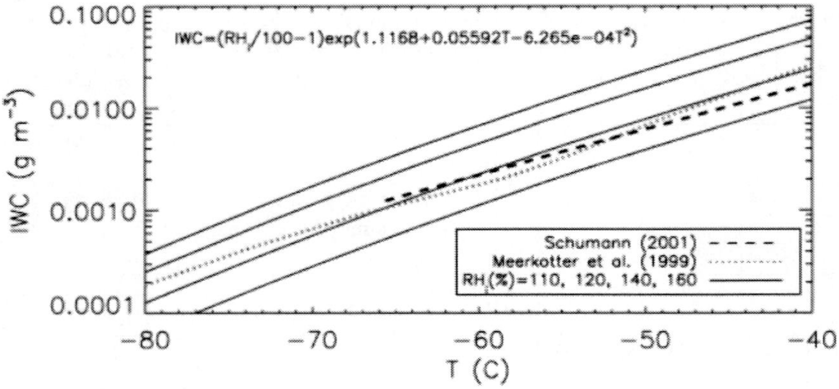

Figure 5. Contrail ice water content as a function of temperature fitted to observations by Schummann (2002), from the model of Meerkotter et al. (1999) and as given by Eq. 4.

To explain the magnitude of the IWCs observed in contrails, Meerkötter et al. (1999) modeled the "potential IWC" as half of the available condensate between the vapor density at the point of ice nucleation (~140-160%, depending on the temperature) and the vapor density at ice saturation (see figure 1). This model is supported by the observations (see figure 5). The Schumann curve fit and Meerkötter et al. model indicate that there is a factor of ten decrease in the observed or potential IWC as temperatures (T) are reduced from -40°C, the warmest possible contrail formation temperature, to -70C.

The potential IWC is a useful empirical representation of the IWC. In practice, however, wake and environmental turbulence produces mixing of contrail and ambient air. The contrail ice crystals are free to draw upon supersaturation in the environment, if present, for growth. Using a large eddy simulation (LES), Lewellen and Lewellen (2001) modeled air motion, moisture and ice crystal size distributions in contrails forming under a range of ambient relative humidities with respect to ice (RHi). Initially, the vortex pair generated by an aircraft descends rapidly for several hundred seconds. The positive buoyancy acquired by the vortex systems' descent through the stratified atmosphere produces Brunt-Vaisala oscillations that are damped by turbulence and mixing between the plume and the environmental air. When the environment is supersaturated with respect to ice, the IWC tends to increase by entrainment of moist air as the volume of the plume expands, its value set by the excess moisture above the ice saturation level. According to Lewellen and Lewellens' model simulations, the ratio of the actual IWC in the plume to its equilibrium value, given by the difference between the vapor density in the environment and its saturation value at the given temperature, is nearly unity throughout the contrail plume. There are obvious exceptions to this estimate, e.g., in the plume center or near the edges .

The representation of potential IWC therefore should account for the ambient humidity. As an approximation, we can take

$$IWC=\rho_a[X_i(RH_i/100-1)], \tag{3}$$

where ρa is the density of air, *Xi* is the saturation mixing ratio with respect to ice at the ambient temperature and *RHi* is in percent. This presumes that there is considerable supersaturation in the environment such that the dominant source of condensed water is the ambient supersaturation and not the aircraft exhaust, as assumed by Appleman (1953) and others since. The curves in figure 5 show the IWC as a function of temperature for the expected range of RHi for ice supersaturated layers in the upper troposphere. For a standard atmosphere this result is well-described by the relationship

$$IWC(g/m^3)= (RH_i/100-1)*\exp(a_o+a_1T+a_2T^2), \tag{4}$$

where a_0=-3.4889, a_1=0.05588 and a_2=6.268x10^{-4}; T here is in $^{\circ}$C.

We have reanalyzed some of the best in-situ contrail data collected to date to explore how well Eq. (3) predicts observations of the IWC within contrails. Some of the most reliable observations come from the 12 May 1996 SUCCESS case study when the DC-8 generated a contrail while flying in a racetrack pattern in highly ice supersaturated, cloud-free air (Heymsfield et al., 1998). Some 20 and 40 minutes after the initial contrail pass the DC-8 returned through the contrail, sampling it in a racetrack pattern. These penetrations occurred long after the times required for the wake vortices to develop oscillations that mixed the contrail plume with the environmental air, i.e. these samples can be considered as taken from the later stage of contrail evolution. The DC-8 then sampled the contrail particles as they grew in the ice supersaturated air (as ascertained from a TDL hygrometer) for almost two hours following contrail formation. The accuracy of the TDL hygrometer was established to be +/-5% based on contrail crossings and wave cloud penetrations at temperatures between -40 and -65C.

The track of the DC-8 during these three penetrations (Pens. 1-3) is shown in figure 6a. The temperature during Pen. 1 was a mean of -50.2 C (figure 6b), 2.20C cooler than during Pen. 2. Pen. 2 was an average of 250 m higher than the generation height, indicating that portions of the contrail had risen at an average rate of 11 cm/s between penetrations. Because the concentration of trace species NO and NOy measured from the DC-8 (Campos et al., 1998) were much above the ambient environmental values (see symbols, bottom of figures 6b-d, with high NO regions identified in Heymsfield et al., 1998 as NO>100 ppb) and the contrail cloud was in the shape of a racetrack, we know with certainty that we were within the contrail generated during Pen. 1. The environment was highly supersaturated, RHi>100%, during the contrail generation run (figure 6c). The RHi values approached the limit for homogeneous ice nucleation at these temperatures of about 160% (see Heymsfield et al., 1997 and others), although ice particles were rarely sampled (see later). Abrupt excursions of the RHi trace during Pen. 2—from high values to nearly ice saturation and visa versa-- clearly indicated turbulent patches of ice mass associated with the contrail plume. Pen. 3 was largely outside of the contrail with only brief encounters during climb and descent through locations of high NO. Following the third penetration the DC-8 sampled crystals descending and growing in the highly ice supersaturated air below the contrail.

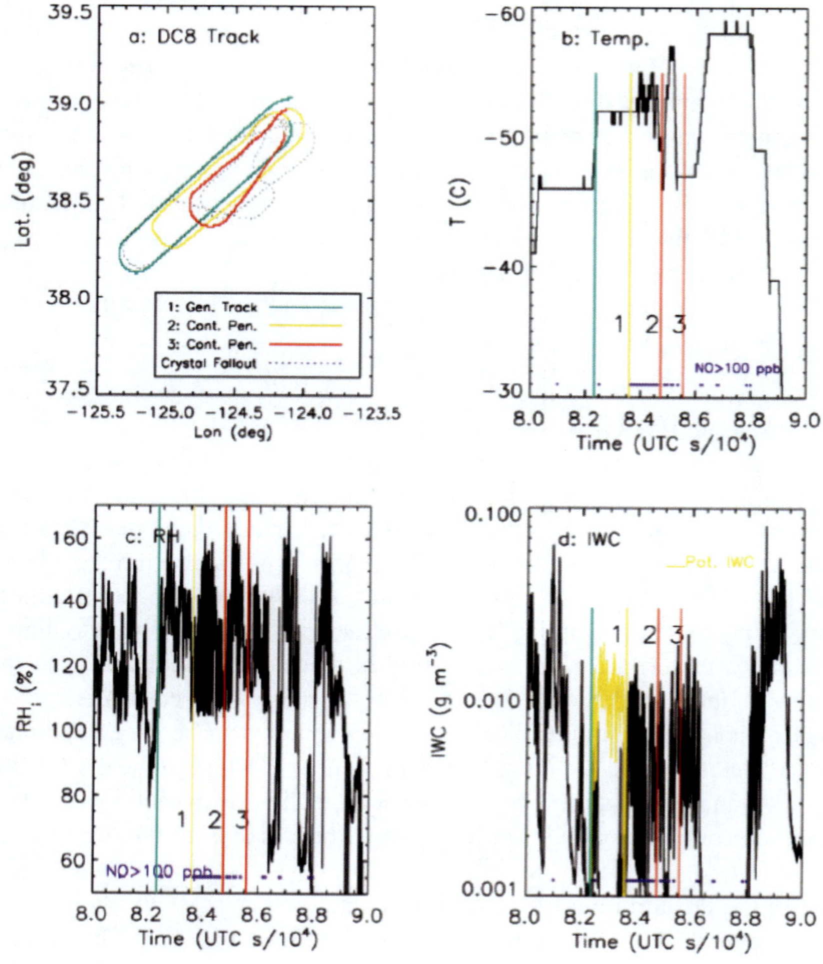

Figure 6. In-situ observations from contrail produced by NASA DC8 aircraft on 12 May 1996 the SUCCESS field campaign. The contrail was generated in region labelled 1 and was subsequently sampled in regions 2 and 3.

The IWC measured by the CVI (Twohey et al., 1997) indicates that there was initially no IWC in the environment (figure 6d). [The CVI should have detected IWC above 0.001 g/m3 because these crystals would have been cirrus crystals, with most of them above the CVI "cut" size of about 5-6 microns]. Abrupt fluctuations were noted in the IWC during Pen. 2 and larger fluctuations during Pen. 3. Particle sizes were primarily above the CVI "cut" size for both penetrations, as discussed in the next subsection.

The potential IWC—derived from Eq. (2), is shown for Pen. 1 in figure 6d. The mean potential IWC for Pen 1 was 0.0086+/- 0.00023 g/m^3 (figure 6d). In the regions of high NO, the IWC measured by the CVI for Pen. 2 was a mean of 0.0074 +/- 0.003 g/m^3. Given that the

IWCs were close to the detection threshold of the CVI and some particles could have been below the particle "cut" size of the CVI, the model and measurements agreed well.

Figure 7 shows an expanded view of the measured and potential IWC and RHI during Pen. 2 where, throughout the region, pockets of high NO were measured. There is consistency between the aircrafts' NO signature (i. e., the contrail plume), detection of IWCs by the CVI and the reduction in RH to near 100%. The tailing off of the measured IWC at the end of each contrail penetration is due to hysteresis of the CVI--where residual IWC remains in the CVI's detection chamber following passage through a cloud. In the contrail-free regions the potential IWC is close to the values in adjacent contrail regions and the RHI is significantly elevated above ice saturation. Although the growth of the ice in the contrail portions developed from near supersaturation at earlier times, it is clear from Pen. 1 that the potential and measured IWCs were comparable.

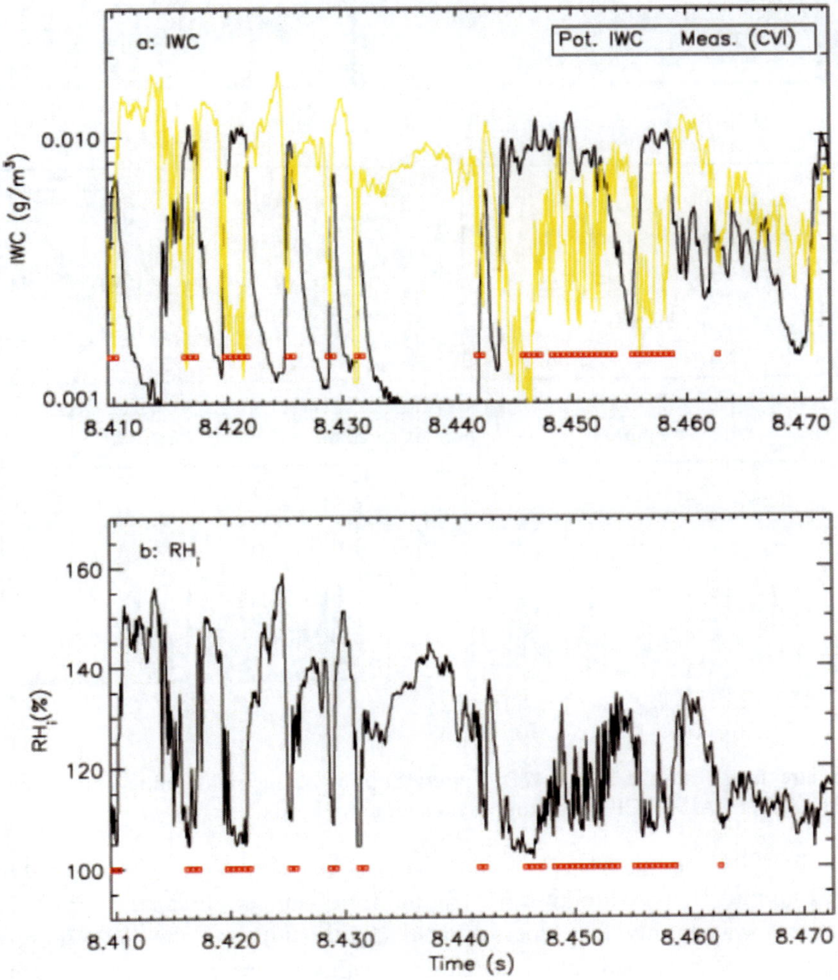

Figure 7. (a) IWC, both measured and potential (as derived from Eq. 1), and (b) RHi as function of time during a portion of Pen. 2.

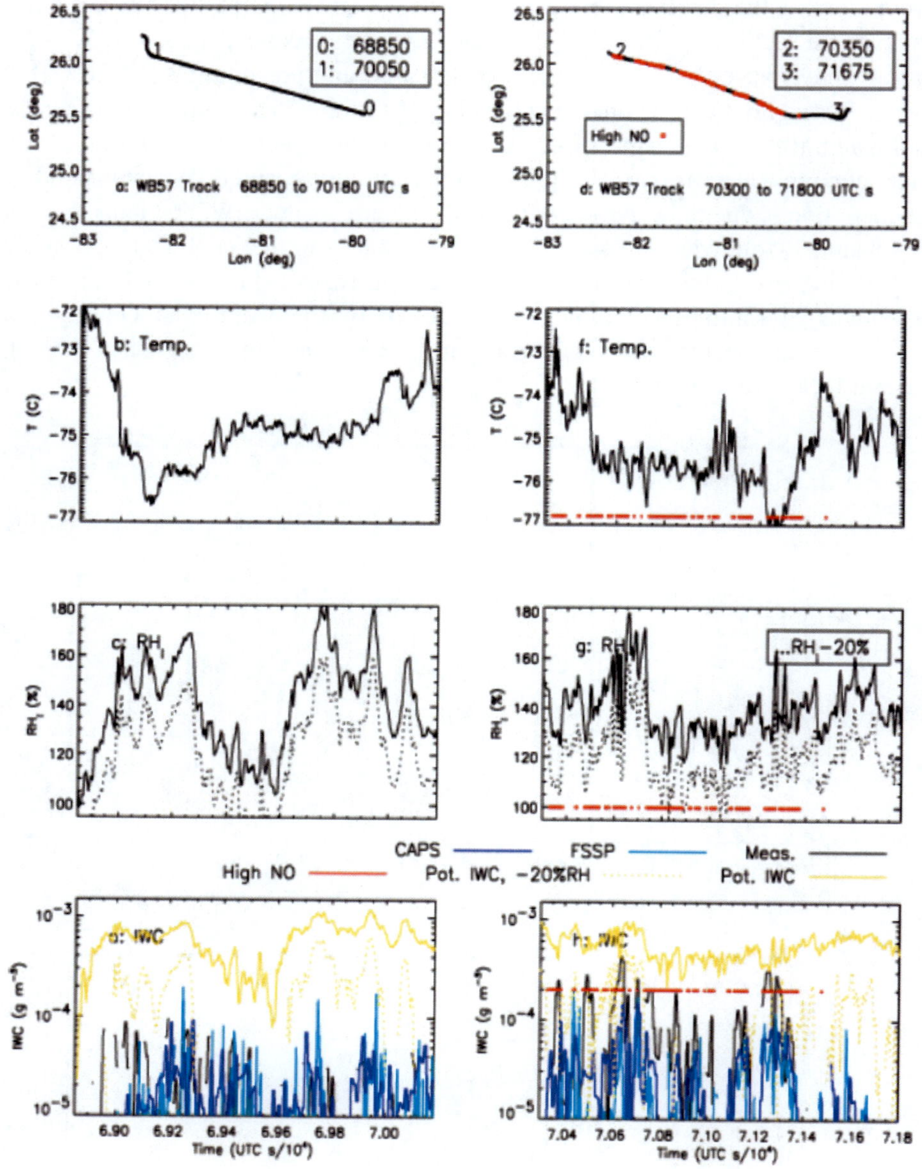

Figure 8. Measurements from the NASA WB57 aircraft during a penetration into a contrail on 13 July 2002 during the CRYSTAL-FACE field program in southern Florida.

On 13 July 2001 during the CRYSTAL-FACE field campaign in southern Florida the NASA WB57 aircraft flew a straight-track (figure 8a) at temperatures near -75C (figure 8b). The environment was highly ice supersaturated (figure 8c), and the WB57F produced a contrail. There has been considerable discussion on the accuracy of the RH measurements during this penetration because the peak values exceeded the threshold for ice production through homogeneous nucleation. A more conservative estimate would be 20% lower than the measured values (dotted line, figure 8c). A low IWC was measured and derived from the

FSSP and CAS PSD during the penetration (figure 8d) and ice concentrations given later are shown to be very low. There was considerable potential IWC.

The WB57 turned and penetrated its contrail (figure 8e), encountering regions of high NO that tagged the contrail presence. The temperature trace on the return leg was a mirror image of the trace from the first penetration (figure 8f). Ice supersaturation was considerably lower than during the initial sampling and when reduced by 20% it agrees with the expectation that RHi is about 100% within a contrail. The measured and derived IWCs are about three times larger than those from the initial penetration and are within 30% of the potential IWC from the first penetration.

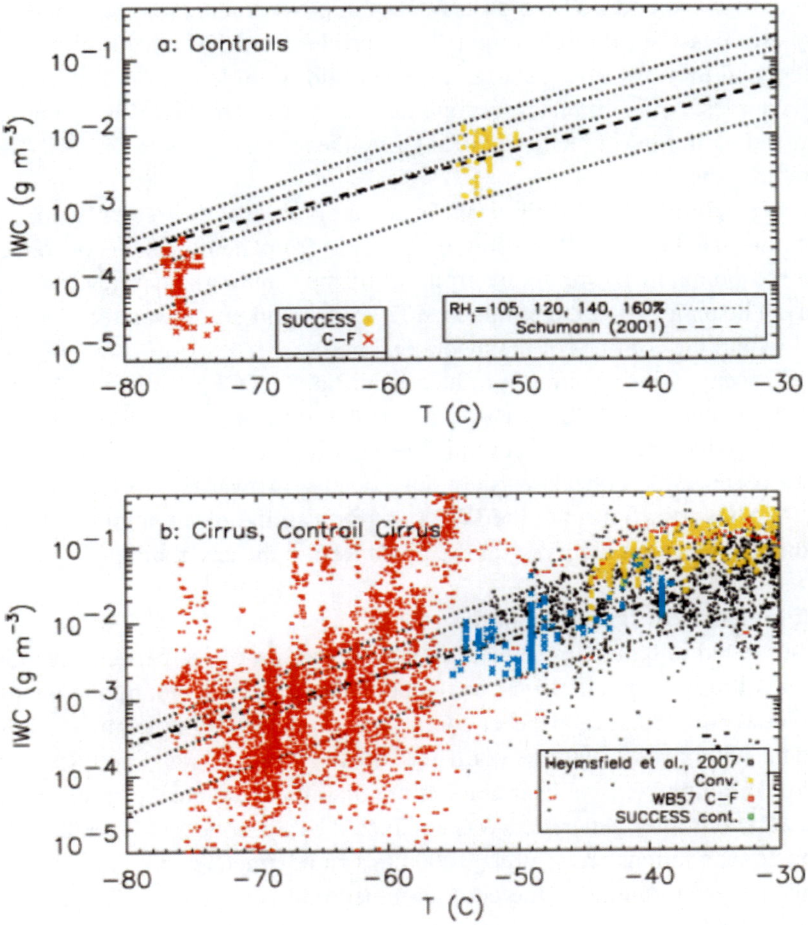

Figure 9. (a) Measurements of the IWC as a function of temperature for the SUCCESS and CRYSTAL-FACE contrail observations, and (b) from measurements in cirrus and in cirrus developing from the SUCCESS contrail.

The IWCs measured during the SUCCESS and CRYSTAL-FACE contrail cases, discussed above for the periods when the aircraft were in regions of high NO following the initial contrail generation run, are shown in figure 9. Given the high RHi noted in each of the cases—nearly 160%, the measured peak IWCs are consistent with Eq. (2). The IWCs are lower than they otherwise may have been because of the five second averaging times used in figure 9 to get a reliable value from the PSD.

These observations for young, non-precipitating contrails can be compared to measurements of the IWC in cirrus and in ice crystals falling out into ice supersaturated air below contrail (i. e., contrail cirrus). Heymsfield (2007, hereafter H07) report on IWCs acquired during nineteen Lagrangian spiral descents from the top to base of mid and low-latitude ice clouds (figure 9b) made during several field experiments where the IWCs were measured directly in most instances. A total of 5000 data points, spanning the temperature range -65 to 0C, are included in this data set. All but four of the spirals are from cirrus clouds generated by in-situ vertical motions; the others, identified in the figure, were outflow generated by deep convection. These observations improve upon earlier estimates of IWC(T) because they are measured directly rather than derived from PSD (with assumptions about crystal masses) and they are from contiguous penetrations of cloud from top to bottom rather than random samples. An additional ice cloud data set—from the WB57F during CRYSTAL-FACE (discussed earlier but for natural cirrus), provides measurements primarily in the -60 to -80C temperature range.

Many of the natural cirrus exhibit IWCs much above the theoretical values, signifying that most of this ice has been transported upwards from below in deep convection. For example, the red points in figure 9b are from all of the Lagrangian spirals through anvils in the H07 study. The majority of the points in H07 are from in-situ generated cirrus and for the most part fall within the confines of the theoretical curves.

For the 12 May 1996 contrail produced during SUCCESS the DC-8 followed the development of ice particles for almost an hour as they fell into the highly supersaturated air (RHi were 120-140%) below the contrail (Heymsfield et al., 1998). This provided an opportunity to observe the crystals as they developed downwards into typical bullet-rosette type shapes. As shown in figure 9b the IWCs conform to the observations for natural cirrus following along the theoretical curves for the given RHi in the environment.

2.1.5.2. Ice Particle Size Distributions

The studies listed in table 1 examined the characteristics of ice particle size distributions in early contrails through to contrail cirrus. Goodman et al. (1998) found the ice particle size distributions some 40 to 80 seconds after contrail generation at a temperature near -61C to be nearly unimodal with a mean volume diameter below 20 microns and a concentration range of 6 to 13 cm^{-3}. A comprehensive examination of contrail crystal characteristics from seconds to greater than an hour after generation was conducted by Schröder et al. (2000). Shortly after generation the concentrations are of order 1000 cm^{-3} and diameters about a micrometer. In the presence of ice supersaturations in the contrail generation zone and for temperatures in the -50 to -55C range, aging over a one hour period leads to larger mean diameters, about eight micrometers, and reduced concentrations due to plume mixing, of about 10-15 cm^{3}. Heymsfield et al. (1998) sampled a contrail for more than an hour from the time of its generation by the NASA DC-8 (see discussion and figures in previous section). Temperatures in the environment were about -52C and RH_i approached 160%. The total ice concentrations within the contrail were 10-100 cm^{-3} and the concentrations of ice crystals >50 μm reached 10-100 l^{-1}. The concentrations for > 50 μm crystals are comparable to those found in the > 50 μm sizes in the Knollenberg (1972) and Gayet et al. (1996) studies. As suggested by Heymsfield et al. (1998) favorable conditions for contrails to develop cirrus-like microphysics include a largely cirrus-free environment, a sustained growth period for crystals in high supersaturated conditions and sustained upward vertical motions leading to a deep layer of high ice supersaturation. Such conditions are likely to have been present in the environment in the study of ice virga developing from contrails (e.g., Knollenberg, 1972 and Atlas, 2006).

A detailed examination of the data from the 12 May 1996 case from SUCCESS provides considerable insight into the microphysical properties of contrails that develop into contrail cirrus and to the issues that are raised about particle probes in Section 2.3.4. Prior to the contrail generation the MASP sampled total concentrations (Nt, > 1 μm^2.) of order 10 cm^{-3} (figure 10a). In the virga falling from the contrail following Pen. 3, where the NO concentrations < 100 ppb, Nt were of order 10-20 cm^{-3}. Few particles were measured during the contrail generation run (Pen. 1). During Pen 2, where NO concentrations indicated prior passage of the DC-8 aircraft, Nt were clearly enhanced, to 20-80 cm^{-3}.

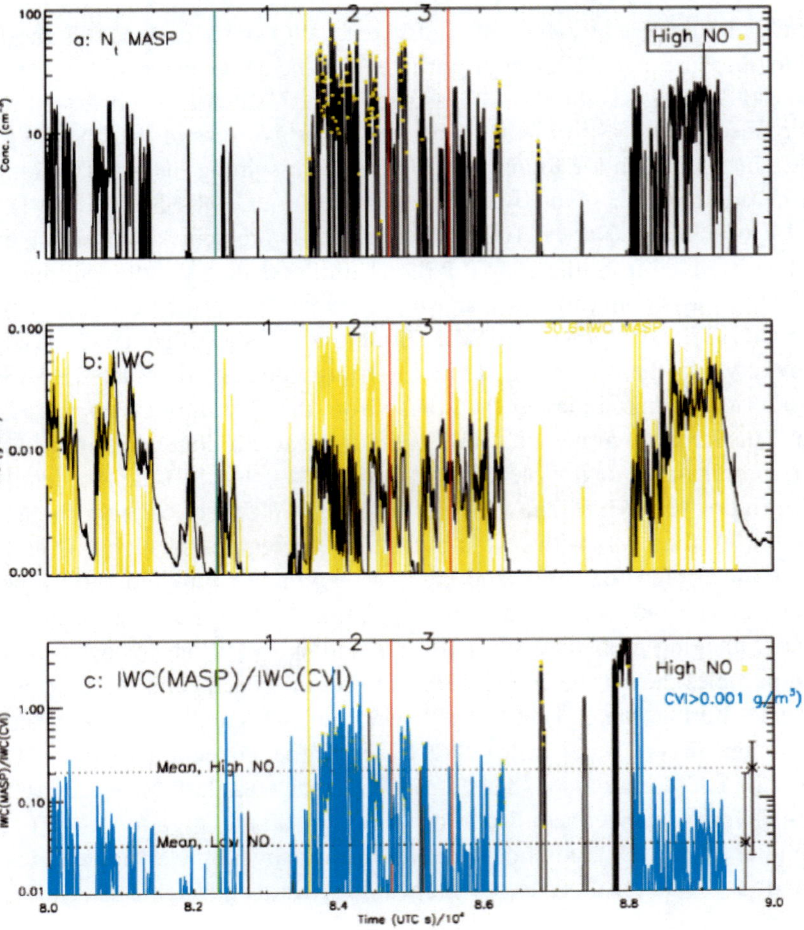

Figure 10. Particle size distribution observations on 12 May 1996 during the SUCCESS field campaign. Regions of high NO, where contrail particles are sampled, are shown.

Ice water contents measured above the CVI's cut size of about 5 μm are several hundredths g/m3 prior to Pen. 1 (figure 10b). In the virga falling from the contrail following Pen. 2 there is a steady increase in the IWC as the DC-8 circled downwards in the contrail cirrus crystals falling into highly ice supersaturated air (see figure 6). The crystals in the contrail were accumulating mass as they developed downwards.

Recent studies and the discussion below in Section 2.3.4 point to a potential high bias in ice concentrations measured in sizes < 50 μm due to shattering of larger particles on the inlets of the probes like the MASP and FSSP. Heymsfield (2007) argues that a nearly constant fraction of the ice mass in large particles swept out by the small particle probes' forward surfaces is shattered and enters the probe's sensing area where it is regis tered as real particles and developed a linear relationships to express these dependencies of the form

$$\text{IWC (FSSP or MASP)}=\text{IWC(small particles, real)}+C^*\text{IWC(large particles)}, \quad (5)$$

where C is an empirically-derived coefficient, IWC(small particles) is the IWC that we are trying to measure and IWC(large particles) is the IWC measured by the 2D probes. Heymsfield (2007) suggests that IWC measured by the CVI can be substituted for IWC(large particles) because IWC(FSSP or MASP) is usually 10% or less of the CVI IWC. The scaled MASP IWCs in figure 10b are found from Eq. (3) by assuming that the IWC(small particles, real) is 0 g/m^3 and that C is given by the mean ratio of IWC(MASP)/IWC(CVI) for the times in figure 10 where the NO < 100 ppb, indicating ambient (contrail-unperturbed) air. IWC(MASP) is derived from the MASP PSD> 1 μm, assuming that the particles are solid ice spheres. What is most noticeable is that where NO < 100 ppb there is excellent agreement between IWC(CVI) and C*IWC(MASP), strongly suggesting that IWC(small particles) is negligible relative to the shattering term. In strong contrast, the scaled MASP IWCs are almost an order of magnitude larger than those from the CVI during Pen. 2, the first contrail penetration. This strongly argues that IWC (small particles) dominates in this case and that shattering is producing a negligible contribution to IWC(MASP). These details are further illustrated in figure 10c, which shows the ratio of the IWC derived from the MASP PSD to those measured by the CVI, with points indicating the regions of high NO and dotted lines indicating the mean ratio for "low" and high NO regions. Figure 11a shows the ratio of the total number concentration measured by the MASP to the total number concentration derived from the 2D-C imaging probe (for sizes 100 μm and above). Noteworthy is the near absence of >100 μm particles during the contrail generation run (Pen. 1) and the first sampling of the contrail particles during Pen. 2. Following Pen. 3, the DC8 sampled the virga falling from the contrail it generated (see time after 88000 sec). The concentrations of 2D-C particles decreased as the DC-8 descended. This is partially attributable to size sorting. Because the smaller crystals have a slower sedimentation velocity than the larger ones—20 versus 80-100 cm/s, the smaller ones are left behind. Furthermore, as shown by Heymsfield et al. (1998), the largest contrail particles for this case developed downwards through aggregation. This would have also reduced the concentrations.

Figures 11b-f show the development of the PSD from the first penetration of the contrail (see times for the 5-sec average samples, top of figure 11a) through to sampling at its lowest levels. Initially, the PSD were narrow (figure 11b). The PSD broadened downwards in the contrail cirrus, with crystals > 1 mm developing.

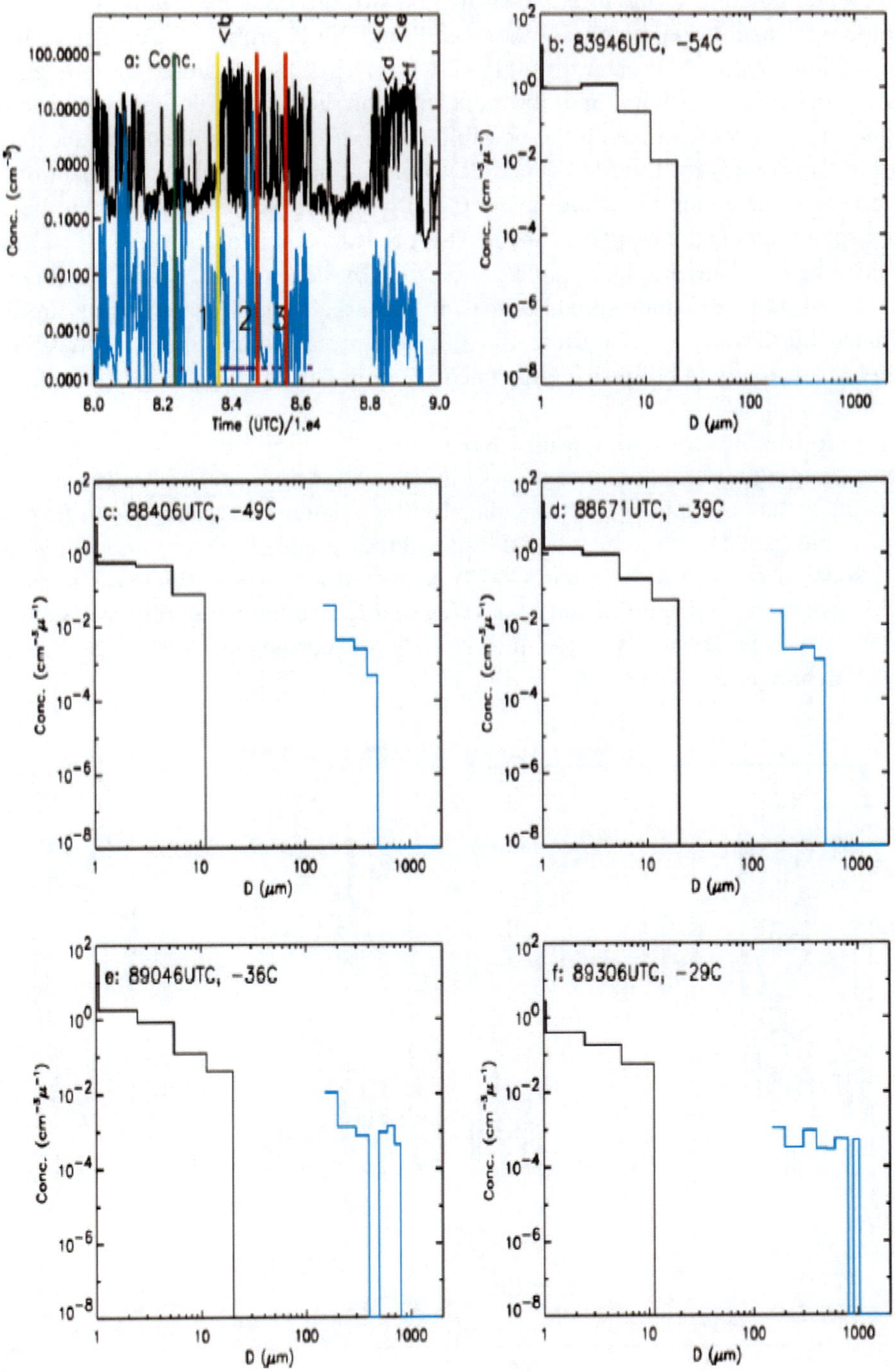

Figure 11. (a) Total concentrations measured by the MASP and 2D-C probes, and (b-f) PSD measured within the contrail plume and in fallout from it, on 12 May 1996 during the SUCCESS field program. In (a), the time of collection of each PSD is shown.

Using a combination of the direct measurements of the IWC from the CVI and the total concentrations measured separately by the MASP and 2D-C probe, we can derive the mean massweighted diameter from Dm=[6/pi*IWC(CVI)/Nt]0.333. Because Nt will always be dominated by MASP particles and the concentration is suspect in regions where large particles are present, we can bound the possible range of Dm and evaluate its uncertainty by deriving Dm separately for the MASP and 2D probes. Further, we must use data only from the CVI above its detection threshold, about 0.004 g/m3. As figure 12 shows, in the contrail generation region Dm is dominated by the MASP particles and is of the order of 10 μm; these estimates can be considered to be reliable. In the zone of fallout of the contrail particles, there is almost an order of magnitude spread in the Dm values. Although these regions are likely to be dominated by the 2D IWCs and therefore those Dm values are likely to be reliable, the shattering issue is likely to produce a large uncertainty in Dm.

2.1.5.3. Formation of Contrail Cloud Layers

A long-standing observation in regions of heavy air-traffic is that contrails tend to appear in groups rather than as single objects (Kuhn, 1970; Carleton and Lamb, 1986; Bakan et al., 1994; Duda et al., 2004). While the contrails spread they gradually merge into an almost solid interlaced sheet, a contrail deck. Kuhn (1970) estimated an average thickness of 500 m for such decks over Barbados and California, a value that is also the measured average thickness of ice supersaturated layers over the mid-latitude meteorological station of Lindenberg, Germany (Spichtinger et al., 2003a).

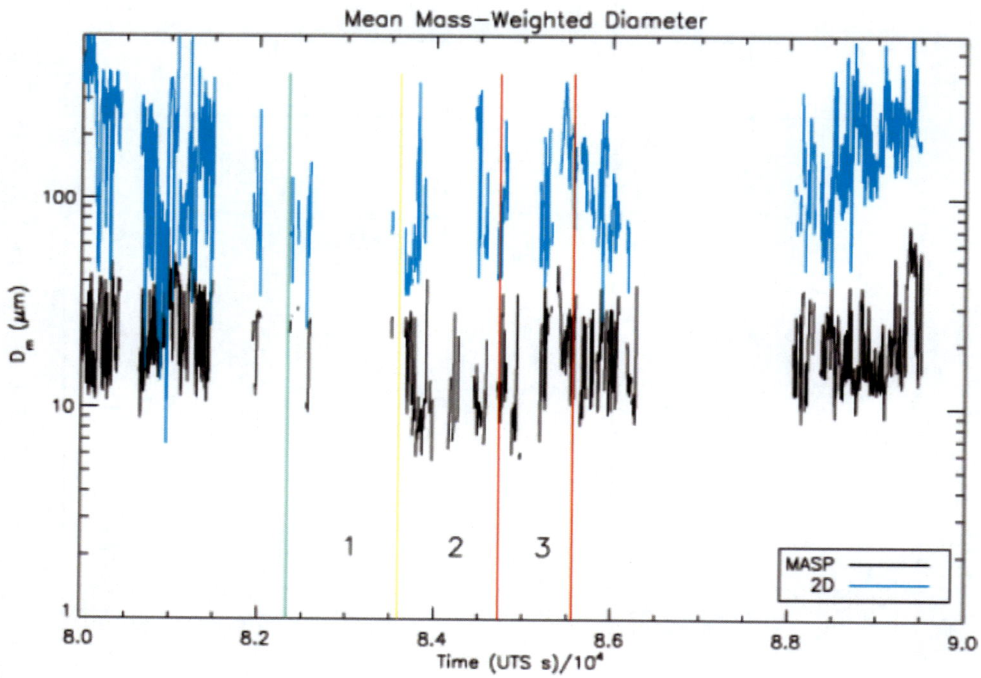

Figure 12. Mean mass weighted diameter as derived from the CVI IWCs, above the probes cut size, and the total concentrations from the MASP and 2DC probes.

Ice supersaturation is often formed by synoptic scale uplifting (e.g. Spichtinger et al. 2005a) which would favor contrail deck development, e.g., Knollenberg (1972) observed an extensive contrail deck developing ahead of a massive upslope snowstorm along the front range. There is only one simple process study of contrail deck development (Gierens, 1998) that only considers the spread dynamics but not the contrail microphysics.

Individual contrails spread horizontally and vertically with rates depending on ambient conditions. Freudenthaler et al. (1995) measured horizontal spreading rates in the range 18 to 140 m/min with a scanning lidar in time frames up to one hour. The average spreading rate determined by Duda et al. (2004) for a contrail outbreak over the Great Lakes region on September 9, 2000 was 45 m/min, i.e. in the middle of the range given by Freudenthaler et al. (1995). Cross sectional areas were observed to increase with rates between 3500 and 25000 m^2/s. Vertical growth rates are often limited by the thickness of the supersaturated layer, in particular at the lower contrail edge, but growth rates up to 18m/min have been measured with the lidar. Vertical growth of contrails is sensitive to the ambient profile of potential temperature (stability) and to radiative heating or cooling within the body of the contrail. Gierens and Jensen (1998) modeled how a contrail can rise 400 m through a very dry layer because the potential temperature at flight altitude was higher than in the layers above leading to strong buoyancy of the plume when it reached the unstable layers. Jensen et al. (1998) showed in other simulations that strong radiative heating of a thick contrail causes a 5 cm/s uplift of the contrail, resulting in a total uplifting of several 100 meters within an hour. Atlas et al. (2006) found convective turrets developing in contrails, probably as a result of radiative heating. In contrast, latent heating seems unimportant for the dynamical evolution of contrails even in cases with substantial supersaturation (hence substantial ice growth). Contrail spreading is controlled mainly by wind shear and ambient humidity. Under conditions of relatively little supersaturation contrail spreading can make them subvisible clouds. Under sufficiently moist conditions (more than 125% RHi) horizontal contrail spreading is affected by processes that control the vertical growth of contrails, the taller a contrail, the more effective the wind-shear. Strong turbulence, e.g., clear air turbulence, with Richardson numbers of $Ri \sim 0.1$ causes 20-fold increase of the vertical diffusivity Dv, compared to a calm background situation (Dürbeck and Gerz, 1996). In contrast, turbulence is unimportant for the horizontal diffusivity Dh. Dürbeck and Gerz (1995, 1996) conducted numerical experiments to determine the typical range of diffusion constants in the free troposphere. Typical values are: Dh [5,20]m^2/s and Dv [0,0.6]m^2/s (in calm situations). In cases with wind shear there is also a slant diffusion parameter Ds, which is typically $0.4\ (Dh{*}Dv)^{0.5}$. Dh increases with atmospheric stability but Dv decreases because turbulent diffusion has to work against gravity. The simulations also show contrail width increasing approximately linearly with time for as long as half a day. One should note here, however, that the simulations used a passive mass-less tracer. The results are not completely applicable for contrail to cirrus transformation when ice crystals are sedimenting.

2.1.5.4. Summary of Contrail and Contrail Cirrus Microphysics

The following summarizes the observations of subsections 1-3:

Total ice crystal concentrations in the initial stages of contrail formation are of order 103-104 cm-3 in the center of the plume.

During the early contrail dispersion phase, which begins some five minutes after contrail generation, N_t is of the order of 10-100 cm^{-3}.

Observations of the ice crystal shapes during the initial contrail formation are sparse. The available observations suggest that when micrometer size they may be irregularly shaped, not spherical. Quasi-spherical, droxtal shapes have also been observed (Schröder et al., 2002).

Contrail crystal shapes evolve to those observed in natural cirrus that develop under the same conditions, i.e. bullet rosettes when the environment is supersaturated with respect to ice.

Direct measurements of ice water contents during the early contrail dispersion phase, when the IWC is dominated by particles five micrometers in diameter or larger (depending upon the temperature), are the most reliable. The IWC measurement, together with the measured ice crystal concentrations, allows a reliable determination of the median volume diameter of the PSD.

A conceptual model for IWC production during the contrail dispersion phase that converts all of the (ice) supersaturated vapor condensate is consistent with a reanalysis of measurements from the SUCCESS and CRYSTAL-FACE field campaigns spanning the -50 to -75C temperature range.

2.2. Present State of Measurements and Data Analysis

2.2.1. Current Understanding of Possible Past Trends in Contrail and Cirrus Coverage and Their Association with Aviation Traffic

A long-standing question in relation to air traffic has been whether aviation increases the average cloudiness and whether it affects other weather parameters like daily sunshine duration and temperature range. More cirrus, formed from contrails, is potentially possible because 10 - 20% of the atmosphere, at typical subsonic flight altitudes, is cloud-free but ice-supersaturated (Gierens et al. 1999).

Boucher (1999) took ground and ship based cloud observations for the period 1982-1991 and grouped them into early (1982-1986) and late (1987-1991) periods. He then correlated the differences, late minus early, of cirrus frequency of occurrence, ΔC (change in cloudiness), in 3°X3° grid boxes with the aviation fuel consumption, F, in the same grid boxes. He found a positive correlation between ΔC and F. The highest ΔC occurred in major air flight corridors, NE USA (+13.3%/decade) and the North Atlantic Flight Corridor (+7.1%/decade). This study concluded that effects of volcanoes, long term changes in relative humidity or climate variations related to the North Atlantic Oscillation (NAO) could not explain the trend in cloudiness or its regional distribution.

Minnis et al. (2001) performed a similar study with the addition of satellite data and found consistency in trends of cirrus and contrails over the USA but not over Europe. This might point to other important influences on cloudiness that are stronger in Europe than in USA.

Zerefos et al. (2003) took other potential factors into account in their study, namely El Niño Southern Oscillation (ENSO), NAO, and the Quasi Biennial Oscillation (QBO). They deseasonalized the cirrus time series and removed the ENSO, NAO, and QBO signals. Possible effects of changing tropopause temperatures and convective activity were removed by linear regression and only the residuals were correlated with air traffic. Cirrus frequency was found to increase, sometimes with statistical significance in regions with heavy air traffic; however, an overall decrease in cirrus frequency was found. Consistent with Minnis et al. (2001), the most significant correlations were found over North America (winter season) and over the NAFC (summer season), whereas the correlations over Europe were insignificant (at a 95% level).

Stubenrauch and Schumann (2005) studied satellite data (1987-1995) for trends of effective high cloud frequency. They introduced a new element in these studies by grouping their data into three classes according to the retrieved upper tropospheric UTH_i (an average of relative humidity over a thick layer in the upper troposphere, say from 200 to 500 hPa). These three classes were grouped as: (1) UTH_i high enough for cirrus formation, (2) UTH_i insufficient for cirrus formation but sufficient for contrail formation and (3) clear sky. This additional classification of the data led to a clear positive trend, +3.7%/decade over Europe and +5.5% over NAFC in effective high cloud amount while the overall trend for all classes combined was weak.

Stordal et al. (2005) found from an analysis of satellite data (1984-2000) that the time series of cirrus coverage C(t) and air traffic density D(t) (flown distance per km^2 per hour) are generally positively correlated. The correlation is inferred from a linear relation: $dC/dt = b\ dD/dt$. Estimated correlations are not strong, partly because other influences mentioned above, have been left in the C(t) time series. They conclude that aviation over Europe produces an extra cirrus coverage of 3 to 5%.

Mannstein and Schumann (2005) also correlated C(t) with D(t), however for 2 months of cirrus data from METEOSAT and actual air traffic data from EUROCONTROL. For relating cirrus cover and traffic density they used a representation that takes overlapping of contrails and saturation effects (e.g. finite size of ice-supersaturated regions) into account: $C(t) = C_i(t) + C_{pot}[1-\exp(-D/D_*)]$, where $C_i(t)$ is cover of natural cirrus, C_{pot} is the potential coverage of persistent contrails (Sausen et al., 1998), and the term in square brackets is the fraction of C_{pot} that is actually covered by contrails. It was shown that the relation between additional cirrus coverage and air traffic density indeed followed roughly the exponential model. The main result of this study was that over Europe aviation is responsible for an additional cirrus coverage of 3% (consistent with the result of Stordal et al.). Unfortunately, it later turned out that the studied data are subject to a serious selection effect which produces an apparent correlation of unknown size (Mannstein and Schumann, 2007).

2.2.2. Range of Radiative Forcing Calculated for a Given Contrail Coverage and What Atmospheric Processes Govern this Range

Once formed, a contrail's direct radiative impact is through scattering of solar radiation and absorption and scattering of longwave thermal radiation. Contrail time of formation, lifetime, size, shape, altitude and crystal optical properties all affect the radiation field. The background atmosphere, especially cloud fields and surface albedo, also affect the radiative forcing by the contrails.

Sensitivity studies have been carried out to explore the importance of many of these factors and to test the impact of using different radiative schemes (e.g. Meerkötter et al., 1999). Contrails' reflection of solar radiation lead to a negative forcing but absorbed/trapped longwave radiation leads to a positive forcing. The overall effect of a contrail, i.e. its net forcing, is expected to be a positive forcing; however, since the net effect is a result of a cancellation of shortwave and longwave terms of roughly equal magnitude, the contrail net forcing is very sensitive to errors in either term. Zhang et al. (1999) have demonstrated that for a given ice water path the net forcing of cirrus clouds is basically determined by ice particle size and that it potentially changes sign from warming to cooling with decreasing particle size. This suggests the possibility that contrails and contrail-induced cirrus, when they are dominated by very small ice crystals, would act climatically opposite to natural cirrus clouds and cool the climate. However, for realistic ice crystal sizes and IWC, contrails are expected to give a net warming effect.

Sensitivity studies like those by Meerkötter et al. (1999), (see figure 13 and table 7) show that the determination of ice crystal size, ice water content and optical depth is key after the contrail coverage is known. These three quantities are all related and the size of the contrail radiative forcing varies more or less proportionally to all of these. Constraining at least two of these quantities is needed before an accurate estimate of contrail forcing can be made. [See Section 2.3.4.]

Table 7. Contrail Radiative Forcing Sensitivity Study From Merkotter Et Al., 1999

			Net forcing in Wm^{-2}		
Case/Parameters	Parameter range	τ	N	FL	M
References		0.52	37.1	37.2	37.2
Different aspherical particles	Spherical - aspherical	0.4	37-22	37-34	37-36
Solar zenith angle	60^0-21^0	0.52	37-48	37-45	37-46
Ice water content (IWC)	7.2-42 mg m^{-3}	0.2-1.0	19-51	18-53	19-52
Particle Radius	5-20 μm	0.85-0.21	41-20	-	-
Surface Temperature	289-299 K	0.52	34-39	36-39	35-40
Optical Depth of Low-level clouds	0-23	0.52	37-40	37-39	37-40
			Net forcing in Wm^{-2}		
Case/Parameters	Parameter range	τ	N	FL	M
Surface albedo	0.05-0.3	0.52	31-40	34-39	34-40
Relative humiduty	Reference - 80 %	0.52	37-31	37-31	37-31
Contrail depth (for fixed ice water path)	200 m -1 km	0.52	37-37	-	37-37
Lower contrail top (for fixed IWC)	11-10 km	0.52	37-31	37-32	37-31
Lower contrail top (increased IWC)	11-10 km	0.52-1.32	37-45	-	37-41

Other sensitivity issues include:

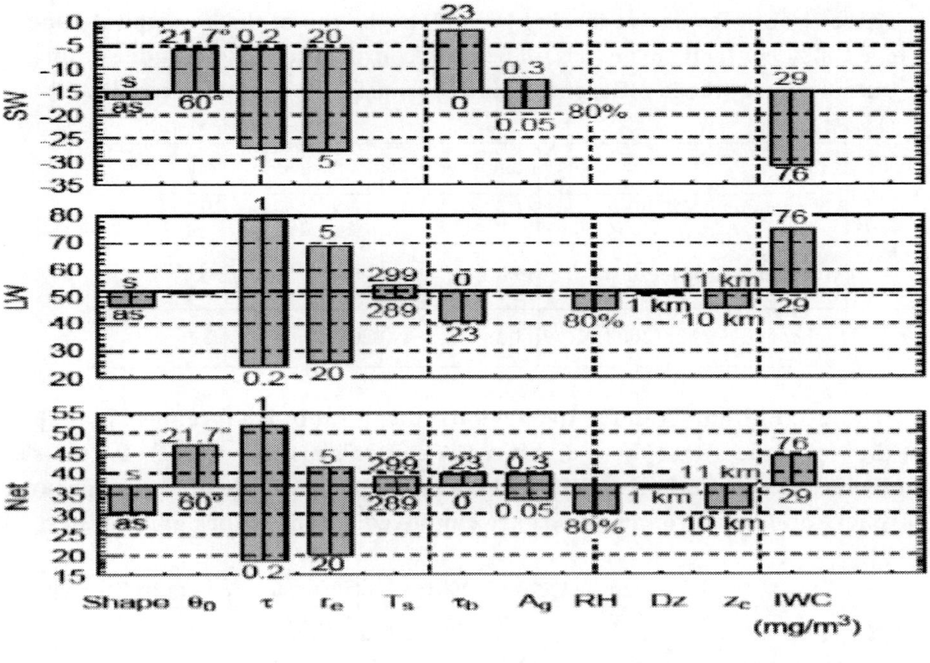

Figure 13. Results of a sensitivity study using model N. The bars indicate the range of variation of shortwave (SW), longwave (LW) and net (Net) flux changes in Wm^{-2} for 100% contrail cover due to the given variation of the parameters: particle shape, solar zenith angle θ_0, IWC or optical depth τ, volume-mean particle radius r_e, (in μm), surface temperature T_s (K), optical depth of lower level cloud τ_b, ground albedo A_g, ambient relative humidity RH (% of liquid saturation), contrail layer depth Dz (km), cloud top level Ze (km), and contrail at 1 km lower level but with increased ice water content IWC (mg m^{-3}).

Radiative Models

Different plane parallel radiation schemes employing different background atmospheres and cloud cover but similar input parameters for contrails give very similar forcings (tables 2 and 3). Note that a comparison of results from previous work (table 2, Models N, L, M, and table 3) indicates that, provided when contrail coverage and optical properties are known the contrail forcing should be readily calculable; however, because they are optically thin, radiation schemes need to account for scattering in the longwave as well as the shortwave to correctly model contrail forcing. Furthermore, all previous estimates of global radiative forcing employ radiation schemes that adopt the plane-parallel approximation and use the same underlying contrail ice particle size distribution proposed by Strauss et al. (1997). This raises the question whether the studies noted in table 3 are truly independent. Given that contrails are thin lines of cloud three dimensional effects and scattering from the sides of contrails can become important, especially when the Sun is low in the sky and the contrail less than a kilometer in width (Gounou and Hogan, 2007). Generally the 3D effects on individual shortwave and longwave forcings are modest (10%); however, as the net forcing is a cancellation of two terms with opposite sign, these authors found that in certain instances the net forcing could double or change sign.

Table 2. Top: Radiative forcing [W/m2] at the top of the atmosphere due to a 100% contrail cover (τvis =0.52) in a continental mid-latitude summer atmosphere (three different radiative transfer schemes are employed)

	Longwave	Shortwave	Net
Meerkotteretal. (1999)	51.5	-22.0	29.5
Myhre and Stordal (2001)	45.6	-25.2	20.4
Stuber and Forster (2007)	44.2	-20.3	23.9

Table 3. Bottom: Annual mean, global mean radiative forcings [W/m2] at the top of the atmosphere due to a 1% contrail cover (τvis =0.3) for all -sky and clear sky conditions (: different radiative transfer schemes, background atmosphere and background clouds are employed. From Stuber and Forster, 2007)

	Myhre and Stordal (2001)		Stuber and Forster (2007)	
	All-sky	Clear-sky	All-sky	Clear-sky
Longwave	0.21	0.27	0.19	0.25
Shortwave	-0.09	-0.15	-0.06	-0.12
Net	0.12	0.12	0.13	0.13

Background Clouds

From an examination of table 3 it would seem that background clouds, although having a large effect on shortwave and longwave terms, have no effect on net radiative forcing and can be ignored in contrail forcing calculations; however, Stuber and Forster (2007) point out that when considering diurnal variations in aviation, excluding clouds leads to a 10% or greater overestimate of the net radiative forcing.

Diurnal Cycle of Air Traffic

As most flights and contrails occur during daylight they contribute more negative radiative forcing than a diurnal average would suggest. Excluding the diurnal cycle can lead to roughly an overall 20% overestimation of the net forcing, although this varies with location depending on the time of day aircraft are typically overhead (Stuber and Forster, 2007).

Ice Crystals Optical Properties

In radiation calculations contrails are typically assumed to be composed of small, spherical, hexagonal column or aggregate ice crystals. Often, and for convenience, they are assumed to have the same shape as those in typical cirrus clouds used in radiation models, only smaller with an effective radius typically about 10 microns. An assumption of aspherical particles of the same effective radius as a spherical particle will slightly reduce the radiative forcing while the uncertainty in crystal habit could lead to errors on the order of 10% (e.g. table 7). An open issue in modeling the radiative properties of ice clouds in general is the lack of accurate light scattering models for the size parameter, defined as crystal size divided by the wavelength of the incoming radiation, between 10 to 100. Below this range approximate solutions like the Discrete Dipole Approximation (Draine and Flatau, 1994) or Finite Time Difference Methods (e.g. Yang and Liou, 1996) are applicable. Larger size parameters are covered by the Geometric Optics Approximation (e.g. Mishchenko and Macke, 1999). The effect on the overall radiative forcing by particles with size parameters in this mid-range can be estimated once experimentally derived optical properties are determined for these aspherical crystals; however, there is a significant lack of observations to constrain the uncertainty that crystals in this size range represent.

Other Effects

Uncertainties in the contrail height and surface albedo can lead to 10% uncertainties in the magnitude of the radiative forcing but these can be reasonably constrained using detailed flight information and a good surface climatology. In summary, whilst several factors could lead to 10-20% uncertainty in the radiative forcing by contrails, it is the contrail coverage and the crystal size that are key to providing an accurate forcing estimate. Global estimates of contrail radiative forcing and their impact will be discussed in Section 2.5.1

2.2.3. How Well Are Aviation-Related Subscale Processes Represented in Large-Scale Global Models?

Global models that predict local contrail formation employ the SAC based on grid cell temperatures and humidities or suitable corrections thereof (Ponater et al., 2002; Rädel and Shine, 2008). The frequently applied concept of potential contrail coverage was introduced by Sausen et al. (1998) because of the inability to simulate contrail development in global models. Potential contrail coverage defines the maximum coverage by contrail clouds in a given region if sufficient air traffic was available and meteorological conditions were favourable for persistence. Therefore, it may be viewed as a proxy for ice supersaturated areas in global models which generally do not resolve ice supersaturation on the grid scale. No effort has been made so far to validate simulated potential contrail coverage.

Another important factor that needs to be parameterized in large-scale models is the mass and number of contrail ice particles that survive the vortex phase. This information is needed as an initialization in global models that track the time evolution of contrails, models that are not yet available. The same holds for any other parameter affecting the contrail-to cirrus transition, such as mesoscale wind shear, vertical winds and relative humidity, because the contrail life cycle is not included in current large-scale models. The latter atmospheric parameters could in principle be used to validate future global models simulating contrail cirrus; however, any aircraft-specific processes related to vortex break-up are difficult to consider because aircraft inventories employed in such models do not provide the necessary information.

Given the above, and that radiative effects of contrails depend on their microphysical evolution, radiation modules of climate models can only be fed with assumed contrail optical properties, in particular the effective ice crystal size. Those are determined in part by the competition of contrail cirrus with natural cirrus for condensable water, which constitutes yet another potentially significant subgrid process that remains unconsidered in current large-scale models. Off-line estimates of contrail radiative forcing have used advanced radiation codes, but lack geophysical variability in optical depth and often rely on simple scaling arguments to compute global contrail coverage.

2.2.4. How Well Have Contrail Microphysical and Optical Properties Been Measured in Past, in-Situ Observational Studies?

The properties of contrail particles can be described by their size distributions, water content, total number concentration, sulfate and carbon chemistry and light scattering phase function. Hydrometeor optical and microphysical properties have been measured directly or derived using five, fundamentally different techniques: impaction, phase change, light scattering from individual particles, light scattering from an ensemble of particles and imaging. Aerosol particle properties are derived from light scattering and incandescence, condensation chambers, counter-flow virtual and wire impactors and aerosol mass spectrometers.

Table 4. Contrail Properties Measurement Techniques

Sensing Techniques	Particle Property	Instruments
Impaction	N: Directly SD: Directly M: Size distribution integration $P\lambda,\theta$: Size distribution integration Re: Size distribution integration σe: Size distribution integration	Wire impactor SEM grid behind Counterflow Virtual Impactor Video Ice Particle Sampler (VIPS)
Phase change of hydrometeors	N: From CVI only M: Directly	Counterflow Virtual Impactor (CVI)3 FISH4
Single particle light scattering	N: Directly SD: Directly M: Size distribution integration $P\lambda,\theta$: Size distribution integration Re: Size distribution integration σe: Size distribution integration	Forward Scattering Spectrometer Probe Models 100 and (FSSP-100, 300)1 Passive Cavity Aerosol Spectrometer Probe (PCASP) Cloud and Aerosol Spectrometer (CAS)1 Multiangle Aerosol Spectrometer (MASP)1 Focussed Cavity Aerosol Spectrometer (FCAS)5
Hydrometeor Ensemble Light Scattering	M: Direct from PVM $P\lambda,\theta$: Partial information from CIN Re: Direct from PVM σe: Direct from CIN	Particle Volume Monitor (PVM)6 Cloud Integrating Nephelometer (CIN)2
Non-Intrusive optical imaging	N: Directly SD: Directly M: Size distribution integration $P\lambda,\theta$: Size distribution integration Re: Size distribution integration σe: Size distribution integration	Cloud Imaging Probe (CIP)1 Cloud Particle Imager (CPI)7 2D Optical Array Probe1 Small ice detector (SID)8
Condensation + light scattering	N: Directly	Condensation Nucleus Counter (CNC)9 Ice Nucleus Counter10

Table 4. (Continued)

Sensing Techniques	Particle Property	Instruments
Incandescence and scattering	N: Directly M: Directly (light absorbing carbon) SD: Directly $P\lambda,\theta$: Size distribution integration Re: Size distribution integration σe: Size distribution integration	Single Particle Soot Photometer (SP2)1
Laser ionization and mass spectrometery	N: Directly M: Directly (dependent on technique)	Particle Analysis by Laser Mass Spectrometry (PALMS)

N = total number concentration,
M = total mass concentration,
PL,\ = Phase function
SD = size distribution,
σ e = Extinction coefficient,
Re = Effective radius.

Table 4 lists the types of airborne measurement methods that have been used to characterize contrails, the properties that they detect and the name of the instrument that incorporates a specific technique. Table 5 lists the majority of measurement campaigns when contrails have been measured, the instruments that were used, the principal investigator who was responsible for the instrument and some of the key publications that describe the instrument and significant results. A description of the most commonly employed techniques for measuring cloud properties is found in Baumgardner (2002) and the majority of aerosol instruments are described by McMurray (2000) and Willeke and Baron (2001). A brief description of all the instruments listed in these tables is given here primarily to highlight the advantages and disadvantages of each of them for measuring contrail properties.

Table 5. Instrumentation on Contrail Measurement Projects

Project Name	Instrumentation	Principal Investigator (responsible for instruments	References
Sulfur I-V	TSI 3067 CN counter N-MASS PCASP FSSP-300 CVI MASP	A. Petzold, C. Brock F. Schröder S. Borrmann J. Ström M. Kuhn, D. Baumgardner	Petzold et al. (1997; 1998a,b; 1999) Brock et al. (2000) Schröder et al (1998, 2000) Schumann et al. (1996; 2002)
ASHOE/MAESA	CN counter (In-House) FCASP MASP	C. Brock J. Wilson D. Baumgardner	Kuhn et al. (1998) Fahey et al. (1995) Jonsson et al. (1995) Baumgardner et al. (1996) Goodmann et al. (1998)

Table 5. (Continued).

Project Name	Instrumentation	Principal Investigator (responsible for instruments	References
SUCCESS	Wire impactors CN CCN IN (CFD) FSSP-300 MASP PCASP CVI PVM VIPS	J. Goodmann D. Hagen L. Radke,W.A. Cooper D.C. Rogers, P. Demott R. Pueschel D. Baumgardner, B. Gandrud J. Anderson C. Twohy H. Gerber A. Heymsfield	Chen et al. (1998), Rogers et al. (1998) Baumgardner and Gandrud (1998) Twohy and Gandrud (1998), Twohy et al. (2007) Gerber et al. (1998)
AEROCONTRAIL	TSI 3067 CN counter PCASP FSSP-300 CVI MASP	A. Petzold, C. Brock F. Schröder S. Borrmann J. Ström M. Kuhn, D. Baumgardner	Toon, O.B. and R.C. Miake-Lye, 1998
SONIC CRYSTAL/FACE	? CN counters (In-House) NMASS PALMS FCAS CAS CIP CPI FSSP-100 VIPS	M. Freeman M. Freeman D. Murphy J. Wilson D. Baumgardner, G. Kok D. Baumgardner, G. Kok P. Lawson P. Lawson A. Heymsfield	Thompson et al. (2000) Jonsson et al. (1995) Baumgardner et al. (2001), Baumgardner et al. (2005) Lawson et al. (2001)
CR-AVE	CN counters (In-House) NMASS (CN In House) PALMS FCAS CAS CIP CVI CPI 2D-S	M. Freeman M. Freeman D. Murphy J. Wilson D. Baumgardner D. Baumgardner C. Twohy P. Lawson P. Lawson	Thompson et al. (2000) Jonsson et al. (1995) Baumgardner et al. (2001), Baumgardner et al. (2005) Twohy et al. (2007) Lawson et al. (2001) Lawson et al. (2006)
CIRRUS	CN counters (In-house) FSSP-100 CIP FISH SP2	G. Kok and D. Baumgardner	Zöger et al. (1999)

Impaction devices are those that are exposed, either intermittently or continuously, to the air stream. The wire impactors have wires with diameters of 70 μm or 200 μm, depending on the size range of particles to be collected (Goodman et al., 1998) and are exposed for periods of several minutes, depending on the ambient concentration of collected particles. The wires are stored for subsequent evaluation of the captured particles with X-ray dispersion analysis (SEM or TEM). The video ice particle sampler (VIPS) uses a moving, 8 mm transparent tape that is coated with silicone oil and exposed to the particle-laden air stream. After exposure the magnified images are recorded digitally as the tape moves in front of two video cameras. The limitation of these types of devices is the break-up of ice crystals larger than approximately 50 μm and the wire impactors have a very limited sample volume

A method for deriving water mass by evaporation is to measure water vapor formed from vaporization of the hydrometeors. The Counterflow Virtual Impactor (CVI) utilizes this technique (Twohy et al, 1997). At the CVI inlet tip cloud droplets or ice crystals larger than some aerodynamic diameter (usually about 5 to 10 ∝m diameter depending on airspeed and density) are separated from the interstitial aerosol and water vapor and are "virtually" impacted into dry nitrogen gas. This separation is possible via a counter-flow stream of nitrogen out the CVI tip, which assures that only larger hydrometeors are sampled. The water vapor and non-volatile, residual nuclei that remain after droplet evaporation are sampled downstream of the inlet with selected instruments. These may include a hygrometer to determine water content, a condensation nucleus counter, an optical particle counter, or particle filters for various chemical analyses. Since droplets or crystals in a large sampling volume converge into a smaller sample stream within the instrument, concentrations within the CVI are significantly enhanced, which leads to a better detection limit. The CVI has the disadvantage that its lower size threshold of approximately 5 μm rejects a large fraction of the ice crystals that are found in young contrails whose median volume diameter is typically less than 5 μm (e.g. Baumgardner and Gandrud, 1998). The Fast In situ Stratospheric Hygrometer (FISH), developed at the Forschungszentrum Jülich (Germany), is another device that has recently been employed to measure ice water content. It measures water vapor from all hydrometeors evaporated in the heated inlet using a hygrometer based on the Lyman-a photofragment fluorescence technique (Zöger et al., 1999.

Optical particle counters detect the light scattered when a particle passes through a focused light beam. Instruments that convert the light amplitude into a size, using Mie scattering theory, are called single-particle spectrometers. Several such instruments for measuring particle sizes and concentrations are the Forward Scattering Spectrometer Probe models 100 and 300 (FSSP-100, - 300), the Passive Cavity Aerosol Spectrometer Probe (PCASP), the Multiangle Aerosol Spectrometer Probe (MASP) and the Cloud and Aerosol Spectrometer (CAS). As droplets in the free air stream pass through a laser beam and scatter light, these instruments collect this light over a solid angle that depends on the instrument and convert the photons into an electrical signal with a photodetector. The MASP and CAS collect light separately over forward and backward angles. Comparison of light measured at two angles provides information on either refractive index (Baumgardner et al., 1996) or ice crystal shape (Baumgardner et al., 2005). The advantage of the instruments is that they are very fast response and measure particle sizes with high resolution. The disadvantage is that they have relatively small sample volumes and the accuracy of the measurement is decreased when measuring non-spherical particles. The information is also ambiguous in certain size ranges as a result of the multivalued nature of the relationship between light scattering and size that stems from the complexity of light refraction.

Light scattering from an ensemble of hydrometeors is the technique used by two of the instruments, one that derives IWC, extinction coefficient, and effective radius from the measurement and the other that derives the asymmetry coefficient and extinction coefficient. The Particle Volume Meter (PVM) measure the diffracted component of scattered light (Gerber et al., 1998). The PVM illuminates an ensemble of hydrometeors with a collimated laser and focuses the near-forward scattered light onto two detectors that are masked with variable-transmission filters. The filters have been designed to provide transmission functions that are mathematically derived to approximate inversions of the integral equations that relate particle surface area (PSA) or LWC to the flux of light scattered by the ensemble of hydrometeors. An instrument that has been specifically designed to measure the asymmetry coefficient, g, is the Cloud Integrating Nephelometer (CIN). The CIN consists of a collimated laser beam that passes through an ensemble of hydrometeors and four detectors that are positioned to measure scattered light in the forward and backward directions The detectors have optical masks that cosine-weight the collected scattered light so that after suitable corrections for the angles over which light is not being collected, the ratio of forward to backward scattered light provides an approximation to the asymmetry factor. The advantage of these techniques is that they measure over larger sample volumes than the single particle instruments and the optical properties are measured directly rather than derived from a size distribution. There remain questions, however, about the sensitivity of the PVM to cloud particles larger than 30 μm (Wendisch et al., 2002) and the robustness and fidelity of the large angle correction for the CIN measurements (Heymsfield et al., 2006).

One of the first methods in cloud physics research of optically measuring hydrometeor size was by imaging onto a linear diode array the shadow of a particle that passed through a laser. The on/off state of the diodes in the array is recorded at a rate proportional to the velocity of the particles passing through the laser and the images can be subsequently reconstructed to show the features of the hydrometeors. This type of instrument, called a two-dimensional optical array probe (2D-OAP), can typically measure in the size range from 10 ∝m to greater than several millimeters, depending on the magnification. This technique has been refined to measure hydrometeors at a higher resolution, down to 2.5 ∝m, with the Cloud Particle Imager (CPI) by using a pulsed laser and a two dimensional photodetector array to capture the particle image (Lawson et al., 2001). In addition, 256 gray levels are measured in the CPI as opposed to the binary levels in the 2D-OAPs. The primary disadvantage of the imaging probes is that the depth of field (DOF) decreases with the square of the particle diameter, e.g. particles with diameters of 20 μm and 10 μm have DOFs of 0.8 mm and 0.2 mm, respectively. Hence, an imaging probe with sufficient resolution to measure contrail particles whose diameters are less than 5 μm would have a very small sample volume.

The perils of measuring ice crystals with optical particle spectrometers have been recognized since the mid 1980s. Studies have identified conclusively that size spectra measured with optical spectrometers that have inlets are contaminated by ice crystal fragments, even in the presence of very low concentrations of larger ice crystals. Gardiner and Hallett (1985) were the first to publicize the effect of ice crystal break-up on the inlet of the FSSP and showed that size distributions in these conditions were broadened and number concentrations were unreasonably large. Later studies by Field et al. (2006) reinforced the conclusions of Gardiner and Hallett by analyzing the distribution of interarrival times of particles measured by the FSSP. McFarquhar et al. (2007) showed that the CAS also suffers from ice fragmentation on its shroud and inlet. In recent projects the flow straightening shroud has been removed to minimize the crystal shattering problem; however, this has not completely eliminated the problem as shown in figure 14. These size distributions are compiled from measurements made with the CAS and CIP during the Tropical Composition, Cloud and Climate Coupling (TC4) project. The images below the size distributions are representative measurements with the CIP that indicate the average size and shape of the ice crystals. Whereas the CAS has an inlet with a diameter of approximately 5 cm, the CIP has no inlet and is minimally affected by crystal break up, although there can be some contamination that can be removed by looking at the interarrival times of particle images in these types of instruments (Korolev and Isaac, 2005). In these figures, we show that when the ice crystals are small, the CAS and the CIP concentrations are relatively well matched in the overlapping size range. As the ice crystals become larger, however, the CAS shows increasingly larger concentrations than the CIP. It is assumed that these are a result of fragments formed from ice crystals shattering on the CAS inlet. Fortunately, given that there are rarely any crystals larger than 20 μm in young contrails, ice crystal shattering is probably not a factor to take into account when analyzing FSSP or CAS measurements in early stages of contrail development. Thereafter, as contrails develop into contrail cirrus, shattering becomes a major source of artifacts (Section 2.1.5.2).

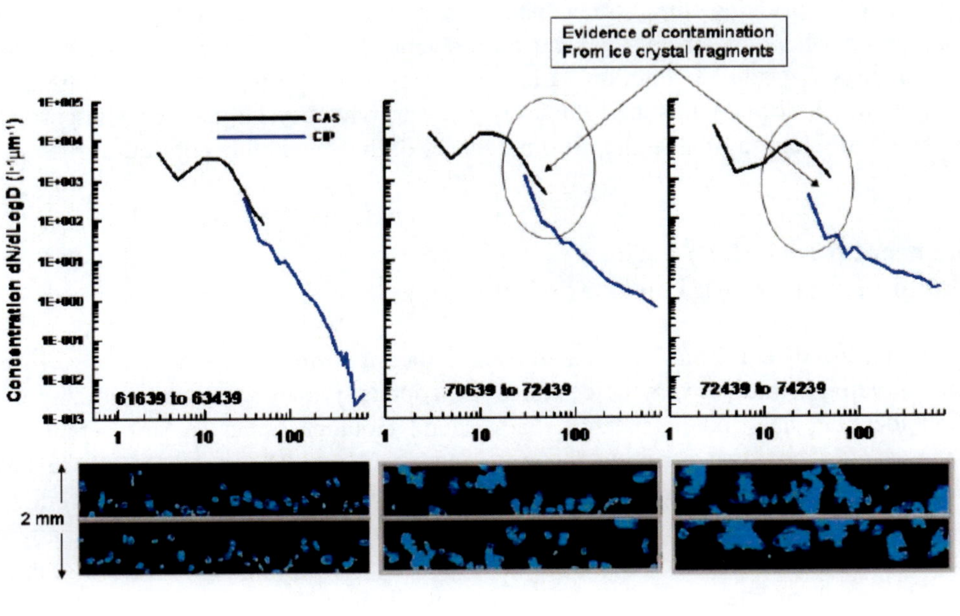

Figure 14. The size distributions shown here are from the Cloud Aerosol Spectrometer (CAS) in black and the Cloud Imaging Probe (CIP) in blue, showing the agreement in the overlapping size ranges when ice crystals are small, as shown in the images from the CIP (below the size distributions), disagreeing when larger ice crystals are present.

2.3. Present State of Modelling Capability / Best Approaches

2.3.1. Representation of Aviation-Related Subscale Processes in Large-Scale Global Models

As introduced in subsection 2.3.3 most contrail-related subscale processes are not represented in current large-scale models, with the notable exception of the thermodynamic conditions for contrail formation, the most advanced approach having been formulated by Ponater et al. (2002).

In their approach contrails are reinitialized every model time step hence, they show no atmospheric life cycle and assumed to share their ice water content with cirrus in proportion to a contrail coverage that was tuned to a specific region using a globally constant tuning coefficient. Contrails are assigned the same optical properties as natural cirrus.

All existing studies address coverage by line-shaped contrails only, constituting only a subset of the total coverage by contrail-cirrus. Even if an advanced global model approach was available there is a scarcity of in-situ and remote sensing observations that could be used for model evaluation, in particular for contrail cirrus older than about 30 min.

2.3.2. Modelling of the Spreading of Ice Crystals Generated by Aircraft

Lidar observations of the vortex phase of contrails and numerical studies show that the early evolution of a contrail is sensitive to ambient conditions and aircraft performance parameters. Recent numerical experiments (Unterstrasser et al., 2007) show that the number of ice crystals surviving the vortex phase is a power-law function of the ambient supersaturation whose parameters depend on ambient temperature, stability and turbulence level. The surviving number fraction varies from less than one percent at ice saturation to 100% at about 130% ambient supersaturation. The surviving crystal number is important for the evolution of microphysical and optical properties of the developing contrail-cirrus.

2.4. Current (or Model-) Estimates of Climate Impacts and Uncertainties

As emphasized in subsection 2.2.2 radiative forcing from contrails depends on many factors, including contrail coverage, ice water path, optical properties, geometry, time of day, size and location, age and persistence, background cloudiness and surface albedo. After coverage, a poorly known quantity, ice water path and optical properties are the largest sources of uncertainty. Here we perform a literature review of previous forcing studies and the weaknesses and strengths of each to explain differences between findings. In particular, all studies seem to show similar ranges and variability but they also all make similar assumptions, such as the restriction to linear contrails and the use of a constant contrail optical depth. A second part of this section will evaluate the role of spreading contrails and aviation induced cirrus and the last two parts examine the climate impact beyond radiative forcing.

2.4.1. Persistent Linear Contrail Radiative Forcing

To date global radiative forcing estimates have employed similar simple diagnostic techniques, i.e. combining estimates of 1) contrail coverage and 2) contrail optical properties to get the contrail radiative forcing.

1. Contrail coverage estimates rely on first estimating regions of the Earth where contrails might form, given suitable background conditions of ice supersaturation, and then using a suitable database of flights to estimate when an aircraft forms a contrail. Assumptions are then needed about contrail lifetime and width so that a suitable estimate of coverage can be made. Given the uncertainty in estimating ice supersaturation a large range might be expected from modelling studies; however, all studies to date have employed a "normalization" step that mutes the effect of any differences in their previous assumptions. This normalization linearly scales the contrail coverage found with that estimated using satellite observations of persistent linear contrails over a particular region. Nearly always the scaling is obtained by normalizing to the 1992 European observations of Bakan et al., (1994). The derived normalization factor can be as large as an order of magnitude, thus having a major impact on results. This also means that all results obtained are actually scaled for line-shaped contrails in 1992; hence, it is likely that this normalization is a major reason for the agreement between otherwise disparate methodologies. It is important to note, however, that even when applying the scaling the modeled global contrail coverage can still differ by a factor of two (table 6) so the other assumptions remain significant as well.
2. Contrail optical depth is either simply assumed (e.g. Stuber and Forster, 2007) or obtained as a diagnostic from a climate model (e.g. Ponater et al, 2002). Until recently GCMs have not included ice supersaturation so estimates of available ice have had to be obtained from parameterisations that assume that ice exists above a relative humidity threshold less than 100%. These schemes are also diagnostic as there is no contrail history, i.e. one timestep does not know about contrails at any previous time steps; therefore, assumptions are also needed about contrail lifetime. In general, following the contrail optical depth used for the IPCC (1999) calculations, the estimate of average optical depth has been reduced, and this, along with reduced estimates of global persistent contrail cover, have been the major reasons for the reduced estimates for the global mean radiative forcing by contrails (Marquart and Mayer, 2002; Mayer et al., 2002, Ponater et al., 2002, Marquart et al., 2003 and table 6).

Results from previous studies are shown in table 6 where the radiative forcing is estimated in the radiative transfer models by combining steps 1) and 2). Radiative forcing can either be estimated within a climate modelling context (e.g. Ponater et al., 2002) or offline (e.g. Myhre and Stordal, 2001, Stuber and Forster, 2007). Offline calculations have the advantage, perhaps, of employing better, observationally-based, background climatology of humidity and temperature to determine ice supersaturation. They can also employ more sophisticated radiative transfer schemes; for example, the Ponater et al., (2002) study using a GCM did not account for scattering in the longwave. On the other hand, offline schemes have the major disadvantage that they have no information of the degree of ice-supersatuation and are therefore unable to estimate the optical depth of contrails. The latest GCMs, such as the European Centre for Medium Range Forecast (ECMWF) model, can now estimate ice supersaturation and it may be possible to obtain a climatology of likely optical thicknesses in the future (see Rädel and Shine, 2008).

Table 6. Contrail radiative forcing studies from Stuber and Forster, 2007

study	Contrail cover	τ	Diurnal cycle air traffic	RF
This study	0.04	Fixed, 0,1	yes	2.0
This study	0.04	Fixed, 0,3	yes	50.
MS2001	0.09	Fixed, 0,3	yes	9.0
This study, scaled	0.09	Fixed, 0,3	yes	11.3
This study	0.04	Fixed, 0,1	no	2.4
Marquart et al. (2003)	0.06	variable, 0,15	no	3.5
This study, scaled	0.06	Fixed, 0,1	no	3.6
Fichter et al. (2005)	0.047	variable, 0,15	no	3.2
This study, scaled	0.047	Fixed, 0,1	no	2.8

The IPCC fourth assessment based their persistent contrail forcing estimate on Sausen et al. (2005) and estimated the contrail forcing for 2005 to be 0.01 Wm-2 with a factor of three uncertainty and a low level of scientific understanding (Forster et al., 2007). Although the forcing best estimate is considerably reduced from the 1999 report, which estimated a forcing about three times larger when scaled by fuel use, the uncertainty remains large and confidence low due to the pronounced flaws in the methodology described above.

2.4.2. Aviation Induced Cloudiness (AIC)

Aviation can potentially have several additional effects on cirrus clouds apart from forming linear contrails and there is some confusion as to precisely which effects are being evaluated in the different literature studies. Firstly, persistent linear contrails can shear and spread into large areas of cirrus cloud. Secondly, aerosol, especially soot, can also act as IN that potentially can alter cirrus cloud evolution (see Section 2.1.4). Studies that have estimated the radiative forcing from such changes likely include all forms of cirrus modification including that from persistent linear contrails. IPCC-1999 did not put a best estimate on this radiative forcing but suggested it could range from 0.0 to 0.04 Wm-2. This was based on a study by Boucher (1999) that looked at differing cirrus trends in regions of high and low aviation traffic (see Section 2.2.1). Since then two further studies have adopted similar techniques (Stordal et al., 2005; Zerefos et al., 2003) as discussed in Section 2.2.1. Here we focus on their radiative forcing estimates.

Cirrus trend estimates that compared high and low aviation use areas were used to estimate radiative forcing by Stordal et al., (2005), assuming standard cirrus optical properties. For the calendar year 2000 a range of 0.01-0.08 Wm-2 was suggested. Minnis et al. (2004) used surface and satellite cloud observations to derive a suggested upper estimate for the AIC radiative forcing of around 0.03Wm-2. The AR4 IPCC report adopted the Stordal et al. (2005) range, neither providing a best estimate nor attributing the AIC forcing to particular causes. Overall it gave the AIC forcing a very low level of scientific understanding. Therefore, in summary, studies to date suggest that the total radiative forcing of aviation on cirrus, including linear contrail formation, is 2-10 times larger than its role solely in the formation of linear persistent contrails; however, the studies have been unable to place a realistic best estimate or even upper bounds on this.

2.4.3. Climate Impact of Contrails on Temperature

Two modeling groups, using three climate models, have examined the climate impact of contrails and AIC beyond that of radiative forcing. Rind et al. (2000) increased cirrus frequency along aircraft flight paths by various amounts in a version of the GISS climate model. For a 1% worldwide increase in cirrus the global surface temperature increased by 0.4 K. The source of this estimate was from an increase in radiative forcing of around 0.1W-2 and a climate sensitivity of 0.9 K/Wm-^{2}; this compared to the models' sensitivity to well mixed greenhouse gases of 1.2 K/Wm-^{2}. This study also found a pronounced hemispheric difference in the climate response. Because most of the forcing was in the Northern Hemisphere, the Northern Hemisphere and especially the Arctic exhibited the greatest response; however, within each hemisphere the geographic variation of temperature response was more-or-less independent of where the aircraft perturbed the cirrus. The maximum temperature response was in the upper troposphere – at a height of 10 km the warming was about twice that at the surface. The recent study with the ECHAM4 model by Ponater et al. (2005) confirmed these findings, i.e. that the response considerably smoothed out the geographical variation of forcing (figure 15). Their findings are also consistent with perturbations of other forcing agents in a similar model (Hansen et al. 2005) and other perturbations within different climate models (Forster et al., 2007).

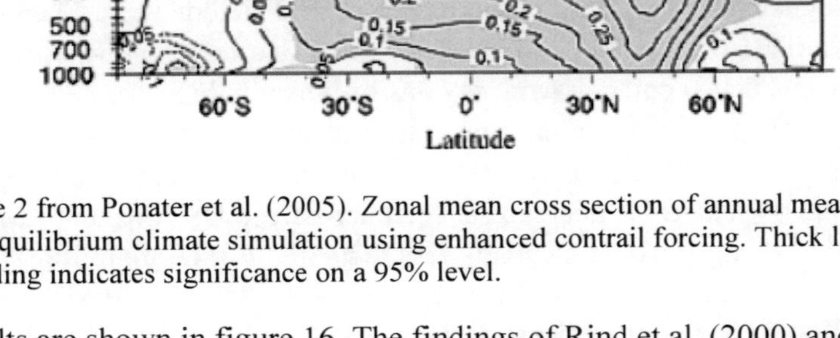

Figure 15. Figure 2 from Ponater et al. (2005). Zonal mean cross section of annual mean temperature response in the equilibrium climate simulation using enhanced contrail forcing. Thick line displays the tropopause. Shading indicates significance on a 95% level.

These results are shown in figure 16. The findings of Rind et al. (2000) and more recently Ponater et al. (2005) suggest that the response of the Earth's climate to contrails and AIC is probably smaller than suggested by a simple evaluation of radiative forcing. The concept of "efficacy" (Hansen et al., 2005; Forster et al., 2007) compares the global mean temperature response for a given amount of forcing, from a particular agent, to an equivalent radiative forcing from carbon dioxide. If the efficacy is larger than 1.0 then the climate will be more sensitive to the forcing agent than it is to perturbations in carbon dioxide. On the other hand if the efficacy is smaller than 1.0 then the climate is not as sensitive to perturbations in the forcing agent. The preliminary study of Rind et al. (2000) suggested an efficacy smaller than 1.0 and Ponater et al. (2005) suggest an efficacy of only 0.6 for contrail changes. These small efficacies are because contrail and/or cirrus perturbations have a relatively larger effect on the upper tropospheric temperatures compared to those at the surface.

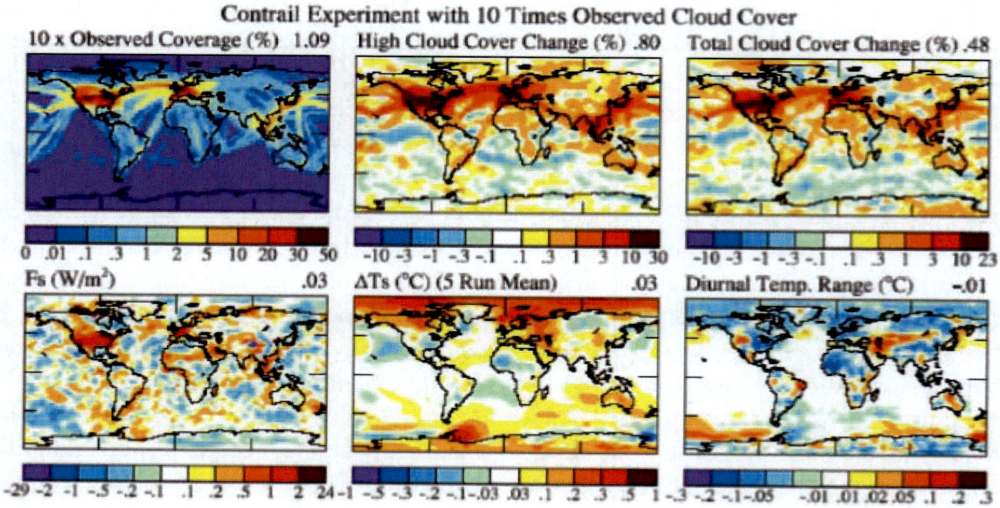

Figure 16. Observed contrail coverage in 1992 from Minnis et al. [2004] and simulated impact of the contrails, increased by a factor of 10, on high cloud cover, total cloud cover, Fs, surface air temperature, and the diurnal range of surface air temperature in years 81-120 of the coupled climate model.

2.4.4. Impact on Diurnal Temperature Range

After the 11 September 2001 terrorist attacks in the U.S. civil aviation flights were grounded for three days. The diurnal temperature range (DTR) during this time was observed to be several degrees larger than days immediately before and after these groundings. The DTR was also higher than similar time periods during different years (Travis et al. 2002). Further work described in Travis et al. (2004) suggests that locations where maximum DTR changes were observed also coincide with regions of typically high contrail frequency and air traffic (figure 16). Daytime maximum temperatures also appeared to be more affected than the night time minimum. Hansen et al. (1995) do suggest a role for high cloud changes in affecting DTR and a clear DTR decrease is seen across the US in figure 15; however, this figure suggests that changes would be small, less than 1 K. Furthermore it has been proposed that unusually clear weather across the U.S. during those three days could have been responsible for the DTR response (Kalkstein and Balling Jr., 2004). Therefore, although a large impact on DTR remains a possibility, more work is certainly needed to investigate these possibilities.

2.5. Interconnectivity with Other SSWP Theme Areas

Strong relation with theme 3–supersaturation and theme 5–climate issues. These will be worked out in the future.

3. Outstanding Limitations, Gaps and Issues that Need Improvement

3.1. Science

3.1.1. Uncertainties in Contrail Formation Conditions

Knowledge of contrail formation conditions is sufficient for most applications including contrail forecasting or climate modeling. Crucial parameters that should be known with high accuracy include the ambient relative humidity and the aircraft propulsion efficiency.

3.1.2. Chemical and Microphysical Mechanisms that Determine the Evolution of Emissions from the Engine Exit to Plume Dispersion

As discussed in Section 2 we think that our current understanding of the properties of young contrails, up to the vortex regime, is sufficient for most applications in models in the near future; nevertheless, there are several aspects of plume dynamics and microphysics that warrant a more complete understanding, in part because of their potential implications for persistent contrail development and subsequent influence on contrail-cirrus formation.

The dynamics of aerosol and ice particles at the point of contrail formation is highly complex according to numerical model simulations (Kärcher, 1998). Some features of the underlying physical and chemical processes are not well covered by observations. For example, the ultrafine liquid particle mode is predicted to take up substantial amounts of nitric acid (HNO_3) during water activation, resulting in the formation of supercooled, ternary ($H_2SO_4/HNO_3/H_2O$) droplets (STS) after ice nucleation that subsequently interact with contrail ice. Nitric acid also interacts directly with contrail ice crystals, potentially forming NAT particles (Kärcher, 1996) such as are observed in polar stratospheric clouds (Dye et al., 1990). Contrails composed of NAT and/or STS particles persist at lower relative humidities than traditional contrails composed of pure water ice particles; however, because the NAT phase is only stable at very low temperatures (< 205 K) at subsonic cruise altitudes, this is mainly an issue for the high latitudes in winter. The global consequences of aviation-induced NAT formation has been discussed in the context of a planned fleet of supersonic aircraft (Peter et al., 1991), but has not received sufficient attention in the case of the subsonic fleet. Besides the potential chemical implications, the global, radiative impact of NAT contrails is probably small, but such processes may enhance cloudiness regionally in the Arctic.

Another factor that could become relevant at low temperatures is the formation of cubic ice. While it is becoming increasingly clear that ice nucleates in cubic form and slowly transforms into hexagonal ice, depending on the chemical composition of the ice-forming aerosol particle precursors (Murray et al., 2007), virtually nothing is known about the relevance of cubic ice in contrail formation. As cubic ice has a higher vapour pressure (by about 15%) than hexagonal ice (Murphy, 2003), it may be speculated that contrails composed of cubic ice have smaller ice particles than traditional contrails for at least part of their lifetime (Gao et al., 2004). As in the case of NAT, the cubic ice issue is likely of limited global importance as it becomes relevant only in regions with low air traffic density; however, given the lack of understanding of the fundamental physics of ice formation at low temperatures and high supersaturations, improving our understanding of this process is important from strictly a scientific perspective. As mentioned above, microphysical and optical properties of persistent contrails and contrail-cirrus are sensitive to conditions and processes during the vortex phase, i.e. about 20 to 120 s of plume ageing. Adiabatic compression and heating due to the vortex's downward migration leads to ice evaporation and the surviving fractions can be as small as one per mille by number. Simulations for one aircraft type (a wide-body) led to the following results (Unterstraßer et al., 2007):

The fraction of ice number and mass that survives the vortex phase has a power law sensitivity to the ambient supersaturation with respect to ice. The dependence is strongest for the highest temperature that allows contrail formation and becomes weaker with decreasing temperature.

Only the ice in the secondary vortex survives at low supersaturation, giving rise to persistent yet very faint contrail.

The stratification of the atmosphere and its turbulence level have a strong impact on the fraction of the surviving ice via their dynamical effect on the sinking vortex pair. Strong turbulence leads to fast vortex decay, whereas weak turbulence allows the vortex pair to travel a long distance downward. Hence, in situations of strong turbulence, more ice is rendered to the atmosphere than in weakly turbulent conditions. The downward travelling distance of the vortex increases with decreasing strength of stratification; hence, more crystals survive in more stable situations and vice versa. Additionally, more ice is detrained into the secondary wake in more stable situations.

The variation of the initial circulation with varying aircraft weight during a flight, details of the spatial distribution and the temperature profile within the vortices have only a minor influence on the surviving ice fraction.

The initial number of ice crystals has an influence on the surviving fraction. The initial number increases with decreasing ambient temperature within a range of about an order of magnitude. At the warmest temperatures that allow contrail formation the surviving fraction is larger when less ice crystals are formed initially; however, the total number of surviving crystals and the surviving ice mass can be larger when more crystals are formed initially.

The relative position of the engines to the wing tips has a small influence on the contrail properties; nevertheless, the aircraft type plays an important role, but merely for the almost trivial reason that different aircraft types burn different amount of fuel per meter of flight path, which can also vary between one order of magnitude (Sussmann and Gierens, 2001).

None of the factors that are listed here, and that are determined from numerical simulations, have been validated with observational studies. The modeled sensitivities seem reasonable from the perspective of the fundamental physics, yet without targeted studies to compare observations with predictions, there will always remain doubt as to the validity of the simulations and hence, the predictions of environmental impact.

3.1.3. How Do the Microphysical and Optical Properties of Natural Cirrus Differ from Naturally Occurring Cirrus?

From detailed particle imagery acquired during the SUCCESS field campaign in 1996, and from the Knollenberg (1972) and Atlas et al. (2006) observations, we surmise that contrail crystal shapes develop to habits similar to those found in natural cirrus that develop under the same conditions when the environment is strongly ice supersaturated. These studies indicate that concentrations of ice crystals >100 microns in cirrus and contrail cirrus are comparable, of the order 10's per liter. The ice water content in contrails and natural cirrus are driven by the ambient supersaturation (e.g., figure 7). More studies are needed to determine whether the IWC's in cirrus and aged contrail cirrus are comparable.

As noted in Section 2.1.5, issues related to the measurement of small ice crystals in the presence of large ones has confounded our ability to distinguish natural from contrail cirrus based on their radiative properties or median volume diameter. Specifically,

Shattering of ice crystals on the inlets of probes that have measured contrail ice crystal concentrations with sizes < 50 μm in maximum dimension, e.g. the FSSP, CAS, MASP, and CPI, can account for ice concentrations of 10's per cm^{-3} or greater under conditions where there are even small concentrations of ice crystals > 50 μm. This occurs some 10-20 minutes following contrail formation, depending upon the temperature and supersaturation.

Because of shattering, previous comparisons of N_t and population mean diameter or median volume diameter of contrail cirrus that has evolved into natural cirrus are likely unreliable.

This issue of measurement contamination from shattered ice fragments renders differentiation of contrail cirrus from natural cirrus, using extinction alone, nearly impossible at the present time.

3.1.4. What Is the Role of Soot Emissions in Altering Cirrus and How Does Soot-Induced Cirrus Relate to Contrail-Cirrus?

While most of the above-mentioned sub-themes address issues associated with contrail-cirrus, it was pointed out in section 2.1.4 that aviation may also affect cirrus via emissions of soot particles. Although the role of soot emissions in contrail formation seems to be reasonably well understood (see 2.1.2 and 2.1.3), the question of cirrus formation or modification induced by soot particles (soot-cirrus) remains largely unresolved.

Many tools are available for modeling the formation of soot-induced cirrus, just as there are for modeling the activation of any atmospheric IN and the photochemical and microphysical mechanisms that might affect the ice nucleation ability of aging soot emissions (Kärcher et al., 2007); however, a number of factors limit our understanding of the potential implications for cirrus modification. First of all, there is a paucity of in-situ observations of processes evolved in the dispersing of plumes and of direct and tractable plumes interactions with actively forming cirrus. Secondly, and equally important, is the lack of laboratory studies to investigate how ice forms on soot-containing particles and how this process may be altered with aging in the upper troposphere.

In-situ ice nucleation measurements of exhaust soot particles, under cirrus formation conditions, have yet to be carried out and current laboratory evidence is inconclusive. Likewise, relatively poor knowledge exists of the IN activity of the ambient aerosol that competes with aircraft soot in ice formation at cirrus temperatures. Therefore, atmospheric validation of the interplay of heterogeneous and homogeneous ice formation in cirrus, gleaned from laboratory studies and numerical modeling, remains an unachieved goal.

As a consequence of these gaps in our knowledge of fundamental physical and chemical atmospheric processes, global model studies that address soot-induced cirrus can only provide preliminary parametric studies exploring possible uncertainties of changes in cirrus properties (Hendricks et al., 2005). The number of cirrus ice crystals could potentially be enhanced or reduced in dispersing aircraft plumes relative to unperturbed cirrus formation conditions, depending on assumed ice nucleation scenarios (Kärcher et al., 2007), with the subsequent consequence probably being one of significant regional or global impact.

Whereas contrail effects are fairly evident, it remains unclear whether soot particles exert an additional effect on radiative forcing (and thus contribute to AIC). However, contrail and soot effects may not act in isolation. Contrails need ice supersaturated air masses to persist; in such an environment, soot particles from aviation could preferentially trigger ice formation if they form ice at significantly lower relative humidities than natural particles.

3.2. Measurements and Analysis

3.2.1. How Can Measurements of Contrail Microphysical and Optical Properties Be Improved?

Reliable measurements of the size distributions of small ice crystals, and the shapes of these crystals, hinder our ability to accurately model the radiative properties of contrails.

3.2.2. Aviation's Share of Cirrus Trends

It is currently not clear how much of the correlation between air traffic and cirrus cloudiness is actually due to a causal relationship. Hence the determination of the radiative forcing of contrail cirrus is fraught with large uncertainties; studies to resolve the differences and to constrain the error margins are certainly needed. All studies suggest that air traffic actually induces additional cirrus clouds, which seems plausible. However it is extremely difficult to demonstrate and prove such a correlation because the variation of cirrus cloudiness due to natural influences is much larger than the possible aviation effect. Hence, to look for the latter is like looking for a signal hidden in strong noise.

3.2.3. Measurement Needs

An array of research and measurements needs is suggested by limitations and gaps in the state of scientific knowledge of soot impacts on ice formation in cirrus.

Fundamental laboratory studies are required to ascertain what makes certain soot particles more active than others and what role contrail and atmospheric processing might play in making exhaust soot more or less active as cirrus ice nuclei.

Direct sampling to test the ice nucleation ability of real exhaust particles during groundlevel emissions studies would be fruitful. Alternately, studies using collected samples of real exhaust emissions for laboratory studies would complement the ground level studies.

New studies of the ice nucleating properties of aircraft exhaust and other ambient ice nuclei measured in-situ for conditions in the cirrus temperature and water vapor regimes are needed. This could be done independently, but would be most useful within the context of a measurement effort to convincingly relate the ice activation properties of aerosols to the microphysical composition of cirrus clouds that form on them. Subsidiary needs for such a study include:

- A fast IN sensor for atmospheric studies that operates at much lower temperatures than present aircraft systems and that can run unattended to take full advantage of various manned or unmanned high altitude aircraft platforms. A number of new ice nuclei sensors for aircraft use are under development as evidenced by participation in the workshop ICIS 2007 (http://lamar.colostate.edu/~pdemott/IN/IN Workshop 2007.htm). It is possible that some of these or other new devices will meet the specifications required for cirrus and contrail studies.
- A need to sample aerosol particles without heating so as not to impact their chemical phase states and states of hydration since potential "preactivation" may be destroyed during sampling through standard inlets.
- Purposeful "seeding" of developing cirrus with aircraft exhaust could be a component of the strategy for such studies.

New studies of cirrus formation would be useful to take advantage of new or improved high resolution measurements of aerosol composition, particle activation to ice, relative humidity, vertical motion, and cloud ice particle size distributions. All of these aspects present varying levels of technical challenge. Nevertheless, instrumentation is steadily becoming available and improving that should be applicable to this task in atmospheric studies.

3.3. Modelling Capability

3.3.1. How Can We Improve Prediction of Persistent Contrails in Weather Forecast?

The upper limit on contrail and contrail-cirrus coverage is largely driven by the upper level humidity structure, i.e. the amount of ice-supersaturated regions (ISSRs) in the upper troposphere. Unfortunately, measurements of relative humidity in these levels by radiosondes (strong negative biases), aircraft (airframe distortions) and satellite instruments (insufficient vertical resolution) are notoriously difficult. This is an area of investigation with strong links to theme 3.

Although the knowledge about ISSRs has increased considerably during the last decade or so, most what has been learned is climatological in nature, i.e., we have compiled statistical information (Gierens et al., 1999, 2000, 2004; Gierens and Spichtinger, 2000; Spichtinger et al., 2002, 2003a,b; Gettelman et al., 2006); however, what we need for forecasting of contrail occurrence and persistence is to know more about single ISSR cases. Two case studies of ISSRs have been conducted by Spichtinger et al. (2005a,b). In one case an ISSR that lasted for at least 24 hours was caused by slow large-scale uplift of an airmass in a frontal system. In the second case the ISSR, lasting only a couple of hours, was caused by small-scale uplift due to a superposition of orographic waves and waves induced by strong curvature of the nearby jet stream.

Weather forecast models, that up to now have ignored ice supersaturation, should be able to predict upper tropospheric ice supersaturation in order to predict contrail formation. Some weather models use higher cut-offs of RHi than 100%; however, as long as there is any cut-off, it will be artificial. A notable exception to the standard weather forecast models is the Integrated Forecast Model of the European Centre for Medium-Range Weather Forecast (ECMWF) which has had upper-tropospheric ice supersaturation included since September, 2006 (Tompkins et al., 2007). Initial evaluation of the skill of the new cloud scheme for contrail prediction, using a confined south England based observation data base, show that it is significantly superior to the old scheme which, like most other models, had a RHi cut-off at ice saturation (Tompkins et al., 2007).

3.3.2. Contrails in Climate Models

The lack in climate models of physical and radiative interaction between contrails and their moist environment (as described in Theme 5) renders a solid determination of global contrail effects on the water budget in the upper troposphere currently impossible. The consequences of a changing background atmosphere, resulting from climate change, cannot be evaluated until simulated contrail cirrus undergo similar interactions as those modeled for natural clouds. Advances in the development of such process-based global models that enable the simulation of these and other relevant contrail processes are the subject of key theme 5 but are also linked to the objectives and priorities of key theme 4.

3.4. Interconnectivity with Other SSWP Theme Areas

This will be done in the future.

4. Recommendations and Prioritization for Tackling Outstanding Issues

This white paper has identified areas where global climate models can improve the treatment of aircraft effects on climate. These include better specification of contrail coverage and day/night differences, representing the whole life cycle of contrails, allowing contrails to interact with their environment, and resolving ice supersaturation and how it is affected by natural and aircraft-induced ice cloud. Rather than focusing on these model-related areas, we will focus on recommendations that focus on measurements that are needed to improve the treatment of contrails and natural ice clouds in global climate models.

Section 2 argues that the net radiative effect of contrails results from a near-cancellation of the shortwave and longwave terms and because of this cancellation between two forcings of roughly equal magnitude the net contrail forcing is very sensitive to any error in either term. Hence, the highest priorities are associated with decreasing the uncertainties associated with evaluating the short and longwave radiative interactions with contrail and cirrus particles.

High Priority – Near term: Field campaigns are needed that employ new technologies to measure the detailed microphysical and chemical structure of aircraft exhaust plumes and contrails during their initial development and subsequent evolution into mature systems that disperse and age.

High Priority – Near term: Develop improved in situ sensors to measure contrail and cirrus particle properties and ice nuclei concentrations and composition.

New measurement capabilities are available that were not in operation during the previous campaigns designed to study contrail physics, chemistry and dynamics. These had to do with probe resolution during the early phases of contrail evolution and in the case of contrail cirrus the inability to identify and separate artifacts produced by shattering of large particles on the probes' inlets from real ice crystals. Secondly, the measurement of size stratified aerosol composition, particularly black carbon and black carbon coated with sulfate was not available during these earlier research programs. Recently, however, there have been aerosol mass spectrometers, single particle soot photometers and fast response optical spectrometers that can distinguish the shapes of very small ice crystals by their depolarization signals. These instruments, deployed on multiple airborne platforms on a program design similar to that of SUCCESS, would provide new information that would constrain cloud parameterizations in models and greatly decrease uncertainties that were associated with previous measurements. There remains, however, a number a serious technological obstacles to be hurdled. There have been some new developments, e.g. of inlet-less optical spectrometers, that hold promise for circumventing the issue of crystal breakup. These are only now being used operationally and detailed evaluation is necessary to assess if they are free of the problems that limited the earlier technology. In addition, direct collections of contrail crystals during the pre-vortex and vortex phases may be necessary as probes that digitally image crystals smaller than about 30 microns, with high enough resolution to delineate the microstructure, have not yet been developed for airborne use.

The A-train satellite constellation, especially CALIPSO, provides an opportunity to study the backscattering behavior of natural and contrail cirrus in coordination with the targeted, airborne research missions that are a high priority recommendation. Especially useful would be to examine the vertical structure of the volume extinction coefficient, deduced from lidar backscatter, and the depolarization ratio, a measure of particle shape, from case-studies and statistically, to determine whether there are fundamental differences between cirrus and contrail cirrus extinction coefficients. The interpretation of these satellite-derived optical properties is highly dependent on validation with in-situ measurements. Once validated, application of the integrated remote and in situ measurements in state-of-the-art radiative transfer models will provide the net radiative forcing of both natural and aircraft induced cirrus for a wide range of synoptical and climatological conditions.

High Priority – Near term: Implementation of a "closure" experiment to evaluate the sensitivity of cirrus cloud formation and evolution to soot particles emitted by aircraft

Most studies of ice nucleation by soot have been focused at temperatures warmer than 235K and have used idealized soot particles of unknown relevance to aircraft exhaust soot. A "closure" experiment suitable to make headway in this area requires high-resolution measurements of aerosol composition, their activation to ice, relative humidity, vertical motion, and cloud ice particle size distribution. Carefully developed research aircraft flight patterns that target specific cloudformation conditions can be used to elucidate the ice-nucleating ability of soot aerosols under certain conditions and their competition with ice formation by ambient aerosols. Specifically, orographic wave clouds provide an excellent environment to characterize the ice-forming activity of soot particles heterogeneously—when environmental temperatures are above -35C, and in competition with homogeneous freezing when below this temperature. Aircraft tracks upwind of these clouds can dispense the soot aerosols and subsequent in-situ and remote sampling of the clouds can discern changes in the ice microphysics. Confirmation of the ice-nucleating behavior of the soot aerosols would involve direct measurement of emitted aerosols using ice nuclei activation instrumentation, and measurements of the composition and ice activation properties of the wave cloud residual aerosols following their sampling by a CVI. Cloud chamber and other fundamental laboratory studies can be used to further elucidate the ice-nucleating properties of soot aerosols and their potential changes with atmospheric aging. Further, laboratory studies could be used in conjunction with carefully designed flight profiles to assess the role of "preactivation" on the ice-forming ability of soot.

High priority – Medium Term: Deploy and acquire sounding of temperature and water vapour from the current generation of radiosondes – the Vaisala RS90 and RS92 that do not have the strong negative biases found in earlier sondes

The upper limit on contrail and contrail-cirrus coverage is largely driven by the amount and spatial extent of ice-supersaturation (ISS) in the upper troposphere. Over the next several years as longterm data sets acquired with these sensors become available, more robust estimates of layering and spatial extent of ice supersaturations will become available for the upper troposphere. Continued acquisition and analysis of data from the MOZAIC project and from satellite-borne instruments (AIRS) will complement the sonde measurements and should be encouraged, although the relative humidities derived from satellites remain highly uncertain.

High priority – Medium Term: Deploy ground-based remote sensors for measuring upper tropospheric water vapor concentrations, specifically Raman lidar, in areas with a high likelihood of ISS in the upper troposphere. Medium priority – Long Term: Equip commercial airliners (like MOZAIC) with humidity probes that are designed especially for use in the upper troposphere (including AMDAR, Aircraft Meteorological Data Reporting to the weather centers) and cloud sensors that detect cirrus layers and contrail plumes

New developments in the capability of meteorological models, such as the ECMWF (European Centre for Medium-Range Weather Forecasts) operational model, to predict ice supersaturated regions means that it may be possible in the near future to predict whether persistent contrails can form in a specific region at a specific time. Unfortunately the meteorological analyses (including the ECMWF one) which serve as initial conditions for forecast runs and for archiving/documenting the atmospheric state still do not represent ISS since radiosonde readings are not operationally corrected and satellite data assimilation suffers from the low vertical resolution in the water vapour bands. Simple cloud sensors would provide a useful contribution to cloud coverage along with basic information on effective radius and number concentration.

5. RECOMMENDATIONS FROM THE CURRENT 'PRACTICAL' PERSPECTIVE

Can our current body of knowledge presented in Section 2 be used in the near term to improve the representation of commercial aircraft influences on climate? Specifically, what can be done using the data collected to date on contrail and cirrus cloud microphysics, their radiative properties, the ice nucleating properties of soot aerosols, and the results of detailed (e. g., LES) modeling of contrail thermodynamics, dynamics and microphysics be used in the near term to improve estimates of the effects of aircraft on climate?

Central to this question is whether the global distribution of ice supersaturation can be better predicted in forecast and climate models and whether it can be evaluated using the existing body of global data on upper tropospheric relative humidity and ice supersaturation. Models will have to carry water vapor and allowing the build up of ice supersaturations rather than condensing and then unloading the condensate from one time step to the next. Ice supersaturation and ice condensate must be carried from one time step to the next and for it to be removed realistically. A physically consistent treatment of ice supersaturation and contrail/cirrus coverage in global models, however, is probably not achievable in the near term, because it requires fundamental changes in cloud parameterization schemes. Meanwhile, substantial progress in global models is still possible in the near future by adapting and validating the subgrid-scale parameterizations of supersaturation and cloud fraction to contrail cirrus that are currently in use to predict natural clouds.

Accurate, high quality information on the vertical structure of water vapor as a function of time of day and season are needed on a global basis for model evaluation and for developing a reliable data base and for evaluating space-borne estimates of water vapor (e. g., AIRS). Improving existing radiosonde water vapor measurements to correct for biases resulting from sensor time lag, 'icing" of the sensor, solar radiative effects, and vertical averaging are now achievable. This is especially true for sonde measurements made with sensors developed and deployed over the past five years that are less influenced by time lag and icing. The correction algorithms should be and could immediately be used in the operational work of the weather centers. This would allow, in the short term, the representation of ice supersaturation in the analyses via data assimilation, not only in the forecasts. With the corrected relative humidity (water vapor) measurements, there is now the ability and also a strong need for performing a comparison of ice supersaturated regions from sondes with weather forecast (e. g., ECMWF) and climate models.

Improvements in the representation of cirrus microphysical and radiative properties in global models are desirable in that they feed back into the upper tropospheric water vapor budget. Improved representations of the scattering properties of natural cirrus ice crystals—scattering models, have recently become available. Improvements in knowledge of cirrus crystal nucleation— homogeneous and heterogeneous processes and how those can be parameterized for use in large scale model, have become available during the past several years. A consideration of recent laboratory studies, if parameterized in models, can be used at least to bracket minimum and maximum potential impacts of soot emissions on cirrus nucleation.

Several flights into contrails at various stages of their development have been acquired but have not been mined. These include observations in the US from MiDCiX and additional observations from CRYSTAL-FACE, among others. Because the contrails in these cases were generated by the research aircraft, they acquired high-quality water vapor and particle microphysical measurements, and several eventually developed into contrail cirrus, analysis of those data would provide information on the properties of contrail cirrus and provide data sets to evaluate contrail microphysical models.

A global database of upper-level cloudiness and information on vertical profiles of extinction through the upper parts of high clouds is now available from the CALIPSO satellite. These data can be scrutinized with a goal of differentiating cirrus from contrail microphysical (extinction) and radiative properties and may help to evaluate how, when and where contrails evolve into contrail cirrus.

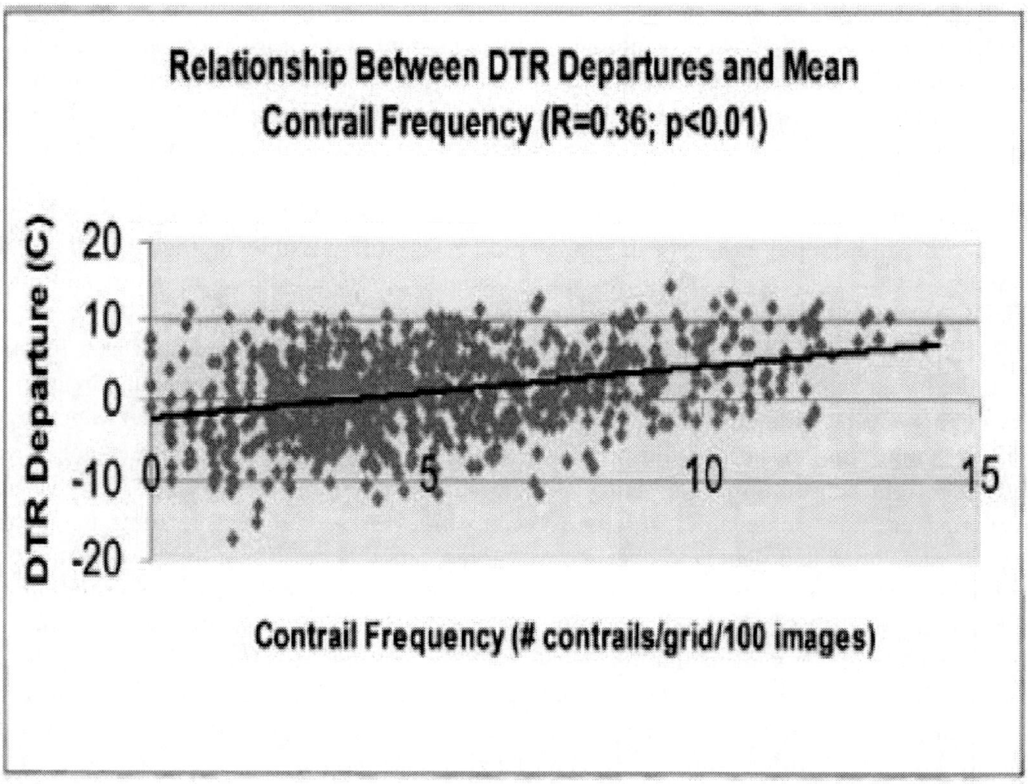

Figure 17. Results of sensitivity study of net radiative fluxes for varying contrail conditions. From Travis et al. 2004.

REFERENCES

Abbatt, J. P. D., Benz, S., Cziczo, D. J., Kanji, Z., Lohmann, U., and Möhler, O., 2006: Solid ammonium sulfate aerosols as ice nuclei: A Pathway for cirrus cloud formation, *Science*, 313, 5794, 1770–1773.

Appleman, H., 1953: The formation of exhaust condensation trails by jet aircraft. *Bull. Am. Meteorol. Soc.*, *34*, 14–20.

Archuleta, C.A., P.J. DeMott, and S.M. Kreidenweis, 2005: Ice nucleation by surrogates for atmospheric mineral dusts and mineral dust/sulfate particles at cirrus temperatures. *Atmos. Chem. Phys.*, 5, 2617–2634.

Atlas, D., Z. Wang, D.P. Duda, 2006: Contrails to cirrus - morphology, microphysics, and radiative properties. *J. Appl. Meteorol. Climatol.*, 45, 5-19.

Bakan, S., M. Betancor, V. Gayler, H. Graßl, 1994: Contrail frequency over Europe from NOAAsatellite images. *Ann. Geophysicae*, 12, 962-968.

Baumgardner, D., J.E. Dye, B. Gandrud, K. Barr, K. Kelly, K.R. Chan, 1996: Refractive indices of aerosols in the upper troposphere and lower stratosphere, *Geophys. Res. Lett*, 23, 749-752.

Baumgardner, D. and. B. Gandrud, 1998: A comparison of the microphysical and optical properties of particles in an aircraft contrail and mountain wave cloud, *Geophys. Res. Lettr.*, 25, 1129-1132.

Baumgardner, D., H. Jonsson, W. Dawson, D. O'Connor and R. Newton, 2001: The cloud, aerosol and precipitation spectrometer (CAPS): A new instrument for cloud investigations, *Atmos. Res.*, 59-60, 251-264.

Baumgardner, D., J.F. Gayet, H. Gerber, A. Korolev, and C. Twohy, 2002: Clouds: Measurement Techniques In-Situ, in the *Encyclopaedia of Atmospheric Science*, Eds. J, Curry, J. Holton and J. Pyle, Academic Press, U.K., ISBN: 0-12-227090-8

Baumgardner, D., H. Chepfer, G.B. Raga, G.L. Kok, 2005: The Shapes of Very Small Cirrus Particles Derived from In Situ Measurements, *Geophys. Res. Lett.,*32, L01806, doi:10.1029/2004GL021300, 2005

Beaver, M. R., Elrod, M. J., Garland, R. M., and Tolbert, M. A., 2006: Ice nucleation in sulphuric acid/organic aerosols: Implications for cirrus cloud formation, *Atmos. Chem. Phys.*, 6, 3231– 3242.

Boucher, O., 1999: Air traffic may increase cirrus cloudiness. *Nature*, 397, 30-31.

Brock, C. A., F. Schröder, B. Kärcher, A. Petzold, R. Busen, M. Fiebig, 2000: Ultrafine particle size distributions measured in aircraft exhaust plumes, *J. Geophys. Res.*, 105(D21), 26555- 26568, 10.1029/2000JD900360.

Busen, R., U. Schumann, 1995. Visible contrail formation from fuels with different sulfur contents. *Geophys. Res. Lett.*, 22, 1357–1360.

Campos, T. L., and 15 coauthors, 1998: Measurements of NO and NOy emission indices during SUCCESS. *Geophys. Res. Ltrs.*, 25, 1713-1716.

Carleton, A.M., P.J. Lamb, 1986: Jet contrails and cirrus cloud: A feasibility study employing highresolution satellite imagery. *Bull. AMS*, 67, 301-309.

Chen, Y., S.M. Kreidenweis, L.M. McInnes, D.C. Rogers, P.J. Demott, 1998: Single particle analysis of ice nucleating aerosols in the upper troposphere and lower stratosphere, *Geophys. Res. Lettr.*, 25, 1391-1394.

Cziczo, D. J., Murphy, D. M., Hudson, P. K., and Thomson, D. S., 2004: Single particle measurements of the composition of cirrus ice residues during CRYSTAL-FACE, *J. Geophys. Res.*, 109, D04201, doi:10.1029/2004JD004032.

Demirdjian, B. , D. Ferry, J. Suzanne, O.B. Popovicheva, N.M. Persiantseva and N.K. Shonija, 2007: Heterogeneities in the microstructure and composition of aircraft engine combustor soot: impact on the water uptake. *J. Atmos. Chem.*, 56, 83, doi: 10.1007/s10874-006-9043-9.

DeMott, P.J., 1990: An exploratory study of ice nucleation by soot aerosols. *J. Appl. Meteor.*, 29, 1072-1079.

DeMott, P. J., Rogers, D. C., and Kreidenweis, S. M., 1997: The susceptibility of ice formation in upper tropospheric clouds to insoluble aerosol components, *J. Geophys. Res.*, 102, 19 575–19 584.

DeMott, P. J., Chen, Y., Kreidenweis, S. M., Rogers, D. C., and Sherman, D. E., 1999: Ice formation by black carbon particles, *Geophys. Res. Lett.*, 26, 2429–2432.

DeMott, P. J., 2002: Laboratory studies of cirrus cloud processes, In *Cirrus*, D.K. Lynch, K. Sassen, D.O.C Starr, G. Stephens Eds., Oxford University Press, New York, 102-135.

DeMott, P.J., A.J. Prenni, C.A. Archuleta, and S.M. Kreidenweis, 2002: Some aerosol effects on cirrus ice formation. *American Meteorological Society Conference on Cloud Physics*, Ogden, UT, June, 2002 (on CD-ROM).

DeMott, P. J., K. Sassen, M. Poellot, D. Baumgardner, D. C. Rogers, S. Brooks, A. J. Prenni, and S. M. Kreidenweis, 2003a: African dust aerosols as atmospheric ice nuclei. *Geophys. Res. Lett.*, 30, No. 14, 1732, doi:10.1029/2003GL017410.

DeMott, P. J., D. J. Cziczo, A. J. Prenni, D. M. Murphy, S. M. Kreidenweis, D. S. Thomson, R. Borys and D. C. Rogers, 2003b: Measurements of the concentration and composition of nuclei for cirrus formation. *Proceedings of the National Academy of Sciences*, 100, No. 25, 14655- 14660.

DeMott, P. J., M.D. Petters, A.J. Prenni, C.M. Carrico, and S.M. Kreidenweis, 2008: Ice nucleation behavior of biomass combustion particles, in preparation for submission to *Atmos. Chem. And Phys.*

Denman, K.L., G. Brasseur, A. Chidthaisong, P. Ciais, P.M. Cox, R.E. Dickinson, D. Hauglustaine, C. Heinze, E. Holland, D. Jacob, U. Lohmann, S Ramachandran, P.L. da Silva Dias, S.C. Wofsy and X. Zhang, 2007: Couplings Between Changes in the Climate System and Biogeochemistry. In: *Climate Change 2007: The Physical Science Basis. Contribution of Working Group I to the Fourth Assessment Report of the Intergovernmental Panel on Climate Change* [Solomon, S., D. Qin, M. Manning, Z. Chen, M. Marquis, K.B. Averyt, M.Tignor and H.L. Miller (eds.)]. Cambridge University Press, Cambridge, United Kingdom and New York, NY, USA.

Diehl, K. and Mitra, S. K., 1998: A laboratory study of the effects of a kerosene burner exhaust on ice nucleation and the evaporation rate of ice crystals, *Atmos. Environ.*, 32, 3145–3151. Draine, B.T., and P.J. Flatau. Discrete dipole approximation for scattering calculations. *J. Opt. Soc. Am. A*, 11:1491-1499, 1994.

Duda, D.P., P. Minnis, L. Nguyen, R. Palikondra, 2004: A case study of the development of contrail clusters over the Great Lakes. *J. Atmos. Sci.,* 61, 1132-1146.

Dürbeck, T., T. Gerz, 1995: Large-eddy simulation of aircraft exhaust plumes in the free atmosphere: effective diffusivities and cross-sections. *Geophys. Res. Lett.*, 22, 3203-3206.

Dürbeck, T., T. Gerz, 1996: Dispersion of aircraft exhausts in the free atmosphere. *J. Geophys. Res.*, 101, 26,007-26,015.

Dye, J.E., B.W. Gandrud, D. Baumgardner, K.R. Chan, G.V. Ferry, M. Loewenstein, K.K. Kelly, and J.C. Wilson, 1990: Observed particle evolution in the polar stratospheric cloud of January 24, 1989. *J. Geophys. Res.*, 17, 413-416.

Dymarska, M., Murray, B. J., Sun, L., Eastwood, M. L., Knopf, D. A., and Bertram, A. K., 2006: Deposition ice nucleation on soot at temperatures relevant for the lower troposphere, *J. Geophys. Res.*, 111, D04204, doi:10.1029/2005JD006627.

Fahey, D.W. et al., 1995. Emission measurements of the Concorde supersonic aircraft in the lower stratosphere. *Science* 270, 70-74.

Field, P. R., A. J. Heymsfield, and A. Bansemer , 2006: Shattering and Particle Interarrival Times Measured by Optical Array Probes in Ice Clouds, *J. Atmos. and Oceanic Tech.,* 23, 1357–1371.

Forster, P., Ramaswamy, V., Artaxo, P., Berntsen, T., Betts, R., Fahey, D.W., Haywood, J., Lean, J., Lowe, D.C., Myhre, G., Nganga, J., Prinn, R., Raga, G., Schulz, M. and Van Dorland, R., 2007: Changes in Atmospheric Constituents and in Radiative Forcing. In: Climate Change 2007: The Physical Science Basis. Contribution of Working Group I to the Fourth Assessment Report of the Intergovernmental Panel on Climate Change. Cambridge University Press, Cambridge, United Kingdom and New York, NY, USA.

Freudenthaler, V., F. Homburg, H. Jäger, 1995: Contrail observations by ground-based scanning lidar: cross-sectional growth. *Geophys. Res. Lett.*, 22, 3501-3504.

Gao, R. S., and coauthors, 2004: Evidence that nitric acid increases relative humidity in lowtemperature cirrus clouds. Science, 303, 516-520.

Gardiner, B. A., and J. Hallett, 1985: Degradation of in-cloud forward scattering spectrometer probe measurements in the presence of ice particles, *J. Atmos. Oceanic Technol.*, 2, 171–180.

Gayet, J. G., G. Febvre, and H. Larsen, 1996: The reliability of the PMS FSSP in the presence of small ice crystals, *J. Atmos. Oceanic Technol.*, 13, 1300– 1310.

Gerber, H., C.H. Twohy, B.W. Gandrud, A.J. Heymsfield, G. M. McFarquhar, P.J. Demott and. D.C. Rogers, 1998: Measurements of wave-cloud microphysical properties with two new aircraft probes, *Geophys. Res. Lettr*, 25, 1117-1120.

Gettelman, A., et al., 2006, The Global Distribution of Supersaturation in the Upper Troposphere from the Atmospheric Infrared Sounder. *J. Climate,* 19, 6089–6103.

Gierens, K.M., 1998: How the sky gets covered with condensation trails. *Meteorol. Z.*, 7, 181-187.

Gierens, K., U. Schumann, 1996: Colors of contrails from fuels with different sulfur contents. J. Geophys. Res. 101, 16731-16736.

Gierens, K., E. Jensen, 1998: A numerical study of the contrail-to-cirrus transition. *Geophys. Res. Lett.* 55, 4341-4344.

Gierens, K., U. Schumann, M. Helten, H.G.J. Smit, A. Marenco, 1999: A distribution law for relative humidity in the upper troposphere and lower stratosphere derived from three years of MOZAIC measurements. *Ann. Geophys.* 17, 1218-1226.

Gierens, K., P. Spichtinger, 2000: On the size distribution of ice-supersaturated regions in the upper troposphere and lowermost stratosphere. *Ann. Geophys.* 18, 499-504.

Gierens, K., U. Schumann, M. Helten, H. Smit, P.H. Wang, 2000: Ice-supersaturated regions and sub visible cirrus in the northern midlatitude upper troposphere. *J. Geophys. Res.* 105, 22743-22754.

Gierens, K. 2003: On the transition between heterogeneous and homogeneous freezing. *Atmos. Chem. Phys.*, 3, 437-446.

Gierens, K., R. Kohlhepp, P. Spichtinger, M. Schroedter-Homscheidt, 2004: Ice supersaturation as seen from TOVS. *Atmos. Chem. Phys.*, 4, 539-547.

Gierens, K., 2007: Are fuel additives a viable contrail mitigation option? *Atmospheric Environment* 41, 4548–4552.

Goodman, J., R.F. Pueschel, E.J. Jensen, S. Verma, G.V. Ferry, S.D. Howard, S.A. Kinne, and D. Baumgardner, 1998: Shape and size of contrail particles, *Geophys. Res. Lettr.*, 25, 1327-1330.

Gorbunov, B., Baklanov, A., Kakutkina, N., Windsor, H. L., and Tuomi, R., 2001: Ice nucleation on soot particles, *J. Aerosol Sci.*, 32, 199–215.

Gounou, A. and Hogan, R. J., 2007: A sensitivity study of the effect of horizontal photon transport on the radiative forcing of contrails, J. Atmos. Sci., 64, 1706–1716.

Haag, W., Kärcher, B., Schaefers, S., Stetzer, O., Möhler, O., Schurath, U., Krämer, M., and Schiller, C., 2003: Numerical simulations of homogeneous freezing processes in the aerosol chamber AIDA, *Atmos. Chem. Phys.*, 3, 195–210.

Haag,W. and Kärcher, B., 2004: The impact of aerosols and gravity waves on cirrus clouds at midlatitudes, *J. Geophys. Res.*, 109, D12202, doi:10.1029/2004JD004579.

Hansen J et al. 2005: Efficacy of climate forcings. J.Geophys.Res., 110:D18104

Heintzenberg, J., K. Okada, and J. Ström, 1996: On the composition of non-volatile material in upper tropospheric aerosols and cirrus crystals. *Atmos. Res.*, 41, 81-88.

Hendricks, J., Kärcher, B., Lohmann, U., and Ponater, M., 2005: Do aircraft black carbon emissions affect cirrus clouds on the global scale? *Geophys. Res. Lett.*, 32, L12814, doi:10.1029/2005GL022740.

Heymsfield, A.J., and R.M. Sabin, 1989: Cirrus crystal nucleation by homogeneous freezing of solution droplets. *J. Atmos. Sci.*, 46, 2252-2264.

Heymsfield, A. J., R. P. Lawson, and G. W. Sachse, 1998: Growth of ice crystals in a precipitating contrail. *Geophys. Res. Lett.*, *25*(9), 1335–1338.

Heymsfield, A.J., C. Schmitt, A. Bansemer, G-J van Zadelho, M. J. McGill, C. Twohy, D. Baumgardner, 2006: Effective Radius of Ice Cloud Particle Populations Derived from Aircraft Probes, *J. Atmos. Oceanic Tech.* 23, No. 3, pp. 361–380.

Heymsfield, A.J., 2007: On Measurements of Small Ice Particles in Clouds, *Geophys. Res. Lettr*, VOL. 34, L23812, doi:10.1029/2007GL030951, 2007

Hung, H.-M., S. T. Malinowski, and S. T. Martin, 2003: Kinetics of heterogeneous ice nucleation on the surfaces of mineral dust cores inserted into aqueous ammonium sulfate particles, *J. Phys. Chem. A,* 107, 1296–1306.

Intergovernmental Panel on Climate Change (IPCC 1999), *Aviation and the Global Atmosphere*. J. E. Penner, D. H. Lister, D. J. Griggs, D. J. Dokken, and M. McFarland (eds.), Cambridge University Press. Cambridge, UK, 1999.

Jensen E.J., and O.B. Toon, 1997: The potential impact of soot particles from aircraft exhaust on cirrus clouds. *Geophys. Res. Lett.*, 24, 249-252.

Jensen, E.J., O.B. Toon, R.F. Pueschel, J. Goodman, G.W. Sachse, B.E. Anderson, K.R. Chan, D. Baumgardner, R.C. Miake-Lye., 1998: Ice crystal nucleation and growth in contrails forming at low ambient temperatures, *Geophys. Res. Lettr.*, 25, 1371-1374.

Jensen et al., 2005: Ice supersaturations exceeding 100% at the cold tropical tropopause: implications for cirrus formation and dehydration. *Atmos. Chem. Phys.*, 5, 851-862.

Jonsson, H.H., J.C. Wilson, C.A. Brock, R.K.Knollenberg, R. Newton, J.E. Dye, D. Baumgardner, S. Borrmann, G.V. Ferry, R. Pueschel, D.C. Woods, M.C. Pitts, 1995: Performance of a Focused cavity aerosol spectrometer for measurements in the stratosphere of particle size in the 0.06 - 2.0 mm-Diameter range, *J. Ocean. Atmos. Tech.*, 12, 115-129.

Kanji, Z. A. and Abbatt, J. P. D., 2006: Laboratory studies of ice formation via deposition mode nucleation onto mineral dust and n-hexane soot samples, *J. Geophys. Res.*, 111, D16204, doi:10.1029/2005JD006766.

Kärcher, B., 1996: Aircraft-generated aerosols and visible contrails. *Geophys. Res. Lett. 23 (15)*, 1933-1936.

Kärcher, B., Th. Peter and R. Ottmann, 1995: Contrail formation: Homogeneous nucleation of H2SO4 / H_2O droplets. *Geophys. Res. Lett. 22,* 1501-1504.

Kärcher, B., T. Peter, U.M. Biermann, and U. Schumann, 1996: The initial composition of jet condensation trails. *J. Atmos. Sci.*, 53, 3066-3083.

Kärcher, B., 1998: Physicochemistry of aircraft generated liquid aerosols, soot, and ice particles, 1. Model description. *J. Geophys. Res. 103 (D14)*, 17111-17128.

Kärcher, B., R. Busen, A. Petzold, F.P. Schröder, U. Schumann, and E.J. Jensen, 1998: Physicochemistry of aircraft generated liquid aerosols, soot, and ice particles, 2. Comparison with observations and sensitivity studies. *J. Geophys. Res. 103 (D14)* , 17129-17148.

Kärcher, B., and U. Lohmann, 2003: A parameterization of cirrus cloud formation: Heterogeneous freezing, *J. Geophys. Res.*, 108, 4402, doi:10.1029/2002JD003220.

Kärcher, B. and Ström, J., 2003: The roles of dynamical variability and aerosols in cirrus cloud formation, *Atmos. Chem. Phys.*, 3, 823–838.

Kärcher, B., Hendricks, J., and Lohmann, U., 2006: Physically based parameterization of cirrus cloud formation for use in global atmospheric models, *J. Geophys. Res.*, 111, D01205, doi:10.1029/JD006219.

Kärcher, B., O. Möhler, P.J. DeMott, S. Pechtl, and F. Yu, 2007: Insights into the role of soot aerosols in cirrus cloud formation, *Atmos. Chem. Phys.*, 7, 4203–4227.

Knollenberg, R.G., 1972: Measurements of the growth of the ice budget in a persisting contrail. *J. Atmos. Sci.*, 29, 1367-1374.

Knopf, D. A., and T. Koop, 2006: Heterogeneous nucleation of ice on surrogates of mineral dust. J. *Geophys. Res.*, 111, D12201, doi:10.1029/2005JD006894.

Koehler, K., M. D. Petters, P. J. DeMott, S. M. Kreidenweis, and O. B. Popovicheva, 2008: Hygroscopicity, cloud condensation nuclei activity and ice nucleation ability of selected combustion soot particles. In preparation for submission to *Atmos. Chem. Phys.*

Koop, T., B, Luo, A. Tsias, and T. Peter, 2000: Water activity as the determinant for homogeneous ice nucleation in aqueous solutions. *Nature*, 406, 611-614.

Koop, T., 2004: Homogeneous nucleation in water and aqueous systems. *Z. Phys. Chem.* 218, 1231-1258.

Korolev, A., and G. A. Isaac, 2005: Shattering during sampling by OAPs and HVPS. part I: Snow particles, *J. Atmos. Oceanic Technol.*, 22, 528–543.

Kuhn, M., A. Petzold, D. Baumgardner, and F.P. Petzold, 1998: Particle composition of a young condensation trail and of upper tropospheric aerosol, *Geophys. Res. Lettr.*, 25, 2679-2682.

Kuhn, P.M., 1970: Airborne observations of contrail effects on the thermal radiation budget. J. Atmos. Sci., 27, 937-942.

Lawson, R. P., A. J. Heymsfield, S. M. Aulenbach, and T. L. Jensen, 1998: Shapes, sizes, and light scattering properties of ice crystals in cirrus and a persistent contrail during SUCCESS, *Geophys. Res. Lett.*, *25*(9), 1331–1334.

Lawson, R.P., Baker, B.A., Schmitt, C.G., and Jensen, T.L., 2001: An overview of microphysical properties of Arctic clouds observed in May and July during FIRE.ACE. *J. Geophys. Res.*, 106, 14,989-15,014.

Lawson, R. P., O'Connor, D., Zmarzly, P., Weaver, K., Baker, B. A., Mo, Q., and Jonsson, H.: The 2D-S (Stereo) probe, 2006: design and preliminary tests of a new airborne, high speed, highresolution particle imaging probe, *J. Atmos. Oceanic Technol.*, 23, 1462-1477.

Lewellen, D. C., and W. S. Lewellen, 2001: The effects of aircraft wake dynamics on contrail development. *J. Atmos. Sci.,* 58, 390–406.

Lin, R-F., D.O.C Starr, P.J. DeMott, R. Cotton, K. Sassen, E. Jensen, B. Karcher, and X. Liu, 2002:

Cirrus parcel model comparison project phase 1: The critical components to simulate cirrus initiation explicitly, *J. Atmos. Sci.*, 59, 2305-2329.

Lohmann, U., 2002: Possible aerosol effects on ice clouds via contact nucleation. *J. Atmos. Sci.*, 59, 647-656.

Lohmann, U. and K. Diehl, 2006: Sensitivity studies of the importance of dust ice nuclei for the indirect aerosol effect on stratiform mixed-phase clouds. *J. Atmos. Sci.*, 63, 968-982.

Mannstein, H. and U. Schumann, 2005: Aircraft induced contrail cirrus over Europe. *Meteorol. Z. 14*, 549-554.

Marquart, S. and Mayer, B., 2002: Towards a reliable GCM estimation of contrail radiative forcing, Geophys. Res. Lett., 29(8), 20-1–20-4, doi:10.1029/2001GL014075.

Marquart, S., Ponater, M., Mager, F., and Sausen, R., 2003: Future development of contrail cover, optical depth and radiative forcing: Impacts of increasing air traffic and climate change, J. Climate, 16, 2890–2904.

McFarquhar, Greg M. Junshik Um, Matt Freer, Darrel Baumgardner, Gregory L. Kok, and Gerald Mace, 2007: Importance of small ice crystals to cirrus properties: Observations from the Tropical Warm Pool International Cloud Experiment (TWP-ICE), *Geophys. Res. Lettr.*, 34, L13803, doi:10.1029/2007GL029865

McMurray, P.H., 2000: A review of atmospheric aerosol measurements, *Atmospheric Environment*, 34, 1959-1999

Meerkötter, R., U. Schumann, D. R. Doelling, P. Minnis, T. Nakajima, and Y. Tsushima, 1999: Radiative forcing by contrails. *Ann. Geophys.*, 17, 1080–1094, 1999.

Meyer, R., Mannstein, H., Meerk¨otter, R., Schumann, U., and Wendling, P., 2002: Regional radiative forcing by line-shaped contrails derived from satellite data, J. Geophys. Res., 107(D10), ACL 17- 1–ACL 17-15, doi:10.1029/2001JD000426

Minnis, P., V. Chakrapani, D. Doelling, L. Nguyen, R. Palikonda, D. Spangenberg, T. Uttal, R. Arduini, and M. Shupe (2001), Cloud coverage and height during FIRE ACE derived from AVHRR data, J. Geophys. Res., 106(D14), 15215-15232.

Minnis P, Ayers JK, Palikonda R. Phan D. 2004: Contrails, Cirrus Trends, and Climate. J. Climate, 17, 1671-1685

Mishchenko, M. I. and A. Macke, 1999: " How Big Should Hexagonal Ice Crystals be to Produce Halos?," Appl. Opt. 38, 1626-1629.

Möhler, O., Stetzer, O., Schaefers, S., Linke, C., Schnaiter, M., Tiede, R., Saathoff, H., Krämer, M., Mangold, A., Budz, P., Zink., P., Schreiner, J., Mauersberger, K., Haag, W., Kärcher, B., and Schurath, U., 2003: Experimental investigations of homogeneous freezing of sulphuric acid particles in the aerosol chamber AIDA, *Atmos. Chem. Phys.*, 3, 211–223.

Möhler, O., Büttner, S., Linke, C., Schnaiter, M., Saathoff, H., Stetzer, O., Wagner, R., Krämer, M., Mangold, A., Ebert, V., and Schurath, U., 2005a: Effect of sulfuric acid coating on heterogeneous ice nucleation by soot aerosol particles. *J. Geophys. Res.*, 110, D11210, doi:10.1029/2004JD005169.

Möhler, O., Linke, C., Saathoff, H., Schnaiter, M., Wagner, R., Mangold, A., Krämer, M., and Schurath, U., 2005b: Ice nucleation on flame soot aerosol of different organic carbon content. *Meteorol. Z.*, 14, 477–484.

Möhler, O., P. R. Field, P. Connolly, S. Benz, H. Saathoff, M. Schnaiter, R. Wagner, R. Cotton, M. Krämer, A. Mangold, and A. J. Heymfield, 2006: Efficiency of the deposition mode ice nucleation on mineral dust particles. *Atmos. Chem. Phys.*, 6, 3007-3021.

Murphy, D.M., 2003: Dehydration in cold clouds is enhanced by a transition from cubic to hexagonal ice. *Geophys. Res. Lett.*, 30(23), 2230, doi: 10.1029/2003GL018566.

Murray, B. J., and A. K. Bertram 2007: Strong dependence of cubic ice formation on droplet ammonium to sulfate ratio. Geophys. Res. Lett. 34, L16810, doi:10.1029/2007GL030471.

Myhre, G. and Stordal, F., 2001: On the tradeoff of the solar and thermal infrared radiative impact of contrails. Geophysical Research Letters, 28, 3119-3122.

Next Generation Air Transportation System, Federal Aviation Administration report to the U.S. Congress, 2004.

Peter, T., C. Marcolli, P. Spichtinger, T. Corti, M. B. Baker, and T. Koop, 2006: When Dry Air Is Too humid. *Science*, 314, 1399-1402.

Peter, T., C. Bruhl, and P. J. Crutzen 1991: Increase in the PSC-formation probability created by high-flying aircraft. *Geophys. Res. Lett.*, 18, 1465-1468.,

Murray, B. J., and A. K. Bertram 2007: Strong dependence of cubic ice formation on droplet ammonium to sulfate ratio. Geophys. Res. Lett. 34, L16810, doi:10.1029/2007GL030471.

Petzold, A., R. Busen, F.P. Schroeder, R. Baumann, M. Kuhn, J. Stroem, D. Hagen, P. Whitefield, D. Baumgardner, F. Arnold, and U. Schumann, 1997: Near field measurements on contrail properties from fuels with different sulfur content, *J. Geophys. Res.,* 102, 29,867-29,880.

Petzold, A., J. Ström, S. Ohlsson and F. P. Schröder, 1998a.: Elemental composition and morphology of ice-crystal residual particles in cirrus clouds and contrails, *Atmos. Res., 49, 21-34*

Petzold A., J. Ström, F. P. Schröder and B. Kärcher, 1998b: Carbonaceous aerosol in jet engine exhaust: emission characteristics and implications for heterogeneous chemical reactions, *Atmos. Environ., 33, 2689-2698.*

Petzold, A., A. Döpelheuer, C. A. Brock, F. Schröder, 1999: In situ observations and model calculations of black carbon emission by aircraft at cruise altitude, *J. Geophys. Res.*, 104(D18), 22171-22182, 10.1029/1999JD900460.

Petzold, A., M. Gysel, X. Vancassel, R. Hitzenberger, H. Puxbaum, S. Vrochticky, E.Weingartner, U. Baltensperger, and P. Mirabel, 2005: On the effects of organic matter and sulphurcontaining compounds on the CCN activation of combustion particles. *Atmos. Chem. Phys.*, 5, 3187–3203.

Poellot, M. R., W. P. Arnott, and J. Hallett, 1999: In-situ observations of contrail microphysics and implications for their radiative impact. *J. Geophys. Res.*, 104, 12,077–12,084.

Ponater, M., Marquart, S., and Sausen, R., 2002: Contrails in a comprehensive global climate model: Parameterization and radiative forcing results, J. Geophys. Res., 107(D13), ACL 2-1–ACL 2-15, doi:10.1029/2001JD000429.

Ponater, M., Marquart, S., Sausen, R., Schumann, U., 2005: On contrail climate sensitivity. Geophysical Research Letters 32, L10706, doi:10.1029/2005GL022580

Popovicheva O. B., Persiantseva N. M., Shonija N. K., P. J. DeMott, K. Koehler, M. D. Petters, S. M. Kreidenweis, V. Tishkova, B. Demirdjian, J. Suzanne, 2007: Water interaction with hydrophobic and hydrophilic soot particles, in preparation for submission to *Phys. Chem. Chem. Phys.*

Rädel G., and K. Shine, 2008. Radiative forcing by persistent contrails and its dependence on cruise altitudes. *J. Geophys. Res.*, (in press).

Richardson, M. S., P. J. DeMott, S. M. Kreidenweis, D. J. Cziczo, E. Dunlea, J. L. Jimenez, D. S. Thompson, L. L. Ashbaugh, R. D. Borys, D. S. Westphal, G. S. Cassucio and T. L. Lersch, 2007: Measurements of heterogeneous ice nuclei in the Western U.S. in springtime and their relation to aerosol characteristics. *J. Geophys. Res.*, 112, D02209, doi:10.1029/2006JD007500.

Rind D., Lonergan P. and Shah K., 2000: Modeled impact of cirrus cloud increases along flight paths. Journal of Geophysical Research 105, 19927-19940

Roberts, P. and J. Hallett, 1968: A laboratory study of the ice nucleating properties of some mineral particulates. *Quart. J. Royal Meteorol. Soc.*, 94, 25–34 (1968).

Rogers, D.C., P.J. DeMott, S.M. Kreidenweiss and Y. Chen, 1998: Measurements of ice nucleating aerosols during SUCCESS, *Geophys. Res. Lettr.*, 25, 1383-1386.

Rogers, D.C., P.J. DeMott, S.M. Kreidenweis and Y. Chen, 2001: A continuous flow diffusion chamber for airborne measurements of ice nuclei, *J. Atmos. Oceanic Technol.*, 18, 725-741.

Rogers, D.C., P.J. DeMott, and S.M. Kreidenweis, 2008: An exploratory field study of the production of ice nucleating aerosol particles by jet aircraft. In preparation for submission to *J. Geophys. Res.*

Sassen, K., and G.C. Dodd, 1988: Homogeneous nucleation rate for highly supercooled cirrus cloud droplets. *J. Atmos. Sci.*, 45, 1357-1369.

Sassen, K., P. J. DeMott, J. M. Prospero, and M. R. Poellot, 2003: Saharan dust storms and indirect aerosol effects on clouds: CRYSTAL-FACE results, *Geophys. Res. Lett.*, 30 (12), 1633, doi:10.1029/2003GL017371.

Sausen, R., K. Gierens, M. Ponater, and U. Schumann, 1998: A diagnostic study of the global distribution of contrails part I: Present day climate. *Theor. Appl. Climatol.*, 61, 127–141.

Sausen, R., Isaksen, I., Grewe, V., Hauglustaine, D., Lee, D., Myhre, G., Kohler, M., Pitari, G., Schumann, U., Stordal, F. and Zerefos, C. , 2005: Aviation radiative forcing in 2000: An update on IPCC (1999). Meteorologische Zeitschrift, 14, 4, 555-561. doi: 10.1127/0941-2948/2005/0049.

Sausen, R., et al., 2005: Aviation radiative forcing in 2000: An update on IPCC (1999). *Meteorol. Z.*, 14, 1–7.

Schiller, C., M. Krämer, A. Afchine, N. Spelten: An in-situ climatology of ice water content in Arctic, mid latitude and tropical cirrus, JGR, in press.

Schmidt, E., 1941: Die Entstehung von Eisnebel aus den Auspuffgasen von Flugmotoren. In: *Schriften der Deutschen Akademie der Luftfahrtforschun.* Verlag R. Oldenbourg, Munich and Berlin, Germany, Vol. 44, pp. 1-15.

Schröder, F. P., B. Kärcher, A. Petzold, R. Baumann, R. Busen, C. Hoell, U. Schumann, 1998: Ultrafine aerosol particles in aircraft plumes: In situ observations, Geophys. Res. Lett., 25(15), 2789-2792, 10.1029/98GL02078.

Schröder, F., C. A. Brock, R. Baumann, A. Petzold, R. Busen, P. Schulte, M. Fiebig, 2000: In situ studies on volatile jet exhaust particle emissions: Impact of fuel sulfur content and environmental conditions on nuclei mode aerosols, J. Geophys. Res., 105(D15), 19941-19954, 10.1029/2000JD900112.

Schröder, F., and Coauthors, 2000: On the transition of contrails into cirrus clouds. *J. Atmos. Sci.,* 57, 464–480.

Schumann U., Ström, J., Busen, R., Baumann, R., Gierens, K., Krautstrunk, M., Schröder, F.P., Stingl, J., 1996. In situ observations of particles in jet aircraft exhausts and contrails for different sulfur-containing fuels. J. Geophys. Res. 101, 6853-6869.

Schumann, U., 2000: Influence of Propulsion Efficiency on Contrail Formation, *Aerospace Science and Technology* 4 391-401.

Schumann, U., R. Busen, and M. Plohr, 2000: Experimental Test of the Influence of Propulsion Efficiency on Contrail Formation, J. Aircraft 37 (2000) 1083-1087.

Schumann, U., 2002: Contrail cirrus, in D. K. Lynch et al. (eds.), *Cirrus*, Oxford University Pre ss, pp. 231–255.

Schumann U., F. Arnold, R. Busen, J. Curtius, B. Kärcher, A. Kiendler, A. Petzold, H. Schlager, F. Schröder, and K.-H. Wohlfrom, 2002: Influence of fuel sulfur on the composition of aircraft exhaust plumes: The experiments SULFUR 1–7, *J. Geophys. Res.*, 107 (D15), doi:10.1029/2001JD000813.

Seifert, M., J. Ström, R. Krejci, A. Minikin, A. Petzold, J.-F. Gayet, H. Schlager, H. Ziereis, U. Schumann, and J. Ovarlez, 2004: Thermal stability analysis of particles incorporated in cirrus crystals and of non-activated particles in between the cirrus crystals: comparing clean and polluted air masses. *Atmos. Chem. Phys.*, 4, 1343–1353.

Shilling, J. E., Fortin, T. J., and Tolbert, M. A., 2006: Depositional ice nucleation on crystalline organic and inorganic solids, *J. Geophys. Res.*, 111, D12204, doi:10.1029/2005JD006664.

Shine, K.P., 2005: Comment on 'Contrails, cirrus, trends, and climate'. *J. Clim.*, 18, 2781–2782.

Spichtinger, P., K. Gierens, W. Read, 2002: The statistical distribution law of relative humidity in the global tropopause region. *Meteorol. Z.* 11, 83-88.

Spichtinger, P., K. Gierens, U. Leiterer, H. Dier, 2003a: Ice supersaturation in the tropopause region over Lindenberg, Germany. *Meteorol. Z.*, 12, 143-156.

Spichtinger, P., K. Gierens, W. Read, 2003b: The global distribution of ice-supersaturated regions as seen by the Microwave Limb Sounder. *Q. J. R. Meteorol. Soc.*, 129, 3391-3410.

Spichtinger, P., K. Gierens, H. Wernli, 2005a: A case study on the formation and evolution of ice supersaturation in the vicinity of a warm conveyor belt's outflow region. *Atmos. Chem. Phys.*, 5, 973-987.

Spichtinger, P., K. Gierens, A. Dörnbrack, 2005b: Formation of ice supersaturation by mesoscale gravity waves. *Atmos. Chem. Phys.*, 5, 1243-1255.

Strauss, B.; Meerkötter, R.; Wissinger, B.; Wendling, P.; Hess, M. (1997): On the Regional Climate Impact of Contrails - Microphysical and Radiative Properties of Contrails and Natural Cirrus Clouds. Annales Geophysicae, Vol. 15, S. 1457 – 1467

Ström, J. and Ohlsson, S., 1998: In-situ measurements of enhanced crystal number densities in cirrus clouds caused by aircraft exhaust, *J. Geophys. Res.*, 103, 11 355–11 361.

Ström, J., M. Seifert, B. Kärcher, J. Ovarlez, A. Minikin, J.-F. Gayet, R. Krejci, A. Petzold, F. Auriol, W. Haag, R. Busen, U. Schumann, and H. C. Hansson, 2003: Cirrus cloud occurrence as function of ambient relative humidity: a comparison of observations obtained during the INCA experiment. *Atmos. Chem. Phys.*, 3, 1807–1816.

Stubenrauch, C.J., and U. Schumann, 2005: Impact of air traffic on cirrus coverage. Geophys. Res. Lett., 32, L14813, doi:10.1029/2005GL022707.

Sussmann, R., K. Gierens, 1999: Lidar and numerical studies on the different evolution of a contrail's vortex system and its secondary wake. *J. Geophys. Res.* 104, 2131-2142.

Stuber, N. and Forster, P., 2007: The impact of diurnal variations of air traffic on contrail radiative forcing, Atmos. Chem. Phys., 7, 3153-3162.

Stordal F, Myhre G, Stordal EJG, Rossow WB, Lee DS, Arlander DW, Svenby T 2005: Is there a trend in cirrus cloud cover due to aircraft traffic? Atmospheric Chemistry and Physics 5:2155 – 2162.

Sussmann, R., K. Gierens, 1999: Lidar and numerical studies on the different evolution of a contrail's vortex system and its secondary wake. *J. Geophys. Res.* 104, 2131-2142.

Sussmann, R., K. Gierens, 2001: Differences in early contrail evolution of 2-engined versus 4-engined aircraft. Lidar measurements and numerical simulations. *J. Geophys. Res.* 106, 4899-4911.

Thomson, D. S., M. E. Schein, D. M. Murphy, 2000: Particle Analysis by Laser Mass Spectrometry WB-57F Instrument Overview, *Aerosol Science and Technology, 3,153 – 169.*

Tompkins, A.M., K. Gierens, G. Rädel, 2007: Ice supersaturation in the ECMWF Integrated Forecast System. *Q. J. R. Meteorol. Soc.*, 133, 53-63.

Toon, O.B. and R.C. Miake-Lye, 1998: Subsonic Aircraft: Contrail and Cloud Effects Special Study (SUCCESS), *Geophy. Res. Lettr.*, 25, 1109-1112.

Travis, D.J., Carleton, A.M., and Lauritsen, R.G., 2002: Contrails reduce daily temperature range. Nature, 418, 601-602.

Travis, D.J., Carleton, A.M., and Lauritsen, R.G., 2004: Regional variations in U.S. diurnal temperature range for the 11-14 September 2001 aircraft groundings: evidence of jet contrail influence on climate. J. Clim., 17, 1123-1134

Twohy, C.H., Schanot, A.J. and W.A. Cooper, 1997: Measurement of condensed water content in liquid and ice clouds using an airborne counterflow virtual impactor, *J. Atmos. Oceanic Technol.*, 14, 197-202, 1997.

Twohy, C.H. and B.W. Gandrud, 1998: Electron microscope analysis of residual particles from aircraft contrail, *Geophy. Res. Lettr.*, 25, 1359-1362.

Unterstraßer, S., K. Gierens, P. Spichtinger, 2007: The evolution of contrail microphysics in the vortex phase. *Meteorol. Z.,* accepted.

Wendisch, M., T.J. Garrett and J.W. Strapp, 2002: Wind tunnel tests of the airborne PVM-100A response to large droplets, *J. Atmos. Oceanic tech.*, 19, 1577-1584.

Willeke, K. and Baron, P.A., 2001: Aerosol measurement: Principles, Techniques, and Applications. Second addition: Van Norstrand Reinhold, N.Y..

Wilson, C. W., A. Petzold, S. Nyeki, U. Schumann, and R. Zellner, 2004: Measurement and Prediction of Emissions of Aerosols and Gaseous Precursors from Gas Turbine Engines (PartEmis): An Overview, *Aerospace Sci. Technol.*, 8, 131–143.

Wuebbles, D., 2006: Workshop on the Impacts of Aviation on Climate Change: A Report of Findings and Recommendations June 7-9, 2006, Cambridge, MA, REPORT N0. PARTNERCOE- 2006-004, Joint Planning & Development Office Environmental Integrated Product Team, 64 pp

Yang, P. and K. N. Liou, 1996: Finite-difference time domain method for light scattering by small ice crystals in three-dimensional space, J. Opt. Soc. Am. A 13, 2072-2085.

Zerefos CS, Eleftheratos K, Balis DS, Zanis P, Tselioudis G, Meleti C, 2003: Evidence of impact of aviation on cirrus cloud formation. Atmos. Chem. Phys., 3:1633-1644 Zhang, Y., Macke, A., and F. Albers, 1999: Effect of crystal size spectrum and crystal shape on stratiform cirrus radiative forcing, Atm. Res., 52, 59-75.

Zobrist, B., Marcolli, C., Koop, T., Luo, B. P., Murphy, D. M., Lohmann, U., Zardini, A. A., Krieger, U. K., Corti, T., Cziczo, D. J., Fueglistaler, S., Hudson, P. K., Thomson, D. S., and Peter, Th., 2006: Oxalic acid as a heterogeneous ice nucleus in the upper troposphere and its indirect aerosol effect, *Atmos. Chem. Phys.*, 6, 3115–3129.

Zöger, M., A. Afchine, N. Eicke, M.-T. Gerhards, E. Klein, D. S. McKenna, U. Mörschel, U. Schmidt, V. Tan, F. Tuitjer, T. Woyke, and C. Schiller, 1999: Fast in situ stratospheric hygrometers: A new family of balloonborne and airborne Lyman- photofragment fluorescence hygrometers, *J. Geophys. Res.*, 104, 1807-1816.

Zuberi, B., Bertram, A. K., Koop, T., Molina, L. T., and Molina, M. J., 2001: Heterogeneous nucleation of aqueous particles induced by crystallized (NH4)2SO4- H_2O, ice and letovicite, *J. Phys. Chem., A* 105, 6458–6464.

Zuberi, B., Bertram, A. K., Cassa, C. A., Molina, L. T., and Molina, M. J., 2002: Heterogeneous nucleation of ice in (NH4)2SO4- H_2O particles with mineral dust immersions, *Geophys. Res. Lett.*, 29, 1504, doi:10.1029/2001GL014289.

In: Aviation and the Environment
Editor: Jon C. Goodman

ISBN: 978-1-60692-320-7

Chapter 5

AVIATION-CLIMATE CHANGE RESEARCH INITIATIVE (ACCRI) SUBJECT SPECIFIC WHITE PAPER (SSWP) ON CONTRAIL/CIRRUS OPTICS AND RADIATION SSWP # V, JANUARY 25, 2008

Steve S. C. Ou and K. N. Liou

Joint Institute for Regional Earth System Science and Engineering and Department of Atmospheric and Oceanic Sciences University of California, Los Angeles, California, USA

EXECUTIVE SUMMARY

In this subject-specific white paper, we present a literature survey of past and current developments regarding the impact of contrails and contrail cirrus on the radiation field of the Earth's atmosphere and climate. A number of recommendations for future long-term and short-term actions that are required to comprehend and quantify this important subject are subsequently outlined.

We first present a survey on the background of the basic problem of aviation's impacts on climate and climate change, followed by a discussion of perspectives based on conclusions of the 1999 Intergovernmental Panel on Climate Change (IPCC) Special Report, and the doubling and tripling growths of aviation industry in the next 20 to 40 years as projected by the Next Generation Air Transportation System, United Nation International Civil Aviation Organization, European Union Nations, and the United Kingdom. In response to the pressing need for further study of the potential impact of aircraft emission on climate and environment, a "Workshop on the Impacts of Aviation on Climate Change" was organized and held in Boston, MA on June 7-9, 2006, and a report on the findings during this workshop was later published.

We then review the definition of contrail and the classification of short-lived and persistent contrails and contrail-induced cirrus clouds. The coverage of contrails and contrail-cirrus clouds (~0.1%) has been found to be much smaller compared to that of naturally formed cirrus clouds (>20%). However, their radiative effects are not negligible and, because of indirect effect and feedback, their potential climatic impact could be substantial, particularly in the vicinity of flight corridors where contrail and contrail-induced cirrus formations are frequent. We point out that the radiative forcing of aviation produced contrails in the past is at least twice as large as the contribution of aircraft CO_2 emissions alone. Finally, estimates of the annual growth rate of cirrus clouds (~0. 1 %/yr) and the global contrail radiative forcing are presented.

State of Science

Persistent contrails and contrail-cirrus that are formed in the upper troposphere and lower stratosphere may play an important role in regulating the radiation balance in the Earth-atmosphere system through the competition between the solar-albedo and greenhouse effects that are determined by the ice crystal microphysical and radiative properties within these clouds. The major issue is whether increasing jet air traffic will enhance the generation of additional cirrus clouds, which can lead to an amplification of global warming caused by the build-up of carbon dioxide and other trace gases in the atmosphere.

We have presented past and current progress in estimating long-term trends of the coverage and frequency of occurrence of contrail and contrail-cirrus clouds. Most works after the publication of the 1999 IPCC report focused on the estimate of long-term frequency trend using meteorological data, satellite observations, and numerical weather model products on global and regional scales, as well as the study of radiative forcings of contrails and contrail cirrus by means of satellite data and radiative transfer calculations. Surface observations and satellite data all show that the trend of cirrus cloud cover increased in the past 50 years and that the formation of cirrus clouds has been more frequent in winter and spring near flight corridors. It is anticipated that this increasing trend will continue as a result of increased aviation-induced contrail cirrus formations that tap hitherto cloud-free supersaturated air.

The 1999 IPCC report estimated that the direct radiative forcings of persistent contrails and contrailinduced cirrus are about 0.02 W m-2 (with a range of uncertainty from 0.005-0.06 W m-2) and anywhere between 0 and 0.04 W m^{-2}, respectively. Estimates of contrail radiative forcings vary from near zero to 0.03 W m^{-2}. A number of GCM results show that surface warming produced by contrails is between 0.2 and 0.3°C/decade. These values must be updated and further assessed in light of new observations and an improved physical understanding of the microphysical and optical properties of contrails and contrail-cirrus.

Lastly, we discuss the issue of aerosol indirect effects on the microphysical and radiative properties of ice clouds, essential to the study of the climatic impact of contrails and contrail-cirrus. The indirect effects are complex and their quantifications require a concerted effort involving laboratory and theoretical research, modeling approach, and *in situ* observations in the atmosphere.

Present State of Measurements, Data Analyses, and Modeling Capability

The long-term contrail and cirrus trends have been compiled using satellite and manual surface observations and ground-based instrument measurements. We report a comprehensive data archive of contrail observations from the surface, compiled by the Global Learning and Observations to Benefit the Environment program. We have also provided a list of satellite remote sensing instruments and retrieval techniques that are applicable to contrails and contrail-cirrus studies, including a discussion of the current capability of ground-based remote sensing instruments.

A number of models for contrail research have been developed, and we identify seven state-of-the- art parameterization programs and models, including parameterization of ice crystal microphysics properties in GCMs, the unified theory of light scattering by ice crystals developed by Liou, Takano and Yang, the LBLE radiative transfer model for satellite remote sensing developed by Takano and Liou, the delta 2/4-stream radiative transfer model for radiative forcing calculations developed by Fu and Liou, the UCLA GCM, the European ECHAM4 global contrail-climate model, and the WRF model for regional study.

Current Estimate of the Uncertainties on the Climatic Impact of Contrails and Contrail-Cirrus

The major debate has focused on the magnitude of radiative forcing and surface warming generated by contrails. Large uncertainties exist in global and regional radiative forcing and surface warming, as determined by observations and modeling studies. This suggests that past and current studies of contrail climate impact are inconclusive and not definitive. However, it is pointed out that the radiative and climatic effects, though small globally, could be substantial on a regional scale, as illustrated by a number of regional modeling studies, a subject requiring further exploration and investigation.

Outstanding Scientific Limitations

Primary sources of data that can be used to estimate the long-term trends in contrail-cirrus and cirrus clouds suffer from uncertainties due to manual operation and high-altitude measurements, limitations in geographical coverage, and low temporal and spatial resolutions. The aerosol indirect effects on the microphysical and radiative properties of cirrus clouds are critical in the discussion of climate and climate change involving contrails and cirrus clouds, but these effects are complex and difficult to quantify by mean of *in situ* observations and/or modeling approaches. Comprehensive and systematic *in situ* measurements of contrail and contrail-cirrus have been extremely limited because of the requirement of high flying aircraft and the development of accurate and durable sampling instruments. Modeling approaches, on the other hand, are limited by insufficient understanding of the physical and chemical processes that control ice formation in the presence of aerosols. It would seem that it is important to reduce these uncertainties before resolving the contemporary issues of the magnitudes of radiative forcing and surface warming.

Due to their narrow geometrical shapes, detection of the freshly formed and young contrails by space-borne sensors and ground-based lidar and radar has been a difficult task. Moreover, satellite contrail detection algorithms using split-window bands suffer from a drawback: cirrus clouds with similar linear shapes can be misidentified as contrails. Further development of the satellite and ground- based remote sensing techniques to infer the microphysical and optical properties of contrails is needed, along with *in situ* observations for validation of ice microphysics and the single-scattering properties.

It appears that current GCMs have had difficulty in predicting supersaturation in the upper troposphere and the lower stratosphere region. Many cloud schemes in GCMs compute cloud fraction based on an empirical function of the grid-mean relative humidity that may not be applicable to stratiform cirrus clouds, which are known to be long-lived and can be transported over many grid boxes of a large-scale model during their lifetime.

In addition to the above uncertainties and limitations, there are other issues related to the study of climatic impact of contrails, including uncertainties in the global distribution of water vapor, aerosols, and thin cirrus; detection and prediction of ice supersaturation; chemistry within emission plumes, contrail-cirrus development; the concentration of small ice crystals; and the physical and chemical properties of heterogeneous ice nuclei from natural and anthropogenic sources.

Prioritization of Research Needs

In situ observations and ground-based remote sensing of contrail cirrus and aircraft emission plumes using high-flying aircraft and accurate and durable sampling instruments are needed for the study of the aerosol and contrail indirect effects on the microphysics and radiative properties, modeling of the microphysical and radiative properties for contrails, and the development of ice crystal single-scattering parameterization. These research activities can be costly, and their planning and preparation can be time- consuming. Laboratory measurements of the optical properties for ice crystal clouds can mitigate the uncertainty in current models and parameterizations. In addition, the airborne and remote broadband and narrow-band radiometric measurements, combined with collocated and coincident ice crystal *in situ* observations can be used to validate atmospheric and surface contrail radiative forcings computed by radiative transfer models. However, we rank the priority for this research category as "low" in regard to cost and time.

For global and regional model studies that address direct and indirect effects involving contrails, understanding of the basic mechanism for ice crystal formation is required to improve parameterization of heterogeneous ice nucleation rates. Data collected from coordinated atmospheric *in situ* measurements of the ice crystal and aerosol properties would assist in the development of physical parameterizations so that the contrail direct and indirect effects could be physically simulated in global models. The estimated cost for modeling efforts would be much smaller than *in situ* measurements, and the required time would also be shorter, perhaps on the order of one to two years. We rank this research category as "medium priority".

An integrated use of satellite observations will improve the dependability of estimating the longterm trends of contrails and contrail-cirrus and complement the study of aerosol indirect effect. Research-grade broadband radiometric observations from satellites can be used directly for the investigation of radiative forcing produced by contrails and contrail-cirrus. Radiative transfer calculations can also utilize satellite-retrieved ice crystal microphysical and optical properties as input. Furthermore, integrated satellite observations can be combined with collocated surface observations, meteorological soundings and ground-based remote sensing measurements to further improve accuracy of the detection of contrails and contrail-cirrus. Therefore, satellite observations would be very useful in advancing our understanding in the climatic effect of contrails. The cost for conducting this line of research would be relatively inexpensive, if a suitable number of focused validation experiments using existing facilities could be configured. We rank this research category as "high priority".

Recommendations for Best Use of Current Tools

We would suggest two current tools for contrails-climate research and development. First, MODIS cloud mask and products, with their superior spatial and spectral resolution, can be used to study longterm trends in the coverage and frequency of contrail-cirrus and cirrus occurrence in conjunction with AVHRR and GOES imager data. Another complementary dataset for estimating contrail long-term trends would be the CALIPSO/CALIOP cloud mask products, which have recently become available. MODIS cloud mask and products can also be analyzed to study aerosol-cirrus and contrail-cirrus indirect effects.

With reference to the modeling aspect, it appears that the best regional model that has been developed so far is the WRF model. We suggest that this model coupled with a spectral radiative transfer and ice microphysics parameterizations be used to simulate the formation, evolution, and dissipation of contrails and contrail cirrus using input from flight track and jet fuel consumption information, and that the simulation results be compared with the independent remote sensing results determined from MODIS and related cloud products.

1. Introduction/Background

Aviation appears to be one of the world's fastest growing sources of greenhouse gases, such as carbon dioxide, water vapor, and nitrogen oxide. The increase in the global surface temperature produced by greenhouse warming has been linked to the occurrence of more frequent extreme weather events such as floods, droughts, hurricanes, and blizzards, leading to catastrophic damages of property and loss of lives (US Environmental Protection Agency 2007). The 1999 Intergovernmental Panel on Climate Change (IPCC) Special Report contains a detailed study of the impact of aviation on the global atmosphere. Major findings from this report include: (1) Aviation produces around 6×10^8 tons of carbon dioxide annually and globally; (2) it accounts for 3.5% of global warming from all human activities in 1990; and (3) aircraft emitted greenhouse gases will continue to rise and could contribute to about 15% of global warming from all human activities by 2050. Since the publication of the IPCC 1999 report, air traffic has been continually growing, particularly in the United States, Europe, and eastern Asia. In fact, the Integrated Plan for the Next Generation Air Transportation System (NGATS) proposed by the Joint Planning and Development Office (JPDO) created by the U. S. Congress demands that air transportation services grow from 2004 to 2025 by three fold (NGATS 2004). A similar projection of the aviation growth has been suggested by the United Nation International Civil Aviation Organization, the European Union, and the United Kingdom (Bows et al. 2005).

In view of the aviation activities projected over the next few decades, it is vitally important that immediate and effective actions be taken to understand the nature of the problem and to assist policy makers in making informed decision to protect the environment from potential threat by the inadvertent modifications of climate. Thus, the potential impact of aircraft emissions on current and future climate of the Earth-atmosphere system has become a serious environmental issue that challenges the aviation industry (e.g., Waitz et al. 2004). In response to this challenge, the NGATS/JPDO and the Partnership for Air Transportation Noise and Emissions Reduction (PARTNER) convened a panel of scientists to participate in a "Workshop on the Impacts of Aviation on Climate Change" in Boston, MA, June 7-9, 2006. The major goal of this workshop was to assess and document the present state of knowledge of the climatic impacts of aviation. A report of findings and recommendations from this workshop was later published (JPDO and PARTNER 2006).

Among the aircraft-emitted greenhouse gases, water vapor contributes to the formation of contrails and cirrus clouds, which effectively transmit solar radiation, but block terrestrial infrared radiation that could produce warming of the Earth-atmosphere system. A contrail or condensation trail is defined by Appleman (1953) as the upper-level ice crystal cloud generated by jet aircraft flying in the upper troposphere and lower stratosphere (UT/LS). Contrails were first observed behind low-flying propeller- driven aircraft in 1915, but have now become a common sight in the skies over the United States and Europe, particularly near airports. They are visible line clouds produced by water vapor emitted from aircraft flying in sufficiently cold air. Emerging from the exhaust of jet engines, water vapor is drastically cooled in the extremely cold environment so that saturation with respect to liquid water can be quickly reached (Schumann 1996). Following the thermodynamic principle as described in Appleman (1953), small water droplets can be formed through heterogeneous nucleation on the emitted soot and sulfuric acid aerosols, which serve as cloud condensation nuclei (CCN). Measurements have shown that saturation with respect to liquid water are usually reached in the fresh plume (age < 0.5 sec) closely behind the aircraft and that contrails would not form if the environment is only ice-saturated (Jensen et al. 1998a; Kärcher et al. 1998; Schumann et al. 2000). Because the environment temperature in UT/LS is generally below -40° C, freshly formed water droplets would then instantly freeze to become contrail ice crystals (Schumann 2002).

In an extremely dry atmosphere, such as the typical condition of UT/LS, contrail ice crystals may not grow to sufficiently large sizes before they undergo complete sublimation. In this case, there would be no visible contrail line behind the aircraft. However, in an adequately moist atmosphere, these ice crystals can continue to grow to a much larger size through water vapor deposition and coalescence processes and become visible at 10-30 m behind the aircraft. In a sub-saturated (with respect to ice) atmosphere, contrail lines only last for a short time period on the order of minutes and these are classified as "short-lived contrails" (Minnis 2002). Two examples of these contrails are shown in figure 1 (a) where a pair of trails forming behind the aircraft gradually dissipated. Some contrails can persist for a much longer time period in an ice-saturated or ice-supersaturated atmosphere and are grouped as "persistent contrails". In an ice-supersaturated atmosphere, emitted soot particles may serve as ice nuclei (IN) upon which natural ice crystals are formed by means of contact or immersion nucleation (Jensen et al. 1998b). The resulting mixture of contrail and natural ice crystals is classified as "contrail-induced cirrus" (hereafter referred to as "contrail cirrus"). Figure 1(b) shows examples of persistent contrail and contrail cirrus. This picture was taken by L. Nguyen, NASA LaRC on January 26, 2001 at eastern Virginia. Distinct crisscrossing persistent contrails are shown along with contrail cirrus at high altitudes and spread to a much wider extent than the younger contrails formed below. No clear-cut age threshold can be detected between short-lived and persistent contrails. Bakan et al. (1994) observed that a group of persistent contrails in the region of flight corridors of heavy air traffic over Europe can merge together and grow into cirrus cloud forms, producing similar radiative characteristics (blocking sunlight) as natural cirrus.

Figure 1. (a) Short-lived contrails and (b) Persistent contrails and contrail cirrus (after Minnis 2002).

Contrails and contrail cirrus transmit, reflect and absorb the incoming solar radiation and, at the same time, transmit and absorb/emit thermal infrared radiation (Liou 1986). It has been noted that they can directly affect climate through these radiative processes (Murcray 1970; Kuhn 1970; Changnon 1981). The net radiative effects of contrails containing nonspherical ice crystals have not been comprehensively quantified, because their composition and structure are poorly understood (Sassen 1997). Although the coverage of contrails and contrail cirrus (~0.1%) is much smaller compared to the coverage of naturally formed cirrus clouds (>20%), their potential climatic impact nevertheless cannot be ignored, particularly near flight corridors where air traffic is heavy and contrail formations are frequent. Figure 2(a) displays an estimated linear contrail coverage over a 2.8° grid resolution based on a parameterization of contrail formation adjusted to match the linear contrails observed from satellites, using air traffic data from 1992 and 10-year global analyses of relative humidity and temperature at selected pressure levels (Sausen et al. 1998; Minnis et al. 2004). Black and white boxes represent the boundaries for the land and ocean air traffic regions, respectively. Figure 2(b) shows the geographical distribution of average total contrail cover computed by Gulberg (2003) using the IFSHAM model. The global mean contrail cover determined from this work is 0.06%, which is somewhat less than the IPCC (1999) estimate of 0.1%. The geographical distributions of contrail coverage from different numerical models shown in figures 2(a) and 2(b) are qualitatively similar and reveal that contrail coverage is largely confined to main flight route and flight frequency. High contrail covers up to 5% are shown to center around the northern United States and western European metropolitan areas.

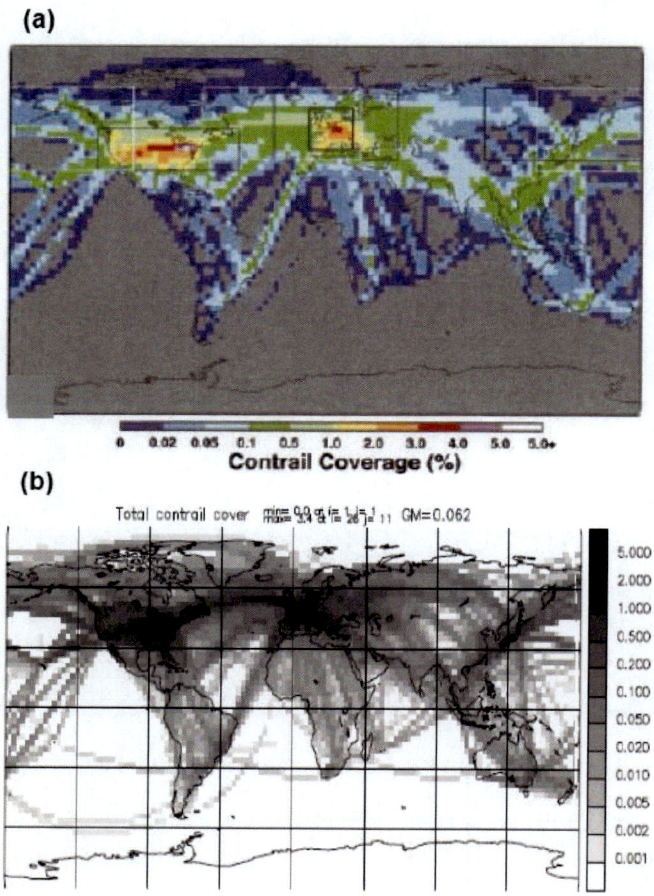

Figure 2. (a) Estimated linear contrail coverage based on a 2.8o grid resolution and a parameterization of contrail formation adjusted to match satellite observations of linear contrails using air traffic data from 1992 and applied to 10 years of global numerical weather analyses of relative humidity and temperatures at selected pressure levels (Sausen et al. 1998 and Minnis et al. 2004). Black and white boxes determine the boundaries for the land and ocean air traffic regions, respectively. (b) Total contrail cover simulated by IFSHAM model (after Gulberg 2003).

Due to numerous factors, including the lack of *in situ* observations and detailed modeling studies, the climatic impact of contrails has not been well understood. In the 1999 IPCC assessment report (IPCC 1999), contrails and their effects have been recognized as one of the largest outstanding uncertainties in the study of air traffic impact on the atmosphere. Moreover, Sausen and Schumann (2007) indicated that even though current civil aviation is only responsible for just 2% of total anthropogenic CO_2 emissions, its impact on environment and climate will be a matter of special concern in the context of anthropogenic global warming since aviation is among the fastest growing economic sectors. It has been stated in a number of assessments (e.g., Shine et al. 1990; Brasseur et al. 1998; Schumann et al. 2001; Ramaswamy et al. 2001; Sausen et al. 2005) that the radiative forcing of current aviation is at least twice as large as the contribution from aircraft CO_2 emissions alone, caused by persistent contrails and contrail cirrus and by the aircraft emitted NO_x, H_2O, and particles.

A significant increase in aviation traffic in recent years has resulted in a noticeable increase in the frequency of occurrence of contrails and contrail cirrus. For example, Minnis et al. (2004) showed a trend in cirrus increase by about 0.1%/yr over the continental USA between 1971 and 1995, and attributed it exclusively to the aviation traffic increase during this period. The radiative and climatic effects of contrails and contrail cirrus appear to have become an important subject for scientific research and in public policy domain. Despite a large degree of uncertainty regarding contrail cover and its ice crystal size and shape, the globally and annually averaged radiative forcings have been estimated. For subsonic aircraft emissions, an estimated positive radiative forcing of 0.02 W m^{-2} with an uncertainty of more than a factor of two was reported for the year 1992 (IPCC 1999). However, for the year 2000, this number was increased to 0.03 W m^{-2} (IPCC 4th Assessment Report, Forster et al. 2007). The mean radiative forcing due to contrails is smaller than that produced by tropospheric aerosols. However, the projected increase in future air traffic could cause the direct climatic effects of contrails comparable to those generated by certain types of tropospheric aerosols.

Under the support of the current FAA program, we have undertaken a survey of available literature and relevant information sources via network websites and put together a focused and in-depth overview of the present knowledge and understanding of scientific principles, uncertainties, and requirements in conjunction with the climatic impacts of contrails and contrail cirrus. In this subject-specific white paper (SSWP), we present the results of our literature survey and provide a number of recommendations for future actions that are required to comprehend and determine the climatic impacts of contrail and contrail cirrus. Section 2 contains a review of the current status of the subject, progress that was made since the IPCC 1999 report, the present state of satellite and ground-based remote sensing as well as modeling capabilities, current estimate of the climatic impact of contrails, and interconnectivity with other SSWP areas. Section 3 lists outstanding limitations, gaps and issues that need improvement. Section 4 prioritizes research needs, followed by recommendations for short-term research in Section 5.

2. A Review of the Climatic Impacts of Contrail and Contrail Cirrus

A. Current State of the Scientific Research

Long-Term Trends in the Coverage and Frequency of Contrail-Cirrus and Cirrus Occurrence

The primary concern in studying the climatic effects of contrails has been the record of the long-term trend of contrails and contrail cirrus. There are four primary data sources that can be used to address this question: manual surface observations of cloud cover, meteorological soundings of temperature and humidity profiles, ground-based measurements by active remote sensors, and satellite data. Each source has its limits. Surface manual observations suffer from insufficient geographical coverage and inaccuracy due to subjective judgment. Humidity soundings display a large degree of uncertainty at high-altitude. Ground-based remote sensing is restricted in geographical coverage. Polar-orbiting and geostationary satellite remote sensing instruments are limited by their temporal and spatial resolution and coverage. We note that surface observations have been continuously used for the compilation of contrail statistics. Machta and Carpenter (1971) first reported secular increases in the amount of high cloud cover in the absence of low or middle clouds at a number of midlatitude stations in the United States between 1948 and 1970. Changnon (1981) analyzed records of monthly sky cover, sunshine and temperature in Midwestern United States (10-state) areas for the period 1901-1977 to discern long-term trends. The sky cover data shows a long-term increase in cloudy days and decrease in clear days since 1901. Figure 3 displays that for a 10-year increment period, the average cloudy days for the south- central area increase from 112 days during the 1901-1910 period to 172 days for during 1968-1977 period. In a separate report to the National Science Foundation, Changnon et al. (1980) further illustrated that high-cloud cover increased from 1951 to 1976 over many Midwestern cities and theorized that such an increase in high clouds could be due to the increase in commercial air traffic. Seaver and Lee (1987) also found more cloud cover, less sunshine and a decrease in the number of clear days over large regions of the United States since 1936.

Liou et al. (1990) analyzed cirrus-cloud cover over Salt Lake City based on surface observations between 1949 and 1994. In this study, the three-hourly weather observations reported by the National Weather Service at Salt Lake City International Airport were used to determine the sky cover information. For each observation, the cloud amount, which is quantified in tenths of the sky coverage, cloud type and visibility were recorded. Figure 4 shows the time series of the mean annual high cloud cover and domestic jet fuel consumption. Based on a student-*t* test, the time series of high cloud coverage can be separated into two periods: 1949-1964 (period 1) and 1965-1982 (period 2). The high- cloud covers for periods 1 and 2 are 11.8% and 19.6%, respectively. The average high-cloud cover for period 2 matches the one for 1965-1969 compiled by Machta and Carpenter (1971). In the time series of domestic jet fuel consumption a sharp increase in the mid-60's occurred corresponding to a substantial increase in high-cloud cover. As shown in figure 4, increased cirrus cloudiness has also been detected in climate data from other stations in the mid-western and northwestern United States that are located in major upper-tropospheric flight corridors (Frankel et al. 1997). Based on correlation between the trends of cirrus cloudiness and jet fuel consumption, increase in cirrus clouds over the last 50 years could be partially attributed to an increase in air traffic (Study of Man's Impacts on Climate 1971).

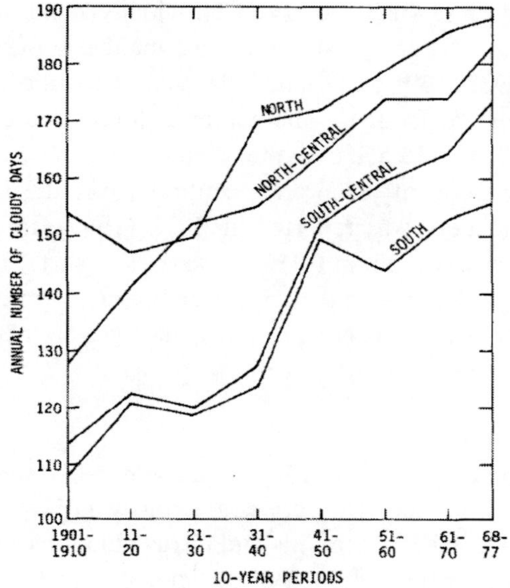

Figure 3. Area mean 10-year values of cloudy days (after Changnon 1981).

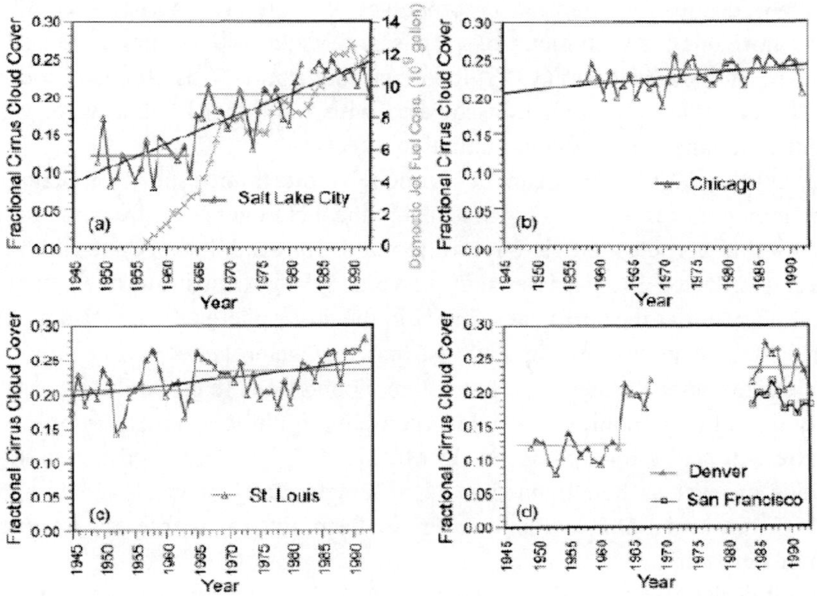

Figure 4. Mean annual high cloud cover over Salt Lake City from 1948 to 1992 and domestic jet fuel consumption (after Liou et al. 1990; Frankel et al. 1997). The two solid lines are the statistical fitting curves for high cloud cover for 1948-1964 and 1965-1992. The statistical fitting curve for the entire period is denoted by the heavy line. Also shown are cirrus cloud covers for several midlatitude cities from 1945 to 1992.

Observations in Germany indicated that the frequency of high clouds during sunny hours increased from 45% in 1954 to 70% in 1995. Over the same period, global radiation during sunshine hours decreased by about 10% (Rebetez and Beniston 1998). A similar increase in high-cloud frequency has also been observed for cloudy conditions (Liepert et al. 1994, Liepert 1997). Boucher (1999) analyzed the surface manual observation reports over North America for the period 1982–1991 and found a decadal increase of 5.6% for the entire region and 13.3% over heavy air traffic areas. The author also reported a global trend of 1.7% per decade over land and 6.2% per decade over the oceans.

A comprehensive analysis of jet aircraft contrails over the United States and Europe using satellite infrared imagery was reported in IPCC (1999). In 1992, aircraft line-shaped contrails were estimated to cover about 0.1% of the Earth's surface on the annually averaged basis but with larger regional values (e.g., 0.5% over central Europe between 1996 and 1997). It is anticipated that global contrail coverage will increase by about 0.5% by 2050 (IPCC 1999).

Aerosol Indirect Effects

Due to the global increase in air traffic, aircraft-emitted water vapor and soot particles mostly composed of black carbon (BC) are continuously infiltrated into the UT/LS, which could cause accelerated increase in contrails and cirrus cloud occurrence. Aerosols affect the atmospheric radiative transfer through their direct interaction with solar radiation (referred to as direct radiative effect) and through their interaction with clouds (referred to as indirect effect). Compared to cloud radiative effect, the aircraft-emitted aerosol direct radiative effect is quite small because of small aerosol optical depth. However, the formation of contrails through heterogeneous ice nucleation processes that involve aerosols could change the vertical and horizontal distributions of clouds and water vapor amount. Based on satellite remote sensing studies, Seinfeld (1998) theorized that some cirrus clouds in fact evolve from contrails. The increase in cloudiness associated with additional IN and water vapor can lead to a substantial enhancement of cloud radiative effects.

Ice crystals in high clouds can be formed by the homogeneous freezing of solution droplets at temperatures below -37°C, and by the heterogeneous freezing of insoluble or partially insoluble particles. BC is one of the major IN candidates (Cantrell and Heymsfield 2005). Aircraft-injected BC particles may serve as IN via deposition nucleation. Laboratory studies have shown that the surrogates for IN in the atmosphere are significant contributors to atmospheric heterogeneous IN populations, and that heterogeneous freezing rates increase with particle size under the same thermodynamic conditions (e.g., Archuleta et al. 2005). BC is generally quite hydrophobic, but could become hydrophilic after exposure to sulfuric acid, and therefore can act as immersion IN. DeMott (1990; 1999) showed in the laboratory that soot particles can act as heterogeneous IN at temperatures between -25°C and -40°C and below -53°C. More laboratory data are now becoming available for characterizing ice nucleation on aerosols.

Aerosol indirect effects on the microphysical and radiative properties of cirrus clouds are important for the study of climatic impact of contrails and contrail cirrus, but these effects are complex and difficult to quantify based on a modeling approach (Seinfeld, 1998). Attempts to mechanistically relate aerosols number density to cloud formation in general circulation models have focused on the initiation of warm/liquid clouds. Much less attention has been given to the study of the potential impacts of aerosols on high-altitude ice clouds for the reasons stated above. Parameterizations for homogeneous and heterogeneous ice nucleation have been developed by various researchers (e.g., Kärcher and Lohmann 2002; DeMott et al. 1997; Gorbunov et al. 2000, Kärcher and Lohmann 2003; Liu and Penner 2005; Kärcher et al. 2006). Some significant steps in quantifying the indirect effect from anthropogenic aerosols have been made by using GCMs. For example, Jones et al. (1994) estimated aerosol indirect effect by performing a series of simulations for the annual mean distribution of low-level cloud droplet effective radius at cloud top using the Hadley Center GCM. Figure 5(a) shows the global distribution of cloud top effective radius, while figure 5(b) displays its instantaneous change due to changes from natural-only aerosols to total aerosol concentration. There is a general decrease in effective radius throughout most of the Northern Hemisphere and over most of the land areas, particularly around major industrial regions.

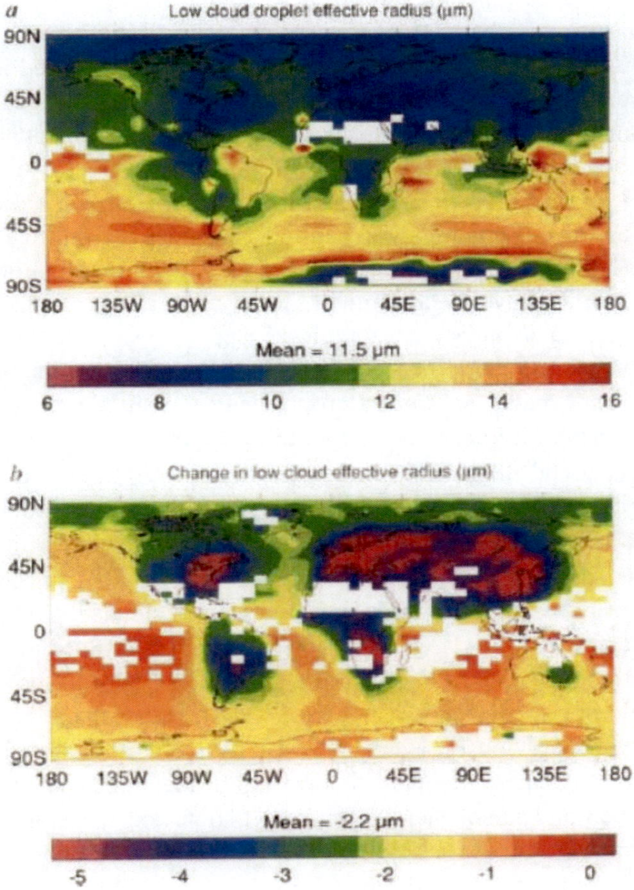

Figure 5. (a) A simulation of annual mean distribution of low-level cloud droplet effective radius at cloud top. The blank areas indicate regions where there were no low clouds during the integration. (b) annualmean composite of the instantaneous change in low cloud droplet effective radius due to changing from natural-only to total aerosol concentration. The blank areas indicate regions where there were no low clouds during any of the sampling periods (after Jones et al. 1994).

Measurements of BC at the level where ice clouds form have been extremely limited due to the requirement of high flying aircraft and limitation of our understanding of the physical and chemical processes controlling ice formation in the presence of aerosols, particularly heterogeneous ice nucleation (Cantrell and Heymsfield, 2005). An adequate understanding of aerosol-cirrus cloud interaction must be derived from *in situ* microphysical measurements. However, it is difficult to isolate and quantify aerosol indirect effects based solely on *in situ* observations, because of measurement uncertainties and sampling considerations, as well as a separation of these effects from the natural variability of meteorological conditions.

In view of various problems encountered in the quantification of aerosol indirect effects based on direct *in situ* observations, an alternative approach to study the aerosol-cirrus and contrail-cirrus indirect effects is through satellite observations of ice clouds and aerosols, making use of an extensive suite of space-based instruments that are currently available along with collocated and coincident *in situ* aerosol measurements. These observations contain rich and valuable information that can be used to investigate the relationship between aerosols and ice cloud formation. Along this line, correlation of the MODIS observed ice crystal effective radius and the level of aerosol loading during the Indian Ocean Experiment (INDOEX) revealed a significant aerosol impact on ice cloud particle size (Chýlek et al. 2006).

Microphysical and Radiative Properties of Contrails and Contrail Cirrus

The radiative forcings of contrails and contrail cirrus depend on their optical properties, which are in turn a function of the ice crystal size and shape distributions. Because *in situ* observations on contrails have been limited, their microphysical properties are largely unknown. Following is a summary of findings based on available *in situ* microphysical measurements. Knollenberg (1972) first used an optical-array spectrometer on board NCAR Sabreliner aircraft and made *in situ* microphysical measurements of ice crystal size distribution, *IWC*, and total ice water budget within its own contrails and the resulting cirrus uncinus clouds. He found that, like cirrus clouds, the *IWC* of contrails depends on temperature, humidity, vertical velocity of air, fall out of ice crystals, and possibly radiative cooling. Konrad and Howard (1974) provided an insightful morphology of contrail cirrus and fallstreaks as viewed by ultra-sensitive radars. From the late 70's to the early 90's, high-altitude *in situ* observations mostly focused on natural cirrus clouds.

In 1996, the Subsonic Aircraft Contrail and Cloud Effects Special Study (SUCCESS) field campaign carried out over Kansas during a 5-week period (April 8-May 10, 1996) provided unique microphysical measurements of the size and shape characteristics of ice crystals that were not previously available. SUCCESS used scientifically-instrumented aircraft and ground-based measurements to investigate the effects of subsonic aircraft on contrails, cirrus clouds and atmospheric chemistry (Toon and Miake-Lye 1998). Airborne platforms used during SUCCESS include a medium-altitude DC-8 and a high-altitude ER-2, both of which were based at the NASA Ames Research Center, Moffett Field, California and a T39 aircraft based at the NASA Wallops Flight Facility, Wallops Island, Virginia. During the SUCCESS observation period, all three NASA aircraft were deployed at the Salina campus of Kansas State University. A series of flights, averaging one every other day during this period, were made near the ARM-SGP site. Flights were also made over the Gulf of Mexico to utilize an oceanic background for remote sensing measurements. In order to achieve experimental objectives, the DC-8 aircraft was used as an *in situ* sampling platform, carrying a wide variety of instruments for sampling gases and particulate matters, and radiometric measurements. Major cloud microphysics measurement instruments included a multi-angle aerosol spectrometer probe (MASP, Baumgardner et al. 1995), a video ice particle sampler (VIPS, Heymsfield and McFarquhar 1996), a cloudscope (Arnott et al. 1995), a PiNephelometer (Lawson et al. 1998), and a FSSP. The T-39 aircraft was used primarily to sample the exhaust from other aircraft. It also carried a suite of instruments to measure particles and gases. The ER- 2 aircraft carried the MODIS Airborne Simulator, which was used as a surrogate for MODIS, so that remote sensing observations could be related to the *in situ* parameters measured by the DC-8 and the T39.

Based on analysis of the data gathered during SUCCESS, Heymsfield et al. (1998) examined the evolution of contrails to precipitation trails using the data collected from various instruments, including PI, VIPS, and a PMS 2D-C imaging probe with a lower detection limit between 50 and 100 ìm. Goodman et al. (1998) used an impaction technique to sample ice crystals in the exhaust trail of a Boeing 757, and found that ice crystals in the contrail of about 1 minute old had a unimodal size distribution, with an equivalent volume radius of less than 10 ìm and an effective radius of about 2 ìm. The crystal habits at the observed temperature of -6 1°C were predominantly hexagonal plates (75%), columns (20%) and few triagonal plates (<5%). Lawson et al. (1998) sampled a persistent contrail generated by the DC-8 during SUCCESS and found that, after 40 minutes, the core of the contrail consisted of mostly small particles ($L = 1 – 20$ ìm) with a concentration larger than 1000 l^{-1}, but the concentration of large particles ($L > 300$ ìm) was less than 10^{-6} l^{-1}. In contrast to the core, the contrail boundary consisted of one order-of-magnitude less small particles, but three order-of-magnitudes more large particles with the shape of columns and bullet rosettes that are typically found in natural cirrus.

In the area of lidar observations of contrail microphysics, Freudenthaler et al. (1996a) found that strong depolarization produced by contrails containing growing particles a few minutes old revealed nonspherical shaped particles. Sassen and Hsueh (1998) analyzed the data from a ground-based polarization lidar during SUCCESS to study contrails and cirrus clouds evolved from contrails. They found that contrail-cirrus is distinctively different from natural cirrus clouds. Contrail-cirrus tends to be thin (~50 -500 m) and can generate coronas indicative of long-lasting small (20 – 30 ìm) particles. Jensen et al. (1 998c) conducted a case study of the persistent contrail evolution in a sheared environment by simulating contrail evolution using a large-eddy simulation model with detailed ice microphysics. Simulation results were compared to satellite and *in situ* measurements of the persistent contrails inferred from the SUCCESS experiment. Using large ambient super-saturations and moderate wind shear in simulation, ice crystals with maximum dimensions greater than 200 ìm were generated within 45 minutes after emission by depositional growth.

Radiative Forcing for Persistent Contrails and Contrail Cirrus

Aircraft emission of water vapor and particles, as well as the creation of contrails, could lead to a change in global cloudiness. A number of atmospheric GCM studies that investigated the impacts of injecting water vapor on creating contrails (e.g., Ponater et al. 1996; Rind et al. 1996) also illustrated the potential importance of these impacts on climate. Persistent contrails are detectable both by surface observation and satellite remote sensing, and their impact on radiative forcing can be evaluated. Fahey et al. (1999) presented the 1992 IPCC estimate of direct radiative forcing from persistent contrails of about +0.02 W m^{-2} with a range of uncertainty from +0.005 to +0.06 W m^{-2}. This estimate is limited to immediately visible, quasi-linear persistent contrails. The radiative forcing associated with contrail formation is a consequence of aircraft activity, and its impact on climate can be directly estimated by various measurement techniques and modeling approaches. Cirrus clouds generally exert a net positive radiative forcing as a result of the domination of longwave greenhouse effect relative to solar albedo effect. Fahey et al. (1999) reported that the 1992 IPCC estimate of the radiative forcing from aircraft- induced cirrus clouds is positive and may be comparable to contrail radiative forcing. The magnitude of this radiative forcing remains very uncertain. A range for the best estimate of the globally averaged radiative forcing due to contrails could fall between 0 and 0.04 W m^{-2}.

The importance of contrails in changing regional and global radiation budgets has been assessed in several modeling studies. Using a one-dimensional radiative transfer model along with specified contrail microphysical properties and atmospheric conditions, Fortuin et al. (1995) estimated that, with 0.5% cloudiness, contrails may produce a radiative forcing at the top of atmosphere (TOA) of -0.15 to 0.3 W m-2 for the Atlantic flight corridor. Minnis et al. (1999) calculated the TOA radiative forcing for the year 1992 with a similar approach and found a net global radiative forcing of 0.01 W m^{-2}. The radiative forcing for heavy air traffic regions is much higher with maximum values reaching 0.71 W m^{-2} over northern France and 0.58 W m^{-2} near New York City. Contrails have important effects on regional climate and for the time period when the upper atmosphere is saturated or supersaturated with respect to ice. Moreover, Fortuin et al. (1995) and Strauss et al. (1997) suggested that the maximum instantaneous radiative forcing directly under a contrail, assuming 100% contrail cover, could have values from -30 W m-2 to 60 W m^{-2}. Such a large radiative forcing could lead to a change in the surface temperature by a few degrees K. Strauss et al (1997) also found that an additional 0.5% contrail cover could cause a warming of 0.05K. Thus, although the global mean magnitude of radiative forcing produced from contrails is relatively small, as compared to the estimated anthropogenic greenhouse effect, contrails could have a significant impact on regional climate.

Climatic Impacts of Contrails and Contrail Cirrus

Non-black, semi-transparent high cirrus clouds are known to produce surface warming, and warming in the lower troposphere caused by the thermal IR fluxes emitted from the cloud. The degree and extent of warming are controlled by the cloud's radiative property and its physical position in the atmosphere as well as feedbacks associated with thermodynamic processes involving cloud formation. Earlier model simulation results show that high clouds above about 8 km produce a warming effect at the surface: the degree of this warming is a function of cirrus cloud optical depth (or emissivity) (Freeman and Liou 1979; Liou and Gebhart 1982). Research efforts pertaining to cirrus clouds and climate have been comprehensively reviewed by Liou (1986) and Liou (2005), both of which also pointed out the importance of cirrus formation from contrails.

Grassl (1990) presented the importance of contrails in the upper troposphere and additional water vapor in the lower stratosphere in conjunction with the radiation budget of the Earth-atmosphere system. Liou et al. (1990) specifically studied the climatic effects of contrail-cirrus by using a two-dimensional cloud-climate model, in view of the fact that the increase in contrail-cirrus has been primarily confined to midlatitudes. This model was a combination of a two-dimensional energy balance climate model (Liou and Ou 1981, 1983, Ou and Liou 1984) and an interactive cloud formation model (Liou et al. 1985) that generates cloud cover and liquid water content based on thermodynamic principles. The effects of contrail cirrus cover on cloud formation and temperature field were investigated by increasing the cloud cover between 20 and 70°N, roughly corresponding to the location of most jet aircraft traffic. A 5% increase in high-cloud cover leads to a substantial amplification in high-cloud cover increase (15%) at 20- 40°N, caused by an increase in specific humidity. Low and middle clouds also increase slightly because of the additional moisture supply. Overall, enhanced downward thermal IR emission from additional high clouds causes a temperature increase in the troposphere of the lower latitudes. Figure 6 shows zonally mean changes in atmospheric and surface temperatures due to increases in high cloud cover of (a) 5%, and (b) 10%. For both experiments, there is a maximum temperature increase in the lower troposphere of the tropics due to a significant increase of humidity in that region. This is in contrast to the results simulated from fixed relative humidity and non-interactive cloud cover, in which the maximum temperature increase occurs in the polar region. Hansen et al. (2005) also produced the maximum temperature increase in the lower troposphere of the tropics for a 4x CO_2 experiment using the GISS GCM. Temperature increases above 5 km are generally reduced with increasing height. The surface albedo feedback effects are also substantially reduced. The temperature increase due to a 5% increase of cloud cover under the condition of interactive cloud cover is less than that of fixed cloud cover, because of the increase in low and middle cloud covers in the former experiment. A 10% increase in high-cloud cover in the perturbation experiment shows a temperature increase of more than a factor of two (relative to a 5% increase in cloud cover) in the troposphere, because additional high clouds are formed due to the humidity feedback effect. In the case of a 5% increase in high-cloud cover, surface temperature increases by about 1 K, but varies with latitude. When the increase in high clouds was doubled, a surface temperature increase of about 2.5 K was obtained in the experiment. In summary, all perturbation experiments involving high-cloud cover increase indicate increases in atmospheric and surface temperature caused by a positive greenhouse feedback from cloud cover and specific humidity.

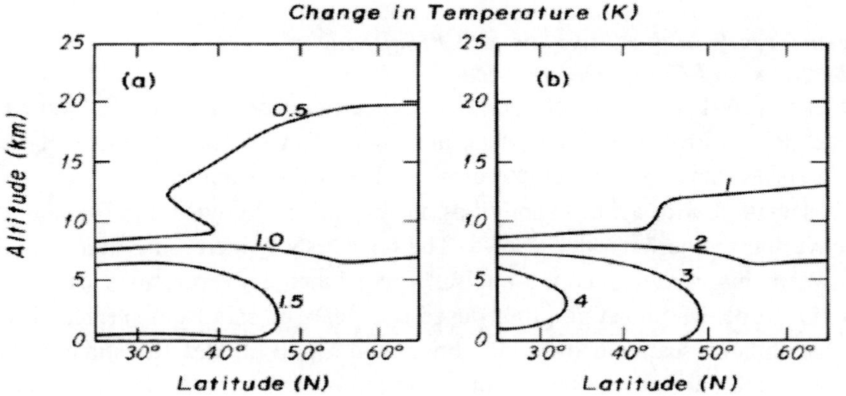

Figure 6. Changes in zonally mean atmospheric and surface temperatures subject to thermal equilibrium due to increases in high cloud cover of (a) 5%, and (b) 10% subject to interaction between humidity and cloud cover, as simulated by the two-dimensional energy balance climate model (after Liou et al 1990).

B. Critical Roles of Contrail and Cirrus Clouds in Climate Processes

Persistent contrails, contrail cirrus, and natural cirrus clouds formed in UT/LS play a significant role in regulating the radiation balance of the Earth-atmosphere system, and so their presence must be recognized as a crucial component in understanding the inadvertent human-induced climate change problem (Liou 1986). Short-lived contrails are not expected to have significant impacts on climate change due to their extremely small coverage and relatively short durations of existence. Persistent contrail and cirrus cloud temperatures are low (< -20°C), and many of them are composed of irregularly shaped ice crystals. Because of their high altitude and cold temperature, they can act as a thermal blanket by absorbing (and therefore trapping) the upward thermal infrared radiation emitted and transmitted from below the cloud, the same as the "greenhouse effect", which warms the Earth- atmosphere system. At the same time, these clouds can also reflect the incoming solar radiation referred to as the "solar albedo effect", which serves to cool down the Earth-atmosphere system. Balance between these competing radiative effects determines the net impact of high clouds on our climate system. The relative importance of the greenhouse vs. albedo effect is dependent on the cloud microphysical and optical properties of clouds (Ackerman et al. 1988; Stephens et al. 1990; Fu and Liou 1993; Ou and Liou 1995), which in turn are governed by atmospheric circulation and water vapor distribution.

In summary, the major issue is whether increase in cirrus clouds related to increasing jet air traffic would enhance or suppress the global warming produced by the build-up of carbon dioxide and other greenhouse gases. Another important issue is whether there are other unknown mechanisms that might have contributed to a global increase in cirrus clouds in recent years. Resolving these issues is vitally important to planning future air traffic operation.

C. Progress Since the IPCC 1999 Report

Long-Term Trends in the Coverage and Frequency of Contrail-Cirrus and Cirrus Occurrence

Chen et al. (2001) estimated contrail occurrence frequency over the Taiwan area based on flight frequency and meteorological data, and found that contrails form more frequently in winter and spring than in summer. Zerefos et al. (2003) examined changes in cirrus-cloud cover in association with aviation activities at busy air traffic corridors based on the ISCCP data set covering the period 1984 – 1998. The results show increasing trends in cirrus-cloud cover between this period over the air traffic corridors of North America, North Atlantic Ocean, and Europe. Minnis et al. (2003) used two years of data from surface observers at 22 military installations scattered over the continental United States to estimate mean hourly, monthly, and annual frequencies of daytime contrail occurrence. During both years, persistent contrails were most prevalent in winter and early spring, but less frequent during summer and occurred simultaneously with cirrus clouds 85% of the time. Although highly correlated with the air traffic fuel consumption, contrail occurrence is also governed by meteorological conditions. Minnis et al. (2004) further collected and analyzed surface observations from 1971 to 1995 and showed that cirrus clouds increased significantly over the northern hemisphere oceans and the United States, while decreasing over other land areas except over

Western Europe, where cirrus coverage was relatively constant. It was pointed out that surface observations are consistent with satellite-derived trends over most areas and that it is most likely that the cirrus trends in the U. S. are correlated with air traffic. The cirrus increase is a factor of 1.8 greater than that expected from the current estimate of linear-contrail coverage, suggesting that a spreading factor of the same magnitude could be used to estimate the maximum contrail effect.

Wylie et al. (1994, 1999, 2005) used NOAA High Resolution Infrared Radiometer Sounder (HIRS) polar-orbiting satellite data from 1979 to 2001, a 22-year record, to determine the frequency of detected high cloud in the upper troposphere (figure 7). The CO_2 slicing method was used to infer cloud amount and height. They estimated that thin cirrus ($\hat{o} < 0.7$) covers about 20% in the mid-latitude region and over 50% in the tropics. High clouds show a small but statistically significant increase in the Tropics and the Northern Hemisphere. The HIRS analysis differed from the International Satellite Cloud Climatology Project (ISCCP, Rossow and Schiffer), which shows a decrease in both total cloud cover and high clouds during most of the 22-year period.

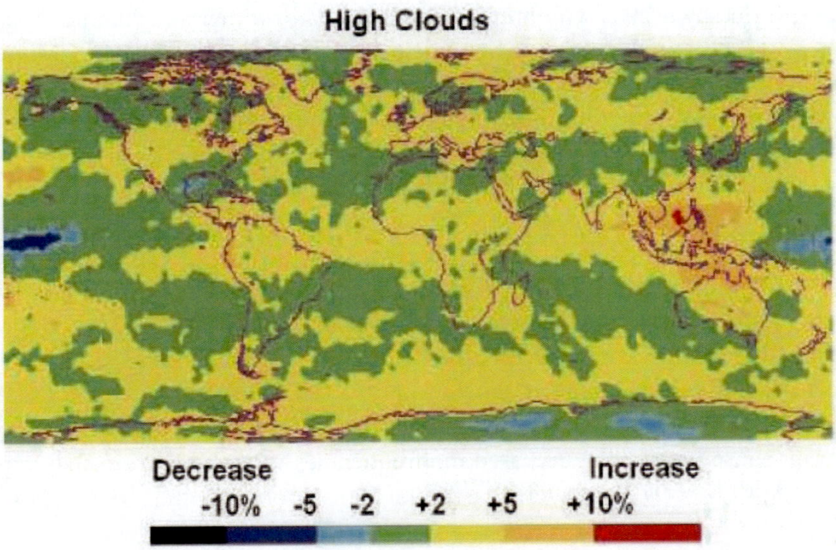

Figure 7. The geographical locations of changes in high-cloud frequency between the 1994-2001 and 1985-1992 periods (after Wylie et al. 2005).

Schumann (2005) presented the formation, occurrence, properties, and climatic effects of contrails. The global cover by lined-shaped contrails and their radiative impact is smaller than that assessed in an international assessment in 1999. To help alleviate uncertainty in the air traffic contribution to cirrus increase, Minnis et al. (2005) analyzed linear contrail coverage over the North Pacific Ocean using the NOAA- 1 6/AVHRR data during a 4-month period in 2002 and 2003. Manual evaluation of the automated contrail detection method revealed that it misclassified, on average, 32 % of the pixels as contrails and missed 15 % of contrail pixels. After a correction for detection errors, the contrail coverage over the domain between 25° and 55°N and between 120° and 150°W varied from a minimum of 0.37 % in February to a maximum of 0.56 % in May. The annual mean coverage, after correcting for the diurnal cycle of air traffic, is 0.31 %, a value very close to earlier theoretical estimates for the region. The average contrail optical depth is 0.24, corresponding to a mean longwave radiative forcing of 14.2 W m^{-2}.

Duda et al. (2003, 2005) estimated contrail frequency and coverage over the contiguous United States (CONUS), using hourly meteorological analyses from the Rapid Update Cycle (RUC) numerical weather prediction model and commercial air traffic data for a 2-month period during 2001. The contrail frequency over the CONUS was computed directly from RUC analyses using several modified forms of the classical Appleman criteria for persistent contrail formation. Various schemes for diagnosing contrails from the RUC analyses were tested. Palikonda et al. (2005) derived linear contrail coverage, optical depth, and longwave radiative forcing from NOAA- 15 and NOAA- 16 daytime AVHRR data over CONUS, southern Canada, northern Mexico, and the surrounding oceans. Contrail coverage averaged 1.17% and 0.65% based on the early-morning NOAA-15 and mid-afternoon NOAA-16 observations, respectively, for the areas and times common to both satellites. The estimated combined maximum coverage for the entire domain was ~1 .05% during February, while a minimum of 0.57% occurred during August. The annual mean optical depth is 0.27, while the monthly value varied by ~ 20% with minima and maxima in winter and summer, respectively. Marquart et al. (2003) used a contrail parameterization in the ECHAM GCM to estimate future contrail coverage. Time slice simulations showed increase in the global annual mean contrail cover from 0.06% in 1992 to 0.14% in 2015 and to 0.22% in 2050. In the northern extratropics, the enhancement of contrail cover is mainly determined by aviation growth, but in the tropics, contrail cover appears to be affected by climate change.

Meyer et al. (2007) presented the contrail coverage over Thailand, Japan and the surrounding area through remote sensing observations. Locally received NOAA/AVHRR satellite data were analyzed by a fully automated contrail detection algorithm. The annual average of the daily mean contrail coverage is 0.13% and 0.25% for the Thailand and Japan regions, respectively, with a maximum value during spring for both regions. Travis et al. (2007) reported a contrail mid-season climatology for the coterminous United States (2000–2002) based on AVHRR data, US jet aircraft flight activity log, and NCEP-NCAR reanalysis data at the tropopause level, and compared the frequencies with those previously reported for an earlier period (1977–1 979) to determine spatial and seasonal contrail frequency changes.

Radiative Forcing of Contrails and Cirrus Clouds

Meerkötter et al. (1999) used three different radiative transfer models and six model atmospheres (McClatchy et al. 1972) to study the instantaneous radiative impacts of contrails and found that a mean contrail cover of 0.1% with average optical depths of 0.2-0.5 would produce about 0.01-0.03 W m-2 daily mean radiative forcings. Duda et al. (2001) used GOES data to study the evolution of solar and longwave radiative forcings in contrail clusters over Midwestern US, Eastern US, Atlantic Ocean, and Hawaii. They showed that observed radiative forcings are less than those from model simulations. Marquart et al. (2003) estimated increase in the global annual mean radiative forcing from 3.5 mW m-2 in 1992 to 9.4 mW m^{-2} in 2015 and to 14.8 mW m^{-2} in 2050. Uncertainties in contrail radiative forcing mainly arise from uncertainties in the microphysical and optical properties such as particle size and shape and optical depth. Sausen et al. (2005) provided an estimate of the various contributions to radiative forcing (RF) from aviation based on results from the TRADEOFF project that was an update of the IPCC (1999). The new estimate of the total RF from aviation for 2000 is approximately the same as that of the IPCC's estimate for 1992 as a consequence of the reduced contrail RF that compensates for the RF increase due to increased aviation traffic from 1992 to 2000. The RF from other aviation-induced cirrus clouds might be as large as the present estimate of the total RF (without cirrus). However, our present knowledge on these aircraft-induced cirrus clouds is too limited to provide a reliable estimate of the associated RF.

Palikonda et al. (2005) derived longwave RF from NOAA- 15 and NOAA- 16 daytime AVHRR data over the CONUS, southern Canada, northern Mexico, and the surrounding oceans. The annual mean optical depth of 0.27 translated to a normalized contrail longwave RF of 15.5 W m^{-2}. The overall daytime longwave RF for the domain is 0.11 W m^{-2}. The normalized longwave RF peaked during summer, while the overall forcing was at a maximum during winter because of greater contrail coverage. Given the U.S. results and using mean contrail optical depths of 0.15 and 0.25, Minnis et al. (2004) estimated that the maximum contrail–cirrus global RF is 0.006–0.025 W m^{-2}, depending on the radiative transfer model used in the calculations. Using contrail results simulated from a GCM, the cirrus trends over the United States are estimated to generate a tropospheric warming of 0.2°–0.3°C/decade. It is noted that the observed tropospheric temperature trend is 0.27°C/decade between 1975 and 1994. The magnitude of the estimated surface temperature change and the seasonal variation of the estimated temperature trends are in general agreement with observations.

D. Present State of Measurements and Data Analysis

Contrail and Cirrus Climatology Based on Analyses of Surface and Satellite Observations

Minnis et al. (2004) demonstrated that global satellite remote sensing can provide long-term climatology datasets for contrails, contrail cirrus, and natural cirrus clouds. However, conventional sensors on the present NASA and NOAA satellites have had difficulty in detecting optically thin cirrus with an optical depth smaller than about 0.1 (Roskovensky and Liou 2003; 2005). A substantial amount of surface data for cloud classification near major airports exists. It appears that analysis of this data, albeit local, could be complementary to satellite observations. To support aviation operation and climate change, one can compile long-term cloud climatology similar to that shown in figure 3 for contrails, contrail cirrus, and natural cirrus near major airport areas using current and future satellite data in combination with surface observations.

Duda et al. (2007) described a comprehensive data archive of surface contrail observations collected by the Global Learning and Observations to Benefit the Environment (GLOBE) program. A primary goal of the GLOBE program is to use detailed written protocols to enable student observers to provide scientifically valuable measurements of environmental parameters (Brooks and Mims 2001). In May 2003, GLOBE initiated a contrail observation protocol to classify observations of contrail occurrence and coverage throughout the CONUS from primary and secondary schools across the country. (See www.globe.gov.). Over 18,500 observations were reported over the region between April 1, 2004 and June 27, 2005, including contrail coverage, contrail number, cloud coverage, cloud type and a classification of contrails into three categories: short-lived, non-spreading persistent contrails, and spreading persistent contrails.

Satellite Remote Sensing of Contrails and Cirrus Clouds

Satellite remote sensing of contrails can provide objective measures to determine the cloud cover induced by contrails globally. High resolution infrared satellite images often provide revealing patterns of contrails, while corresponding visible images are less clear (Joseph et al. 1975; Lee 1989). Carleton and Lamb (1986) showed that the occurrence of contrails can be determined by DMSP high-resolution visible bands (0.6 km) and infrared bands (1.0 km). From a pilot study, they found that contrails tend to occur frequently in association with natural cirrus clouds and tend to cluster in groups. Duda and Minnis (2002) reported GOES results for dissipating contrails over southeast Virginia and Chesapeake Bay. Duda et al. (2004) examined the development of widespread persistent contrails over the western Great Lakes on October 9, 2000 using the GOES data. Table 1 summarizes the current and future satellite observations that are relevant to the remote sensing of contrails and cirrus clouds. A more detailed description of each instrument follows.

i *Advanced Very High Resolution Radiometer (A VHRR).* The AVHRR has been onboard NOAA polar-orbiting satellites for a number of years. It is a radiation-detection imager that can be used for remotely determining cloud cover and surface and cloud temperatures. The latest instrument version was the 6-channel AVHRR/3 on board NOAA-15 launched in May, 1998. The AVHRR/3 is an imaging system in which a small field-of-view (1.3 milliradians by 1.3 milliradians) is scanned across the Earth from one horizon to the other by a continuous 360 degree rotation of a flat scanning mirror. There are 1.362 samples per IFOV (instantaneous field-of-view). A total of 2048 samples are obtained per channel per Earth scan covering the area from the scan angles of $\pm 55.4^{o}$ with reference to the nadir. The channel characteristics of AVHRR/3 are as follows: Ch.1 (, = 0.58 – 0.68 ,m, for daytime cloud and surface mapping), Ch.2 (, = 0.725 – 1.00 ,m, for characterizing land and water), Ch.3A (, = 1.58 – 1.64 ,m, operating only during daytime for detecting snow and ice); Ch.3B (, = 3.55 – 3.93 ,m, operating only during nighttime for nighttime cloud mapping and sea-surface temperature), Ch. 4 (, = 10.3 – 11.3 ,m, for nighttime cloud mapping and sea-surface temperature), and Ch.5(, = 11.5 – 12.5 ,m, for sea-surface temperature).

ii *High Resolution Infrared Radiation Sounder (HIRS).* The HIRS instrument has also been onboard NOAA polar-orbiting satellites and provides multispectral data from 1 visible channel (0.69 ,m), 7 shortwave channels (3.7-4.6 ,m) and 12 longwave channels (6.7-15 ,m) using a single telescope and a rotating filter wheel containing 20 individual spectral filters. The IFOV for each channel is approximately 0.7^{o} which, from a spacecraft altitude of 833 km, encompasses a circular area of 10 km at its nadir on the Earth. It is almost impossible for HIRS to detect contrails because of the large footprint, but the CO_2 slicing method applied to HIRS data appears to be capable of estimating the effective emissivity and temperature of cirrus clouds that are within the HIRS footprint.

iii *Geostationary Operational Environmental Satellites (GOES)/Imager.* GOES satellites are located around a fixed position above the Earth and provide continuous monitoring of about 1/3 of Earth's spherical surface. The geosynchronous plane is about 35,800 km (22,300 miles) above the Earth. The GOES I-M Imager is a 5-band (1 visible, 4 infrared) imaging radiometer designed to sense radiant and solar reflected energy from sampled areas of the Earth. The channel characteristics of

GOES Imager are as follows: Ch.1 (= 0.55 – 0.75 m, Instantaneous Geographic Field of View at nadir (IGFOV) = 1 km, for daytime cloud and surface mapping), Ch.2 (= 3.8 – 4 m, IGFOV = 4 km, for characterizing land and water), Ch.3 (= 6.5 – 7 m, IGFOV = 8 km, for measuring precipitable water); Ch. 4 (= 10.2 – 11.2 m, IGFOV = 4 km, for nighttime cloud mapping and sea-surface temperature), and Ch.5(= 11.5 – 12.5 m, IGFOV = 4 km, for sea-surface temperature). Compared to AVHRR, the GOES IR Imager's spatial resolution is lower, and thus it is not ideal to detect fresh and short-lived contrails. As demonstrated by Minnis et al. (1998), however, this imager can be of some use for detecting persistent contrails and contrail-cirrus.

Table 1. Current and future satellite observations for remote sensing of contrails and cirrus clouds

Satellite Instrument	Measurements relevant to contrails and cirrus
NOAA/AVHRR	Detection of contrail and cirrus clouds, aerosol and cirrus cloud ptical depths, ice crystal effective radius, cloud-top parameters
NOAA/HIRS	Detection of cirrus clouds, cloud effective missivity and cloud-top ressure
GOES/Imager	Detection of contrail and cirrus clouds, aerosol and cirrus cloud ptical depths, cloud-top parameters
Terra/Aqua/MODIS	Detection of contrail and cirrus clouds, aerosol and cirrus cloud ptical depths, ice crystal effective radius, cloud-top parameters
CALIPSO	Aerosol and cloud vertical profiles
CloudS at	Vertical profile of IWC
IceS at/GLAS	Vertical Structure of Cloud
Terra/MISR	Aerosol optical depth and height
NPOESS (NPP)/VIIRS	Detection of contrails and cirrus clouds, Aerosol and cirrus cloud ptical depths, ice crystal effective radius, cloud-top parameters
JMA/MTSAT-R/Imager	Detection of contrails and cirrus clouds
EUMESAT/ESA/Mete sat-9/Imager	Detection of contrails and cirrus clouds
EUMESAT/ESA/Meto /AVHRR	Detection of contrail and cirrus clouds, aerosol and cirrus cloud ptical depths, ice crystal effective radius, cloud-top parameters

iv *Moderate Resolution Imaging Spectroradiometer (MODIS).* The MODIS has been on both Terra and Aqua satellites that were launched in December, 1999 and May, 2001, respectively. Both Terra and Aqua are in sun-synchronous polar orbits with daytime equator crossings at 10:30 am and 1:30 pm LTC, respectively. Aqua is the leading platform of the NASA A-Train, a constellation of polar-orbiting satellite platforms flying in formation. MODIS has a 1 km^2 IFOV mapping to a swath of approximately 2330 km to achieve near complete global coverage every day. The MODIS cloud product contains both physical and radiative cloud properties, including cloud mask, cloud-particle phase (ice vs. water, clouds vs. snow), cloud-top temperature/pressure/height, effective cloud-particle radius, and cloud optical depth. Because of high spatial resolution and multi-spectral-band characteristics, MODIS can be effectively used to detect contrails and contrail cirrus. Terra/MODIS and Aqua/MODIS now have 8- and 5-year datasets, respectively, but a systematic compilation of contrail statistics using MODIS data has not yet been conducted.

v *CloudSat and Cloud-Aerosol Lidar and Infrared Pathfinder Satellite Observation (CALIPSO).* Both CloudSat (Stephens et al. 2002) and CALIPSO (Winker et al. 2004) were launched on 28 April 2006. In the A-Train, CloudSat and CALIPSO lag Aqua by 1 to 2 minutes and are separated from each other by 10 to 15 seconds. The close proximity between these two platforms offers a unique opportunity for almost exact collocated and coincident observations of global cloudy areas. CloudSat's sensor consists of a 94 GHz radar referred to as the cloud profiling radar (CPR). With a sampling rate of 6 profiles/sec, the CPR generates a vertical profile for every 1.1 km along the flight track. Each profile has 125 vertical "bins", while each bin is about 240 m thick. The footprint covers a rectangular area of 1.4 km by 2.5 km. The backscattering reflectivity measurements from CloudSat/CPR provide the cloud liquid and ice water content profiles, with a 500-m vertical resolution from the surface to 30 km along with an effective FOV of 1.4 (across track) ×3.5 (along track) km^2. The CALIPSO is equipped with a dual-wavelength (532 nm and 1064 nm) polarization sensitive lidar. Its vertical and horizontal resolutions are 30-60 m and 333 m, respectively. It will provide the vertically-resolved information on aerosol distribution, extinction coefficient, hydration state, and discrimination of large and small particles. It will also offer an improved cloud masking of aerosol data and the opportunity to assess possible aerosol biases in cirrus cloud detection. Because of the cross-track coverage of CALIPSO and CloudSat, search for contrails using both instruments would be limited.

vi *Ice, Cloud, and land Elevation Satellite (ICESat)/ Geoscience Laser Altimeter System (GLAS).* ICESat is the benchmark Earth Observing System mission for measuring ice sheet mass balance, cloud and aerosol heights, as well as land topography and vegetation characteristics. The GLAS onboard ICESat is a diode-pumped Q-switched Nd:YAG laser operating in the near infrared (1064 nm) and visible (532 nm) wavelengths. It is a facility instrument designed to measure ice-sheet topography and associated temporal changes, as well as cloud and atmospheric properties. Dessler et al. (2006a, b) have used GLAS data to produce global statistics of thin cirrus and cloud-top height.

vii *Multi-angle Imaging SpectroRadiometer (MISR).* The MISR sensor is aboard the Terra satellite. This instrument has the unique capability to determine the altitude of aerosol layers in the atmosphere. The MISR sensor uses nine cameras pointed at fixed angles to observe reflected and scattered sunlight. In each of the nine MISR cameras, images are obtained in four spectral bands corresponding to four different colors: blue, green, red, and near-infrared. The center wavelength of each of these bands is 446, 558, 672, and 867 nm. The FOV is 17.6× 17.6 km^2 and the return period is 9 days. Validation of the MISR aerosol optical depth data over North America using AERONET has shown that the products from this instrument are of high quality and unbiased. The aerosol layer height can also be derived from MISR data. MISR aerosol products can be used for the quantitative detection of aerosol indirect effect on cirrus cloud formation.

viii *National Polar-orbiting Operational Environmental Satellite System (NPOESS)/ Visible-Infrared Imager-Radiometer Suites (VIIRS).* The VIIRS is being developed as a part of the NPOESS platform to satisfy the operational requirements for the global remote sensing of atmospheric and surface properties. Its design is similar to MODIS in terms of spectral characteristics, but has a smaller number of spectral image bands (16). One of the prime applications of VIIRS channels would be the remote sensing of cloud properties, including optical depth, particle size, cloud-top temperature, cloud cover/layers and cloud height, termed as cloud environmental data records. The first VIIRS onboard the NPOESS Preparatory Platform (NPP) is scheduled for launch in the 2009 time frame.

ix *Multi-functional Transport Satellite (MTSAT).* The MTSAT-1R is a geostationary platform operated by the Japan Meteorological Agency to fulfill meteorological and aviation functions covering East Asia, Western Pacific Ocean, and Australia. The geosynchronous plane is about 35,800 km (22,300 miles) above the Earth at 135° E, 140° E (the operational position for meteorological function) or 145° E. The MTSAT-1R Imager is a 5-band (1 visible and 4 infrared) imaging radiometer designed to sense radiant and solar reflected energy from sampled areas of the Earth. The channel characteristics of GOES Imager are as follows: VIS (, = 0.55 – 0.9 m, Instantaneous Geographic Field of View at nadir (IGFOV) = 1 km,), IR1 (, = 10.3 – 11.3 m, IGFOV = 4 km), IR2 (, = 11.5 – 12.5 m, IGFOV = 4 km); IR3 (, = 6.5 – 7 m, IGFOV = 4 km), and IR4 (, = 3.5 – 12.5 m, IGFOV = 4 km).

x *Meteosat-9*. The Meteosat Second Generation 2 (MSG-2) is a geostationary platform operated by the European Organisation for the Exploitation of Meteorological Satellites (EUMETSAT) and the European Space Agency, renamed as Meteosat-9 after its launch on 21 December 2005. Its purpose is to monitor the atmospheric and surface condition over Europe, Africa, and Eastern Atlantic Ocean. The geosynchronous plane is about 35,800 km (22,300 miles) above the Earth at 0° Longitude. The Meteosat-9 carries the Spinning Enhanced Visible and InfraRed Imager (SEVIRI), a 12-band spectroradiometer imaging suite. The 12 bands are: 1 High Resolution Visible band, 3 visible bands (, = 0.6, 0.8 and 1.6 m)and 11 IR bands (, = 3.9, 6.2, 7.3, 8.7, 9.7, 10.8, 12.0 and 13.4 m).

xi *Metop-A*. This is the first of three satellites of the EUMETSAT Polar System (EPS), launched on 19 October 2006, and was operational on 15 May 2007. The Metop-A is a polar orbiter with the equator-crossing time at 0930 LTC. It carries the US-made AVHRR (see *(i)* for details).

Satellite Remote Sensing Techniques Applicable to Contrails and Cirrus Clouds

A number of satellite remote sensing techniques have been developed to detect the presence of contrails and cirrus clouds, and to retrieve their microphysical and optical properties. The detection/retrieval products can be further applied to determine the aerosol indirect effect on cirrus cloud formation. Table 2 summarizes these remote sensing techniques. A more detailed description of each instrument follows.

i *A VHRR Split-window Pattern Recognition.* Schumann and Wendling (1990) introduced a pattern recognition method for the detection of contrails using AVHRR split-window (10.7 and 12 .m bands) data. Contrails have also been identified by their linear shape using images from visible reflectance and infrared brightness temperature (Palikonda et al. 2004). A drawback of this method is that natural cirrus with similar linear shape could be mistakenly identified as contrails.

ii *A VHRR Multi-Spectral Method.* Ou et al. (1996) developed a multi-spectral numerical scheme to identify pixels containing cirrus clouds overlapping low clouds using AVHRR channels based on their spectral characteristics. This scheme has been applied to the AVHRR data collected over the FIRE-II IFO area during nine overpasses within seven observational dates. Results from the cloud typing program have been verified using the co-located and coincident ground-based radar and lidar return images, balloon-borne replicator data and the NCAR Cross-chain Loran Atmospheric Sounding System humidity soundings on a case-by-case basis.

Table 2. Satellite remote sensing techniques for detection and retrieval of contrails and cirrus clouds

Satellite Remote Sensing echnique	Application to contrails and cirrus louds	References
AVHRR Split-window attern Recognition	Detection of line-shaped contrails	Schumann and Wendling (1990), etancor-Gothe and Grassl (1993), Iannstein et al. (1999), Palikonda 2004)
AVHRR Multi-Spectral Iethod	Detection of cirrus clouds verlapping low clouds	Ou et al. (1996)
GOES Imager Detection echnique	Detection of persistent contrails nd contrail cirrus	Minnis et al. (1998)
MODIS Cloud Iask/Phase Program	Detection of cirrus clouds	Ackerman et al. (2002), Platnick et l. (2003), King et al. (2004)
MODIS 1.38 ìm etection Method	Detection of contrails and cirrus louds	Roskovensky and Liou (2003) Gao t al. (1993), King et al. (1996), Iutchison and Choe (1996)
HIRS and MODIS CO_2 licing Method	Retrieval of cirrus cloud effective missivity and cloud-top Pressure	Smith and Platt (1978), Menzel et l. (2002)
AVHRR Split-window etrieval Method	Retrieval of cloud optical depths nd ice crystal size of thin cirrus louds and contrails	Parol et al. (1991), Betancor- Gothe nd Grassl (1993), Duda and pinhirne (1996), Duda et al. (1998)

Satellite Remote Sensing echnique	Application to contrails and cirrus louds	References
AVHRR and IPOESS/VIIRS Thermal R Window Retrieval Method	Retrieval of cloud optical depths, e crystal size and cloud-top mperature of thin cirrus clouds nd contrails	Ou et al. (1993, 1995, 1998a, b, 002, 2003), Rao et al. (1995), Wong t al. (2007)
AVHRR, MODIS, and IPOESS/VIIRS Visible-Iear-IR Window Retrieval Method	Retrieval of cloud optical depths, e crystal size of thin cirrus louds, and contrails	Hansen and Pollack (1970); womey and Cocks, (1982, 1989); Iakajima and King (1990), King et al. (1996, 1997), Ou et al. 1999, 2003), olland and Liou (2001), olland et al. (2000), Platnick et al. 2003), Roskovensky and Liou
AIRS Hyperspectral etrieval Method	Retrieval of cloud optical depths nd ice crystal size of cirrus clouds	Yue et al. (2007)

iii *GOES Imager Detection Technique.* Minnis et al. (1998) summarized a method for detecting contrails using GOES 0.65, 3.9, 11, and 12 μm data. A contrail is detected either as a distinct or other geometrical cold feature in the IR imagery or by using the BTD between thermal IR window bands. Once identified, a box is drawn around the contrail area and all pixels with brightness temperatures less than a threshold and BTD < 2K are flagged as contrails.

iv *MODIS Cloud Mask/Phase Program.* The MODIS cloud mask/phase programs use several cloud detection tests to indicate a level of confidence that the MODIS is observing a clear sky scene, and to assess the likelihood of a pixel being obstructed by clouds (Ackerman et al. 2002). Fourteen of the MODIS 36 spectral bands are utilized to maximize reliable cloud detection. Their products are generated globally for both daytime and nighttime overpasses with a 1 km-pixel resolution. Because cloud cover can occupy a pixel to varying extents, the MODIS Cloud mask program was designed to allow for varying degrees of clear sky confidence. The MODIS cloud mask/phase programs identify several conceptual domains according to surface type and solar illumination, including land, water, snow/ice, desert, and coast for both daytime and nighttime overpasses.

v *MODIS 1.38 m Detection Method.* The MODIS cloud mask products, which include data from the 1.38-m channel, have shown that the global cirrus-cloud coverage is less than that presented by Wylie et al. (1999). We note that MODIS products have not adequately utilized the 1.38-m channel reflectance. This channel is particularly useful for detecting thin cirrus due to its high sensitivity to upper tropospheric clouds and a nearly negligible sensitivity to low-level reflectance (Gao et al. 1993; King et al. 1996; Hutchison and Choe 1996). Specific 1.38-m reflectance threshold levels can be utilized to detect thin cirrus that has previously been undetectable by downward looking satellite imagery. Roskovensky and Liou (2003) developed a new cloud-detection scheme that utilizes 1.38-m reflectance to detect thin cirrus clouds. In this new method, the threshold is dependent on neighboring cloud type, water vapor concentration, and viewing geometry.

vi *HIRS and MODIS CO_2 Slicing Method.* The CO_2 slicing method is designed to determine the cloud- top pressure and the cloud effective emissivity based on the principle that the ratio of cloud signals (defined as the difference between cloudy and the clear radiances) for the two spectral bands is a function of cloud-top pressure only, which can then be evaluated by matching the ratio values derived from satellite measurements and from radiative transfer calculations using the cloud-top pressure and atmospheric temperature and humidity profiles. The CO_2 Slicing Method is most applicable to high- level clouds because of the strong sensitivity of the ratio value in the 15-μm CO_2 band to cloud-top pressure at high altitudes.

vii *A VHRR Split-Window Retrieval Method.* This method was designed to determine the cloud optical depth and mean particle size based on the principle that correlation of the split-window BTD and Ch. 5 brightness temperature (T_5) depends on both optical depth and mean particle size. For this method to work, it is necessary to know cloud-top and surface temperatures and pre-computed BTD and T_5 based on a prescribed cloud microphysical model.

viii *A VHRR and NPOESS/VIIRS Thermal IR Window Retrieval Method.* Ou et al. (1993) developed a physical retrieval scheme using radiance data from AVHRR 3.7μm and 10.9μm bands to infer nighttime cirrus cloud parameters, including cloud temperature, optical depth, and mean effective ice crystal size based on the theory of radiative transfer and microphysics parameterizations. To aaply this IR retrieval algorithm to daytime conditions, a numerical scheme to remove the solar component in the 3.7μm radiance has been developed (Rao *et al.*, 1995). Analysis of the effects of error sources on retrieval results reveal that the maximum error in the 3.7μm solar component is less than 10 %.

ix *A VHRR, MODIS, and NPOESS/VIIRS Visible-Near-IR Look-Up Table Method.* This method was designed to determine cloud optical depth and mean particle size based on the principle that the reflection function of clouds at a non-absorbing band in the visible wavelength region is primarily a function of cloud optical depth, whereas the reflection function at a water (or ice) absorbing channel in the near-infrared (e.g., 1.61μm band) is primarily a function of cloud particle size (King *et al.* 1997). This principle was initially applied to the determination of water cloud optical depth and effective droplet radius during daytime. The approach has been discussed by Hansen and Pollack (1970), Twomey and Cocks (1982 and 1989), and Nakajima and King (1990) using visible and near-IR radiometers from an aircraft platform. Ou et al. (1999) applied this principle to the retrieval of cirrus cloud optical depth and mean particle size using AVHRR 0.67 and 3.7 μm data. The same principle was applied to the MODIS Airborne Simulator (MAS) 0.657 and 1.609 μm band reflectances by Rolland *et al.* (2000; 2002), to the MAS 0.657, 0.74, 0.86, and 1.87μm (surrogate of the MODIS 1.38 μm) band reflectances by Roskovensky and Liou (2005), and to the MODIS 0.65, 0.86, 1.38, and 1.64 μm band reflectances by Roskovensky and Liou (2006) to evaluate cirrus cloud and aerosol parameters. This multi-channel technique has been incorporated into both the MODIS cloud retrieval program (King et al. 1997) and the NPOESS/VIIRS cloud optical property retrieval code (Ou et al. 2002; 2003).

x *AIRS Hyperspectral Thin Cirrus Retrieval Method.* This method was based on a thin cirrus cloud thermal infrared radiative transfer model constructed by combining the Optical Path Transmittance (OPTRAN, Mcmillin et al. 1995) model, developed for a speedy calculation of transmittances in clear atmospheres, and a thin cirrus cloud parameterization using a number of observed ice crystal size and shape distributions (Yue et al. 2007a). Numerical simulations show that cirrus cloudy radiances in the 800–1130 cm^{-1} thermal infrared window are sufficiently sensitive to variations in cirrus optical depth and ice crystal size and shape if appropriate habit distribution models are selected *a priori* for analysis. The parameterization model has been applied to the Atmospheric Infrared Sounder (AIRS) on board the Aqua satellite to interpret clear and thin cirrus spectra observed in the thermal infrared window. Five clear and 29 thin cirrus cases at nighttime over and near the ARM program Tropical Western Pacific (TWP) Manus Island and Nauru Island sites have been chosen for this study. A χ^2-minimization program was employed to infer the cirrus optical depth and ice crystal size and shape from the observed AIRS spectra. Independent validation shows that the AIRS-inferred cloud parameters are consistent with those determined by collocated ground-based millimeter-wave cloud radar measurements.

Ground-Based Remote Sensing of Contrails and Cirrus

Observations of contrails by lidar dated back as early as the late 80's. During the International Ice Experiment at German Bay (Raschke et al. 1990), a lidar developed by Morl et al. (1981) mounted on an aircraft was used to scan contrails from below (Schumann and Wendling, 1990). Later, Sassen et al. (1 989a, 1 989b), Freudenthaler et al. (1 996b) and Gayet et al. (1996) used different lidar systems with a number of lidar wavelengths for detection and characterization of the contrail properties. Sassen (1997) presented a variety of persistent-contrail measurements employing polarization lidar and radiometric observations in Salt Lake City, Utah and gathered new information on contrails in a geographical area previously identified as being affected by relatively heavy air traffic. This dataset includes the hourly and monthly frequency of occurrence; the height, temperature, and relative humidity statistics; visible and infrared radiative impacts; the microphysical properties evaluated from *in situ* data, and the contrail optical phenomenon such as halos and coronas.

Figure 8 presents an image of a 45-min old contrail generated by commercial jet aircraft flying in a flight corridor north of the ARM-SGP site on May 2, 1996 during SUCCESS. This image was obtained from a high-resolution (1.5 m by 0.1 sec) polarized diversity lidar deployed at Lamont, Oklahoma.

Contrail images, similar to the one shown in figure 8, contain abundant information regarding contrails' fine structure. It has been suggested that small particles typical of those in persistent contrails may favor albedo cooling over greenhouse warming, dependent on such factors as the geographic distribution and patterns of the day vs. night aircraft usage.

Atlas et al. (2006) discussed the morphology of contrails, their transition to cirrus uncinus, and their microphysical and radiative properties on the basis of the coincidental occurrence of a cluster of nearly parallel contrails, and the availability of collocated and concurrent observations by photography, satellite, automated ground-based lidar, and a freshly available database of aircraft flight tracks. Each contrail was observed sequentially by a lidar and tracked backward to the time and position of the originating aircraft track using the appropriate wind field. This lidar also provided particle fall speeds and estimated ice particle size, extinction coefficient, optical depth, and ice water path.

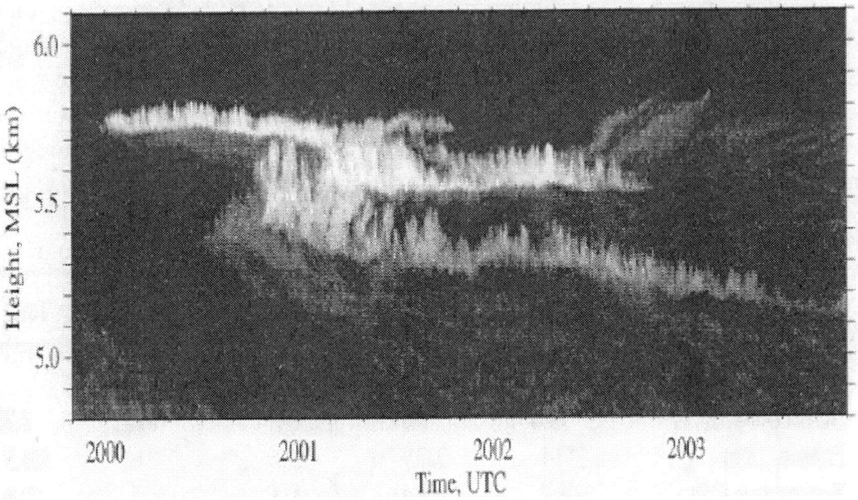

Figure 8. High-resolution image of spreading contrails, resembling cirrocumulus and natural cirrus probed in the 1.06mm polarization diversity lidar channel at the ARM-SGP site on May 2, 1996 during SUCCESS field experiment (after Sassen 1997).

Systematic and continuous ground-based observations by lidar and other remote sensing instruments have been conducted by the Atmospheric Radiation Measurement (ARM) Program, a program established in 1989 that has been sponsored by the DOE Office of Science and managed by its Office of Biological and Environmental Research. One of the primary objectives of the ARM program is to improve scientific understanding of the fundamental physics governing the interaction between clouds and radiative feedback processes in the atmosphere. The ARM Program establishes and operates field research sites to study the effects of clouds on climate and climate change, and to improve their physical parameterization in GCMs. Three primary locations—the Southern Great Plains (SGP), Tropical Western Pacific (TWP), and North Slope of Alaska—were identified as representing the range of Earth's climate conditions. Each site has been heavily instrumented to gather a massive amount of climate data. Among these sites, the SGP site is particularly relevant to contrail observation because of its location near the flight corridors. Relevant instruments deployed at its Central Facility include the micropulse lidar (MPL), millimeter-wave cloud radar (MMCR), Raman lidar, total sky imager, Vaisala ceilometer, AERI, and various radiometers. Details of each instrument are given in the ARM website (http://www.arm.gov/instruments/instclass.php?id=cloud).

Among these instruments, the MPL is the most suitable for contrail and cirrus observation because of the strong sensitivity of the laser beam to small ice particles that are typical of contrails and thin cirrus. The MPL can detect cloud and aerosol signals between the ground level and 20 km with a vertical resolution of 0.03-0.3 km. The MMCR can detect cirrus clouds composed of particles with maximum dimensions larger than 100 m. The difference in cirrus detection between MPL and MMCR is illustrated by Comstock et al (2002), as shown in figure 9. Nevertheless, MMCR data collected at the ARM sites and during field campaigns in the past decade has been extensively used to study cirrus cloud characteristics (Mace et al. 1998a, 1998b, 2002, 2005) and to compile cirrus cloud climatology (Mace et al. 2006). MMCR can detect cloud signal between the ground level and

20 km with a vertical resolution of 0.05-0.1 km. For the retrieval of cirrus microphysical properties, several promising algorithms have been developed some of which can be applied to contrails. Matrosov et al. (1992) estimated layer- averaged ice cloud particle characteristic sizes and concentrations as well as the integrated ice water path from simultaneous ground-based radar and infrared radiometric measurements.

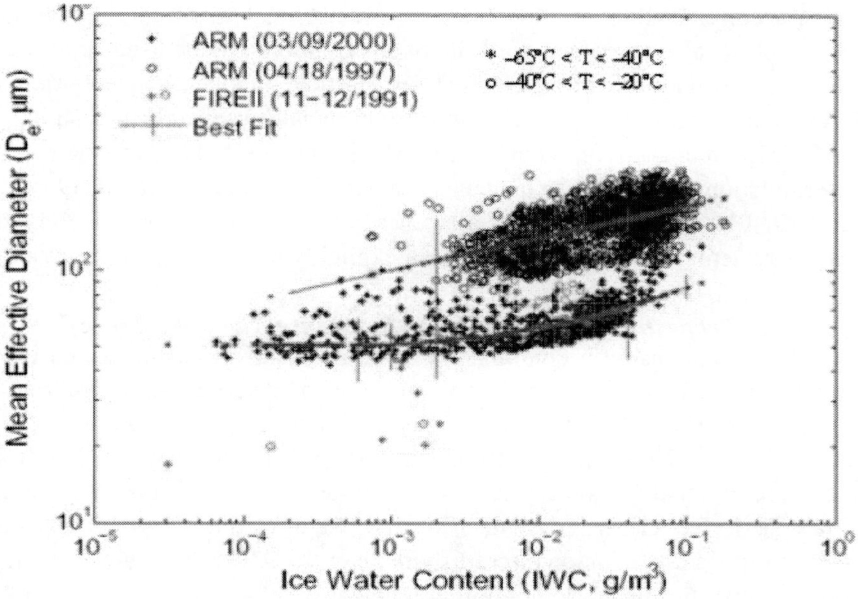

Figure 9. The IWC and mean effective size correlation for midlatitudes cirrus based on 4066 aircraft observations during ARM and FIRE intensive cirrus cloud field campaigns. The solid curves denote the best fitting with vertical bar representing standard deviations.

E. Present State of Modeling Capability

Parameterization of Ice Crystal Microphysics Properties in GCM

A number of GCMs used temperature to determine ice crystal size (Donner et al., 1997; Kristjansson et al., 2005; Gu and Liou, 2006). This approach is rooted in earlier ice microphysics observations from aircraft and attests to the fact that small and large ice crystals are related to cold and warm temperatures in cirrus cloud layers. Ou and Liou (1995) developed a parameterization equation relating cirrus temperature with a mean effective ice crystal size, De, based on a large number of midlatitude cirrus microphysics data presented by Heymsfield and Platt (1984). Ou et al. (1995) reduced large standard deviations in the size-temperature parameterization by incorporating a dimensional analysis between ice water content (IWC) and De. Using CEPEX data, McFarquhar et al. (2003) developed a D_e parameterization as a function of IWC for use in a single column model to understand the impact of tropical ice clouds on radiation fields.

Liou et al. (2007) recently analyzed the ice crystal size distribution data obtained from *in situ* aircraft measurements during a number of field experiments, including the ARM Intensive Cloud Observation Programs that were conducted over Oklahoma during April

1997 and March 2000, and the First ISCCP Regional Experiment (FIRE) II that was carried out over Kansas during November-December 1991 (Liou and Gu, 2006; Yue et al. 2007b). The IWC and D_e for radiation calculation are correlated on the basis of their mathematical definitions (Fu and Liou 1993). Excellent correlations between D_e and IWC have been found in the datasets by dividing the observed air temperature into two groups: -20°C to - 40°C and - 40°C to -65°C (figure 9). IWC and temperature are prognostic variables in most climate models. Thus, a corresponding D_e for radiation calculations can be determined using these correlations. Analysis also reveals that a Gamma distribution may be used to fit the observed ice crystal size distributions for calculating ice particle number concentration from the predicted IWC and temperature. These results are especially useful for evaluating ice cloud radiative effects in numerical simulations in which aerosol fields are not resolved. The empirical formulation allows us to calculate radiative transfer interactively with the ice microphysics used in numerical models, as well as to calculate effective ice crystal size in simulations where aerosol indirect effects are not explicitly considered.

Modeling Optical Properties for Contrails for Input into Radiative Transfer Models

Ice crystal size and shape in contrails and contrail cirrus are complex and intricate. Ice crystal images collected by the optical probe and replicator aboard aircraft during a limited number of field experiments have shown that contrails consist predominantly of bullet rosettes, columns, and plates with sizes ranging from about 1 to 100 m. Liou et al. (1998) presented four representative ice crystal size distributions in contrails and contrail cirrus clouds (figure 10). Ice crystal size distributions were obtained by FSSP onboard the University of North Dakota Citation aircraft flying over the ARM SGP site on April 18, 1994, re-penetrating its own 6-minute old contrail at a height of 13 km and a temperature of - 65.9 °C. The sampled contrail contains a substantial number of small ice crystals on the order of 10 m. Ice crystal size distributions were also sampled over the ARM-SGP area by the Citation from near the top (13.4 km and -69.4°C) of an optically thin cirrus cloud that had contrails embedded in it. The two ice crystal size distributions over northeast Oklahoma on May 4, 1996 were measured by the replicator system developed by Arnott et al. (1994) mounted on the DC-8 aircraft, which tailed a Boeing 757 at a distance of 11.5 km and a time lag of 50 sec. The ambient temperature and dew point are -61.1°C and - 62.9°C, respectively. Based on the SUCCESS replicator data, contrails contain a combination of bullet rosettes (50%), hollow columns (30%), and plates (20%). Using these shape factors, the mean effective sizes for the four ice crystal size distributions are 4.9, 9.8, 15.9, and 13.3 m.

To compare the ice crystal size distributions for contrails and natural cirrus clouds, figure 11 shows six representative distributions that were obtained from aircraft observations presented by Heymsfield and Platt (1984), Takano and Liou (1989) and the FIRE-IFO microphysical data. They are denoted as cold Ci, -60° C, Cs, FIRE-I IFO 1 Nov, FIRE-I IFO 2 Nov, and Ci Uncinus. The ice crystal sizes span from about 5 to 2000μm, which is much wider than the range of contrail size distributions. The mean effective sizes vary from 24 to 124 m, much larger than the contrail size distributions shown in the previous figure. Ice crystal shapes range from bullet rosettes, solid and hollow columns, and plates to aggregates, exhibiting a greater variety than contrail ice crystal shapes.

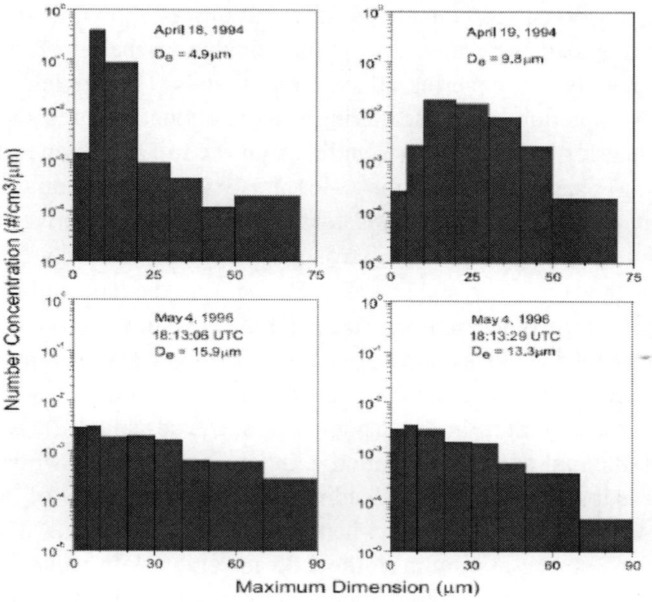

Figure 10. Discretized ice crystal size distributions for a contrail and a cold cirrus (~ 6 min duration) measured by FSSP on board the University of North Dakota Citation on April 18 and 19, 1994 (upper panels); and for contrail cirrus (~ 50 sec duration) measured by the replicator system mounted on the NASA's DC-8 that tailed a Boeing 757 during SUCCESS on May 4, 1996 (after Liou et al. 1998).

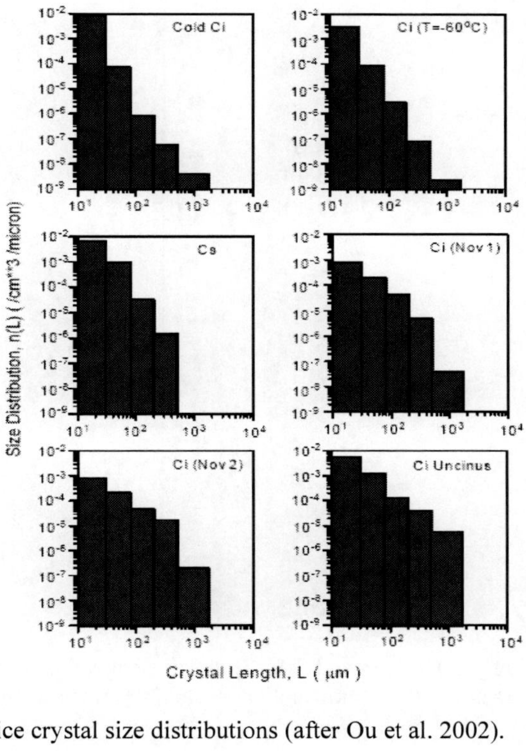

Figure 11. Six discretized ice crystal size distributions (after Ou et al. 2002).

Using the four observed ice crystal size distributions as shown in figure 10, Liou et al. (1998) carried out the scattering and absorption calculations based on a unified theory for light scattering by ice crystals covering all sizes and shapes. The single-scattering parameters in terms of the phase function, single-scattering albedo, extinction coefficient, and asymmetry factor were computed for 200 solar wavelengths from 0.2 to 5 μm. Figures 12(a) and (b) show the single-scattering phase functions for 0.46 and 3.5 μm. Substantial differences in the backscattering part of the phase functions for the four mean effective sizes at the 3.5 μm wavelength are noted. Because ice is a strong absorber at this wavelength, the scattered energy strongly depends on ice crystal size. For D_e = 4.9 μm, the halo and backscattering peaks are lower than those fore other smaller sizes. Figure 13 shows the extinction coefficient, single-scattering co-albedo and asymmetry factor based on a shape model of 50% bullet rosettes, 30% hollow columns, and 20% plates. The extinction coefficients show little variation, except for a minimum in the 2..5 μm region, the so-called Christiansen effect. This effect occurs when the real part of the refractive index approaches 1, while the corresponding imaginary part is substantially larger, leading to the domination of absorption in light attenuation. This is particularly evident when ice particles are small. The single-scattering albedo also displays a strong minimum in the 2.85 μm region with values much less than 0.5. When absorption is strong, the scattered energy is primarily contributed by diffraction in the forward directions. For this reason, maximum values of the asymmetry factor are noted around 3 μm.

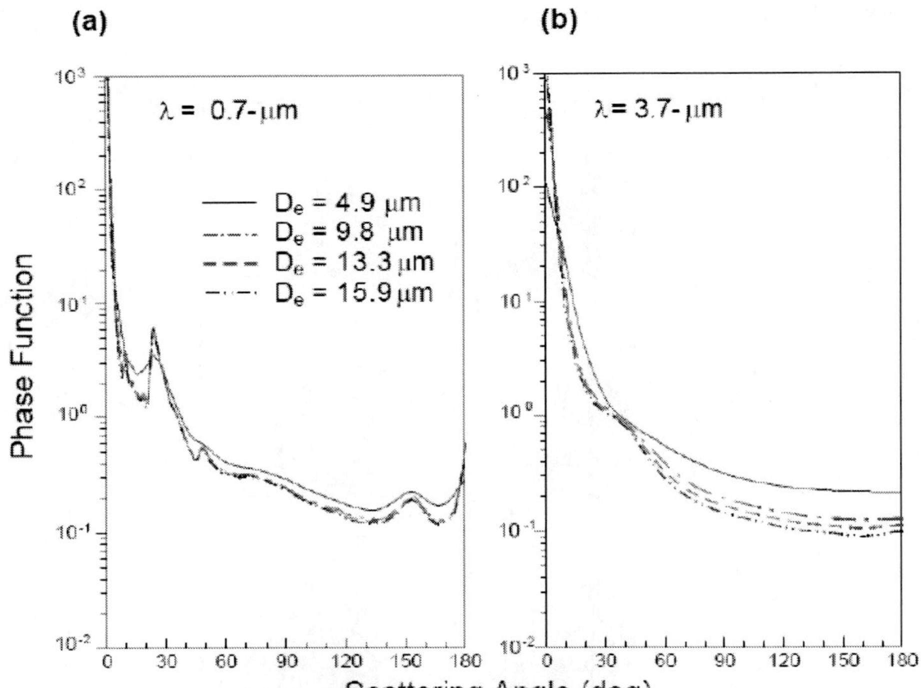

Figure 12. Phase functions for (a) 0.7 μm and (b) 3.7μm wavelengths using a contrail cirrus model consisting of 50% bullet rosettes, 30% hollow columns, and 20% plates (after Liou et al. 1998).

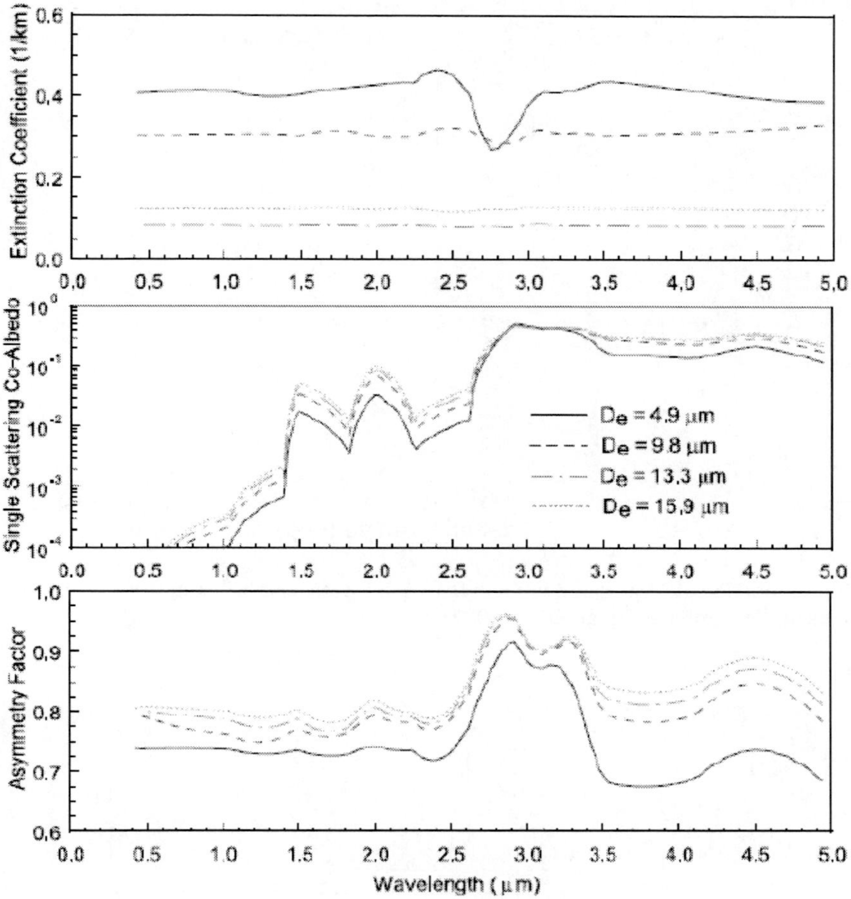

Figure 13. Extinction coefficient (top), single-scattering co-albedo (1 – ω, middle), and asymmetry factor (bottom) as functions of wavelength from 0.2 to 5 μm. The minima for the extinction coefficient and the maxima for single-scattering co-albedo and asymmetry factor located at 2.85 μm are due to the wellknown Christiansen effect (after Liou et al. 1998).

To compare the single-scattering properties between contrails and cirrus clouds, figures 14 (a) and (b) show the phase functions for three representative size distributions involving cold Ci, cirrostratus, and cirrus uncinus for 0.672 and 3.7 m (Ou et al. 2002; 2003). For the non-absorbing 0.672 m wavelength, the overall phase function feature is not sensitive to variation in size distribution. The 22° and 46° halo features produced by two refracted rays are distinct in addition to the forward diffraction peak. Between about 150° and 160° scattering angles, there is another peak for all sizes produced by rays undergoing double internal reflections. Side-scattering is larger for smaller ice crystals. For the 3.7m wavelength, the halos and backscattering peaks disappear due to strong absorption. Also, the strength of the forward scattering associated with diffraction varies with size distribution in this case. Figure 15 shows the extinction coefficient, single-scattering co-albedo, and asymmetry factor for the three ice crystal size distributions and for wavelengths between 0.2 and 5 m. Extinction coefficients are nearly constant. Both the single-scattering co-albedo and asymmetry factor generally increase with increasing wavelengths and D_e.

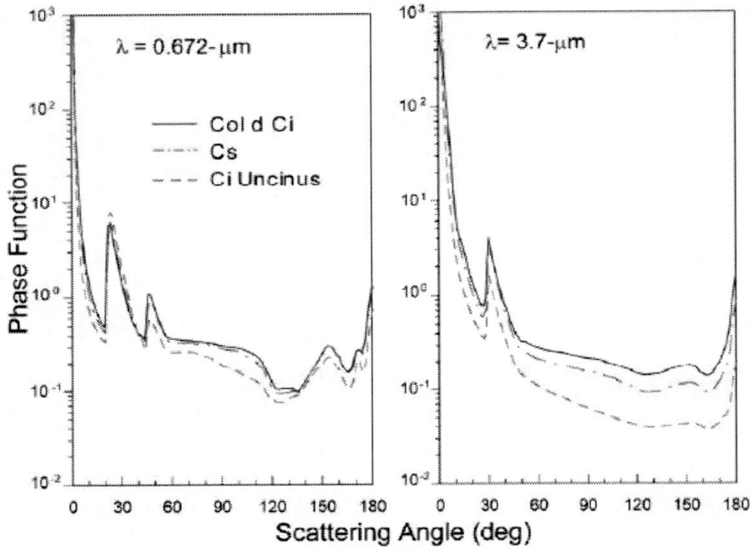

Figure 14. Phase functions for (a) 0.672 μm and (b) 3.7μm wavelengths using Cold Ci, Cirostratus, and Cirrus Uncinus models (after Ou et al. 2002; 2003).

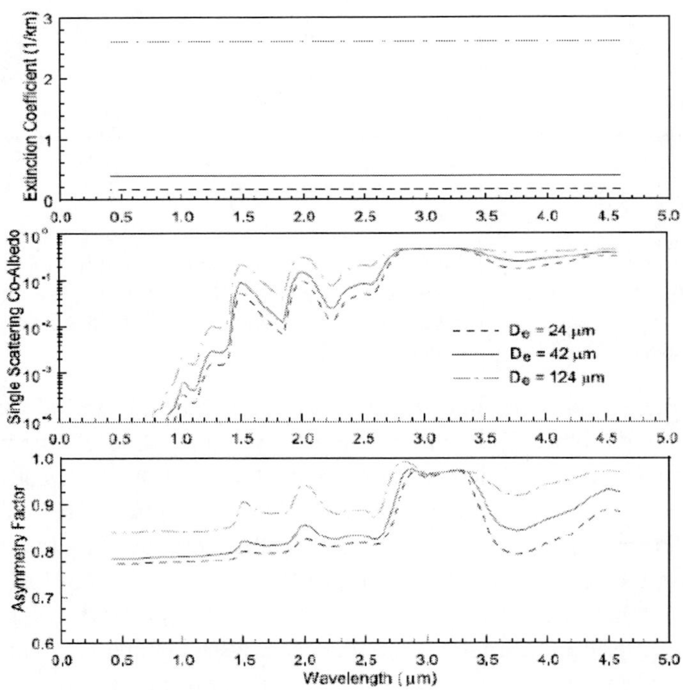

Figure 15. Extinction coefficient (top), single-scattering co-albedo (1 – ω, middle), and asymmetry factor (bottom), as functions of wavelength from 0.2 to 5 μm for three representative cirrus cloud size distributions. The minima for the extinction coefficient and the maxima for single-scattering co-albedo and asymmetry factor located at 2.85 μm are due to the well-known Christiansen effect (after Ou et al. 2002; 2003).

Radiative Transfer Model for Application to Satellite Remote Sensing – LBLE Model

The LBLE radiative transfer model uses the adding-doubling method including full Stokes parameters developed by Takano and Liou (1989a, b) for vertically inhomogeneous atmospheres. The input parameters required to drive LBLE are generated by several pre-processors, including solar insolation, spectral band wavenumber, solar and viewing zenith angles, relative azimuthal angle, spectral surface albedo and emissivity, atmospheric temperature and humidity, and aerosol profiles. Input cloud configuration parameters include phase, base height, thickness, optical depth, and mean effective particle size.

The 1996/2000 HITRAN line-by-line absorption coefficients were used to develop the correlated k- distribution method for spectral radiative transfer. The correlated-*k* coefficients for H_2O covering the spectral region from 2,000 to 21,000 cm^{-1} (0.5-5 μm) were derived following a numerical approach in which efficient and accurate parameterizations for the calculation of pressure- and temperature- dependent absorption coefficients were developed on the basis of the theoretical values at three reference temperatures and 19 reference pressures. Absorption due to O_3 and O_2 bands follows Beer's law. In the original LBLE, the entire solar spectrum was divided into a total of 380 intervals with $\Delta\nu$ = 50 cm^{-1}. For each spectral interval, the inverse of the cumulative probability function k (g) is evaluated at 30 g values, where $0<g<1$. The model vertical domain was divided into 51 layers (Δp = 20 mb for each layer, except for the bottom layer, where Δp = 13 mb). For wavelengths between 3.5 and 5 μm, thermal emission contribution was accounted for in the solar flux transfer by adding the thermal emission component. Comparing the visible, near-IR and IR clear radiances computed from LBLE with those computed from MODTRAN shows differences within 10%, due to the different multiple- scattering treatment in the two models. Single-scattering parameters for ice clouds, which include the asymmetry factor (g), the extinction coefficient (β_e) and the absorption coefficient [$(1 - \tilde{\omega})\beta_e$], were computed from the geometric ray-tracing method assuming randomly-oriented hexagonal ice crystals. The phase function values at discrete scattering angles were computed using a 200-term Lengendre polynomial expansion of the phase function with the δ-function transmission and diffraction-peak truncations. The resulting single-scattering parameters, cumulative *k*-distribution functions, phase functions, and other auxiliary data were combined and built into a radiative transfer program.

Using the preceding LBLE code, Liou et al. (1998) showed from spectral curves that cloud reflection increases as ice crystal sizes become smaller, but the cloud absorption increase is only evident for wavelengths longer than about 2.7 μm (figure 16). The ice crystal shape has a substantial effect on cloud reflection and absorption for a given size. More complex ice particles reflect more solar radiation. For comparison, figure 17 shows bidirectional reflectances for a number of representative ice crystal size distributions.

Radiative Transfer Model for Radiative Forcing Calculations – Fu-Liou Model

Fu and Liou (1992; 1993) developed a state-of-the-art radiative transfer model that can be used for radiative forcing calculations associated with contrails and contrail-induced cirrus clouds. For the sake of computational efficiency and a high degree of accuracy, this radiation scheme was modified to use the δ4-stream approximation (Liou et al. 1988) for solar flux calculations, and the δ2/4-stream approximation (Fu et al. 1997) for IR flux calculations. The

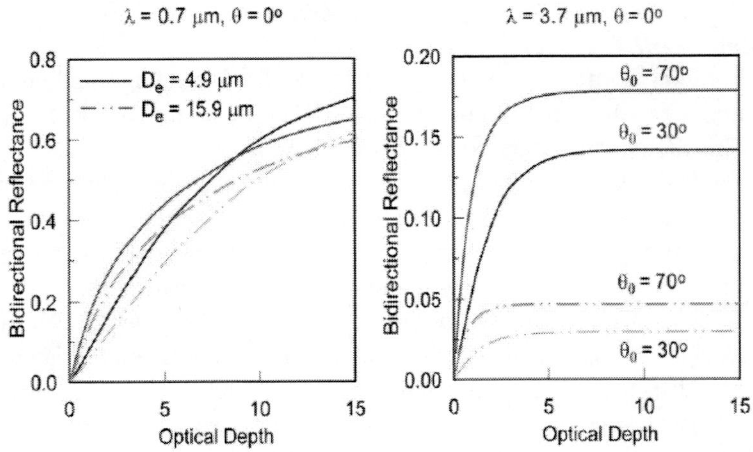

Figure 16. Bidirectional reflectances for two contrail size distributions, two solar zenith angles and two wavelengths as functions of optical depth, computed by LBLE model after (Liou et al. 1998).

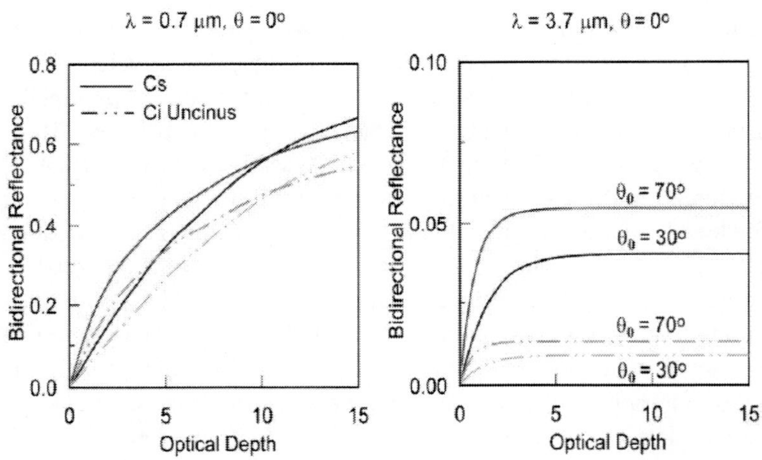

Figure 17. Bidirectional reflectances for two cirrus size distributions, two solar zenith angles and two wavelengths as functions of optical depth, computed by LBLE model (after Ou et al. 2002; 2003).

spectral integration has been carried out using the correlated *k*-distribution method developed by Fu and Liou (1992). The solar and IR spectra were divided into 12 and 6 bands, respectively according to the location of major gaseous absorption bands. Absorption due to water vapor, ozone, carbon dioxide and oxygen is accounted for in the solar spectrum. In the IR bands, absorption and emission included water vapor, ozone, carbon dioxide, methane, nitrogen oxide, and chlorofluorocarbons.

Calculations of the cloud radiative effects follow the procedure presented by Gu et al. (2003, 2006) in which the spectral extinction coefficient, the asymmetry factor, and the single-scattering albedo are determined in terms of cloud water content and the mean effective particle size for nonspherical ice crystals and liquid droplets. Calculations of the single-scattering properties of clouds require information about the particle shape and size distributions, and the indices of refraction as a function of wavelength. In typical GCM calculations, the mean effective radius of ice particles is prescribed with a constant value (e.g., 75 m). The single-scattering properties for cirrus cloud particles were first parameterized based on 12 *in situ* measured composite ice crystal size distributions (Fu and Liou 1993). The extinction coefficient, single-scattering albedo, and asymmetry factor are dependent on wavelength and the cloud position, and are parameterized in terms of IWC and mean effective ice crystal size. Figure 18 shows the cirrus cloud radiative forcing as functions of ice water content for four different ice crystal sizes.

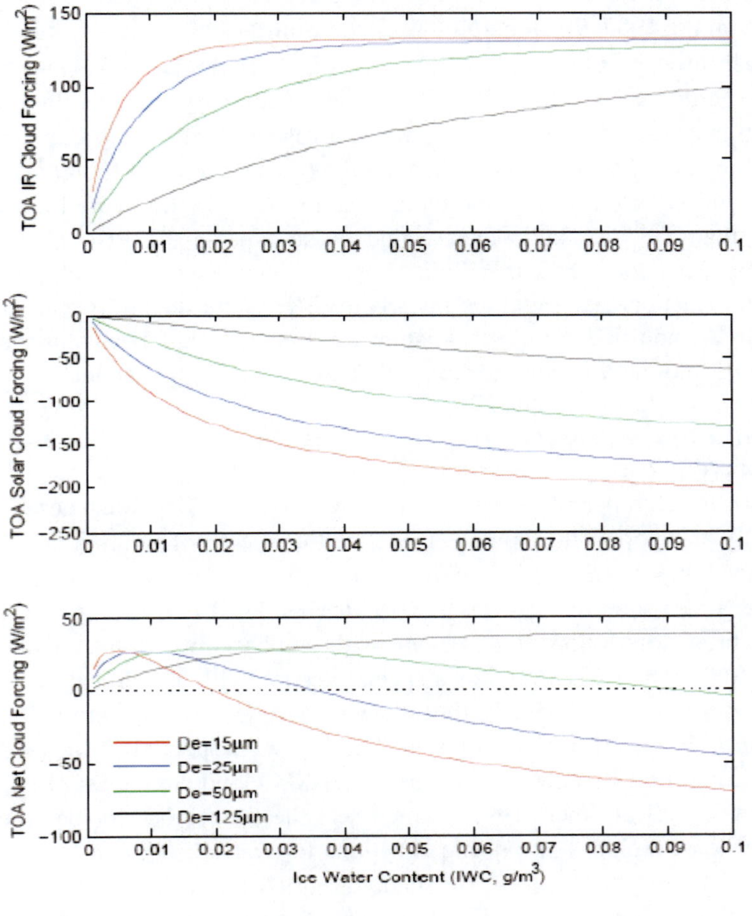

Figure 18. Cirrus cloud radiative forcings as function of ice water content for four size distributions computed by Fu-Liou model (after Fu and Liou 1993).

Global Climate Model – The UCLA AGCM

The UCLA AGCM is a state-of-the-art grid-point model of the global atmosphere. The model prognostic variables are the horizontal wind, potential temperature, mixing ratios of water vapor, cloud liquid water and ice water, planetary boundary layer (PBL) depth, surface pressure, land surface temperature, and snow depth over land. The horizontal finite differencing of the primitive equations is based on a staggered Arakawa "C" grid scheme, while the vertical coordinate employed is the modified sigma coordinate developed by Suarez et al. (1983). For the time integration of prognostic variables, a leapfrog time-differencing scheme is used with a Matsuno step regularly inserted. The PBL is parameterized as a well-mixed layer of variable depth (Li et al. 2002). Parameterization of cumulus convection and its interaction with the PBL follows Arakawa and Schubert (1974). The geographical distribution of sea surface temperature (SST) is prescribed based on a 31-yr (1960–90) climatology corresponding to the Global Sea Ice and Sea Surface Temperature dataset (GISST) version 2.2. Daily values of the surface conditions are determined from the monthly mean values by linear interpolation. Ozone (O_3) mixing ratios are prescribed as a function of latitude, height, and time based on the 1985–90 climatology (Li and Shine 1995). A low-resolution version of AGCM has been used by Gu et al. (2003), which is 4° (latitude) x 5° (longitude) with 15 layers from the earth's surface to 1 hPa. Two cloud "types" are generated by the model. The first is free atmosphere clouds, whose main sources are grid- scale supersaturation and cumulus detrainment. In the current UCLA AGCM, a grid box of the free atmosphere is assumed to be entirely cloudy (i.e., the cloud fraction is 1) if the total cloud water mixing ratio is larger than 10^{-10} kg kg^{-1}. The second cloud type is the PBL clouds, which are generated at the PBL top when there are above the condensation level. PBL clouds are assigned a cloud fraction that increases linearly with pressure thickness to become 1 at and above 12.5 mb. A modified Fu-Liou radiative transfer code has been implemented in the UCLA AGCM in conjunction with the clouds that are formed in the model.

Global Contrail-Climate Model – The ECHAM4 GCM

Version 4 of the European Center/Hamburg GCM (ECHAM4, Roeckner et al. 1996) has been specifically developed to simulate the climatic effects of contrail by Ponater et al. (2002). ECHAM4 has been applied to numerous climate sensitivity and climate change experiments (e.g., Feichter et al. 1997; Roeckner et al. 1999; Bengtsson et al. 1999). Following is a brief description of ECHAM4 with emphasis on parameterization for cirrus clouds and contrails. The ECHAM4 was a special version with 39 vertical layers that offer a vertical resolution of about 700 m in the upper troposphere and lower stratosphere where contrails occur. The horizontal resolution chosen was spectral T30 (about 670 km isotropically) with a time step of 30 min. The ECHAM4 cloud parameterization scheme was described in detail by Roeckner (1995), which sets the framework for parameterization of contrails. The scheme includes prognostic equations for water vapor and cloud water mixing ratios. It follows the original concept of Sundqvist (1978) who introduced a diagnostically determined fractional cloud cover as a function of relative humidity. Condensation and evaporation of cloud water, auto-conversion from cloud to rain, and evaporation of rain drops are parameterized. Parameterization of the contrail formation was based on the thermodynamic principle that predicts threshold values for contrail-forming temperature and humidity (Schumann 1996). Because only persistent contrails are assumed to contribute to climate change (Ponater et al. 1996), an environment that is supersaturated with respect to ice is required so that contrails may exist for a longer time, up to several hours, before they gradually transform into cirrus clouds or eventually disappear (e.g., Gierens and Jensen 1998; Minnis et al. 1998). Finally, the dependency of actual contrail coverage on air traffic density was included by introducing the local amount of aircraft fuel consumption as a linear weighting factor to calculate the actual contrail coverage from contrail coverage (Sausen et al. 1998).

The radiative effect of clouds in the ECHAM4 is represented by the single-scattering albedo, the asymmetry factor, and the optical depth in the solar spectrum, and by the emissivity in the thermal infrared. These values were parameterized in terms of cloud water content and particle effective radius. For ice clouds, the calculation of solar radiative key parameters was based on the Lorenz-Mie theory assuming spherical ice particles (Rockel et al. 1991). The ice crystal effective radius was parameterized in terms of ice water content according to the data given in Heymsfield (1977) and McFarlane et al. (1992). The single-scattering albedo, asymmetry factor, and optical depth were fitted to the spectral radiation scheme, as documented in Boucher and Lohmann (1995). In order to account for the nonsphericity of ice particles, the asymmetry factor was empirically reduced by a factor of 0.91 (Roeckner 1995). Emissivity in the thermal infrared spectrum was approximated by an exponential function of ice water content and effective crystal radius developed by Stephens et al. (1990).

The optical properties of contrails were calculated exactly the same as those of natural cirrus clouds. However, known differences between the optical properties of cirrus and contrails are accounted for by making specific assumptions for contrail optical depth and particle size. Radiative fluxes and heating rates in the ECHAM4 were calculated using the radiation parameterizations developed by Fouquart and Bonnel (1980) and Morcrette (1991) for the solar and infrared spectrum, respectively. Evaluation of the cloud radiative effect in ECHAM4 has been reported by Lohmann and Roeckner (1996) and Chen and Roeckner (1996, 1997). Wild et al. (1998) indicated that the shortwave radiative budgets simulated by ECHAM4 are closer to observed values than those generated by other GCMs. However, this favorable result could be incidental rather than a result of a superior physical formulation.

Regional Climate Model –
The WRF Model

The WRF model (Skamarock et al. 2005) is a next-generation mesoscale forecast model and data assimilation system that will advance both the understanding and prediction of mesoscale weather and accelerate the transfer of research advance into operation. It is designed to be a flexible, state-of-the-art, portable code efficient in a massively parallel computing environment. A modular single-source code is maintained that can be configured for both research and operation. It is a fully compressible, nonhydrostatic model and its vertical coordinate system is a terrain-following hydrostatic pressure coordinate. The grid staggering follows the Arakawa C-grid and the model uses the Runge-Kutta 2nd- and 3rd-order time integration schemes that offer numerous physics options. The principal components of WRF include data initialization, WRF-VAR (the data assimilation system), dynamic solvers, and physics packages. A version of the WRF, called the Advanced Research WRF (ARW), was developed at NCAR for research and development purposes. It has all the physical parameterizations required to produce simulation results.

Several cloud microphysics schemes have been implemented in WRF: (1) The Kessler scheme is a simple warm cloud scheme that includes water vapor, cloud water, and rain; (2) The Purdue Lin's scheme considers six classes of hydrometeors, including water vapor, cloud water, rain, cloud ice, snow, and graupel. All parameterization production terms are based on those developed by Lin et al. (1983) and Rutledge and Hobbs (1984) with modifications including saturation adjustment and ice sedimentation; (3) The NCEP simple ice scheme follows Hong et al. (1998) with a modification involving the ice sedimentation effect. Three categories of hydrometers are included: vapor, cloud water/ice, and rain/snow. The cloud ice and cloud water are considered to be in the same category but distinguished by temperature; (4) The NCEP mixed phase scheme is similar to the NCEP simple ice scheme. However, rain and snow (cloud ice and cloud water) are in different categories. It allows supercooled water to exist and the gradual melting of snow as it falls (Hong et al. 1998); (5) The Eta microphysics scheme explicitly predicts the cloud water/ice mixing ratio. Liquid and frozen precipitations are derived diagnostically from cloud mixing ratio and are assumed to fall to the ground in a single time step. (6) The Eta grid-scale cloud and precipitation scheme predicts changes in water vapor and total condensate that are advected in the model. The density of precipitation ice is estimated from information on the total growth of ice by vapor deposition and accretion of liquid water. Sedimentation is treated by partitioning the time-averaged flux of precipitation into a grid box. The mean size of precipitation ice is assumed to be a function of temperature following the observational results. Mixed- phase physics are considered at temperatures warmer than -10° C, whereas ice saturation is assumed for cloudy conditions at colder temperatures. (7) The Thompson et al. (2004) microphysical parameterization scheme includes six classes of moisture species, plus the ice number concentration as prognostic variables. Key improvements on the ice microphysics have been implemented in the primary ice nucleation and auto-conversion processes. Except for the Kessler scheme, all other schemes consider the ice-phase process.

For demonstration purposes, figure 19 shows results from an illustration simulation corresponding to a specific dust storm and ice cloud case that occurred in the East Asia region.

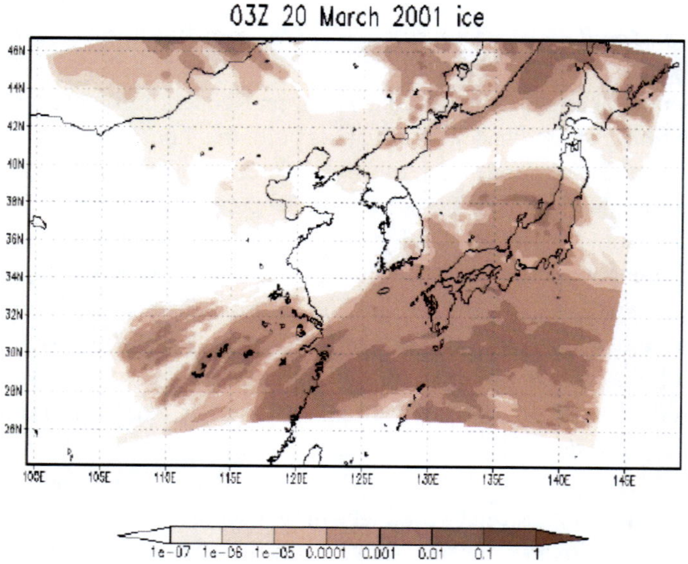

Figure 19. A simulation of ice water content field using WRF-ARW over the region corresponding to figure 1(d) on March 19-20, 2001.

The model simulations were made with the current version (2.1.2) of WRF-ARW using the initial and boundary conditions from the GFS (Global Forecast System) "final analysis" product, which is a sequence of six hourly global gridded fields from the GFS data assimilation system. Seifert and Beheng's (2006) two- moment microphysics scheme was used that predicts mass content and number density for each of the five forms of condensed water: cloud droplets, ice crystals, snow, graupel and raindrops. The model simulations were started at 18Z on 19 March 2001, about 9 hours prior to the satellite overpass time.

F. Current Estimates of Climatic Impacts and Uncertainties

Radiative forcing by the line-shaped contrails has been estimated by Fahey et al. (1999). Rind et al. (2000) used a global circulation model coupled with a mixed-layer ocean model to show that a 1% increase in global cirrus cloud cover with an optical depth of 0.33 leads to a 0.43 K global warming. Ponater et al. (2005) found a smaller climate feedback from contrails than that from CO_2 increase. In their global contrail-climate model, the equilibrium responses of surface temperature due to changes in radiative forcing are 0.43 $K/(Wm^{-2})$ and 0.73 $K/(Wm^{-2})$ for contrail and CO_2 increases, respectively. For a scenario involving a global contrail-cover increase from 0.06% in 1992 to 0.15% in 2015, the mean radiative forcing increases from 3.5 mW/m^2 to 9.8 mW/m^2. The computed transient global mean surface temperature increases by about 0.0005 K. According to Schumann (2005), the climatic impact of contrail cirrus could not be accurately estimated, because factors other than changes in the radiative forcing due to the presence of contrail and cirrus may also impact climate.

A much stronger climatic impact has been presented by Minnis et al. (2004) who analyzed the cirrus cloud cover trend over the continental U.S. between 1971 and 1995. The average 1%/yr increase in cirrus cloud cover is attributed exclusively to air traffic increase during this period. Assuming an optical depth of 0.25, this increase in high cloud cover results in a global mean radiative forcing of 25 mW/m^2 and a tropospheric temperature response of 0.2 to 0.3 K/decade in the region of the forcing, which would provide a practical explanation for the observed warming over this area between 1975 and 1994. However, Shine et al. (2005) and Ponater et al. (2005) pointed out several simplified assumptions employed in this study and reported temperature changes two orders of magnitude smaller.

In respomse to Shine et al. (2005), Minnis (2005) noted that the regional, non-equilibrium responses in climate models are highly uncertain and the estimated tropospheric response is comparable to the instantaneous changes expected in the presence of contrails. Since Minnis et al. (2004) used the contrail temperature response efficacy given by Rind et al. (2000) for the GISS climate model used by Hansen et al. (2005), the GISS model must have changed in the interval between the two model studies. In addition to questionable regional responses, many uncertainties exist in the treatment of high clouds in global models. For example, the GISS model substantially underestimates high cloud coverage in the midlatitude (Zhang et al., 2005) and significantly overestimates ice water path (Waliser et al. 2007). These types of uncertainties suggest that none of the current studies of contrail climate impact are conclusive and more definitive observations and models are needed. Specific information summarizing the needs and how research could be coordinated, funded and integrated should be important action items for future discussion.

Travis et al. (2001) claimed observable increases in the daily temperature range (DTR) due to reduced contrails in the three-day period of 11–14 September 2001, when air traffic over parts of the U.S. was reduced. They reported that DTR was 1 K above the 30-year average for the grounding period, which was interpreted as evidence that jet aircraft do have an impact on the radiation budget over the USA. However, the statistical significance of the data may not be strong enough to conclude that the above-than-average DTR is solely caused by missing contrails, since for unknown reasons, the DTR in 1982 was also nearly 1 K above the average (Travis et al. 2002). Hence, Travis et al. (2004) further analyzed the spatial variations of the DTR and of minimum and maximum temperatures, and estimated the contrail cover that would have occurred under normal traffic conditions. The potential contrail cover appears to be related to the observed variation in DTR. However, a quantitative model which relates the DTR to change in contrail cloud cover is not provided, and other reasons may be responsible for observed DTR variations. Kalkstein and Balling (2004) analyzed the air-mass and weather conditions in relation to observed temperature range over the USA for a short period after 9 September 2001. Theyoo, found a higher-than-average DTR.

Long-term responses to aviation-induced contrails and contrail cirrus have been estimated by inserting small percentages of cirrus into a general circulation model (GCM) at various time steps along the air traffic fly routes, and then letting the model run until equilibrium (Rind et al. 2000). The GCM results account for many of the feedbacks and the redistribution of the radiative energy in the system. Gu et al. (2004) used the UCLA atmospheric GCM to study the effects of cirrus cloud inhomogeneity on the atmospheric thermal structure with an interactive cirrus cloud parameterization. It is not clear at this point whether the regional climatic effects of contrails and contrail cirrus can be captured by GCMs, an area, which requires intensive literature survey and research.

The radiative and climatic effects, though small on a global scale, could be significant on a regional scale. A regional study by Strauss et al. (1997) shows that a 1% increase in local cirrus cloud cover (optical thickness 0.28) leads to a local surface temperature increase of about 0.1 K. Wang et al. (2001) studied effects of contrails on the radiative forcing and climate impact around Taiwan using the State University of New York at Albany regional climate model. The effects are calculated based on the contrail coverage, radiative properties as functions of particle effective radius, and solar and infrared optical depths as simulated from the National Taiwan University contrail model (Chen et al. 2001). Both short-lived and persistent contrails are considered. For persistent contrails with diurnal variation, the daily mean solar and infrared radiative forcings at the top of atmosphere are calculated to be 5.8 and 2.1 W m^{-2}, respectively, while the radiative forcings at surface are 4.9 and 0.19 W m^{-2}. Radiative forcings for short-lived contrails are smaller than those for persistent contrails.

Overall, it is difficult to determine the net warming effect on climate caused by the presence of contrails and cirrus clouds. The extent of feedbacks between ice cloud microphysics and radiation is also not well understood, leading to numerous uncertainties and gaps, some of which are described below.

G. Interconnectivity with Other SSWP Theme Areas

Formation, Evolution, and Persistence for Contrails and Contrail Cirrus

Duda et al. (2007) evaluated the use of high-resolution meteorological data from two operational numerical weather analyses (NWA) to diagnose and predict contrail formation using a variety of contrail observation database. Monthly contrail coverage statistics derived from surface and satellite observations were compared to the NWA-derived humidity, vertical velocity, wind shear, and atmospheric stability. The relationship between contrail occurrence and the NWA-derived statistics was analyzed to determine the atmospheric conditions under which persistent contrail formation is favored. Humidity is the most important factor that determines whether contrails are short-lived or persistent. Persistent contrails are more likely to appear when vertical velocities are positive and to spread when the atmosphere is less stable. Although artificial upper limits on the upper tropospheric humidity within NWA prevent a quantitative agreement of the model data with the contrail formation theory, logistic regression or similar statistical methods may improve the prediction of contrail occurrence.

The study by Duda et al. (2007) and other recent investigations showed that for common periods, surface and satellite data agree in the general direction of the trends but not in magnitude. Ensuring that the trends are due to air traffic requires knowledge of the concomitant trends in upper tropospheric humidity (UTH), a parameter that has not been measured either adequately or consistently for any length of time. Recent efforts to separate natural humidity effects and anthropogenic impacts have had limited success. Furthermore, confidence in the results remains tepid because of uncertainties in humidity record and differences between surface and satellite observations. The humidity issue has been skirted, to a certain degree, by focusing on the relationship between cirrus trends or amounts and upper tropospheric air traffic. Such studies generally agree that cirrus coverage is greater in areas where air traffic occurs, but they do not answer the question regarding the suppression of cirrus in other areas. Definitive answers to these questions would depend on understanding humidity variability in both areas with special emphasis on ice supersaturated regions.

Contrails and Contrail Cirrus Specific Microphysics

After initial homogeneous and/or heterogeneous nucleation involving suitable aerosol particles and atmospheric conditions, ice crystal growth is governed by diffusion processes and the subsequent actions by means of collision and coalescence. These physical processes are complicated by the nature of the ice crystal's hexagonal and irregular shape. For cirrus clouds, nucleation of ice particles can occur via the heterogeneous process involving insoluble IN. At temperatures below about –3 8°C, nucleation occurs by the homogeneous freezing of liquid solution droplets. For cirrus forming *in situ*, homogeneous freezing occurs in increasingly concentrated solution droplets as temperature decreases below –36°C. Heterogeneous ice nucleation in particles that are partially or fully soluble can potentially cause cirrus formation at warmer temperatures and, for temperatures below –3 8°C, lower RH. Heterogeneous ice- nucleation mechanisms (modes) most relevant to UTLS conditions include immersion freezing (ice nucleation induced by an IN previously immersed in a liquid aerosol droplet), contact freezing "inside- out" (freezing initiated when a solid IN immersed in a liquid aerosol particle collides with the drop surface from the inside), and deposition nucleation (direct nucleation of ice from vapor at a solid particle surface), which may occur in rare cases when IN reach the upper troposphere in a dry state. Recent research demonstrates the relevance of organics in heterogeneous ice nucleation (Zuberi et al., 2001; Zobrist et al., 2006; Shilling et al., 2006). Not all of these modes have been shown to operate efficiently at cold temperatures. The concentration and nucleation relative RH of IN mainly determines their impact on cirrus cloud properties (DeMott et al., 1997; Kärcher and Lohmann, 2003).

Based on *in situ* observations of the microphysical properties of upper-tropospheric contrails and cirrus clouds by FSSP and replicator during more than 15 airborne missions over central Europe, Shröder et al.(2000) investigated the development of contrails into cirrus clouds on the timescale of 1 hour in terms of a representative set of number densities, size distributions and surface area distributions of aerosols and cloud elements, with special emphasis on small ice crystals (diameter20 m). They found that ontrails are dominated by high concentrations (100 cm^{-3}) of nearly spherical ice crystals with mean diameters in the range 1–10 m. Young cirrus clouds, which mostly contain small regularly shaped ice crystals in the range of 10–20 m diameter and typical concentrations of 2–5 cm^{-3}, have been observed. Measurement results are compared to simple parcel model calculations to identify parameters relevant to the contrail–cirrus transition. Observations and model estimates suggest that contrail growth is only weakly, if at all, affected by preexisting cirrus clouds.

The basic uncertainty associated with ice nucleation processes is that they occur within short time scales (often only within seconds) and are rather localized (in sufficiently supersaturated patches of air). For this reason, it is extremely difficult to determine their relative importance in *in situ* measurements, or to even determine the basic nucleation mode. It is possible to isolate different ice nucleation pathways in the laboratory, but the question arises whether the employed IN particles are representative of atmospheric particles, an issue particularly important for aircraft because real engine soot and its processing cannot easily be represented in laboratory measurements.

Past studies of IN compositions have identified clay particles and mineral dust as important atmospheric IN. Lidar studies have documented the strong cloud-glaciating effect of dust particles from both Asian and Saharan sources. Laboratory studies using surrogates for airborne crystal and mineral dust particles predict the strong ice-nucleation efficiency of such particles throughout cirrus cloud forming temperatures.

Aircraft soot particles must compete with efficient IN in dust layers, but dust aerosols are highly variable in time and space and it remains uncertain how many dust particles are actually present in aircraft flight corridors. In the absence of dust, measurements in aging aircraft plumes face the difficulty of distinguishing between soot particles from aircraft exhaust and those from other sources (biomass burning, forest fires). Recent findings that certain organics might cause precipitation of ice-nucleating crystalline solids in liquid particles render efforts to disentangle the roles of various particle types in ice formation even more complicated.

3. Outstanding Limitations, Gaps and Issues that Need Improvement

A. Science

Long-Term Trends in Contrail Cirrus and Cirrus

Four primary sources of data can be used to estimate the long-term trends in contrail-cirrus and cirrus clouds: surface observations of cloud cover, meteorological sounding of humidity profiles, ground-based active remote sensors, and satellite observations. Each source has its limit. Surface observations suffer from inaccuracy problems, humidity soundings have large uncertainty for high-altitude measurements, ground-based remote sensing is restricted in geographical coverage, and polar- orbiting and geostationary satellite remote sensing is limited by temporal and spatial resolution, respectively.

Aerosol-Cirrus and Contrail-Cirrus Indirect Effects

The aerosol and contrail indirect effects on the microphysical and radiative properties of cirrus clouds are critical, but these effects are complex and difficult to quantify by either *in situ* observations or modeling approach. Extensive and systematic *in situ* measurements of contrail and contrail cirrus have been extremely limited because of the requirement of high flying aircraft and the development of accurate and durable sampling instruments. Modeling approaches, on the other hand, are limited by an insufficient understanding of the physical and chemical processes that control the ice formation in the presence of aerosols. Satellite remote sensing is a viable alternative to capture the indirect effects, but research in this area is in its embryonic stage.

Microphysical and Radiative Properties on Contrails and Cirrus

Because of the lack of *in situ* measurements, contrails' microphysical and radiative properties are largely unknown. Large-scale field campaigns like SUCCESS did provide microphysical observations of contrails, but it is not known how representative these observations are. It is quite clear that a selection of a number of representative *in situ* observations similar to SUCCESS must be conducted in order to obtain necessary and sufficient understanding and knowledge on contrail properties for climate impact assessment.

Radiative Forcing of Contrails

The global average value of the radiative forcing has been increased from 0.02 Wm^{-2} in 1992 to 0.03 Wm^{-2} in 2000. However, uncertainty associated with these average values is two to three times greater. For regional radiative forcing, particularly near air traffic corridors, the range of radiative forcing is between -0.15 and 0.7 Wm^{-2} based on contrail coverage of 0.5%. Model simulation of the contrail radiative forcing using 100% contrail coverage produces a range of -30 to 70 Wm^{-2}. Further studies are clearly needed to narrow down the uncertainty range.

Climatic Impacts of Contrails and Contrail Cirrus

It is well recognized based on physical principle that contrails and contrail cirrus could cause the global temperature to increase, as discussed above. However, many disagreed on the magnitude of this warming. The global average value of temperature increase due to an increase in contrail coverage is between near 0 K/decade (Ponater et al. 2005) to 0.2 – 0.3 K/decade (Minnis et al. 2004). For regional temperature, Struass et al. (1997) showed that a 1% increase in local cirrus cover can lead to about 0.1 K local temperature rise.

B. Measurements and Analyses

Satellite Remote Sensing

Satellite remote sensing of contrails is limited by both spatial resolution and coverage as well as observation frequency. The polar-orbiting satellites visit the same local spot once or twice a day. For the scanning radiometer on board a satellite, the viewing zenith angle may be too far from the nadir, rendering observations useless. Moreover, because polar-orbiting satellites are usually 700 – 900 km above the Earth, the sensor coverage for one sweep is only about 2000 – 3000 km. The geostationary platform is equipped with spin-scan sensors and can scan the same spot every 30 minutes. However, the sensor can only cover about 1/3 of the Earth's surface. Further, because the geostationary satellite is located about 40,000 km away from the Earth, the sensor pixel resolution is on the order of 4 km, too large to detect the narrow contrails.

For the reasons stated above, it has been extremely difficult to detect freshly formed and young contrails that are narrow by low-spatial-resolution space-borne sensors, including NOAA/HIRS, Aqua/AIRS, and GOES/Imager/IR-bands. Before 2000, NOAA/AVHRR and DMSP/OLS were the only space-borne meteorological satellite instruments that could detect contrails with linear shape. As mentioned above, the contrail detection algorithms using AVHRR split-window bands suffer from a drawback that cirrus clouds with similar linear shape can be misidentified as contrails. With the launch of high-spatial-resolution sensors such as Aqua/MODIS and future NPOESS/VIIRS, the detection of contrails is expected to be improved in view of a better image resolving power. Still, an effective contrail detection algorithm remains to be developed for the accurate and effective determination of contrail coverage. In light of the preceding discussion, the remote sensing of microphysical and optical properties of contrails is still in its embryonic stage. It is quite clear that a coordinated *in situ* observation of the microphysical and single-scattering properties is needed to narrow down numerous uncertainties presented above.

Ground-Based Remote Sensing

The ground-based remote sensing by lidar and radar is limited to spatial density of the instrument deployed and the areas that the lidar pulse and radar emission cover. Thus, it would be difficult to use them to detect freshly formed and young contrails that are geometrically narrow. If a lidar can capture a contrail, then it can determine its vertical extent as it drifts over the observation site. However, radar would be unable to detect freshly formed or young contrails containing small particle sizes.

C. Modeling Capability

Modeling the Optical Properties for Contrails and Cirrus for Radiative Transfer Calculations

Because of limitations in the *in situ* observation of contrails and the lack of understanding of their microphysics properties, it has been difficult to construct representative models for their microphysical and optical properties for the purpose of radiative transfer calculations. Further concerted observation and model analysis research on the contrail properties must be carried out to improve our current understanding.

Radiative Transfer Model

In addition to uncertainties in the microphysical and optical models, the spatial inhomogeneity of the contrail composition is another source of error in radiative transfer calculations for remote sensing and radiative forcing applications. However, Gounou and Hogan (2007) found that the 3D inhomogeneity effect of contrails is not important.

Global and Regional Climate Modeling

Process studies of the evolution of aerosols and their role as INs will be required in order to produce effective emission indices for input to global models because the characteristics of emitted particles may have changed significantly by the time emissions have been placed onto global model grid scales. In this regard, global models and contrail/cirrus studies need to establish the essential parameters that can properly incorporate aviation aerosols and their effects into atmospheric calculations.

Most climate models have not been capable of predicting supersaturation in the UT/LS region. Saturation adjustment schemes are normally employed to remove excess water vapor above saturation obtained within one time step due to cooling or ice water transport. This type of approach is well justified for warm (liquid phase) clouds. However, ice cloud formation requires tens of percent of ice supersaturation. An accurate knowledge of ice supersaturation is crucial to quantify both the direct and indirect effects of aviation on cirrus cloudiness. High-resolution regional models are needed to accomplish this important task.

Many cloud schemes in GCMs compute cloud fraction based on an empirical function of grid-mean RH that may not be applicable to stratiform cirrus clouds. These clouds are known to be long-lived and can be transported over many grid boxes of a large-scale model during their lifetime. For cirrus and long-lived contrail-cirrus, a prognostic description of its cloud cover would appear to be more appropriate. In many cases, only ice water mass is predicted in global models. This hampers the introduction of physically-based links to ice crystal nucleation to distinguish between many different types of cirrus and to track the indirect effect of aircraft-produced aerosols. Also, in most global models cirrus is treated as one class of clouds in terms of their radiative properties. However, contrail-cirrus are composed of a large number of small (diameter ~1 0-30 ım) ice crystals as compared to particle sizes that have been observed in natural cirrus clouds.

It would be very difficult to provide reliable global real-time forecasts of contrail-cirrus to support control strategies or to project with some confidence the impact of aviation-induced cloudiness in future climate change scenarios without resolving some of the first order problems. While some progress has been made in the past 10 years to deal with ice supersaturation and to parameterize ice crystal nucleation in cirrus clouds, coordinated interdisciplinary research commitment must be made to gather a suitable set of dynamical, microphysical, and radiative components related to cirrus clouds in global models. This must be carried out in concert with laboratory, *in situ*, and remote sensing data analyses that can provide guidance in developing the parameterization schemes of subgrid-scale processes for use in GCMs.

Global Distribution and Properties of Supersaturation, Aerosols, and Thin Cirrus

Even if the degree and frequency of occurrence of supersaturation and the composition and size distributions of aerosol and cirrus cloud particles can be simulated in GCMs, it would be a difficult task to validate these parameters. We have previously addressed the problems of detecting and verifying ice supersaturation measurements from satellites. The global inventory is not available to yield quantitative information about the aerosol budget in the UTLS region particularly for soot-containing aerosols. Current global model aerosol validation exercises (e.g., the AEROCOM initiative) strongly focus on lower tropospheric aerosols. Cloud climatology such as those presented by ISCCP does not include high clouds with optical depths less than about 0.2, but such thin cirrus clouds are common in regions where aircraft activities occur. Stratospheric aerosol and subvisual cloud climatology, especially those from the Stratospheric Aerosol and Gas Experiment (SAGE) and the Halogen Occultation Experiment (HALOE), have limited observations in the upper troposphere region.

D. Interconnectivity with Other SSWP Theme Areas

Detection and Prediction of Ice Supersaturation

Recent studies have shown that for the same period, the surface and satellite data agree in the general direction of trends but not in magnitude. Ensuring that trends are due to air traffic requires knowledge of the concomitant trends in upper tropospheric humidity (UTH), a parameter that has not been measured either adequately or consistently for any length of time. The humidity issue has been skirted, to certain degree, by focusing on the relationships between cirrus trends or amounts and upper tropospheric air traffic. Such studies generally agree that cirrus coverage is greater in areas where air traffic occurs, but they do not answer the question regarding the suppression of cirrus in other areas. Answers to these questions appear dependent on understanding humidity variability in both areas, with special emphasis on ice supersaturated regions. While it is not possible to return to the past and reconstruct a more accurate UTH record, it is recommended that improved methods for measuring UTH and supersaturation be standardized and applied consistently on a global basis in the future. Development of innovative methods to unscramble the natural and anthropogenic effects should be continued. Furthermore, both surface and satellite observations should be sustained in order to detect cirrus changes as air traffic patterns evolve over the coming years.

Contrails and the expansion of contrails into cirrus clouds occur in a supersaturated environment. However, global distributions of supersaturation in the upper troposphere where aviation-produced cirrus is likely to occur are not well known. MLS (Microwave Limb Sounder), AIRS (Atmospheric Infrared Sounder) and TOVS (TIROS Operational Vertical Sounder) observations may provide improved data for supersaturation analysis in the upper troposphere. However, current retrievals that are adequate for water vapor and temperature measurements under sub-saturated conditions may not be sufficient for supersaturated cases.

Existing satellites do not have the horizontal or vertical resolution to accurately define the frequency and extent of supersaturated regions. Required in the future is a remote sounding instrument that measures both temperature and humidity with good vertical resolution and/or can detect RH directly.

Chemistry within Emission Plumes

Current global models have treated aircraft emissions as well mixed within the grid box, but ignoring the plume processing of emissions. The effects of nonlinear plume processes (both chemical and microphysical) have not been evaluated in depth in the context of global chemical transport models. None of the currently available emission inventories considers the effect of plume processing on species or particle mixing ratios (e.g., NOx to NOy repartitioning, volatile and soot aerosol number concentrations and size) that eventually enter global simulations. Two types of aerosols are known to exist in aircraft plumes: the first is associated with soot particle emission and has a number emission index of ~10^{14}-10^{15} /kg-fuel; the second is due to the formation of volatile particles induced by chemiions (e.g., Eichkorn et al. 2002) and has a number emission index of ~1016-1017/kg-fuel. The aviation-generated particles may perturb the abundance and properties of climate-relevant particles in the upper troposphere. To properly assess this perturbation and the associated climatic effect, further research is needed to understand the properties, transformation, and fate of aircraft-generated particles. Aviation aerosols are composed of water, sulfuric acid, organics, and soot. The composition of particles may affect their role to act as IN. In this regard, it is necessary to characterize the dependence of particle composition on engine operation conditions and fuel properties as well as the relative contribution of organics versus sulfur to the mass of particles of different sizes as a function of time in dispersing aircraft plumes. The aircraft-generated particles interact with background aerosols through coagulation and mixing, and will eventually become part of ambient aerosols. In addition, photochemistry will provide additional condensable material (e.g., sulfuric acid from emitted SO2).

Research is lacking on how the properties (number concentration, surface area, composition, and mixing state) of ambient aerosols are perturbed at the presence of jet engine emissions under various conditions. In this regard, a detailed investigation of the microphysical (condensation and coagulation) and chemical processes (oxidation of precursor gases) governing the evolution of aviation aerosols in the time scale of days to weeks after emission is required. Research is also needed to define the abundance and properties of ambient aerosols as well as gaseous aerosol precursor concentrations in the troposphere. Both theoretical modeling and in situ measurements are needed to advance our knowledge about the perturbation of climate-relevant particles in the upper troposphere by aviation emissions.

Contrail-Cirrus Development

The development of cirrus clouds from contrails and the resulting radiative effects are poorly characterized in current climate models and have been studied only on a limited basis from satellites and detailed cloud-scale models. The role of wake dynamics in determining immediate contrail ice particle concentrations has not been fully explored. For example, the interactions between the wakes of four versus two engines could, perhaps, dissipate young contrails through induced subsidence, even in nominally super-saturated conditions. Otherwise, once a contrail forms in supersaturated conditions, it will continue to grow and spread. However, knowledge of the impact of the type and numbers of primary and secondary emission particles on the number of ice crystals and hence particle growth potential and precipitation is inadequate. Such factors, along with wind shear, local vertical humidity profiles, and wake turbulence will determine how contrails grow vertically and horizontally and whether they dissipate in a few minutes or hours. The resulting vertical distribution of particles and their sizes determine the contrail-cirrus optical depth and effective particle size that govern their radiative effects.

A number of modeling studies have examined the transformation of young contrails to cirrus clouds and its sensitivity to the number of nucleating particles and wind shear, but the effects of realistic emission particle distributions, induced turbulence, radiation interactions, and the mesoscale environment have not yet been examined in a meaningful way. Such modeling studies are clearly needed but will remain theoretical exercises until relevant variables can be measured simultaneously. While early field campaigns have been conducted to accomplish that goal, the amount of useful data is insufficient to confidently model and understand the processes determining the microphysical and optical properties of contrails and contrail-cirrus in a wide range of atmospheric conditions.

Cloud-to-regional scale measurements and modeling are necessary steps in building a dependable set of tools to determine the contrail-cirrus impacts on climate. They form the basis for modeling the effects on global scale. However, knowledge of the global distribution of contrail-cirrus optical properties and coverage still remains uncertain. To date, satellite-based estimates of contrail particle sizes, optical depths, altitudes, and coverage are confined to only a few regions, seasons, and years. The most studied area is Western Europe followed by the United States. A more comprehensive climatology of aircraft- induced cirrus properties and radiative effects is needed, at least, for those areas where air traffic is significant or will become increasingly significant in the near future, e.g., eastern and southern Asia. The climatology should include several years that differ in upper tropospheric humidity in order to determine variability over the actual range of conditions that occur over time scales greater than a decade. A database of this type will serve as the basis for understanding the direct impact of contrails and contrail- cirrus and for guiding and validating global climate models that include this new class of ice clouds.

In Situ Measurements of Aerosol Composition and Small Ice Crystals

It is difficult to make *in situ* measurements of both the aerosol composition and small ice crystals. The low mass loading of particles provides a challenge to instrumentation even if they can adequately measure particle composition at lower altitudes. Mass spectrometric data indicates that at least in some regions, a majority of the particle mass in the upper troposphere is carbonaceous (e.g., include organic and elemental carbon). These data do not extend to the most numerous small particles below ~100 nm in diameter, nor is there information on the type of organic molecules. These organics could significantly change the freezing behavior of particles, affecting the evolution of contrail-cirrus and cirrus.

Significant problems exist with the measurement of small ice particles. Cirrus ice crystals can range from a few to hundreds of ìm or more in diameter. Measuring this range requires several instruments and the agreement between them in the same size range has not always been good. Although the basic size modes in cirrus are not in as much question as for mixed phase clouds, it appears questionable whether any of the existing probes can obtain accurate ice crystal sizes and shapes that are both important for evaluating cloud radiative properties. Existing instrumentation appears difficult to measure the shape of very small (diameter <20 ìm) crystals that have often been found in contrails and contrail cirrus clouds. Ice can form much more easily on some particles in the atmosphere. The measurement of these ice nuclei in both contrail-cirrus and the background atmosphere is crucial to understanding how particles emitted from aircraft compete with background particles in the formation of new or modification existing cirrus. There are many fewer IN than other aerosol particles that requires improved instrumentation to measure their number and properties. The radiative properties of cirrus clouds and contrail-cirrus also depend on the vertical distribution of their microphysical properties. An examination of those radiative properties using remote sensing instruments (e.g., lidar, radar, radiometers, and interferometers) is often obscured by the presence of lower level clouds.

Properties of Heterogeneous Ice Nuclei from Natural and Anthropogenic Sources

The formation of cirrus clouds is characterized by a competition between freezing particles for the available water vapor. Because of this competition, the ice-nucleating behavior of particles from aviation depends on the ice nucleation properties of particles from other anthropogenic and natural sources. The chemical composition of IN in the free troposphere is important to understand both the details of their freezing behavior and sources. A special case is elemental or soot-like carbon. The available data indicate that their ice nucleating behavior depends on source and processing in the atmosphere such as the addition of sulfate or organics. For example, elemental carbon from biomass burning probably has different ice-nucleating properties than aviation soot. None of the laboratory studies of ice nucleation has used authentic aviation soot. Sulfates and organics have been shown to affect ice nucleation ability, but the role of organics that condense in the plume behind a jet engine has not been studied in cruise conditions.

Measurements carried out in wave clouds have shown that at temperatures close to –3 8°C, certain aerosol particles can nucleate ice at lower supersaturations (Field et al. 2001). This preconditioning effect is neither theoretically understood nor well explored experimentally. Short-lived contrails that form in sub-saturated air could lead to pre-conditioning of exhaust soot particles, as it is known that contrail ice crystals mainly form on emitted soot particles. After sublimation, these modified soot particles could facilitate ice formation in the atmosphere, increasing the relative importance of indirect effect. Conversely, if the soot is not so transformed by this conditioning, it may never be as effective as ambient IN, even though these soot particles previously served as nuclei for contrail particles. Whether or not soot particles are effective IN and to what extent contrail processing changes their properties is an important question.

4. Prioritization of Research Needs for Tackling Outstanding Issues

A. Airborne *In Situ* and Remote Observations and Ground-Based Remote Sensing of Contrail Cirrus and Aircraft Emission Plumes

Impacts

Airborne *in situ* observations and ground-based remote sensing of contrail, contrail-cirrus, and aircraft emission plumes are needed. As mentioned in Section 3, the aerosol and contrail indirect effects on the microphysical and radiative properties of cirrus clouds are critical in the analysis of climatic impacts of contrails, but these effects are difficult to quantify by *in situ* measurements. Nevertheless, with the availability of high-flying aircraft and accurate and durable sampling instruments, ice crystal and aerosol properties can be obtained to gain further understanding of ice cloud and aerosol interaction.

Measurements of the basic ice crystal properties can also help to build up the microphysical and radiative properties models for contrails. In fact, a comprehensive archive of ice crystal microphysical and spectral radiative properties models for cirrus clouds has been constructed based on *in situ* microphysics measurements collected during several field campaigns, including FIRE-I, FIRE-II, ARMCloud-IOP-2000, TRMM, CRYSTAL-FACE (Baum et al. 2005a,b; 2007; Young et al. 2000; 2005). Prototype models have been developed by Liou et al. (1998), as shown in figure 13. Similar procedures can be followed for contrails and contrail-cirrus.

Further measurements of ice crystal microphysics properties can also assist in the development of parameterization of ice crystal single-scattering properties in terms of ice crystal size parameters and ice water content for incorporation into a radiative transfer model for radiative forcing calculations. Preliminary parameterizations between ice crystal mean effective size and ice water content have been developed by Liou et al. (2007) based on limited *in situ* measurements of ice crystal size distributions as shown in figure 9.

Uncertainty Reduction in Climate Impact Estimate

A direct measurement of the optical properties for ice crystal clouds can help to reduce uncertainty in current models and parameterizations. Advanced instrument design (e.g., the nephelometer developed by Barkey et al. 2002) can be used to validate the light scattering properties computed from theoretical models. In addition, the airborne and remote broadband and narrow-band radiometric measurements combined with collocated and coincident ice crystal in situ observations can be used to validate atmospheric and surface contrail radiative forcings computed by radiative transfer models. Such validation efforts have been made by Ou et al (1995).

Practical Application and Achievability

Following the suggestion of JPDO and PARTNER (2006), a series of coordinated regional-scale campaigns need to be carefully designed and executed to measure appropriate variables governing the formation and dissipation of contrail and contrail-cirrus. Because the properties and life times of contrails and contrail-cirrus clouds are highly variable, a large number of flights must be executed for measurement in order to gather statistically meaningful samples. These regional-scale experiments would preferably be conducted over long-term ground-based measurement sites and aircraft sampling flights should be coordinated with the overpasses of satellite platforms listed in table 1.

Estimated Cost and Timeline

Estimated cost and timeline for coordinated research efforts depend on the number of participants, instruments, and platforms involved. Based on past experiences, a regional-scale field campaign involving a large number of participants (more than 100), instruments, and multiple aircraft platforms (e.g., SUCCESS) have been very costly. Thus, it is necessary to achieve an optimal balance between scientific objective and cost. The timeline from the start of scientific and logistic planning, instrument development, and to the field deployment and operation could stretch over several years.

Prioritization

A coordinated research effort involving *in situ* and ground-based observations of contrails and contrail cirrus is needed to reduce uncertainty in the assessment of climatic impacts of contrails and contrail-cirrus produced by aviation effects and is theoretically ranked "high priority". However, considering the possible cost and time involved and the intricate scientific and logistic planning, we must add a "low priority" tag to such an effort.

B. Global and Regional Model Studies Addressing Contrail Direct and Indirect Effects

Impacts

In recent years, development in cloud modeling has included prognostic equations for the prediction of IWC for high-level clouds formed in GCMs and climate models. This is a milestone accomplishment from the standpoint of incorporating a physically-based cloud microphysics scheme in these models, and at the same time, it is also essential from the prospective of studying cloud-radiation interactions. For the study of contrails and contrail cirrus, it is also necessary to develop prognostic equations for computing IWC in contrails.

Cloud particle size is also an independent parameter that affects radiation transfer. For example, for a given IWC in clouds, smaller particles would reflect more sunlight than larger counterparts, an effect that has been recognized by Twomey et al. (1984) and Liou and Ou (1989) in conjunction with aerosol- cloud indirect effects. Incorporating a fully interactive ice microphysics based on the first principle in a GCM appears to be a challenging but an extremely difficult computational task. Innovative ice crystal size parameterization based on theory and observation must be developed for GCM applications.

Global and regional model studies addressing contrail direct and indirect effects have been limited by insufficient understanding of the physical and chemical processes that control the ice formation in the presence of aerosols. As pointed out by Kärcher et al. (2007), the inclusion of indirect aerosol effects in global models is at its infancy. At present, simplified parametric models of indirect effects of soot and dust aerosols using simplified assumptions of ice nucleation thresholds appear feasible for global model studies. A better grasp of the basic mechanism for ice crystal formation is required to improve the parameterization of heterogeneous ice nucleation rates. Data collected from coordinated atmospheric *in situ* measurements of the ice crystal and aerosol properties would help in the development of physical parameterization so that the contrail direct and indirect effects in global models can be physically simulated.

Practical Use and Achievability

Following the suggestion of JPDO and PARTNER (2006), the direct effects of persistent contrails on cirrus cloud cover and radiative forcing can be studied even in the framework of conventional climate models, if appropriate parameterizations can be developed to separate contrail cirrus as a distinct class of ice clouds in a manner that is consistent with the built-in physics in GCMs. Even without explicit calculation of ice supersaturation, such studies will be more realistic than existing GCM estimates and lead to improved prediction of the hitherto poorly quantified global contrail climate impact.

Estimated Cost and Timeline

The estimated cost for modeling efforts is much smaller than the efforts involving in situ measurements. The former timeline would also be shorter, perhaps on the order of one to two years.

Prioritization

Because of the important potential impact and the potential benefit of uncertainty reduction in the estimate of climatic impact of contrails and contrail cirrus and because of its practical applications in the study of aviation effects on climate change, the need for global and regional model studies addressing contrail direct and indirect effects is ranked "high priority". Such studies are also less expensive and require less time than coordinated efforts for in situ measurements.

C. Synergistic Satellite Remote Sensing of Contrails and Contrail Cirrus

Impacts

A combination use of satellite observations listed in table 1 will improve the dependability of the estimate of long-term trends in contrails and contrail cirrus. As mentioned in Section 3, there are limits in the capability of the four primary sources of data: surface observations, meteorological soundings, ground-based remote sensing, and satellite observation. Only satellite observations cover the whole Earth. However, polar-orbiting and geostationary satellite measurements are limited by temporal and spatial resolutions, respectively. Therefore, it is recommended that a concerted effort be carried out to gather data from currently operating polar-orbiting and five geostationary satellites and integrated these datasets to establish the global long-term estimates of contrails and contrail cirrus with high resolutions in both space and time.

In addition, integrated satellite observations making use of an extensive suite of space-based instruments currently in operation, assisted by collocated and coincident *in situ* measurements, can also be used in the study of aerosol-cirrus indirect effect. These observations contain rich and valuable information that can be used to investigate the relationship between aerosol and ice cloud formation. Meaningful statistical and physically based methodologies need to be developed to systematically quantify the indirect effect.

Integrated satellite observations can also be used to assist in the investigation of radiative forcing due to contrails and contrail cirrus in two aspects. First, the radiative flux measurements from research- grade broadband radiometric sensor, such as CERES/Aqua, may be used to directly evaluate contrail radiative forcing, provide that contrail areas can be reliably detected by using collocated imager data. Second, contrail radiative forcing can also be computed using a radiative transfer model with the ingestion of satellite-retrieved ice crystal microphysical properties. Both approaches are expected to improve our estimates of contrail radiative forcing.

Uncertainty Reduction in Climate Impact Estimate

Integrated satellite observations can be combined with collocated surface observations, meteorological soundings and ground-based remote sensing measurements to further improve the accuracy of the detection of contrails and contrail cirrus. This improvement will add reliability to the satellite-estimated long-term trends in contrails and contrail-cirrus.

Integrated satellite observations can also be used to determine improved parameterizations of aerosol indirect effects and combined with model-derived parameterizations for incorporation into GCM studies.

Practical Application and Achievability

Following the suggestion of JPDO and PARTNER (2006), synergistic satellite observation would be a high priority to determine the optical and microphysical properties of contrails and contrail-cirrus, including the cloud-top temperature/height, optical depth and effective particle size, three basic cloud parameters that are essential to the study of cloud radiative forcing. The space-borne polar-orbiting and geostationary radiometric measurements facilitate a unique approach to investigate the characteristics of the global distribution and temporal evolution of contrails, respectively. In particular, the passive sensors (e.g. MODIS) and active sensors (CALIPSO/CALIOP and CloudSat/CPR) in the A-Train constellation offer unprecedented opportunity to explore aviation-induced and modified high-level contrails and contrail-cirrus. The research objective would be to systematically analyze the characteristics of the global and temporal distributions of contrails and contrails-cirrus. To achieve this goal, robust and efficient new analysis programs must be developed to detect contrails and contrail cirrus and to retrieve their microphysical and optical properties.

Estimated Cost and Timeline

Along the line of modeling study efforts, the estimated cost on the basis of *per annum* for analyzing synergistic satellite observations is much smaller than the coordinated efforts of *in situ* field measurements. The timeline for long-term estimates can stretch into decades. Other analysis efforts, including the detection of indirect effects and uncertainty reduction require only 3 to 5 years.

Prioritization

For uncertainty reduction in the assessment of climatic impact of contrails and contrail-cirrus and for understanding of the aviation effects on climate change, fusion of the current and future satellite data containing contrails and contrail-cirrus is ranked "high priority". Such an approach is also less expensive and requires less time than coordinated efforts for in situ field measurements. Note that success of this effort relies upon innovative developments involving reliable and effective contrail detection programs.

5. Recommendations for Best Use of Current Tools for Modeling and Data Analysis

Long-Term Trends in the Coverage and Frequency of Contrail-Cirrus and Cirrus Occurrence

In conjunction with the modeling study, it is necessary to have global contrail and cirrus data sets for validation and analysis. Satellite derived cirrus cloud parameters that have the best quality in terms of spatial resolution and reliability have been the MODIS cloud mask and products and along with AVHRR and GOES imager data, they can be used to study long-term trends in the coverage and frequency of contrail-cirrus and cirrus occurrence. The MODIS products are available from 2000 for Terra and from 2002 for Aqua. Another potential dataset for estimating contrail long-term trends is the CALIPSO/CALIOP cloud mask products, which are recently made available for the period since its launch in April, 2006. However, it is difficult for CALIPSO/CALIOP to intercept contrails given the fact that the lidar aboard has no cross track scanning capability. To match the CALIPSO/CALIOP track with MODIS pixels requires labor-intensive search. Once the collocation is done, CALIPSO/CALIOP observations can be used to validate MODIS contrail detection results.

Aerosol-Cirrus and Contrail-Cirrus Indirect Effects

In view of various problems encountered in the quantification of aerosol indirect effects solely through direct *in situ* observations, an alternative approach to study the aerosol-cirrus and contrail-cirrus indirect effects would be through the satellite observation of ice clouds and aerosols. In this conjunction, MODIS cloud retrieval products can be used to study long-term trends in the coverage and frequency of contrail-cirrus and cirrus occurrence. Parameterizations of aerosol-cloud processes can be derived by using the correlations between MODIS retrieved cloud and aerosol microphysical and radiative properties.

Climatic Impacts of Contrails and Contrail Cirrus

The global climatic effect of contrails and contrail cirrus has been estimated to be relatively small. However, their regional climatic impact could be substantial and significant. A regional climate model including a physically-based radiative transfer model interactive with ice microphysics must be developed to understand radiative forcing issues and provide assessment on this important area. It appears that the best regional model that has been developed so far is the WRF model. We suggest that this model coupled with a spectral radiative transfer and ice microphysics parameterizations be used for simulating the formation, evolution, and dissipation of contrails and contrail cirrus using input of the flight track and jet fuel consumption information, and that the simulation results be compared with the independent remote sensing results determined from MODIS and other satellite cloud products.

REFERENCES

Ackerman, S., K. Strabala, P. Menzel, R. Frey, C. Moeller, L. Gumley, B. Baum, S. W. Seeman, and H. Zhang, 2002: Discriminating clear-sky from cloud with MODIS. *MODIS Algorithm Theoretical Basis Document (MOD35)*, 115pp.

Ackerman, T. P., K. N. Liou, F. P. J. Valero, and L. Pfister, 1988: Heating rates in tropical anvils. *J. Atmos. Sci.*, 45, 1606 – 1623.

Appleman, H., 1953: The formation of exhaust contrails by jet aircraft. *Bull. Amer. Meteor. Soc.*, 34, 14- 20.

Arakawa, A. and W. H. Schubert, 1974: Interaction of a cumulus cloud ensemble with the large-scale environment, Part I. *J. Atmos. Sci.*, 31, 674–701.

Archuleta, C. M., P. J. DeMott, and S. M. Kreidenweis, 2005: Ice nucleation by surrogates for atmospheric mineral dust and mineral dust/sulfate particles at cirrus temperatures. *Atmos. Chem. Phys.*, 5, 2617-2634.

Arnott, W. P., Y. Y., Dong, and J. Hallett, 1994: Role of small ice crystals in radiative properties of cirrus: A case study, FIRE II, November 22, 1991. *J. Geophys. Res.*, 99, 1371- 1381.

Atlas, D., Z. Wang, and D. P. Duda, 2006: Contrails to cirrus – morphology, microphysics and radiative properties. *J. Appl. Meteor. Clim.*, 45, 5-19.

Bakan, S., M. Betancor, V. Gayler and H. Grassl, 1994: Contrail frequency over Europe from NOAAsatellite images. *Anna. Geophys. Atmos. Hydro. and Space Sci.*, 12, 962-968.

Barkey, B., M. Bailey, K. N. Liou, and J. Hallett, 2002: Light-scattering properties of plate and column ice crystals generated in a laboratory cold chamber. *Appl. Opt.*, 41, 5792-5796.

Baum, B. A., A. J. Heymsfield, P. Yang, S. T. Bedka 2005a: Bulk scattering properties for the remote

sensing of ice clouds. Part I: Microphysical data and models. *J. Appl. Meteor.,* 44, 1885–1 895. Baum, B. A., P. Yang, A. J. Heymsfield, S. Platnick, M. D. King, Y.-X. Hu, and S. T. Bedka, 2005b:

Bulk scattering properties for the remote sensing of ice clouds, Part II: Narrowband Models. *J. Appl. Meteor.,* 44, 1896–1911.

Baum, B. A., P. Yang, S. L. Nasiri, A. K. Heidinger, A. J. Heymsfield, and J. Li, 2007: Bulk scattering properties for the remote sensing of ice clouds. Part III: High resolution spectral models from 100 to 3250 cm^{-1}. *J. Appl. Meteor. Clim.,* 46, 423–1350 434.

Baumgardner, D., J. E. Dye, B. Gandrud, R. G. Knollenberg, 1995: The multi-angle aerosol spectrometer probe: a new instrument for airborne particle research. AMS 9th Symposium on Meteorological Observations and Instrumentation, March 27-3 1, 1995, Charlotte, N. C., 1995.

Bengtsson, L., E. Roeckner, and M. Stendel, 1999: Why is the global warming proceeding much slower than expected? *J. Geophys. Res.,* 104, 3865–3876, 1999.

Betancor-Gothe, M., and H. Grassl, 1993: Satellite remote sensing of the optical depth and mean crystal size of thin cirrus and contrails. *Theo. Appl. Clim.*, 48, 101-113.

Boucher, O., 1999: Influence of air traffic on cirrus occurrence. *Nature*, 397, 30-3 1.

Boucher, O. and U. Lohmann, 1995: The sulfate-CCN-cloud albedo effect: A sensitivity study with two general circulation models. *Tellus*, 47B, 28 1-300.

Bows, A., P. Upham, and K. Anderson, 2005: Growth Senarios for EU and UK Aviation: Contradictions with climate policy. *Report for Friends of the Earth Trust Ltd.* Tyndall Center for Climate Change (North), University of Manchester. 91pp.

Brasseur, G., R. Cox, D. Hauglustaine, I. Isaksen, J. Lelieveld, D. Lister, R. Sausen, U. Schumann, A. Wahner, P. Wiesen, 1998: European scientific assessment of the atmospheric effect of aircraft emissions. *Atmospheric Environment*, 32, 2329–2418.

Brooks, D. R., and F. M. Mims III, 2001: Development of an inexpensive handheld LED-based Sun photometer for the GLOBE program, *J. Geophys. Res.*, 106, 4733-4740.

Cantrell, W., and A.J. Heymsfield, 2005: Production of ice in tropospheric clouds: A review. *Bull. Amer. Meteor. Soc.*, 86, 795-807.

Carleton, A. M., and P. J. Lamb, 1986: Jet contrails and cirrus clouds: a feasibility study employing high-resolution satellite imagery. *Bull. Amer. Meteor. Soc.*, 67, 301-309.

Changnon, S. A., R. G. Semonin, and W. M. Wendland, 1980: *Effects of contrail cirrus on surface weather conditions in the Midwest – Phase 1.* Final Report to National Science Foundation, Illinois Institute of Natural Resources, and University of Illinois, 141pp.

Changnon, S. A., Jr., 1981: Midwestern cloud, sunshine and temperature trends since 1901: Possible evidence of jet contrail effects. *J. Appl. Meteor.*, 20, 496-508.

Chen, C.-T., and E. Roeckner, 1996: Validation of the Earth radiation budget as simulated by the Max Planck Institute for Meteorology general circulation model ECHAM4 using satellite observations of the Earth radiation budget experiment. *J. Geophys. Res.*, 101, 4269–4287.

Chen, C.-T., and E. Roeckner, 1997: Cloud simulations with the Max Planck Institute for Meteorology general circulation model ECHAM4 and comparison with observations. *J. Geophys. Res.,* 102, 9335– 9350.

Chen, J.-P., W.-H. Lin, and R.-F. Lin, 2001: Estimation of contrail frequency and radiative effects over the Taiwan area. *Terr. Atmos. Oceanic Sci.,* 12, 155-178.

Chýlek, P. et al., 2006: Aerosol indirect effect over the Indian Ocean. *Geophys. Res. Lett.*, 33, L06806, doi: 10.1 029/2005GL025397.

Comstock, J. M., T. P. Ackerman, and G. G. Mace, 2002: Ground-based lidar and radar remote sensing of tropical cirrus clouds at Nauru Island: Cloud statistics and radiative impacts. *J. Geophys. Res.*, 107, D23, 4714, doi:10.1029/2002JD002203.

DeMott, P. J., 1990: An exploratory study of ice nucleation by soot aerosols. *J. Appl. Meteorol.*, 29, 1072-1079.

DeMott, P. J., D. C. Rogers, S.M. Kreidenweis, 1997: The susceptibility of ice formation in upper tropospheric clouds to insoluble aerosol components. *J. Geophys. Res.*, 102, 19,575-19,584.

DeMott, P. J., Y. Chen, S. M. Kreidenweis, D. C. Roger, and D. E. Shermann, 1999: Ice formation by BC particles. *Geophys. Res. Lett.*, 26, 2429-2432.

Dessler, A. E., S. P. Palm, W. D. Hart, and J. D. Spinhirne, 2006a: Tropopause-level thin cirrus coverage revealed by ICESat/Geoscience Laser Altimeter System. *J. Geophys. Res.,* 111, D08203, doi: 10.1 029/2005JD006586.

Dessler, A. E., S. P. Palm, W. D. Hart, and J. D. Spinhirne, 2006b: Tropical cloud-top height distributions revealed by the Ice, Cloud, and land elevation Satellite (ICESat)/Geoscience Laser Altimeter System. *J. Geophys. Res.,* 111, D12215, doi:10.1029/2005JD006705.

Donner, L., C. J. Seman, B. J. Soden, R. S. Hemler, J. C. Warren, J. Ström, and K. N. Liou, 1997: Large-scale ice clouds in the GFDL SKYHI general circulation model. *J. Geophys. Res.,* 102, doi: 10.1029/97JD01488, 21745-21768.

Duda , D. P., and J. D. Spinhirne, 1996: Split window retrieval of particle size and optical depth in contrails located above horizontally inhomogeneous ice clouds. *Geophys. Res. Lett.,* 23, 3711- 3714. Duda, D. P. and P. Minnis, 2002: Observations of Aircraft Dissipation Trails from GOES. *Mon. Wea. Rev.,* 130, 398-406.

Duda , D. P., P. Minnis, and L. Nguyen, 2001: Estimates of cloud radiative forcing in contrail clusters using GOES imagery. *J. Geophys. Res.,* 106, 4927 – 4937.

Duda, D. P., P. Minnis, P. K. Costulis, R. Palikonda, 2003: CONUS contrail frequency estimated from RUC and flight track data. *Proceedings of European Conference on Aviation,Atmosphere, and Climate.*

Duda, D. P., P. Minnis, L. Nguyen, and R. Palikonda, 2004: A Case Study of the Development of Contrail Clusters over The Great Lakes. *J. Atmos. Sci.,* 61, 1132-1146.

Duda, D. P., P. Minnis, D. P. Garber, and R. Palikonda, 2005: CONUS contrail frequency and coverage estimated from RUC and flight track data. *Meteorol. Z.,* 14, 537-548.

Duda, D. P., R. Palikonda, and P. Minnis, 2008: Relating satellite and surface detection of contrails to numerical weather output. *Geophys. Res. Lett.* (in press).

Eichkorn, S., K.-H. Wohlfrom, and F. Arnold, 2002: Massive positive and negative chemiions in the exhaust of an aircraft jet engine at ground level: Mass distribution measurements and implications of aerosol formation. *Atmospheric Environment*, 36, 11, 1821-1825.

Fahey, D. W., U. Schumann, S. Ackerman, P. Artaxo, O. Boucher, M.Y. Danilin, B. Kärcher, P. Minnis, T. Nakajima, O.B. Toon, 1999: Aviation produced aerosols and cloudiness, in: J.E. Penner, D.H. Lister, D.J. Griggs, D.J. Dokken, M. McFarland (Eds.), Aviation and the Global Atmosphere, A Special Report of IPCC (Intergovernmental Panel on Climate Change), Cambridge Univ. Press, Cambridge, UK, , pp. 65–120.

Feichter, J., U. Lohmann, and I. Schult, 1997: The atmospheric sulfur cycle and its impact on the shortwave radiation. *Clim. Dyn.,* 13, 235– 246.

Forster, P., V. Ramaswamy, P. Artaxo, T. Berntsen, R. Betts, D.W. Fahey, J. Haywood, J. Lean, D.C. Lowe, G. Myhre, J. Nganga, R. Prinn, G. Raga, M. Schulz and R. Van Dorland, 2007: Changes in Atmospheric Constituents and in Radiative Forcing. *In: Climate Change 2007: The Physical Science Basis. Contribution of Working Group I to the Fourth Assessment Report of the Intergovernmental Panel on Climate Change* [Solomon, S., D. Qin, M. Manning, Z. Chen, M. Marquis, K.B. Averyt, M.Tignor and H.L. Miller (eds.)]. Cambridge University Press, Cambridge, United Kingdom and New York, NY, USA.

Fortuin, J. P., F. R. Van Dorland, W. M. F. Wauben, and H. Kelder, 1995: Greenhouse effects of aircraft emissions as calculated by a radiative ransfer model. *Ann. Geophys.,* 13, 413-418.

Fouquart, Y., and B. Bonnel, 1980: Computations of solar heating of the Earth's atmosphere: A new parameterization. *Beitr. Phys. Atmos.,* 53, 35 – 62.

Frankel, D., K. N. Liou, S. C. Ou, D. P. Wylie, and P. Menzel, 1997: Observations of cirrus cloud extent and their impacts to climate, *Proceedings for the Ninth Conference on Atmospheric Radiaiton. Amer. Meteor. Soc.*, February 2-7, 1997, Long Beach, CA, pp 414 – 417.

Freeman, K. P., and K. N. Liou, 1979: Climate effects of cirrus clouds. *Adv. Geophys.,* 21, 23 1-287. Freudenthaler, V., F. Homburg, and H. Jager, 1 996a: Contrail observations by ground-based scanning lidar: Cross-sectional growth. *Geophys. Res. Lett*., 22, 3501 – 3504.

Freudenthaler, V., F. Homburg, and H. Jager, 1996b: Optical parameters of contrails from lidar measurements: Linear depolarization. *Geophys. Res. Lett*., 23, 3715 – 3718.

Fu, Q., and K. N. Liou, 1992: On the correlated *k*-distribution method for radiative transfer in non- homogeneous atmospheres. *J. Atmos. Sci.,* 49, 2139 – 2156.

Fu, Q., and K. N. Liou, 1993: Parameterization of the radiative properties of cirrus clouds. *J. Atmos. Sci.,* 50, 2008-2025.

Fu, Q., K.N. Liou, M. Cribb, T.P. Charlock, and A. Grossman, 1997: Multiple scattering in thermal infrared radiative transfer. *J. Atmos. Sci*., 54, 2799-28 12.

Gao, B.-C., A. F. H. Goetz, and W. J. Wiscombe, 1993: Cirrus cloud detection from airborne imaging spectrometer data using the 1.38-ìm water vapor band. *Geophys. Res. Lett.,* 20, 301–304.

Gayet, J. F., G. Febvre, G. Brogniez, H. Chepfer, W. Renger, and P. Wendling, 1996: Microphysical and optical properties of cirrus and contrails: Cloud field study on 13 October 1989. *J. Atmos. Sci.,* 53, 126-13 8.

Gierens, K., and E. Jensen, 1998: A numerical study of the contrail to cirrus transition. *Geophys. Res. Lett*., 25, 4341–4344.

Goodman, J., R. F. Pueschel, E. J. Jensen, S. Verma, G. V. Ferry, S. D. Howard, S. A. Kinne, and D. Baumgardner, 1998: Shape and size of contrails ice particles. *Geophys. Res. Lett.,* 25, 1327-1330.

Gorbunov, B., A. Baklanov, N. Kakutkina, H. L. Windsor, and R. Toumi, 2001: Ice nucleation on soot particles. *J. Aerosol Sci.,* 32(2), 199-2 15.

Gounou, A., and R. J. Hogan, 2007: Asensitivity study of the effect of horizontal photon transport on the radiative forcing of contrails. *J. Atmos. Sci.,* 64, 1706-1716.

Grassl, H., 1990: Possible climatic effects of contrails and additional water vapor. In *Air Traffic and the Environment Background, Tendencies and Potential Global Atmospheric Effects.* U. Schumann (ed), Springer-Verlag, Berlin, pp 124-137.

Gu, Yu, and K. N. Liou, 2006: Cirrus cloud horizontal and vertical inhomogeneity effects in a GCM. *Meteror. Atmos. Phys*., 91, 223–235.

Gu, Y., J. Farrara, K. N. Liou, and C. R. Mechoso, 2003: Parameterization of cloud/radiation processes in the UCLA general circulation model. *J. Climate*, 16, 3357-3370.

Gu, Y., K. N. Liou, Y. Xue, C. R. Mechoso, W. Li, and Y. Luo, 2006: Climatic effects of different aerosol types in China simulated by the UCLA general circulation model. *J. Geophys. Res.,* 111, D15201, doi: 10.1029/2005JD0063 12.

Gulberg, A., 2003: A study of contrails in a general circulation model. Proceeding of the AACConference, June 30 to July 3, 2003, Friedrichshafen, Germany, 261 – 265.

Hansen, J. E., and J. B. Pollack, 1970: Near-Infrared light scattering by terrestrial clouds. *J. Atmos. Sci.,* 27, 265-281.

Hansen, J., Mki. Sato, R. Ruedy, L. Nazarenko, A. Lacis, G.A. Schmidt, G. Russell, I. Aleinov, M. Bauer, S. Bauer, N. Bell, B. Cairns, V. Canuto, M. Chandler, Y. Cheng, A. Del Genio, G. Faluvegi, E. Fleming, A. Friend, T. Hall, C. Jackman, M. Kelley, N. Kiang, D. Koch, J. Lean, J. Lerner, K. Lo, S. Menon, R. Miller, P. Minnis, T. Novakov, V. Oinas, Ja. Perlwitz, Ju. Perlwitz, D. Rind, A. Romanou, D. Shindell, P. Stone, S. Sun,

N. Tausnev, D. Thresher, B. Wielicki, T. Wong, M. Yao, and S. Zhang, 2005: Efficacy of climate forcings. *J. Geophys. Res.* 110, D18104, doi: 10.1 029/2005JD005776.

Heymsfield, A. J., 1977: Precipitation development in stratiform ice clouds: A microphyical and dynamical study. *J. Atmos. Sci.,* 34, 367– 381.

Heymsfield, A. J., and G. M. McFarquhar, 1996: On the high albedos of anvil cirrus in the tropical Pacific warm pool: Microphysical interpretations from CEPEX and from Kwajalein, Marshall Islands. *J. Atmos. Sci.,* 53, 2424-2451.

Heymsfield, A. J., and C. M. R. Platt, 1984: A parameterization of the particle size spectrum of ice clouds in terms of the ambient temperature and the ice water content. *J. Atmos. Sci.,* 41, 846-855.

Heymsfield, A. J., R. P. Lawson, and G. W. Sachse, 1998: Growth of ice crystals in a precipitating contrail. *Geophys. Res. Lett.,* 25, 1335-1338.

Hong, S.-Y., H.-M. H. Juang, and Q. Zhao, 1998: Implementation of prognostic cloud scheme for a regional spectral model. *Mon. Wea. Rev.,* 126, 262 1–2639.

Hutchison, K. D., and N. J. Choe, 1996: Application of 1.38 ìm imagery for thin cirrus detection in daytime imagery collected over land surfaces. *Int. J. Remote Sens.,* 17, 3325–3342.

IPCC, 1999: *Aviation and the Global Atmosphere*. Penner et al., eds., Ch. 3, *Aviation-Produced Aerosols and Cloudiness*. D. W. Fahey and U. Schumann, eds. pp. 63-120. Cambridge Univ. Press, Cambridge, U.K.

Jensen, E. J., O. B. Toon, S. Kinne, G. W. Sachse, B. E. Anderson, K. R. Chan, C. H. Twohy, B. Gandrud, A. Heymsfield, and R. Miake-Lye, 1998a: Environmental conditions required for contrail formation and persistence. *J. Geophys. Res.* 103, 3929-3936.

Jensen, E. J., O. B. Toon, A. tabazadeh, G. W. Sachse, B. E. Anderson, K. R. Chan, C. W. Twohy, B. Gandrud, S. M. Aulenbach, A. Heymsfield, J. Hallet, and B. Gary, 1998b: Ice nucleation processes in upper tropospheric wave-clouds observed during SUCCESS. *Geophys. Res. Lett.,* 25, 1363 – 1366.

Jensen, E. J., A. S. Ackerman, D. E. Stevens, O. B. Toon, P. Minnis, 1998c: Spreading and growth of contrails in a sheared environment. *J. Geophys. Res.* 103, 31557 - 31567.

Joint Planning and Development Office and Partnership for AiR Transportation Noise and Emissions Reduction, 2006: *A Report of Findings and Recommendations for workshop on the Impacts of Aviation on Climate change*, June 7-9, 2006, Boston, MA., 63pp.

Joseph, J. H., Z. Levin, Y. Mekler, G. Ohring, an dJ. Otterman, 1975: Study of contrails observed from the ERST-1 satellite imagery. *J. Geophys. Res.,* 80, 366-372.

Jones, A., D.L. Roberts, and A. Slingo, 1994: A climate model study of the indirect radiative forcing by anthropogenic sulphate aerosols. *Nature*, 370, 450-453.

Kalkstein, A. J., and R.C. Balling, Jr, 2004: Impact of unusually clear weather on United States daily temperature range following 9/11/2001. *Climate Res.*, 26, 1–4.

Kärcher, B. and U. Lohmann, 2002: A parameterization of cirrus cloud formation: Homogeneous freezing of supercooled aerosols. *J. Geophys. Res.* 107, doi: 10.1029/2001JD000470.

Kärcher, B. and U. Lohmann, 2003: A parameterization of cirrus cloud formation: Heterogeneous freezing. *J. Geophys. Res.* 108, doi: 10.1029/2002JD003220.

Kärcher, B. , R. Busen, A. Petzold, F. P. Schroder, U. Schumann, and E. Jensen, 1998: Physicochemistry of aircraft-generated liquid aerosols, soot, and ice particles, 2, Comparison with observations and sensitivity studies. *J. Geophys. Res.,* 103, 17129-17147.

Kärcher, B., J. Hendricks, and U. Lohmann, 2006: Physically based parameterization of cirrus cloud formation for use in global atmospheric models. *J. Geophys. Res.,* 111, D01205, doi: 10.1 029/2005JD0062 19.

Kärcher, B., O. Möhler, P. J. DeMott, S. Pechtl, and F. Yu, 2007: Insights into the role of soot aerosols in cirrus cloud formation. *Atmos. Chem. Phys.,* 7, 4203–4227.

King, M. D., W. P. Menzel, P. S. Grant, J. S. Myers, G. T. Arnold, S. E. Platnick, L. E. Gumley, S. C. Tsay, C. C. Moeller, M. Fitzgerald, K. S. Brown, and F. G. Osterwisch, 1996: Airborne scanning spectrometer for remote sensing of cloud, aerosol, water vapor and surface properties. *J. Atmos. Oceanic Technology*, 13, 777-794.

King, M. D., S.-C. Tsay, S. E. Platnick, M. Wang, and K. N. Liou, 1997:. Cloud retrieval algorithms for MODIS: Optical thickness, effective particle radius, and thermodynamic phase. *MODIS Algorithm Theoretical Basis Document, No. ATBD-MOD-05,* MOD-06-Cloud Product, version 5, 79pp. (http ://modis.gsfc.nasa.gov/data/atbd/atbd _mod05 .pdf).

King, M. D., S. Platnick, P. Yang, G. T. Arnold, M. A. Gray, J. C. Riedi, S. A. Ackerman, and K. N. Liou, 2004: Remote Sensing of Liquid Water and Ice Cloud Optical Thickness and Effective Radius in the Arctic: Application of Airborne Multispectral MAS Data. *J. Atmos. Oceanic Technol.*, 21, 857- 875.

Knollenberg, R. G., 1972: Measurements of the growth of the ice budget in a persisting contrail. *J. Atmos. Sci.*, 29, 1367-1374.

Konrad, T. G., and J. C. Howard, 1974: Multiple contrail streamers observed by radar. *J. Appl. Meteor.,* 13, 563-572.

Kristjánsson, J. E., T. Iversen, A. Kirkevåg, Ø. Seland, and J. Debernard, 2005: Response of the climate system to aerosol direct and indirect forcing: Role of cloud feedbacks. *J. Geophys. Res*., 110, D24206, doi: 10.1 029/2005JD006299.

Kuhn, P. M., 1970: Airborne observations of contrail effectson the thermal radiation budget. *J. Atmos. Sci.,* 27, 937 – 942.

Lawson, R. P., A. J. Heymsfield, T. L. Jensen, and S. M. Aulenbach, 1998: Shapes, sizes, and light scattering properties of ice crystals in cirrus and a persistent contrail during SUCCESS. *Geophys. Res. Lett.,* 25, 133 1-1334.

Lee, T. F., 1989: Jet contrail identification using the AVHRR infrared split window. *Journal of Applied Meteorology,* 28, 99399.

Li, D.-M., and K. P. Shine, 1995: A 4-dimensional ozone climatology for UGAMP models. UGAMP Internal Rep. 35.

Li, J.-L. F., M. Köhler, J. D. Farrara, and C. R. Mechoso, 2002: The impact of stratocumulus cloud radiative properties on surface heat fluxes simulated with a general circulation model. *Mon. Wea. Rev.,* 130, 1433–1441.

Liepert, B., P. Fabian, and H. Grassl, 1994: Solar radiation in Germany – observed trends and an assessment of their causes. I. regional approach. *Beitrage zur Physik der Atmosphare*, 67, 15-29.

Liepert, B., 1997: Recent changes in solar radiation under cloudy conditions. *Int. J. Climatology,* 17, 158 1-1593.

Lin, Y.-L., R. D. Farley, and H. D. Orville, 1983: Bulk parameterization of the snow field in a cloud model. *J. Clim. Appl. Meteor.*, 22, 1065 – 1092.

Liou, K. N., 1986: The influence of cirrus on climate processes: a global perspective. *Mon. Wea. Rev.,* 114, 1167-1199.

Liou, K. N., 2005: Cirrus clouds and climate. *McGraw-Hill 2005 Yearbook of Science & Technology,* 51 – 53.

Liou, K. N., and S. C. Ou, 1981: Parameterization of infrared radiation transfer in cloudy atmospheres. *J. Atmos. Sci.*, 38, 2707-2716.

Liou, K. N., and K. L. Gebhart, 1982: Numerical experiments on the thermal equilibrium temperature in cirrus cloudy atmospheres. *J. Meteor. Soc. Japan*, 60, 570-582.

Liou, K.N., and S.C. Ou, 1983: Theory of equilibrium temperatures in radiative-turbulent atmospheres. *J. Atmos. Sci.*, 40, 2 14-229.

Liou, K. N., and S.C. Ou, 1989: The role of cloud microphysical processes in climate: An assessment from a one-dimensional perspective. *J. Geophys. Res.*, 94, 8599–8607.

Liou, K. N. and Y. Gu, 2006: Radiative transfer in cirrus clouds: Some thought on its connection to climate and general circulation models (*INVITED*). *2006 AGU Fall Meeting,* San Francisco, California.

Liou, K. N., S. C. Ou, and P. J. Lu, 1985: Interactive cloud formation and climatic temperature perturbations. *J. Atmos. Sci.,* 42, 1969 – 1981.

Liou, K. N., Q. Fu, and T. P. Ackerman, 1988: A simple formulation of the delta-four-stream approximation for radiative transfer parameterizations. *J. Atmos. Sci.,* 45, 1940-1947.

Liou, K. N., S. C. Ou, and G. Koenig, 1990: An investigation on the climatic effect of contrail cirrus. In *Air Traffic and the Environment Background, Tendencies and Potential Global Atmospheric Effects.* U. Schumann (ed), Springer-Verlag, Berlin, pp 154-169.

Liou, K. N., P. Yang, Y. Takano, K. Sassen, T. Charlock, and W. Arnott, 1998: On the radiative properties of contrail cirrus. *Geophys. Res. Lett.,* 25, 1161-1164.

Liou, K. N., Y. Que, Y. Gu, and G. MacFarguhar, 2007: On the correlation between ice water content and ice crystal size and its application to radiative transfer and general circulation models (Submitted to *Geophys. Res. Lett.*).

Liu, X. and J. E. Penner, 2005: Ice nucleation parameterization for global models. *Meteorol. Z.,* 14, 499- 514.

Lohmann, U., and E. Roeckner, 1996: Design and performance of a new cloud microphysics scheme developed for the ECHAM4 general circulation model. *Clim. Dyn.,* 12, 557– 572.

Mace, G. G., C. Jakob, and K. P. Moran, 1998a: Validation of hydrometeor prediction from the ECMWF model during winter season 1997 using millimeter wave radar data. *Geophys. Res. Lett.,* 25, 1645– 1648.

Mace, G. G., T. P. Ackerman, P. Minnis, and D. F. Young, 1998b: Cirrus layer microphysical properties derived from surface-based millimeter radar and infrared interferometer data. *J. Geophys. Res.,* 103, 23,207 – 23,216.

Mace, G. G., E. E. Clothiaux, and T. P. Ackerman, 2001: The composite characteristics of cirrus clouds: Bulk properties revealed by 1-year of continuous cloud radar data.. *J. Clim.,* 14, 2185–2203.

Mace, G. G., A. J. Heymsfield, and M. R. Poellot, 2002: On retrieving the microphysical properties of cirrus clouds using the moments of the millimeter-wavelength Doppler spectrum. *J. Geophys. Res.,* 107, doi:10.1029/2001JD001308.

Mace, G. G., Y. Zhang, S. Platnick, M. D. King, P. Minnis, and P. Yang, 2005: Evaluation of cirrus cloud properties derived from MODIS data using cloud properties derived from ground-based observations collected at the ARM SGP site. *J. Appl. Meteor.,* 44, 221 - 240.

Mace, G. G., S. Benson, and E. Vernon, 2006: Cirrus clouds and the Large-scale atmospheric State: Relationship revealed by six years of ground-based data. *J. Clim.,* 19, 3257–3278.

Machta, L., and T. Carpenter, 1971: Trends in high cloudiness at Denver and Salt Lake City. In *Man's Impact on the Climate*, W. H. Matthews, W. W. Kellogg, and G. D. Robinson, eds., MIT Press, Boston, pp. 410-415.

Mannstein, H., R. Meyer, and P. Wendling, 1999: Operational detection of contrails from NOAAAVHRR data. *Int. J. Rem. Sens.,* 20, 1641-1660.

Marquart, S., M. Ponater, F. Mager, and R. Sausen, 2003: Future development of contrail cover, optical depth, and radiative forcing: Impacts of increasing air traffic and climate change. *J. Climate,* 16, 2 890–2904.

Matrosov, S. Y., T. Uttal, J. B. Snider, and R. A. Kropfli, 1992:Estimates of ice cloud parameters from ground-based infrared radiometer and radar measurements. *J. Geophys. Res.,* 97, 11,567–11,574.

McClatchy, R. A., R. W. Fenn, J. E. A. Selby, F. E. Volz, and J. S. Garing, 1972: *Optical Properties of the Atmosphere*. Air Force Cambridge Research Laboratories. Bedford, MA. 80pp.

McFarlane, N. A., G. J. Boer, J.-P. Blanchet, and M. Lazare, 1992: The Canadian Climate Centre second-generation general circulation model and its equilibrium climate. *J. Clim.,* 5, 1013– 1044.

McFarquhar, G. M., S. Iacobellis, and R. Sommerville, 2003: SCM simulations of tropical ice clouds using observationally based parameterizations of microphysics. *J. Climate,* 16, 1643-1664.

McMillin, L. M., L. J. Crone, M. D. Goldberg, and T. J. Kleespies, 1995: Atmospheric transmittance of an absorbing gas. 4. OPTRAN: a computationally fast and accurate transmittance model for absorbing gases with variable mixing ratios at variable viewing angles. *Appl. Opt.*, 34, 6269–6274.

Meerkötter, R., U. Schumann, D. R. Doelling, P. Minnis, T. Nakajima, Y. Tsushima, 1999: Radiative forcing by contrails. *Ann. Geophys.,* 17, 1080-1094.

Menzel, W. P.,B. A. Baum, K. I. Strabala and R. A. Frey, 2002: Cloud top properties and cloud phase algorithm theoretical basis document, *Algorithm theoretical basis document, ATBD_MOD_04, NASA GSFC*. (http://modis.gsfc.nasa.gov/data/atbd/atbd _mod04.pdf).

Meyer, R., R. Buell, C. Leiter, H. Mannstein, S. Pechtl, T. Oki, and P. Wendling, 2007: Contrail observations over Southern and Eastern Asia in NOAA/AVHRR data and comparisons to contrail simulations in a GCM. *Int. J. Remote Sens.,* 28, 2049-2069, doi:10.1080/01431 160600641707, 2007.

Minnis, P., 2002: Contrails. *Encyclopedia of Atmospheric Sciences*, Academic Press, London, J. Holton, J. Pyle, and J. Curry, Editors, pp. 509-520.

Minnis, P., 2005: Reply. *J. Clim.,* 18, 2783-2784.

Minnis, P., D. F. Young, D. P. Garber, L. Nguyen, W. L. smith, Jr., and R. Palikonda, 1998: Transformation of contrails into cirrus during SUCCESS. *Geophys. Res. Lett.,* 25, 1157-1160. Minnis, P., U. Schumann, D. R. Doelling, K. M. Gierens, and D. W. Fahey, 1999: Global distribution of contrail radiative forcing. *Geophys. Res. Lett.,* 26, 1853 – 1856.

Minnis, P., J. K. Ayers, M. L. Nordeen, and S. P. Weaver, 2003: Contrail frequency over the United States from surface observations. *J. Climate*, 16, 3447 – 3462.

Minnis, P., J. K. Ayers, R. Palikonda, and D. Phan, 2004: Contrails, cirrus trends, and climate. *J. Climate*, 17, 1671-1685.

Minnis, P., R. Palikonda, B. J. Walter, J. K. Ayers, and H. Mannstein, 2005: Contrail properties over the eastern North Pacific from AVHRR data. *Meteorol. Z.,* 14, 5 15-523.

Morcrette, J.-J., 1991: Radiation and cloud radiative properties in the European Centre for Medium Range Weather Forecasts forecasting system, *J. Geophys. Res.*, 96, 9121–9132.

Morl, P., M. E. Reinhardt, E. Renger, and R. Schellhase, 1981: The use of the airborne lidar system "ALEX F 1" for aerosol tracing in the lower atmosphere. *Beitr. Phys. Atmos.,* 54, 403-410.

Murcray, W. B., 1970: On the possiblility of weather modification by aircraft contrails. *Mon. Wea. Rev.,* 98, 745 – 748.

Nakajima, T. and M. D. King, 1990: Determination of the optical thickness and effective particle radius of clouds from reflected solar radiation measurements. Part I: Theory. *J. Atmos. Sci.*, 47, 1878-1893. Next Generation Air Transportation System, 2004: *Federal Aviation Administration report to the US Congress.*

Ou, S. C., and K. N. Liou, 1984: A two-dimensional radiation-turbulence climate model: I. Sensitivity to cirrus radiative properties. *J. Atmos. Sci.*, 41, 2289-2309.

Ou, S. C., and K. N. Liou, 1995: Ice microphysics and climatic temperature perturbations. *Atmos. Res*., 35, 127–138.

Ou, S. C., K. N. Liou, W. M. Gooch, and Y. Takano, 1993: Remote sensing of cirrus clouds parameters using AVHRR 3.7 and 10.9 ìm channels. *Appl. Opt.,* 32, 2171-2180.

Ou, S. C., K. N. Liou, and Coauthors, 1995: Remote sounding of cirrus cloud optical depths and ice crystal sizes from AVHRR data: Verification using FIRE-II-IFO measurements. *J. Atmos. Sci*., 52, 4143-4158.

Ou, S. C., K. N. Liou, and B. A. Baum, 1996: Detection of multilayer cirrus cloud systems using AVHRR data: Verification based on FIRE-II-IFO composite measurements. *J. Appl. Meteor*., 35, 178-19 1.

Ou, S. C., K. N. Liou, and T. R. Caudill, 1998a: Remote sounding of multilayer cirrus cloud systems using AVHRR Data collected during FIRE-II-IFO. *J. Appl. Meteor*., 37, 241-254.

Ou, S. C., K. N. Liou, P. Yang, P. Rolland, T. R. Caudill, J. Lisowski, and B. Morrison, 1998b: Airborne retrieval of cirrus cloud optical and microphysical properties using ARES 5.1-5.3 and 3.7 ìm channel data. *J. Geophys. Res.,* 103, 2323 1-23242.

Ou, S. C., K. N. Liou, M. D. King, and S. C. Tsay, 1999: Remote sensing of cirrus cloud parameters based on a 0.63-3.7 ìm radiance correlation technique applied to AVHRR data. *Geophys. Res. Lett.*, 26, 243 9-2440.

Ou, S. C., K. N. Liou, and Y. Takano, G. J. Higgins, A. George, and R. Slonaker, 2002: *VIIRS cloud effective particle size and cloud optical depth algorithm theoretical basis document.* Algorithm Theoretical Basis Document, Version 5, Raytheon, Lanham, MD, .176pp.

Ou, S. C., Y. Takano, K. N. Liou, G. J. Higgins, A. George and R. L. Slonaker, 2003: Remote sensing of cirrus cloud optical thickness and effective particle size for the National Polar-orbiting Operational Environmental Satellite System Visible/Infrared Imager Radiometer

Suite: sensitivity to instrument noise and uncertainties in environmental parameters. *Appl. Opt.*, 42, 7202-72 14.

Palikonda, P., D. N. Phan, V. Chakrapani, and P. Minnis, 2004: Contrail coverage over the USA from MODIS and AVHRR data. *Preprints, European Conf. on Aviation, Atmosphere, and Climate*. Friedrichschafen at Lake Constance, Germany, Institut für Physik der Atmosphäre, DLR.

Palikonda, R., P. Minnis, D. P. Duda, and H. Mannstein, 2005: Contrail coverage derived from 2001 AVHRR data over the continental United States of America and surrounding areas. *Meteorol. Z.,* 14, 525-536.

Parol, F., J. C. Buriez, G. Brognies, and Y. Fouquart, 1991: Information content of AVHRR channels 4 and 5 with respect to particle size. *J. Appl. Meteor.*, 30, 973-984.

Platnick, S., M. D. King, S. A. Ackerman, W. P. Menzel, B. A. Baum, J. C. Riedi, R. A. Frey, 2003: The MODIS cloud products: algorithms and examples from Terra. *IEEE Trans. Geos. Rem. Sens.*, 41, 459-473.

Ponater, M., S. Brinkop, R. Sausen, and U. Schumann, 1996: Simulating the global atmospheric response to aircraft water vapour emissions and contrails-a first approach using a GCM. *Annales Geophysicae*, 14, 941-960.

Ponater, M. S. Marquart, and R. Sausen, 2002: Contrails in a comprehensive global climate model: Parameterization and radiative forcing results. *J. Geophys. Res.,* 107, D13,4164, doi: 10.1029/2001 JD000429.

Ponater, M., S. Marquart, R. Sausen, and U. Schumann, 2005: On contrail climate sensitivity. *Geophys. Res. Lett.*, 32, L1 0706, 10.1 029/2005GL0225 80.

Ramaswamy, V., O. Boucher, J. Haigh, D.A. Hauglustaine, J. Haywood, G. Myhre, T. Nakajima, G.Y. Shi, S. Solomon, 2001: Radiative forcing of climate change. In: Houghton, J.T., et al. (Eds.), *Climate Change 2001: The Scientific Basis Intergovernmental Panel on Climate Change.* Cambridge University Press, Cambridge, New York, 349–416.

Rao, N. X., S. C. Ou, and K. N. Liou, 1995: Removal of solar component in the AVHRR 3.7 ìm radiances for the retrieval of cirrus cloud parameters. *J. Appl. Meteor.*, 34, 482-499.

Raschke, E. J., J. Schmetz, J. Heintenberg, R. Kandel, and R. W. Sawnders, 1990: The international Cirrus Experiment (ICE) – A joint European effort. *ESA J.,* 14, 193 -199.

Rebetez, M., and M. Beniston, 1998: Changes in sunshine duration are correlated with changes in daily temperature range this century : an analysis of Swiss climatology data. *Geophys. Res. Lett.,* 25, 3611-3613.

Rind, D., P. Lonergan, and K. Shah, 1996: Climatic effect of water vapor release in the upper troposphere. *J. Geophys. Res.,* 101, 29395-29406

Rind, D., P. Lonergan, K. Shah, 2000: Modeled impact of cirrus cloud increases along aircraft flight paths. *J. Geophys. Res.,* 105, 19927–19940.

Rockel, B., E. Raschke, and B. Weyres, 1991: A parameterization of broad-band radiative transfer properties of water, ice and mixed clouds. *Beitr. Phys. Atmos.,* 64, 1– 12.

Roeckner, E., 1995: Parameterization of cloud radiative properties in the ECHAM4 model. In *World Climate Research Program Workshop Proceedings, WMO/TD 713,* World Meteorol. Org., Geneva, 105–116.

Roeckner, E., K. Arpe, L. Bengtsson, M. Christoph, M. Claussen, L. Dűmenil, M. Esch, M. Giorgetta, U. Schlese, and U. Schulzweida, 1996: The atmospheric general circulation model ECHAM-4: Model description and simulation of present-day climate. *Report No. 218, Max-Planck-Institut für Meteorologie*, Hamburg, Germany, 90pp.

Roeckner, E., L. Bengtsson, J. Feichter, J. Lelieveld, and H. Rodhe, 1999: Transient climate change simulations with a coupled atmosphere-ocean GCM including the tropospheric sulfur cycle. *J. Clim.*, 12, 3004– 3032.

Rolland, P., and K. N. Liou, 2001: Surface variability effects on the remote sensing of thin cirrus optical and microphysical properties. *J. Geophys. Res.,* 106, 22965-22977.

Rolland, P., K. N. Liou, M. D. King, S.-C. Tsay, and G. M. McFarquhar, 2000: Remote sensing of optical and microphysical properties of cirrus clouds using Moderate-Resolution Imageing Spectroradiometer channels: Methodology and sensitivity to physical assumptions. *J. Geophys. Res.,* 105, 11721–11738.

Roskovensky, J. K., and K. N. Liou, 2003: Detection of thin cirrus from 1.38 μm/0.65 μm reflectance ratio combined with 8.6–11 μm brightness temperature difference. *Geophys. Res. Lett.,* 30, 1985, doi: 10.1029/2003GL01 8135.

Roskovensky, J. K., and K. N. Liou, 2005: Differentiating airborne dust from cirrus clouds using MODIS data. *Geophys. Res. Lett.,* 32, L12809, doi:10.1029/2005GL022798.

Roskovensky, J. K., and K. N. Liou, 2006: Simultaneous determination of aerosol and thin cirrus optical depths over oceans from MODIS Data: Some Case Studies. *J. Atmos. Sci.,* 63, 2307-2323.

Rossow, W. B., and R. A. Schiffer, 1999: Advances in understanding clouds from ISCCP. *Bull. Am. Meteor. Soc.*, 80, 2261-2287

Rutledge, S. A., and P. V. Hobbs, 1984: The mesoscale and microscale structure and organization of clouds and precipitation in midlatitude cyclones. XII: A diagnostic modeling study of precipitation development in narrow cloud-frontal rainbands. *J. Atmos. Sci*., 20, 2949–2972.

Sassen, K. 1997: Contrail-cirrus and their potential for regional climate change. *Bull. Amer. Meteor. Soc.*, 78, 1885 – 1903.

Sassen, K., and C. Hsueh, 1998: Contrail properties derived from high-resolution polarization lidar studies during SUCCESS. *Geophys. Res. Lett.,* 25, 1165-1168.

Sassen, K., D. O'C Starr, and T. Uttal, 1989a: Mesoscale and microscale structure of cirrus clouds: Three case studies. *J. Atmos. Sci.,* 46, 371 – 396.

Sassen, K. M. Griffin, and G. C. Dodd, 1989b: Optical scattering and microphysical properties of subvisual cirrus clouds, and climatic implications. *J. Appl. Meteor.,* 28, 9 1-98.

Sausen, R., K. Gierens, M. Ponater, and U. Schumann, 1998: A diagnostic study of the global distribution of contrails, part I, Present day climate. *Theor. Appl. Climatol.,* 61, 127– 141.

Sausen, R. and U. Schumann, 2007: Climate Impact of Aviation. *ASD Focus,* 1, 4-5.

Sausen, R. and Co-Authors, 2005: Aviation radiative forcing in 2000: An update on IPCC (1999). *Meteorol. Z.*, 14, No. 4, 555-561.

Schröder, F., B. Kärcher, C. Duroure, J. Ström, A. Petzold, J.-F. Gayet, B. Strauss, P. Wendling, and A. Thomas, 2000: On the transition of contrails into cirrus clouds. *J. Atmos. Sci.,* 57, 464 – 480. Schumann, U., 1996: On conditions for contrail formation from aircraft exhausts. *Meteor. Z.,* 5, 4-23. Schumann, U., 2001: Climate sensitivity to contrails for increasing propulsion efficiency. In: Schumann, U., and Amanatidis, G.T.

(Eds.), *Aviation, Aerosols, Contrails and Cirrus Clouds (A2C3)*. European Commission, Brussels, 271–274pp.

Schumann, U., 2002: Contrail cirrus. In: *Cirrus*, D. K. Lynch, K. Sassen, D. O'C. Starr, G. Stephens (Eds.), Oxford Univ. Press, Oxford, UK., 23 1-255.

Schumann, U., 2005: Formation, properties and climatic effects of contrails. *C. R. Physique*, 6, 549–565.

Schumann, U., and Wendling, 1 990:Determination of contrails from satellite data and observational results. In *Air Traffic and the Environment Background, Tendencies and Potential Global Atmospheric Effects*. U. Schumann (Ed), Springer-Verlag, Berlin, pp 13 8-153.

Schumann, U., H. Schlager, F. Arnold, J. Ovarlez, H. Kelder, Ø. Hov, G. Hayman, I. S. A. Isaksen, J. Staehelin, and P. D.Whitefield, 2000: Polution from aircraft emissions in the North Atlantic flight corridor: Overview on the POLINAT projects. *J. Geophys. Res.*, 105, 3605 – 3631.

Seaver, W. L., and J. E. Lee, 1987: A statistical examination of sky cover changes in contiguous United States. *J. Climate Appl. Meteor.*, 26, 88-95.

Seifert, A., and K. D. Beheng, 2006: A two moment cloud microphysics parameterization for mixed- phase clouds. Part I: Model description. *Meteor. Atmos. Phys.*, 92, 46-66.

Seinfeld, J. H., 1998: Clouds, contrails, and climate. *Nature*, 391, 837-83 8.

Shilling, J.E., T.J. Fortin, M.A. Tolbert, 2006: Depositional ice nucleation on crystalline organic and inorganic solids. *J. Geophys. Res.*, 111, D12204, doi:10.1029/2005JD006664.

Shine, K.P., R.G. Derwent, D.J. Wuebbles, J.-J. Morcrette, 1990: Radiative forcing of climate. In: Houghton, J. T., et al. (Eds.), *Climate change: The IPCC Scientific Assessment.* Cambridge University Press, Cambridge, New York, pp. 41–68.

Shine, K. P. 2005: Comments on "Contrails, Cirrus Trends, and Climate. *J. Clim.*, 18, 278 1-2782. Skamarock, W. C., J. B. Klemp, J. Dudhia, D. O. Gill, D. M. Barker, W. Wang, and J. G. Powers, 2005: *A Description of the Advanced Research WRF Version 2*, NCAR Technical Note, NCAR/TN- 468+STR., NCAR, Boulder, Colorado, 88pp.

Smith, W. L., and C. M. Platt, 1978: Comparison of satellite-deduced cloud heights with indications from radiosonde and ground-based laser measurements. *J. Appl. Meteor.*, 17, 1796-1802.

Stevens, D. E., and C. S. Bretherton, 1996: A forward-in-time advection scheme and adaptive multilevel flow solver for nearly incompressible atmospheric flow. *J. Comput. Phys.*, 129, 284-295.

Stephens, G. L., S. Tsay, P. W. Stackhouse Jr., and P. J. Flatau, 1990: The relevance of the microphysical and radiative properties of cirrus clouds to climate and climate feedback. *J. Atmos. Sci.*, 42, 2682-2694.

Stephens, G. L., D. G. Vane, R. J. Boain, G. G. Mace, K. Sassen, Z. Wang, A. J. Illingworth, E. J. O'Connor, W. B. Rossow, S. L. Durden, S. D. Miller, R. T. Austin, A. Benedetti, C. Mitrescu, and the CloudSat Science Team, 2002: The CloudSat mission and the A-Train. *Bull. Amer. Meteor. Soc.*, 83, 1771-1790.

Strauss, B., R. Meerkotter, B. Wissinger, P. Wendling, and M. Hess, 1997: On the regional climatic impact of contrails: microphysical and radiative properties of contrails and natural cirrus clouds. *Ann. Geophysicae*, 15, 1457-1467.

Study on Man's Impact on Climate, 1971: *Inadvertent Climate Modification.* MIT Press, 308 pp.

Suarez, M. J., A. Arakawa, and D. A. Randall, 1983: The parameterization of the planetary boundary layer in the UCLA general circulation model: Formulation and results. *Mon. Wea. Rev.,* 111, 2224–2243.

Sundqvist, H., 1978: A parameterization scheme for non-convective condensation including prediction of cloud water content. *Q. J. R. Meterol. Soc.,* 104, 677–690.

Takano, Y., and K. N. Liou, 1989a: Solar radiative transfer in cirrus clouds. Part I. Single-scattering and optical properties of hexagonal ice crystals. *J. Atmos. Sci.*, 46, 3-19.

Takano, Y., and K. N. Liou, 1989b: Solar radiative transfer in cirrus clouds. Part II. Theory and computation of multiple scattering in an aniso-tropic medium. *J. Atmos. Sci.*, 46, 20-3 6.

Takano, Y. and K.N. Liou, 1995: Radiative transfer in cirrus clouds. III. Light scattering by irregular ice crystals. *J. Atmos. Sci.*, 52, 818-837.

Thompson, G., R. M. Rasmussen, and K. Manning, 2004: Explicit forecasts of winter precipitation using an improved bulk microphysics scheme. Part I: Description and sensitivity analysis. *Mon. Wea. Rev.*, 132, 5 19–542.

Toon, O. B., and R. C. Miake-Lye, 1998: Subsonic aircraft: Contrail and cloud effects special study. *Geophys. Res. Lett.,* 25, 1109 - 1112.

Twomey, S.A., M. Piepgrass, and T. Wolfe, 1984: An assessment of the impact of pollution on global cloud albedo. *Tellus*, 36B, 356–366.

Travis, D. J., A.M. Carleton, and R.G. Lauritsen, 2001: Jet aircraft contrails: Surface temperature variations during the aircraft groundings of September 11–13, 2001. In: *10th Conference on Aviation, Range & Aerospace Meteorology, 13–16 May 2002, Portland, Oregon,* American Meteorological Society, paper J1 .1.

Travis, D. J., A. M. Carleton, R.G. Lauritsen, 2002: Contrails reduce daily temperature range, brief communications. *Nature*, 418, 601.

Travis, D. J., A. M. Carleton, R.G. Lauritsen, 2004: Regional variations in U.S. diurnal temperature range for the 11–14 September 2001 aircraft groundings: Evidence of jet contrail influence on climate. *J. Clim.*, 17, 1123–1134.

Travis, D. J., A. M. Carleton, J. S. Johnson, and J. Q. DeGrand, 2007: US jet contrail frequency changes: influences of jet aircraft flight activity and atmospheric conditions. *Int. J. Clim.,* 27, 621- 632.

Twomey, S. and T. Cocks (1982). Spectral reflectance of clouds in the near-infrared: Comparison of measurements and calculations. *J. Meteor. Soc. Japan,* 60, 583-592.

Twomey, S, and T. Cocks (1989). Remote sensing of cloud parameters from spectral reflectance in the near-infrared. *Beitr. Phys. Atmos.,* 62, 172-179.

Waitz, I., J. Townsend, J. Cutcher-Gershenfeld, E. Greitzer, and J. Kerrebrock, 2004: *Report to the United States Congress: aviation and Environment, A National Vision, Framework for Goals and Recommended Actions*. PARTNER, MIT, Cambridge, MA. 52pp.

Waliser D., Li F., Jiang J., Wu D., Stephens G., Vane D., Tompkins A., Chern J., Khairoutdinov M., Woods C., 2007: Exploiting MLS and CloudSat Data to Support Weather and Climate Model Development and Validation: Applications to Upper Tropospheric Cloud Processes, A-Train Lille 2007 Symposium, Lille, France, October 22-25, O06.

Wang, W.-C., W. Gong, and J.-P. Chen, 2001: SUNYA Regional Model simulation of radiative forcing and climate impact due to contrails over regions around Taiwan. *Terr. Atmos. Oceanic Sci.,* 12, 179- 194.

Wang, Z., and K. Sassen, 2002a: Cirrus cloud microphysical property retrieval using lidar and radar measurements. Part I: Algorithm description and comparison with in-situ data. *J. Appl. Meteor.,* 41, 218-229.

Wang, Z., and K. Sassen, 2002b: Cirrus cloud microphysical property retrieval using lidar and radar measurements. Part II: Midlatitude cirrus microphysical and radiative properties. *J. Atmos. Sci.,* 59, 2291-2302.

Wild, M., A. Ohmura, H. Gilgen, E. Roeckner, M. Giorgetta, and J.-J. Morcrette, 1998: The disposition of radiative energy in the global climate system: GCM calculated versus observational estimates. *Clim. Dyn.,* 14, 853– 869.

Winker, D. M., W. H. Hunt, and C. A. Hostetler, 2004: Status and Performance of the CALIOP lidar. *Proc. SPIE*, 5575, 8-15.

Wong, E., K. Hutchison, S. C. Ou, and K. N. Liou, 2007: Cirrus cloud top temperatures retrieved from radiances in the National Polar-Orbiting Operational Environmental Satellite Syste - Visible Infrared Imager Radiometer Suite 8.55 and 12.0 ìm bandpasses. *Appl. Opt.*, 46, 1316-1325.

Wylie, D., W. P. Menzel, and K. I. Strabala, 1994: Four years of global cirrus cloud statistics using HIRS. *J. Climate*, 7, 1972–1986.

Wylie, D. P., and W. P. Menzel, 1999: Eight years of high cloud statistics using HIRS. *J. Climate*, 12, 170–1 84.

Wylie, D., D. L. Jackson, w. P. Menzel, and J. J. Bates, 2005: Trends in global cloud Cover in two decades of HIRS observations. *J. Climate*, 18, 3021-303 1.

Yang, P., and K. N. Liou, 1996a: Finite-difference time domain method for light scattering by small ice crystals in three-dimensional space. *J. Opt. Soc. Amer.,* A13, 2072-2085.

Yang, P., and K. N. Liou, 1996b: Geometric-optics-integral-equation method for light scattering by non- spherical ice crystals. *Appl. Opt.*, 35, 6568-6584.

Yang, P., H. Wei, H.-L. Huang, B. A. Baum, Y. X. Hu, G. 1640 W. Kattawar, M. I. Mishchenko, and Q. Fu, 2005: Scattering and absorption property database for nonspherical ice particles in the near- through far-infrared spectral region. *Appl. Opt.,* 44, 55 12–5523.

Yue, Q., K. N. Liou, S. C. Ou, B. H. Kahn, P. Yang, and G. G. Mace, 2007a: Interpretation of AIRS data in thin cirrus atmospheres based on a fast radiative transfer model. *J. Atmos. Sci.*, 64, 3827-3 842.

Yue, Q., K. N. Liou, and Yu Gu, 2007b: Investigation of the radiative forcings of thin cirrus in the tropical atmosphere using AIRS/ARM data. *2007 ARM Science Team Meeting,* March 26-30, 2007, Monterey, California.

Zhang, M. H., W. Y. Lin, S. A. Klein, J. T. Bacmeister, S. Bony, R. T. Cederwall, A. D. Del Genio, J. J. Hack, N. G. Loeb, U. Lohmann, P. Minnis, I. Musat, R. Pincus, P. Stier, M. J. Suarez, M. J. Webb, J. B. Wu, S. C. Xie, M.-S. Yao, and J. H. Zhang, 2005: Comparing clouds and their seasonal variations in 10 atmospheric general circulation models with satellite measurements. *J. Geophys. Res.*, 110, 10.1029/2004JD005021.

Zerefos, C. S., K. Eleftheratos, D. S. Balis, P. Zanis, G. Tselioudis, C. Meleti, 2003: Evidence of impact of aviation on cirrus cloud formation. *Atmos. Chem. Phys*., 3, 1633 – 1644.

Zobrist, B. , Marcolli, C., Koop, T., Luo, B. P., Murphy, D. M., Lohmann, U., Zardini, A., Krieger, U. K., Corti, T., Cziczo, D. J., Fueglistaler, S., Hudson, P. K., Thomson, D. S. and Peter, T., 2006: Oxalic acid as a heterogeneous ice nucleus in the upper troposphere and its indirect aerosol effect. *Atmos. Chem. and Phys. Disc.,* 6, 3571-3609.

Zuberi, B., A.K. Betram, T. Koop, L.T. Molina, and M.J. Molina, 2001: Heterogeneous freezing of aqueous particles induced by crystallized (NH4)2SO4, ice, and letovicite. *J. Phys. Chem. A*, 105, 645 8-6464.

In: Aviation and the Environment
Editor: Jon C. Goodman

ISBN: 978-1-60692-320-7

Chapter 6

AVIATION-CLIMATE CHANGE RESEARCH INITIATIVE (ACCRI) SUBJECT SPECIFIC WHITE PAPER (SSWP) ON CONTRAILS/CIRRUS OPTICS AND RADIATION SSWP # VI, JANUARY 22, 2008

Ping Yang**, *Andrew Dessler** and *Gang Hong*
Department of Atmospheric Sciences Texas A&M University;
College Station, Texas 77843, USA

EXECUTIVE SUMMARY

The effect of aircraft emissions on the climate of Earth is one of the most serious long-term environmental issues facing the aviation industry (IPCC, 1999; Aviation and the Environment – Report to the United States Congress, 2004). Aviation emissions, including gases and particles in the upper troposphere and lower stratosphere, have both direct and indirect climate effects. The direct effect is principally the emission of carbon dioxide, a powerful greenhouse gas. The indirect effects include the changes in ozone due to emissions in nitrogen oxides, the effects of aerosol emissions and water vapor on clouds, and the effects associated with contrails and contrail-induced cirrus clouds.

As stated in *the Executive Summary of the Workshop on the Impacts of Aviation on Climate Change,* June 7-9, 2006, Boston, MA (hereafter, the *Workshop Executive Summary*, http://web.mit.edu/aeroastro/partner/reports/climatewrksp-rpt-0806.pdf), "The effects of aircraft emissions on the current and projected climate of our planet may be the most serious long-term environmental issue facing the aviation industry... The only way to ensure that policymakers fully understand trade-offs from actions resulting from implementing engine and fuel technological advances, airspace operational management practices, and policy

* pyang@ariel .met.tamu
* adessler@tamu.edu
* hong@ariel.met.tamu.edu

actions imposed by national and international bodies is *to provide them with metrics that correctly capture the climate impacts of aviation emissions.*"

Cloud radiative forcing, defined as the difference of shortwave and longwave radiative fluxes between cloudy and clear-sky conditions, is *a common metric* to quantify the effect of clouds on climate. In addition, the optical properties of cloud particles are fundamental physical quantities that are connected to cloud radiative forcing. Significant uncertainties exist in our present understanding of the radiative forcing and optical properties of contrails and contrail cirrus clouds, as explicitly stated in the *Executive Summary.*

With the support of the Volpe National Transportation System Center under the solicitation "Aviation-Climate Change Research Initiative (ACCRI)" (DTRT57-07- 2003), this subject white reports on the 5 th 5th key area identified in the solicitation, i.e., *Climate Impacts of Contrails and Contrail-Cirrus*. Specifically, this report provides a review of our current knowledge of

1. the optical properties of individual ice crystals in contrails and contrail- cirrus clouds from modeling and measurement (including laboratory analog measurements) perspectives;
2. the bulk optical properties of contrails and contrail-cirrus clouds, including the effect of ice crystal size-habit distributions; and
3. the radiative forcing of contrails a nd contrail-cirrus cloud systems.

In this report, the current state of science on the climate impacts of contrails and contrail-induced cirrus is reviewed, including an analysis of the uncertainties. In so doing, we identify ten key problems:

The uncertainties in contrail and contrail-induced cirrus coverage estimated from different detection algorithms and measurements.

The uncertainties in optical thicknesses of contrail and contrail-induced cirrus clouds from different analyses. These differences result in distinct differences in the radiative forcing of contrails and contrail-induced cirrus clouds.

Ice particle sizes vary from contrails to contrail-induced cirrus clouds. But the measurements for ice particle sizes are strongly dependent on the detection approaches, particularly for small ice particles.

Different ice habits have been found in contrails and contrail -induced cirrus clouds. In turn, different mixtures of ice habits must be used in the study of the radiative forcing of contrails and contrail-induced cirrus clouds. A better understanding of the single-scattering properties of ice crystals in contrails and contrail-induced cirrus clouds is necessary.

Shortwave radiative forcing varies substantially with solar zenith angle. The diurnal variation of air traffic is an important factor for the radiative forcing calculations, but is often neglected in the calculation of the impact of contrails and contrail-induced cirrus.

New techniques should be developed to accurately detect contrails and cont rail- induced cirrus clouds. This is critical for understanding the microphysical and optical properties of these clouds.

Our current understanding of the effects of aerosols emitted by aircraft on cirrus cloud formation and the interaction between aerosols and cirrus clouds needs to be improved.

Furthermore, the chemical effects of black carbon and sulphate particles on the microphysical properties (e.g., the refractive indexes) of ice particles in contrails and contrail-induced cirrus clouds are poorly known and need to be improved.
There are no studies on the contrails embedded within natural cirrus clouds. This is quite important since the aircraft flight heights are often at the heights where ice clouds frequently occur.
Representation of contrails and contrail-induced cirrus clouds in global atmospheric models needs to be improved. There are shortages of observations for validation of GCM results, and the representation of the aerosol-cirrus interaction in GCMs needs substantial improvements. Furthermore, there is an urgent need to develop radiation schemes that are suitable for use in GCMs.
More and better measurements of supersaturation in the atmosphere.

To reduce the above uncertainties in estimating the radiative forcing of contrails and contrail-induced cirrus clouds, we recommend the following prioritized research areas:

Area 1. Development of a database of the single-scattering properties of ice particles and aerosols in contrails and contrail -induced cirrus clouds. Although various habits of ice crystals have been found in contrails and contrail-induced cirrus clouds, there is not a single-scattering database specified for studies involving contrails and contrail-induced cirrus clouds. The particle size distributions for contrails tend to have smaller ice particle sizes in comparison with those for natural cirrus clouds. A contrail single-scattering database would provide the basis for retrieving the microphysical and optical properties of contrails and contrail-induced cirrus clouds and studying their radiative forcing, which could be of great benefit to the community of the scientists who study different aspects of the contrail-ice cloud dependencies.
Area 2. Develop new parameterization schemes for shortwave and longwave radiation calculations in GCMs that are suitable for contrails and contrail-induced cirrus clouds.
Area 3. Improve detection of contrails and contrail-induced cirrus clouds from multiple satellite sensors, including passive radiometric measurements from imagers/interferometers and active lidar and radar sensors. Specifically, we note the availability and maturity of data products from the NASA A-Train suite of satellites. This effort is to provide the microphysical, macrophysical, and optical properties of contrails and contrail-induces cirrus clouds that are essential to the study of the radiative forcing of contrails and contrail-induced cirrus clouds.
Area 4. Make more use of in situ measurements and laboratory experiments to understand the formation of contrails and contrail-induced cirrus clouds, the accurate formation criteria of the formation of contrails, and the physical and chemical processes involved in the transition from contrail to cirrus clouds. Especially, improvement on our knowledge about the effect of water vapor and temperature on contrail formation is a prioritized area.
Area 5. Further use of measurements to improve theoretical models of the interaction between aerosols and cirrus clouds. Global model studies should strive to include both the direct and indirect effects, as well as the effects of the direct and indirect effects on high clouds.

Recommendations for best use of current tools for the prioritized research areas:

A combination of the FDTD (or DDA) method and the geometric optics method (GOM) is recommended for the development of the single-scattering properties of ice particles and aerosols in contrails and contrail-cirrus clouds. The FDTD and DDA methods are accurate approaches applicable to small particles. The GOM method is an approximation that is valid for large ice crystals. Several studies reported in the literature have shown that this combination of the FDTD and GOM methods is appropriate in the study of the optical properties of nonspherical ice crystals within natural cirrus clouds.

A new parameterization of the bulk radiative properties of contrails and contrail- cirrus clouds is necessary for the radiative transfer schemes used in several popular climate models such as the NCAR Community Atmosphere Model (CAM). The parameterization can be done on the basis of the single-scattering properties of contrails and contrail-induced cirrus clouds, which are derived from the FDTD (or DDA) and GOM. Furthermore, the validation of the parameterization can be carried out by comparing the model simulations of radiation fluxes and measurements (i.e., the CERES data and DOE-ARM ground radiometric measurements).

Existing satellite sensors, including the narrow bands (MODIS, GOES), broad bands (CERES), high-resolution spectra (AIRS), active (CALIPSO), multi-viewing angles (MISR), polarized (POLDER) sensors, provide an unprecedented opportunity to observe contrails and contrail-cirrus clouds. Moreover, the MODIS band 26 centered at 1.375 çm is effective for detecting thin and high clouds including contrails. We recommend to synergistically use the existing satellite-based retrieval products to quantify the extent of contrails and contrail-cirrus clouds from a global perspective.

To improve the current understanding of the formation of contrails and contrail- cirrus clouds, more in-situ observations regarding the relationship between contrails and ambient parameters are recommended. The prediction of water vapor supersaturation has recently been incorporated into some European models such as the ECWMF, ECHAM4, and IFSHAM models. Modeling efforts are also recommended to understand the influence of water vapor and temperature on contrail formation. Furthermore, in situ measurements can be used to validate and improve the representation of supersaturation in models.

The interaction between aerosols and contrail/cirrus clouds is still an open area. More in-situ measurements and laboratory experiments regarding the correlation of aerosols and contrails are recommended. Furthermore, modeling on the basis of cloud resolving models including the interaction of aerosols and contrail/cirrus clouds may also provide an efficient way to study the interaction between aerosols and contrail/cirrus clouds, which is complementary to the recommended in-situ measurements and laboratory experiments.

The costs and timeline for the aforementioned five prioritized research areas are estimated as follows:

Area 1: 2 FTE for 3 years
Area 2: 2 FTE for 2.5 years
Area3: 3FTEfor3years
Area4: 3FTEfor3years
Area 5: 4 FTE for 3 years

1. Introduction and Background

The first task of the Aviation-Climate Change Research Initiative (ACCRI) is to survey and document our current understanding of the impact of aviation on climate in seven key areas. One of these areas is the climate impact of contrails and contrail-cirrus clouds (i.e., contrails that occur within widespread cirrus cloud decks). These features, formed in the wakes of aircraft, provide one of the most visible anthropogenic effects in the atmosphere. They are often observed in the skies, especially near airports in the United States and Europe. Generally, contrails are composed of ice particles and are unique because they tend to have narrow widths and linear shapes, at least initially.

A contrail typically has a relatively short lifetime when formed in a subsaturated environment — e.g., descending air of a high pressure system — and unlikely to have a significant perturbation on climate. However, in supersaturated air — e.g., ascending air of a low pressure system — a contrail may be quite persistent, and can quickly (minutes to hours) spread into an extended cirrus deck. Contrails with longer lifetimes and larger horizontal extent may affect both the radiation budget and climate in a manner similar to natural cirrus clouds except that their microphysical properties are different (Gayet et al., 1996).

Cloud radiative forcing, defined as the difference of the radiative fluxes between cloudy and clear sky at the top of the atmosphere, *is a straightforward metric to estimate the effects of specific cloud on climate* (Ramanathan et al., 1989). It is therefore useful to document the current understanding of the radiative forcing of contrails and contrail- cirrus clouds.

Fu and Liou (1993) suggested that contrails have radiative effects similar to cirrus clouds because they are ice clouds. Thus, they may have significant regional climate effects. IPCC (1999) reported a best estimate of approximately 0.02 Wm^{-2} based on the study of Minnis et al. (1999) of the global distribution of contrail radiative forcing. While contrails may have large regional effects, the magnitude of the radiative forcing tends to reduce when averaged globally. Recently, the global radiative effects of contrails have been further studied by using different datasets, models and methods (Marquart et al., 2003; Fichter et al., 2005). These studies estimated the radiative impact due to linearly shaped contrails. Duda et al. (2001) investigated the evolution of solar and longwave radiative forcing in contrail clusters over the continental United States and Hawaii by using the GOES data and suggested that the microphysical properties (e.g., ice particle shape) of contrails may have an important effect on radiative forcing. Moreover, air traffic annually increases by 2-5%, according to Minnis et al. (1999, 2004).

All these studies indicate that the accuracy of contrail radiative forcing depends on the extent, persistence, and microphysical and optical properties (i.e., the optical thickness, ice particle size, and shape distributions) of contrails. In addition, almost all reported studies of contrails radiative forcing are focused on linearly shaped contrails that can be easily discriminated from natural cirrus in satellite images. Not included in these estimates is the influence of diffuse contrails (contrails that had spread over time) or the influence of aircraft emissions on naturally-occurring cirrus.

Ice particles in contrails and contrail-cirrus clouds are nonspherical particles. Goodman et al. (1998) collected samples of contrail ice particles using an impaction technique. Their results show that the ice habits at a temperature of approximately -6 1°C included hexagonal plates (75%), columns (20%) and a few triangular plates (<5%). The nonsphericity of these ice particles must be taken into account in the modeling of the radiative properties of contrails and contrail-cirrus clouds because the radiative impact can be quite different for clouds composed of spherical versus nonspherical particles. For example, Liou et al. (2000) demonstrated that the approximation of nonspherical ice particles as "equivalent" ice spheres for the single-scattering and radiative transfer processes can substantially underestimate ice cloud albedo. Moreover, the single- scattering properties associated with realistic ice particle morphologies must be used for a correct interpretation of other bulk optical properties of cirrus clouds. Consequently, it is important that the nonsphericity of ice particles be accurately modeled in the radiative transfer computations involving contrails or contrail-cirrus clouds.

2. Review of Current Knowledge of the Optical Properties and Radiative Forcing of Contrails and Contrail-Induced Cirrus Clouds

On any given day, clouds cover approximately 70% of the Earth's surface. Of this total, approximately 30% of the cloud cover occurs at high altitudes where the clouds are composed exclusively of ice particles. Substantial effort has been made in the last three decades to understand and determine the fundamental scattering and absorption properties of ice particles in these high altitude clouds. Early research efforts to account for the nonsphericity of ice crystals assumed that these clouds were composed of long circular cylinders. From the late 1970s to the 1990s, the geometric optics method, by means of the ray -tracing technique, was used extensively to investigate the single - scattering properties of relatively simple nonspherical ice particles (e.g., Takano and Liou, 1989a, b; Macke, 1996). Over the past decade, other methods such as the T-matrix method and the finite-difference time domain method have also been applied to the study of the optical properties of nonspherical ice particles (Mishchenko and Sassen, 1998; Yang and Liou, 2006). Laboratory and in situ measurements of the optical properties of nonspherical ice particles were also reported in the literatures (e.g., Gayet et al. 1996; Barkey et al. 2000). The research results from these efforts have been used in various applications in conjunction with the study of cirrus clouds, e.g., the parameterization of the radiative properties of ice clouds for use in climate models. *Given the wealth of recent progress (e.g., field campaigns and scientific publications) in the understanding of the microphysical and optical properties of ice clouds, it is necessary to first provide a condensed survey of the current scientific understanding of the scattering properties of non-spherical particles.*

2.1. Optical Properties of Ice Particles in Contrails and Contrail-Induced Cirrus Clouds

2.1.1. Size Distributions and Habits of Ice Particles in Contrails and Contrail-Induced Cirrus Clouds

A. Current State of Science

In-situ observations have demonstrated that ice crystals within contrails and contrail-induced cirrus have diverse size distributions. As an example, figure 1 shows a representative selection of ice crystal number distributions to illustrate the contrail evolution. In general, both small and large ice crystals are found in contrails (e.g., Gayet et al., 1996; Sausen et al., 1998; Minnis et al., 1998; Ström and Ohlsson, 1998; Meerkötter et al., 1999; Schröder et al., 2000; Ponater et al., 2002). During the transition of contrails into cirrus clouds, the ice particles become larger and the number of particles decreases (Schröder et al., 2000). Schröder et al. (2000) investigated the microphysical properties of contrails and contrail - induced cirrus from in situ observations performed during more than 15 airborne missions over central Europe. They found that the observed contrails were dominated by high concentrations (> 100 cm $^{-3}$) of ice crystals with mean diameters in the range of 1-10 çm. Larger ice crystals in the range 10-20 çm with typical concentrations 2 -5 cm $^{-3}$ were found in young contrail-induced cirrus, which mostly contain regularly shaped ice crystals.

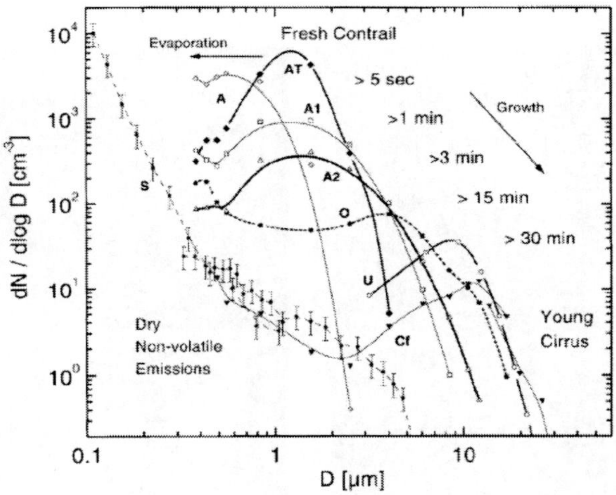

Figure 1. Representative selection of particle concentrations illustrating the transition of contrails into cirrus clouds. AT, A, A1, A2, O, and U are for contrail cases, CF is for young cirrus cloud, and S is for dry exhaust jet aircraft emission. Adapted from Schröder, et al. (2000).

Liou et al. (1998) reported two size distributions measured during the Subsonic Aircraft: Contrail and Cloud Effects Special Study (SUCCESS) on May 4, 1996. In comparison with two size distributions measured by FSSP for a contrail and a cold cirrus on April 18 and 19, 1994, it was shown that the maximum particle sizes in contrails were less than 100 çm. But the typical sizes of ice crystals in natural cirrus clouds range from several tens to hundreds microns.

Large ice crystals in contrails and contrail-induced cirrus are essentially nonspherical, as found from in-situ observations (e.g., Gayet et al., 1996; Sussmann, 1997; Liou et al., 1998; Goodman et al., 1998). Figure 2 shows the examples of scanning microscope images of small crystals (Goodman et al., 1998), including the hexagonal plates, columns, and triagonal plates. From microphotographs taken from airborne particle sampling in a young contrail (aged 3-4 min) at an altitude between 8 and 9 km and a temperature between –49^{0}C and –53^{0}C, Weickmann (1945, 1949) found a few large (~100 çm) hollow prisms typical of highly ice-supersaturated conditions. Strauss (1994) reported that the ice crystals in a y oung contrail had a droxtal habit (see figure 10) with sizes ranging from 1 to 5 μm. These observations were made with an ice replicator instrument. Sussmann (1997) presented photographs of a 120^0 parhelion and a 22^0 parhelion within persistent contrails. It was found that the contrails consisted of a considerable number of hexagonal plates orientated horizontally and with diameters between 300 and 2000 çm. This confirms that a subset of the particle population in persistent contrails may be composed of oriented plates and columns that grow at least as regularly as the most regular crystals found in natural cirrus. Immler and Schrems (2003) found that the southern hemispheric cirrus clouds tend to have larger particles than the northern hemispheric cirrus clouds. Furthermore, the southern hemispheric cirrus clouds have higher density of column-like particles whereas the northern hemispheric cirrus clouds seem to be dominated by plates, as found from lidar measurements. This probably reveals that the influence of human activities on the formation of cirrus clouds, which include the aviation emissions that are more intense over the northern hemisphere than over the southern hemisphere.

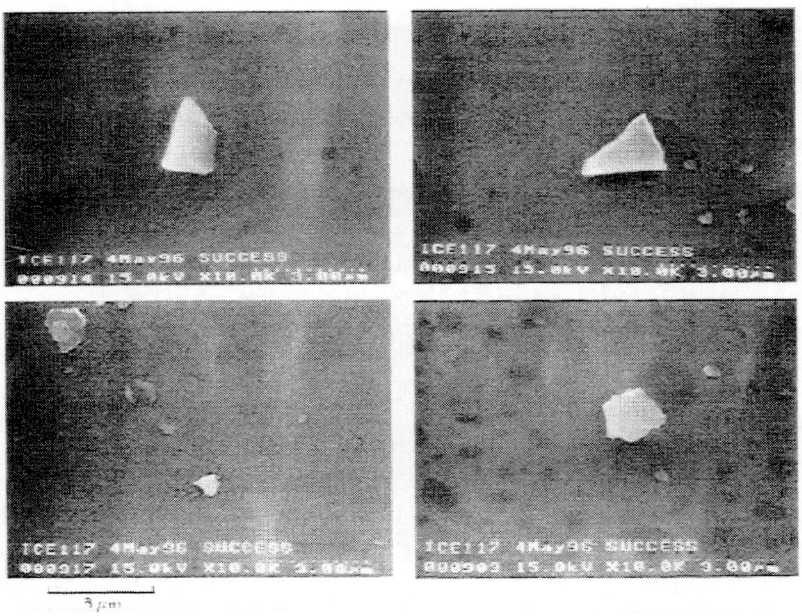

Figure 2. Scanning microscope images of small crystals (Goodmann et al., 1998). b. Critical role of size distributions and habits of ice crystals.

The particle size distribution and habit distribution are two microphysical parameters that are most important to the determination of the bulk optical properties of contrails or contrail-induced cirrus clouds. From the SUCCESS replicator data, Liou et al. (1998) showed that the habits of ice crystals within contrails are approximately 50% bullet rosettes, 30% hollow columns, and 20% plates. They also calculated the bulk single- scattering properties of contrails and cold cirrus shown in figure 3. The effective size (Foot, 1988) for a given size distribution is defined as follows:

$$D_e = \frac{3}{2}\frac{\int_{D_1}^{D_2} Vn(D)dD}{\int_{D_1}^{D_2} An(D)dD}, \tag{1}$$

where V and A are the volume and projected area of an ice crystal with a maximum dimension of D, respectively. The quantity n indicates the size distribution. The lower and upper limits of ice crystal sizes are D_1 and D_2, respectively. Note that effective size defined in Eq. (1) is different from that in Liou et al. (1998) by a factor of 3/2. The definition given by Eq. (1) is consistent with that used by the Moderate Resolution Imaging Spectroradiometer (MODIS) operational cloud products (King et al. 2004). Figure 3 shows the asymmetry factor, single-scattering albedo, and extinction coefficient for four ice clouds (Liou et al., 1998). It is evident from figure 3 that the bulk optical properties of ice crystals are quite sensitive to the effective size. Note that the differences in the extinction coefficient are caused by the differences in both the effective particle size and total number concentration of ice crystals.

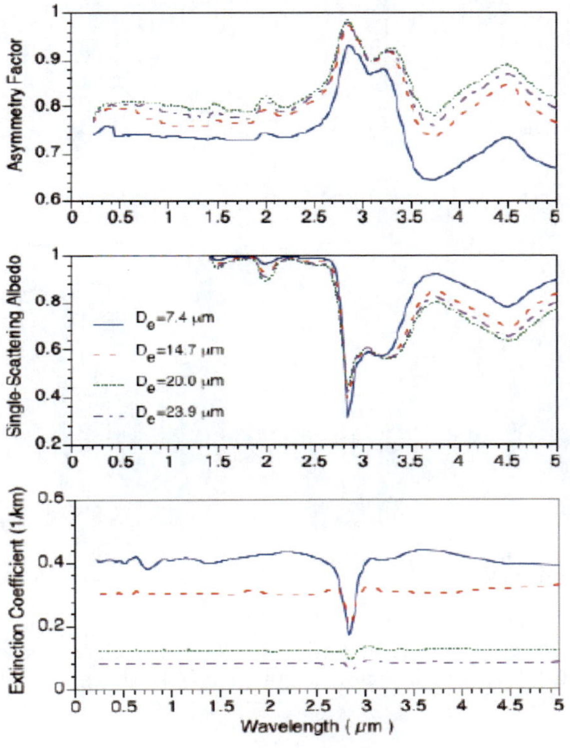

Figure 3. Asymmetry factor, single -scattering albedo, and extinction coefficient as functions of wavelength from 0.2 to 5 çm. The minima located at 2.85 çm are the well-known Christiansen effect. Adapted from Liou et al. (1998). The values of the effective sizes shown in the figure have been converted to the definition given by Eq. (1).

Goodman et al. (1998) collected ice crystal samples in the cont rail using an impaction technique. Their results show that the crystal habits at a temperature of -61°C included hexagonal plates (75%), columns (20%) and a few triagonal plates (<5%). The scattering properties of the ice crystals in contrails and natural cirrus are different and result in differences in their radiative forcing. Figure 4 illustrates the effect of ice particle sizes and habits of contrails, contrail-induced cirrus, and natural cirrus on the solar and infrared radiation. The influences of particle effective size and habit on radiation are stronger for solar radiation than for infrared radiation, particularly for thicker cirrus clouds. Furthermore, the influences of particle effective size and habit on infrared radiation are observed primarily for thin cirrus clouds

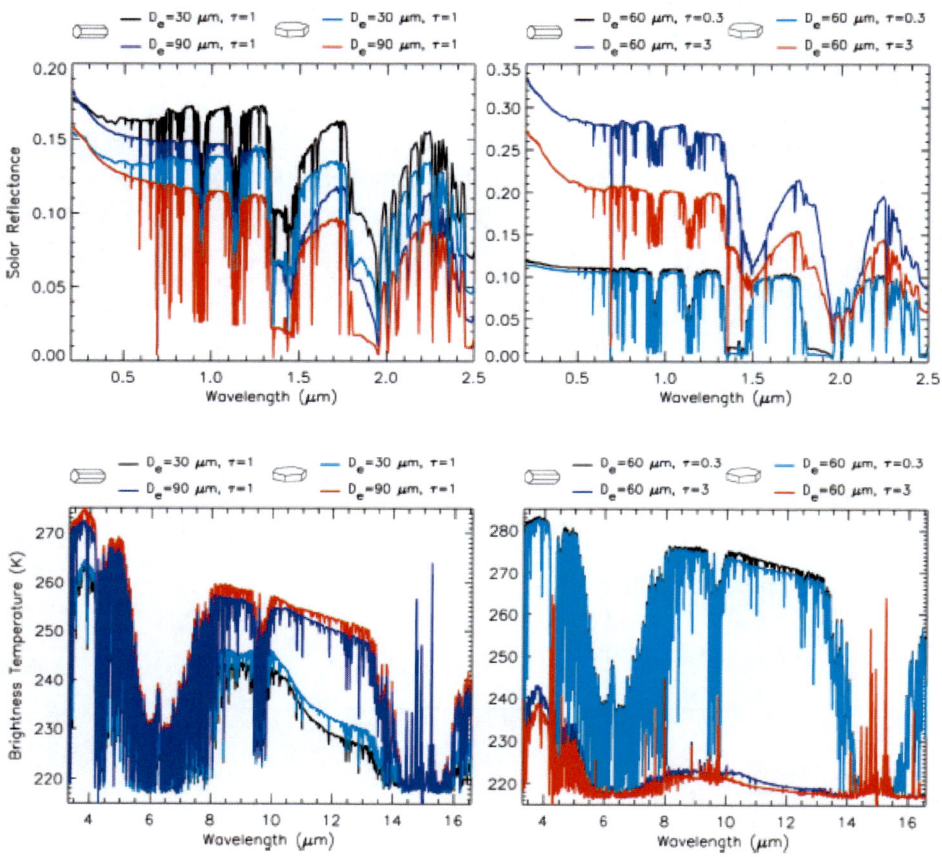

Figure 4. Effects of shape, ice particle size, and optical thickness on solar reflectance spectra (upper panels) and infrared brightness temperature spectra (bottom panels) (Hong et al., unpublished).

2.1.2. Measurements and Simulations of the Optical Properties of Ice Crystals

A. Current State of Science

As discussed above, the microphysical and optical properties of nonspherical ice crystals are important to the understanding of the radiative impact of contrails. Figures 5 and 6, adapted from Liou et al. (2000), illustrate the importance of the ice particle nonsphericity in the case of cirrus clouds. It is expected that the nonsphericity of ice crystals would be important as well in the case of contrails and contrail-induced cirrus clouds. Specifically, figure 5 shows the cirrus albedo for the solar spectrum, calculated by assuming particles are plates/columns or spheres, in comparison with measurements from the FIRE field campaign held in Wisconsin during 1986. The shaded area indicates the range of the results corresponding to equivalent spheres. It is evident that neglecting the nonsphericity can lead to significant errors in the radiative effect of cirrus clouds.

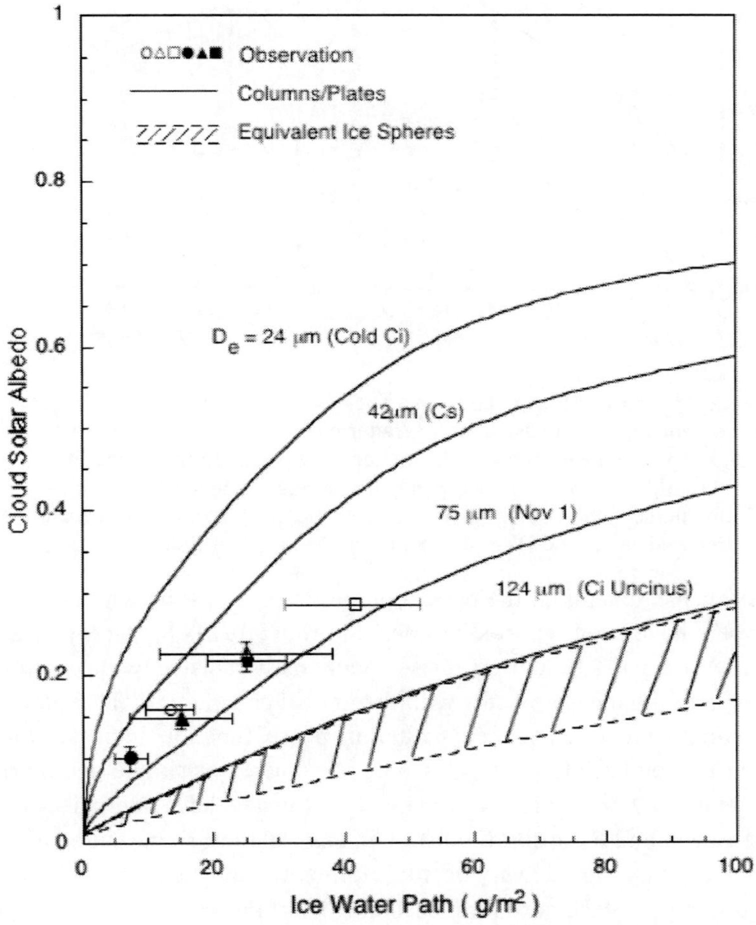

Figure 5. Solar albedo as a function of ice water path determined from broadband flux observations from aircraft for cirrus clouds that occurred during the FIRE experiment, Wisconsin, November-December, 1986 (Stackhouse and Stephens 1991). The solid lines represent theoretical results computed from a line-by-line equivalent solar model using observed ice crystal sizes and shapes for a range of mean effective ice crystal sizes. The dashed lines are corresponding results for equivalent spheres. Adapted from Liou et al. (2000).

Figure 6 shows the sensitivity of surface temperature to high cloud coverage or the ice water path associated with these clouds based on a 1D cloud and climate model (Liou et al. 2000). For present climate conditions, the simulations based on "equivalent spheres" for nonspherical ice crystals lead to a 0.4-K overestimation of the surface temperature. Thus, the nonsphericity of ice crystals is significant for climate studies.

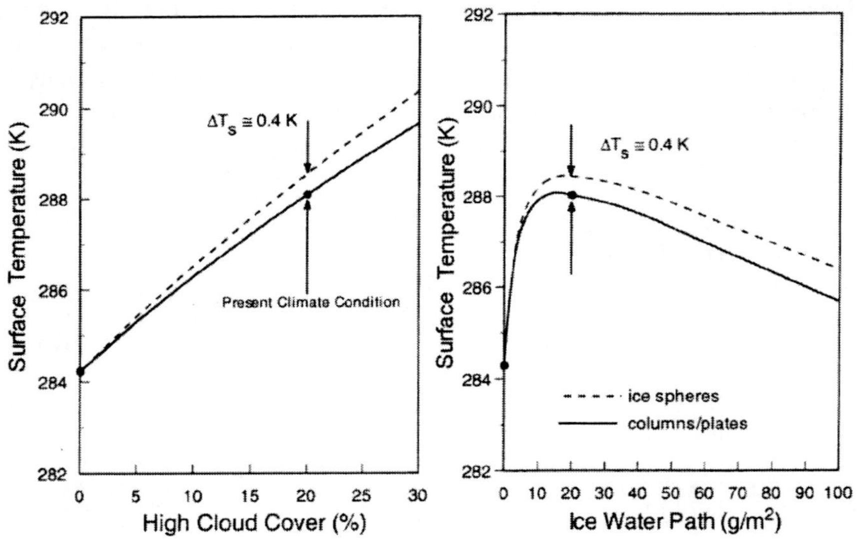

Figure 6. Surface temperatures determined from a one-dimensional cloud and climate model using a radiative transfer parameterization based on the scattering and absorption properties of hexagonal columns/plates and equivalent ice spheres. The model has a present climate condition corresponding to a surface temperature of 288K involving a typical cirrostratus located at 9 km with a geometric thickness of 1.7 km, an ice water content of 10^{-2} g m $^{-3}$, and 20% fractional coverage. Perturbations are performed for both cloud cover and IWP. Adapted from Liou et al. (2000).

Laboratory measurements of the optical properties of nonspherical ice crystals generated in cloud chambers have been reported (Sassen and Liou, 1979a,b; Barkey et al. 2000). Gayet et al. (1998) measured the scattering phase functions associated with contrails and natural cirrus and compared the measurements with theoretical calculations. They concluded that, "in contrails and natural cirrus, measured scattering phase function indicates major difference with those used in cloud models which assume ice spheres or simple geometric shape of ice crystals." The scattering properties of individual ice crystals have been measured by Bacon et al. (1998) and Bacon and Swanson (2000). Although laboratory measurements of the optical properties of ice crystals provide very useful information, they are limited in terms of spectral coverage, angular range (necessary for measurements of the scattering phase function), and an incomplete set of the single-scattering properties (i.e., the phase function, extinction cross sections, and single-scattering albedo are not simultaneously measured). Thus, for many practical applications (e.g., the parameterization of the bulk radiative of ice clouds for applications to climate models), a theoretical approach is generally used to infer the single -scattering properties for a wide variety of ice habits.

At present, four numerical methods are usually applied to the computation of the optical properties of nonspherical ice crystals: the geometric optics method (the so-called ray-tracing technique), the T-matrix method, the finite-difference time domain (FDTD) method, and discrete dipole approximation (DDA) method.

The early applications of the principles of geometric optics to the scattering of light by nonspherical ice crystals can be traced back to the studies of Jacobowitz (1971) and Wendling et al. (1979) for 2D and 3D hexagonal ice crystals, respectively. In 1980s and 1990s, the ray-tracing technique was used extensively to compute the optical properties of nonspherical ice crystals with various shapes (Cai and Liou, 1982; Takano and Jayaweera, 1985; Takano and Liou, 1989a,b; Macke, 1993; Hess and Wiegner, 1994; Macke et al., 1993, 1996, Yang and Liou, 1998). Modified geometric optics methods have also been developed (Muinonen, 1989; Yang and Liou, 1995, 1996b, and 1997) to overcome some shortcomings in the conventional ray-tracing technique.

Figure 7 illustrates the physical basis of the conventional ray-tracing technique. When the size of an ice crystal is much larger than the incident wavelength, the incident radiation can be thought of as being composed of a bundle of localized waves or rays. Snell's law and the Fresnel formulas can be employed to trace the propagation of the incident ray and calculate the magnitude and phase of the electric field vector associated with the ray. After propagations of all the incident rays are traced, the angular distribution of scattered energy in conjunction with the incident rays can be obtained. Additionally, diffraction phenomenon also contributes to the scattered energy, which can be accounted for in terms of the classical Fraunhifer diffraction theory.

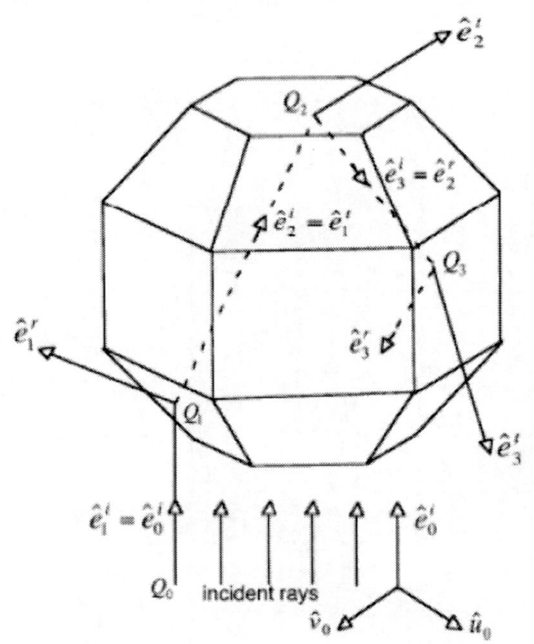

Figure 7. A conceptual diagram for the principle of the ray-tracing technique for computing the single-scattering properties of a particle that is much larger than the incident wavelength. Adapted from Yang and Liou (2006).

The T-matrix method (Waterman 1971, Mishchenko and Travis, 1994) is a computationally efficient method for deriving the optical properties of small and moderately-sized particles. This method is usually applied to axially symmetric particles (e.g., spheroid and circular cylinders) although this method, in principle, can be applied to an arbitrary particle shape. At present, it is quite challenging to implement the T- matrix method for complex ice crystal shapes such as bullet rosettes and aggregates. The T-matrix method was applied to investigate the depolarization of lidar returns associated with contrails

(Mishchenko and Sassen, 1998). The technical details of the method are described in Mishchenko et al. (2000).

The FDTD method pioneered by Yee (1966) is a powerful numerical method to solve electromagnetic scattering problems. In principle, the FDTD method solves Maxwell's equations in the time domain. Figure 8 illustrates the basic principles of the FDTD method. In the FDTD method, a finite spatial region containing a scattering particle is discretized in terms of a grid mesh and the time-dependent Maxwell curl equations are replaced with their finite-difference analogs. To suppress the artificial reflection due to the truncation of the computational domain, an absorbing boundary condition is applied at the edges of the domain. Then, a plane wave is introduced into the computational domain. The interaction of the incident wave and the scattering particle can be simulated by the finite-difference analogs of the Maxwell equations in the time domain. The signals regarding the interaction between the particle and the incident wave in the time domain can be transformed into their counterparts in the frequency domain. After the near field is obtained in the frequency domain, the near field can be mapped to the far field to derive the single-scattering properties of the particle.

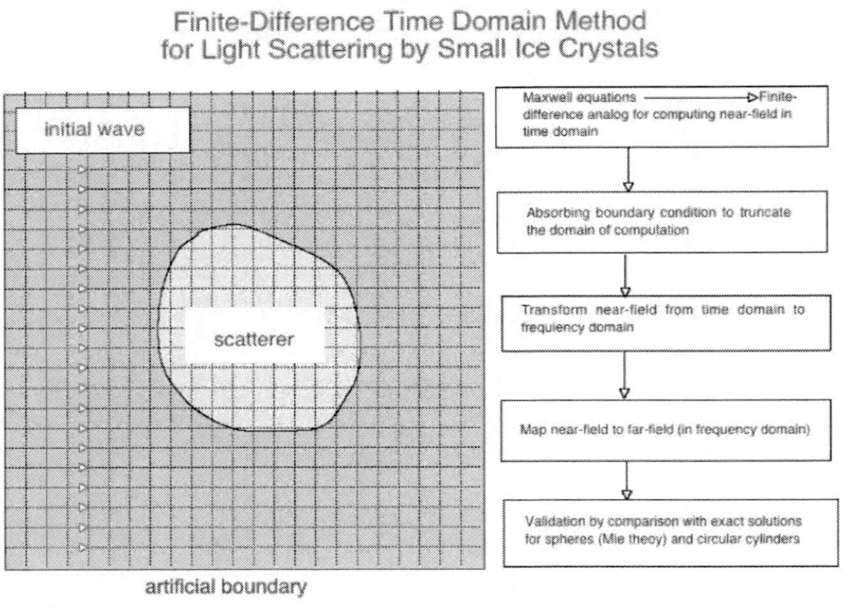

Figure 8. Schematic diagram illustrating the basic principle of the finite-difference time domain method.

The DDA method was developed originally by Purcell and Pennypacker (1973). In the DDA computation, a scattering particle is approximated by a number of electric dipoles, as shown in figure 9. Each dipole responds to local field that is the superposition of the incident field and the field induced by the other dipoles. The governing equations for the coupled dipoles are a set of linear equations. The solutions to the linear equations provide the field scattered by the dipoles, which can be used to compute the optical properties of the scattering particles. A DDA computational code developed by Drain and Flatau (1994) has been released for non-commercial applications.

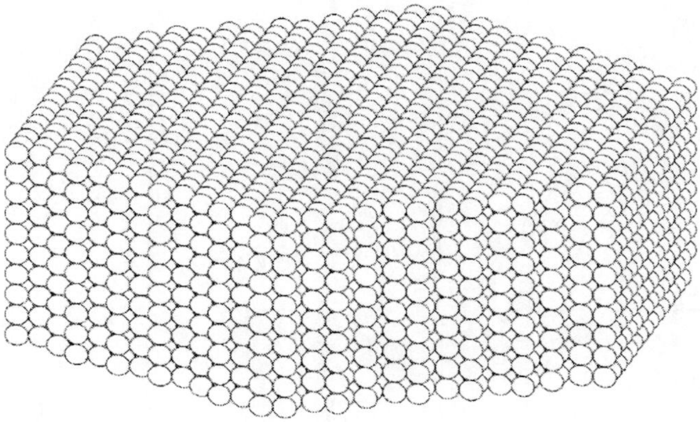

Figure 9. Representation of a hexagonal plate ice crystal for the DDA computation.

As examples, figures 10 and 11 show the scattering phase functions of randomly - oriented large and small ice crystals, computed from the ray-tracing technique and the FDTD method, respectively. For large aggregates shown in figure 10., the surfaces of the particles are assumed to be rough. Note that surface roughness for the aggregates results in a smoother phase function in comparison with the pristine crystals such as the columns and plates. The phase function describes the scattered energy of unpolarized incident radiation. It is evident from figure 10 that the phase functions of large ice crystals show strong scattering peaks with the exception of the aggregates. The peaks at scattering angle 22° and 46° correspond to the well-known 22° and 46° halos. Such peaks are not observed for small ice crystals.

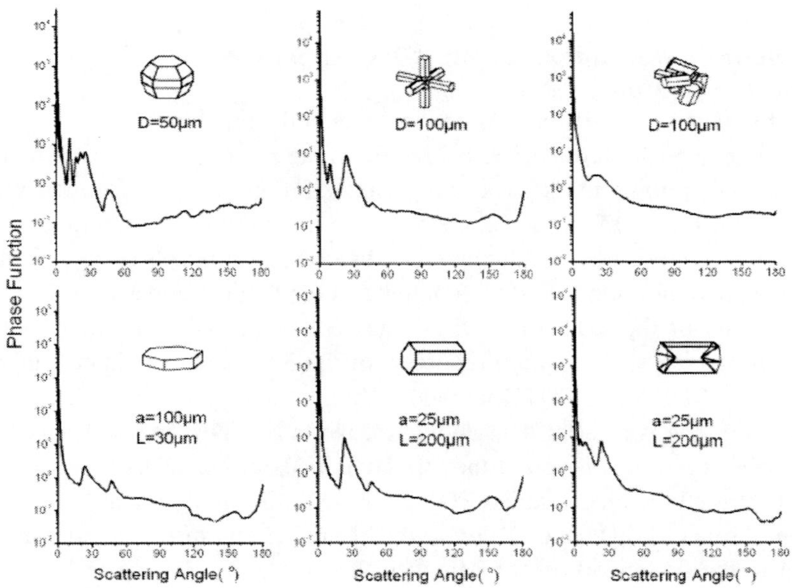

Figure 10. Comparison of the phase functions computed from the geometric optics method for six ice crystal habits (top row, left to right: droxtals, bullet rosettes, aggregates; bottom row, left to right: plates, solid columns, and hollow columns). Aggregates have surface roughness included in the light scattering computations. Adapted from Yang and Liou (2006).

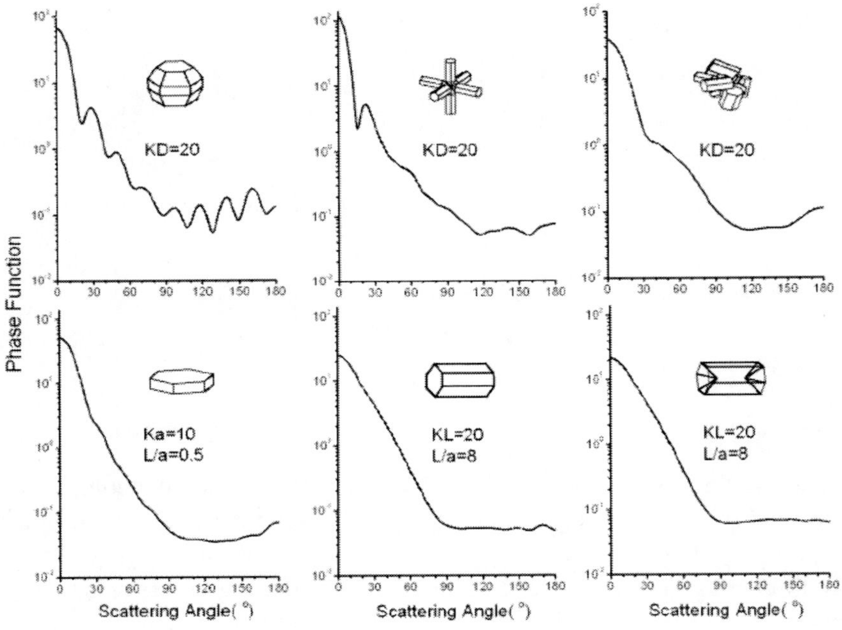

Figure 11. Comparison of the phase functions computed from the FDTD method for ice ice crystal shapes that are commonly observed in ice clouds. The parameter, D, is the maximum dimension for a droxtal, a bullet rosette, or an aggregate ice crystal. For plates and columns, a denotes the half-width and L is the length (for columns) or thickness (for plates). K=2/is the wavenumber. Adapted from Yang and Liou (2006).

B. Present State of Modeling Capability/Best Approach for Light Scattering Computation

The performances of different numerical methods rely strongly on the shape, size, orientation, and composition (or refractive index) of the particle. The most practical approach at present for computing the optical properties of nonspherical ice crystals may be a combination of FDTD, DDA, T-matrix methods and the ray-tracing technique.

FDTD and DDA are two methods that are applicable to arbitrarily shaped inhomogeneous particles. On a desktop computer, the practical upper limit of size parameter for the application of the FDTD (or DDA) method to the scattering of light by randomly oriented small ice crystals is on the order of 20-30. A comparison study of parallel implementations of DDA and FDTD methods (Yurkin, 2007) shows that DDA is faster when the refractive index m is smaller than about 1.4, while FDTD is more efficient for larger m. According to the spectral variation of the refractive index of ice, DDA may be suitable for the UV and near infrared regions, whereas FDTD for the far infrared region.

Different from FDTD and DDA, which are exact methods for light scattering computation in the framework of the Maxwell theory, the ray-tracing technique is based on the geometric optics, an approximate and asymptotic electromagnetic theory. This method is quite inaccurate if the size parameters are less than 50. Thus, substantial errors may be incurred for moderate sizes where the solutions of the FDTD (or DDA) method and the ray-tracing technique are emerged because the edge effect is not included in the ray-tracing technique. The advantage of this method is that the internal field or near field can be obtained quickly, although not accurate enough. In order to improve the accuracy of optical properties in the region of moderate size parameters, a possible hybrid method with two steps may be used. In the first step, the ray-tracing technique is employed to estimate the internal field within the particle on each grid. This process may not take much CPU time. In the second step, the approximate internal field is going to be substituted into DDA equations as the initial value. As the iteration is executed, the accuracy of the internal field may be improved with the edge effect considered.

2.2. Radiative Forcing Contrails and Contrail-Induced Cirrus Clouds

A. Current State of Science

As one of the most visible anthropogenic effects in the atmosphere, contrails and contrail-cirrus are a common sight in the skies over regions with heavy air traffic. Figure 12 shows the effect of persisting contrails on the diurnal temperature range (Travis et al., 2002). This range tends to be reduced by the occurrence of contrails, which affect the transfer of solar and infrared radiation, as demonstrated by measurements taken around 9/11/01, when all civil and commercial aircraft were temporarily grounded.

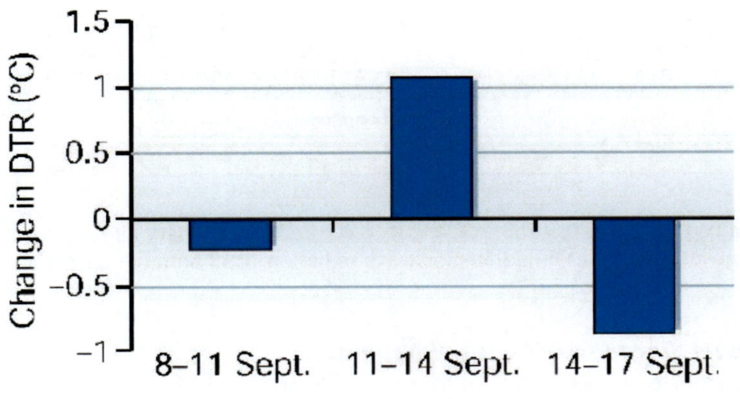

Figure 12. Differences between the average diurnal temperature ranges and the normal values derived from 1971-2000 climatology data for the indicated three-day periods in September 2001 (Travis et al., 2002). All commercial aircraft traffic was grounded in 11-14 September.

Duda et al. (2003) reported that the values of the radiative forcing of global linearly-shaped contrails derived by Minnis et al. (1999) differ from that derived by Ponater et al. (2002) by nearly two orders of magnitude. Figure 13 shows the radiative forcing of contrails calculated by Ponater et al. (2002). In addition, almost all existing studies about contrail radiative forcing focus on linearly-shaped contrails that can be discriminated from natural cirrus in satellite images. These results represent the minimum impact because they do not include the cirrus shields that develop from expanding contrails (Minnis et al. 1998; Duda et al. 2001).

The expansion of contrails into larger -scale cirrus provides another significant effect of contrails on climate, and requires further observational and theoretical modeling studies. We need to investigate the effects of contrail/cirrus coverage and their mean optical thickness to estimate their radiative forcing accurately. The contrail and contrail- induced cirrus radiative forcing depends on the microphysical, macrophysical, and optical properties of contrail and contrail-induced cirrus, as well as the temporal characteristics (i.e., longevity) of these clouds. These uncertainties are discussed in section 3.

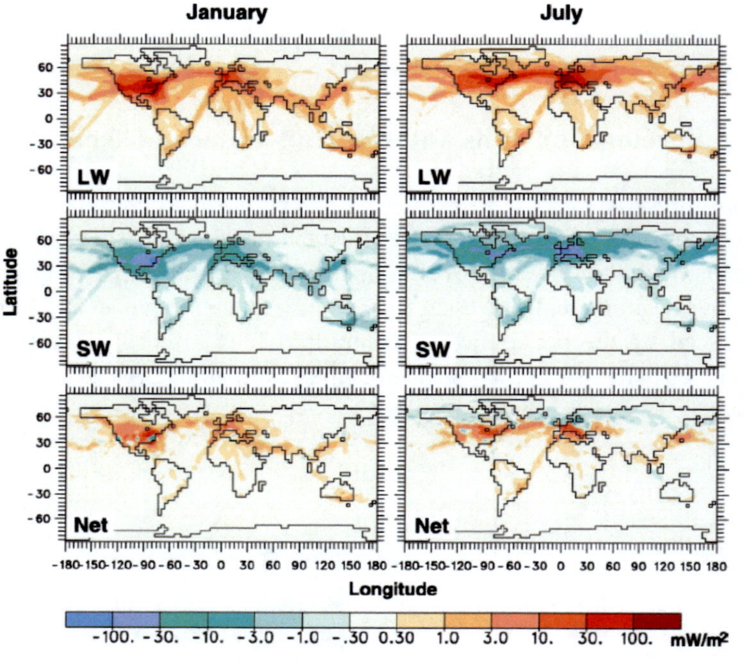

Figure 13. Radiative forcing at the tropopause due to contrails for January and July (Ponater et al., 2002). Top (longwave radiation), middle (shortwave radiation), and bottom (net radiation).

b. Advancements Since the IPCC 1999 Report

The radiative effects of contrails may be similar to those of natural cirrus clouds because both are ice clouds (Fu and Liou, 1993). Contrails may have significant regional climatic effects (Liou et al., 1990). However, the climate effects which are measured by the radiative forcing of contrails and contrail-induced cirrus are extremely uncertain. In IPCC (1999), a best estimate based on the study of Minnis et al. (1999) of the global distribution of the radiative forcing of contrails observed in 1992 is approximately 0.02 Wm^{-2}. Recently, the global radiative effects of contrails have been further investigated using different datasets, models, and methods (Myhre and Stordal, 2001; Marquart et al., 2003; Fichter et al., 2005; Stuber and Forster, 2007). From these new investigations, the global mean and annual mean radiative forcing due to line-shaped contrails of Minnis et al. (1999) were overestimated. Sausen et al. (2005) reported a value of 0.01 Wm^{-2} as the best estimation of contrail radiative forcing in 2000, which is only half of the value provided in IPCC (1999). Figure 14 shows the annual mean net radiative forcing at the top of the atmosphere for the contrails observed in 1992 with an optical thickness of 0.3 (Minnis et al., 1999). Evidently, the maximum values of contrail radiative forcing are observed over the United States and Europe.

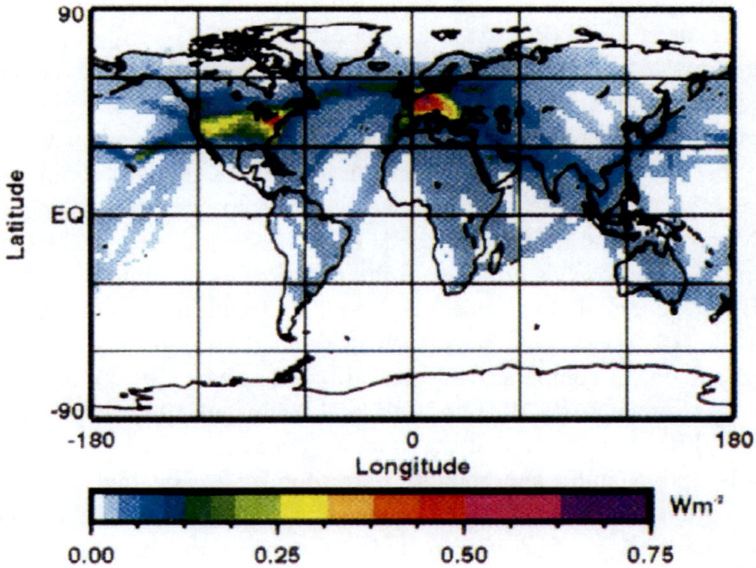

Figure 14. Annual mean net radiative forcing at the top of atmosphere for estimate contrails with an optical thickness of 0.3 in 1999. Adapted from Minnis et al. (1999).

3. Outstanding Limitations, Gaps and Issues that Need Improvement

3.1. Improvements Are Needed on the Computation of the Optical Properties of Ice Crystals

An efficient method at present for computing the light scattering properties of nonspherical ice crystals may be a combination of FDTD, DDA, T-matrix and the ray- tracing technique. The computational efficiency of existing computer codes (FDTD and DDA, in particular) needs to be substantially improved for practical applications.

3.2. Uncertainties in Estimates of Radiative Forcing of Contrails and Contrail-induced Cirrus Clouds

3.2.1. Contrail and Contrail-Induced Cirrus Covers

The estimates of contrail and contrail-induced cloud covers are important since air traffic increases 2-5% annually (Minnis et al., 1999). However, global contrail and contrail-induced radiative forcing is difficult to estimate since the global mean contrail and contrail-induced coverage is poorly known. Sausen et al. (1998) estimated the global mean cover by linearly-shaped contrails to be about 0.1%. The value of 0.02 Wm^{-2} for the global and annual mean radiative forcing by line-shaped contrails (Minnis et al., 1999) was based on this value of contrail cover. Ponater et al. (2002) reported that the annual average for visible contrail coverage amounts to 0.07%. In regions with high air traffic density, time mean coverage of more than 2% has been found (Ponater et al., 2002).

The uncertainty in contrail coverage estimates has been illustrated by the results from different detection methods (Duda et al., 2003). There are two methods generally used to infer contrail coverage: One approach uses a parameterization in a numerical weather prediction model to diagnose contrails based on the ambient conditions (Sausen et al., 1998; Ponater et al., 2002; Duda et al., 2003). The other approach is based on satellite imagery analysis (Bakan et al., 1994; Mannstein et al., 1999; Palikonda et al., 1999; Meyer et al., 2002). The contrail coverage from these approaches provides different estimates of contrail coverage and spatial distribution. But in general, both approaches depend on the air traffic density. Boucher (1999) and Fahey et al. (1999) showed that cirrus occurrence and coverage tend to increase in regions of high air traffic compared with the rest of the globe (IPCC). Figure 15 shows the trends in cirrus coverage from 1971 to 1995 and estimated 1992 linear contrail coverage (Minnis et al., 2004). It is clear that the largest concentrated increases occurred over the northern Pacific and Atlantic and roughly corresponded to the high air traffic densities. Moreover, the regional and local climate response to the regional and local air traffics with high aviation emissions is different from the global mean radiative forcing that is highly important to the global mean climate change.

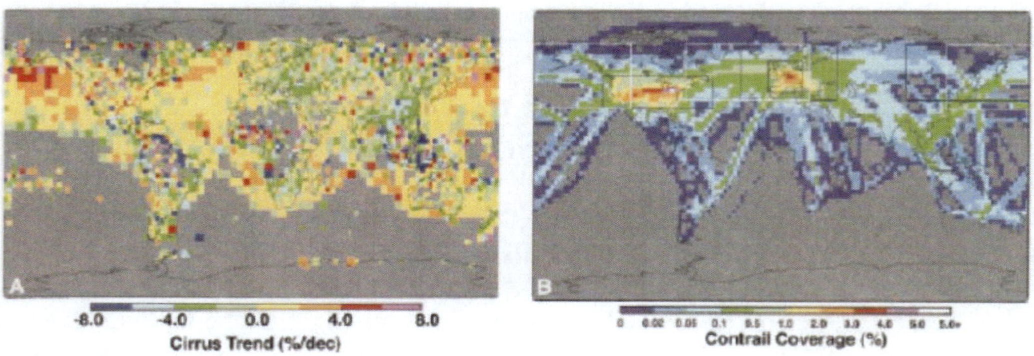

Figure 15. (a) Trends in cirrus coverage from 1971 to 1995 and (b) estimated 1992 linear contrail coverage. Adapted from Minnis et al. (2004)

Stordal et al. (2005) estimated the trends in cirrus cloud cover on the basis of 16 years of data from the International Satellite Cloud Climatology Project (ISCCP, Rossow and Schiffer, 1999). The trends were then spatially correlated with aircraft density to determine the variations in cirrus cloud cover due to aircraft traffic. Cirrus cloud amount increases were reported to accompany an increase in aircraft in the period of 1984-1999. They found that the strongest influence on cirrus clouds occurs in the regions with highest aircraft traffic. However, they also documented that the relationship between cirrus cloud cover and aircraft density was uncertain and they could not draw firm conclusions or quantify the effect with high certainty. Figure 16 shows the correlation coefficients between trends in cirrus cloud amount from ISCCP and aircraft traffic density (Stordal et al., 2005). The correlations are moderate and many other factors may also have contributed to variations in cirrus cloud amount. Their conclusions were based on monthly mean data, and they suggested the use of daily data to improve their results.

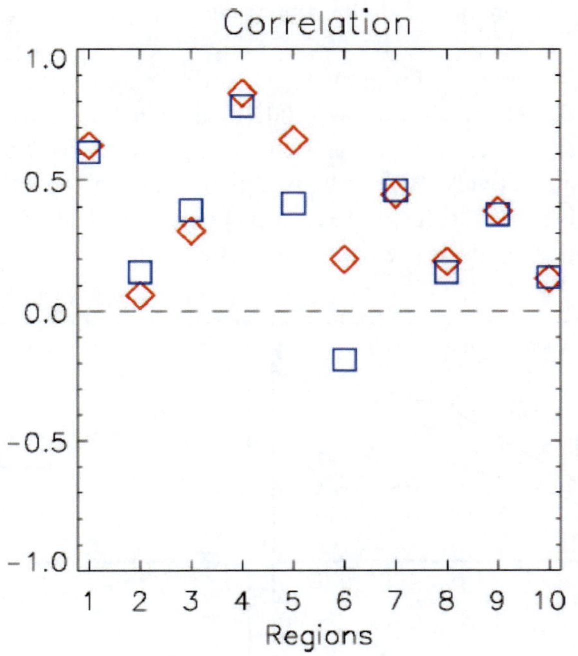

Figure 16. The correlation coefficients between trends in cirrus cloud amount from ISCCP and aircraft traffic density over chosen ten regions. Diamond and square indicate two different methods to determine the cirrus clouds. Adapted from Stordal et al. (2005).

3.2.2. Contrail and Contrail-Induced Cirrus Optical Thicknesses

The optical thickness determines in large part of the radiative forcing of contrails and contrail-induced cirrus. The optical thickness of contrails is typically between 0.1 and 0.5 (e.g., Sassen, 1997; Minnis et al., 1998), and can sometimes be as low as 3.0 10 (Schröder et al., 2000). Larger values of optical thickness may approach or exceed 1 at warmer temperatures (up to -30¥C) (Schumann and Wendling, 1990; Gayet et al., 1996). Different mean contrail optical thicknesses have been used to study the radiative forcing of contrails globally and regionally. For example, an optical thickness of about 0.1 was assumed by Stuber and Forster (2007) for a global mean; 0.15 was used by Ponater et al. (2002) for a global mean; 0.2 was a value given over the life cycles of contrails that developed into cirrus clouds (Duda et al., 2004). Minnis et al. (2002) estimated an optical depth of 0.26 for a large area of contrails from initiation to dissipation; Palikonda et al. (2004) found a similar value of 0.26 using an automated analysis data from two satellites during all of 2001; 0.3 was used by Myhre and Stordal (2001) for a global mean; and 0.52 was used by Meerkötter et al. (1999), which combined with a global mean contrail cover of 0.1%, leads to 0.01 to 0.03 Wm $^{-2}$ daily and annual mean radiative forcings.

The dependence of radiative forcing on the assumed contrail optical thickness is shown in figure 17 (Meerkötter et al., 1999). At the top of atmosphere, the longwave forcing is larger than the shortwave forcing. Thus, the radiative forcing of contrail and contrail-induce cirrus is positive and generally increases with increasing optical thicknesses. However, at the surface, the longwave forcing is smaller than the shortwave forcing, which results in negative radiative forcing.

As optical thickness increases for naturally-occurring cirrus, the net radiative forcing at the top of atmosphere decreases. Hong et al. (2007 b) analyzed three years of MODIS data, it was found that the mean optical thickness and effective particle size of tropical ice clouds are about 8 and 50 çm, respectively. The radiative forcing at the top of the atmosphere and surface by ice clouds is shown in figure 18 as a function of optical depth. The figure shows the impact of optical thicknesses on the radiative forcing.

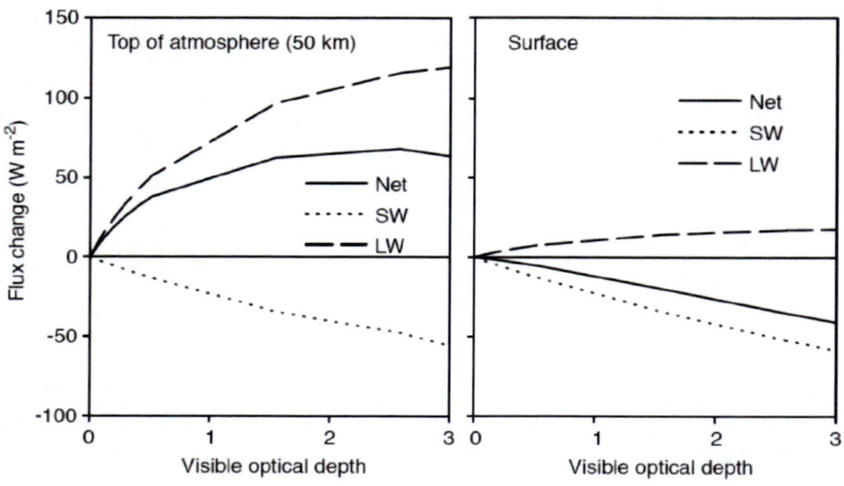

Figure 17. Computed shortwave and longwave net fluxes for 100% contrail cover at the top of atmosphere and at the surface as a function of optical thickness of contrails Adapted from Meerkötter et al. (1999).

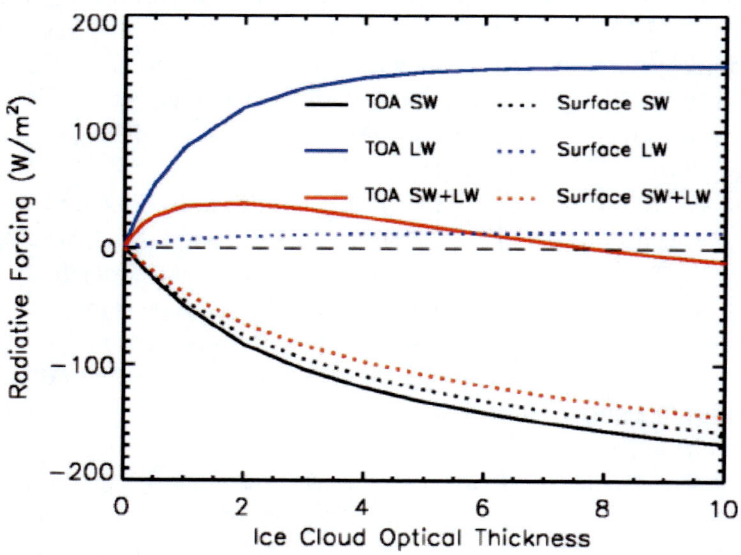

Figure 18. Computed shortwave and longwave net fluxes for 100% cirrus cover at the top of atmosphere and at the surface as a function of optical thickness of cirrus (Hong et al., unpublished).

The contrail optical thickness also varies geographically, which affects the distribution of the radiative forcing. Figure 19 shows the spatial and interannual variability of the optical thickness at pressures of 200 and 250 hPa (Ponater et al., 2002). The optical thickness of contrails in the extratropical regions is considerably smaller in winter (mostly less than 0.1) than in summer.

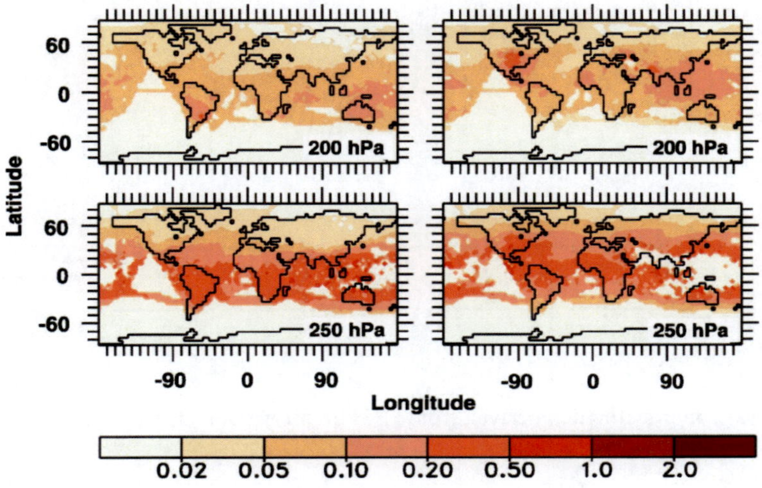

Figure 19. The geographical and seasonal distributions of optical thickness for visible contrails at 200 and 250 hPa. Left is for January and right for July. Adapted from Ponater et al. (2002).

3.2.3. Ice Particle Sizes in Contrail and Contrail-Induced Cirrus

Significant problems exist with the measurement of small ice particles as reported by Heymsfield et al. (2006) and Garrett (2007). The ice particle sizes in contrail and contrail-induced cirrus clouds can range from a few microns to hundreds of microns or more. Measuring the ice particle sizes over this range requires several instruments (e.g., Schröder et al., 2000), such as the forward scattering spectrometer probe (FSSP), the optical array probe (OAP), high -volume precipitation spectrometer (HVPS), cloud virtual impactor (CVI), Hallett-type repicator (REP), cloud intergrating nephelometer (CIN), and most recently the small ice detector (SID).

The agreement between measurements from the different instruments has not always been good. Specially, the measurement of small ice particles strongly depends on the instrument (e.g., Heymsfield et al., 2006; Garrett, 2007). Heymsfield et al. (2006) compared the in situ direct and indirect measurements of extinction coefficient, as well as measured values from lidar. They reported that the direct-extinction measurements used by Garrett et al. (2003) are overestimated by a factor of 2 to 2.5.

Figure 20 shows the extinction coefficient estimated from the FSSP and from the particle probes, as well as values measured directly by the CIN probe. Heymsfied et al. (2006) speculated that the source of the errors is from large ice particles that shatter on the housing of the instrument aperture, and subsequently intersect the sample volume. Shattering does not affect ice mass density but increases total surface area, thereby affecting the inference of extinction coefficient. Heymsfield et al. (2006) concluded that indirect measurements of ice particle sizes are the most accurate.

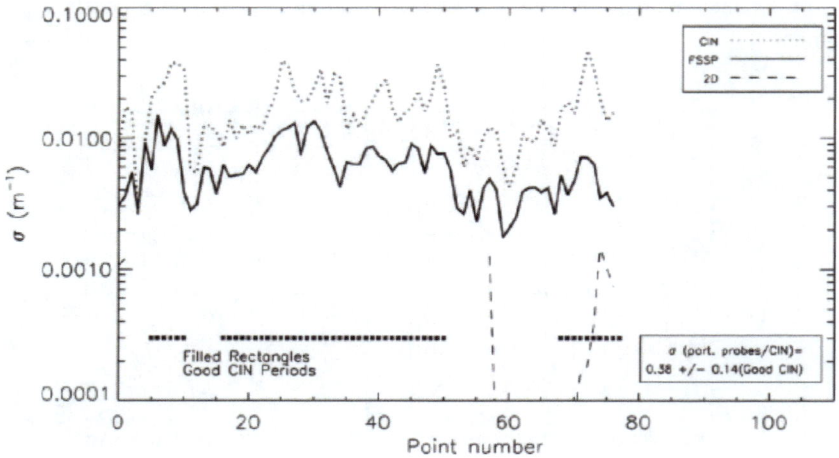

Figure 20. Comparison of extinction derived from FSSP to measurements from the CIN probe. Adapted from Heymsfield et al. (2006).

Garrett (2007) reassessed the analysis and conclusions of Heymsfied et al. (2006). It was there found that the discrepancy between the CIN and particle probe measurements of extinction coefficient during CRYSTAL-FACE was not from shattering. The measurements of ice particle sizes and optical thickness of cirrus clouds derived using a CIN probe were in agreement with those derived from a variety of passive remote sensors employing different retrieval algorithms. It is further confirmed that measurements from different FSSP-type probes can differ greatly in their estimation of ice water content and extinction coefficient although they operate on the same basic principles. Noel et al. (2007) compared extinction coefficients retrieved in ice clouds from lidar observations using a CALIPSO-like algorithm to in-situ measurements from the CIN during CRYSTAL-FACE. The results show a very good agreement between both instruments, as evident from figure 21. Despite these achievements, the microphysical and optical properties of small ice particles need to be further investigated.

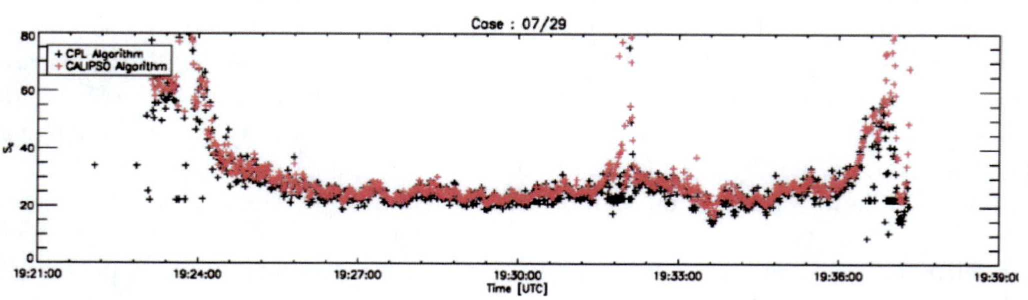

Figure 21. Lidar ratio from the cloud physical lidar (black) CALIPSO algorithm (pink) Adapted from Noel et al. (2007).

3.2.4. Diurnal Variations of Air Traffic

Solar radiative forcin g varies substantially with solar zenith angley (Myhre and Stordal, 2001). Figure 22 from Myhre and Stordal (2001) shows how the solar radiative forcing varies with solar zenith angle for different values of surface albedo with a fixed mid-latitude summer atmosphere. The strongest negative forcing was found at high solar zenith angles between 75°-80° (Meerkötter et al., 1999; Myhre and Stordal, 2001). The strong forward scattering of larger ice particles leads to a decrease in backscattering at low solar zenith angles (Haywood and Shine, 1997; Myhre and Stordal, 2001). Myhre and Stordal (2001) also investigated the diurnal variation in the global shortwave radiative forcing performed for 1% homogeneous global contrail cover, with results shown in figure 23. The forcing during sunrise and sunset substantially differs from that at noon.

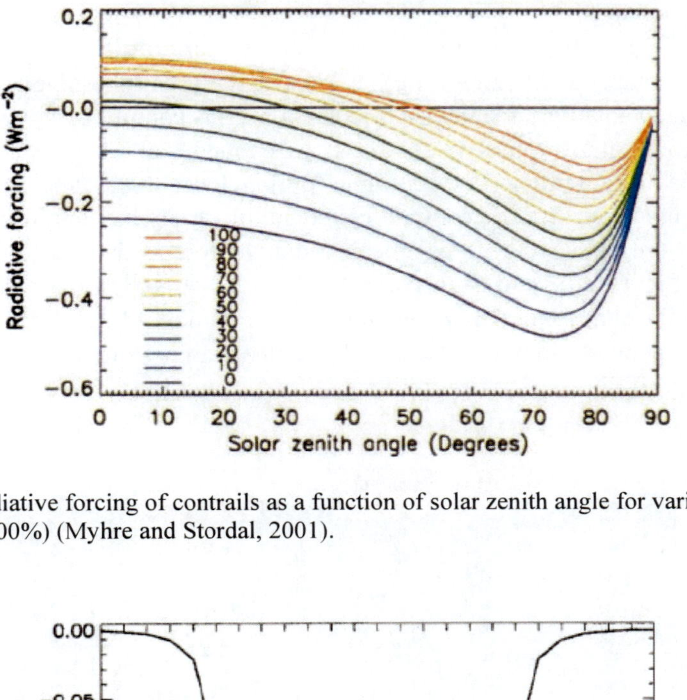

Figure 22. Solar radiative forcing of contrails as a function of solar zenith angle for various values of surface albedo (0-100%) (Myhre and Stordal, 2001).

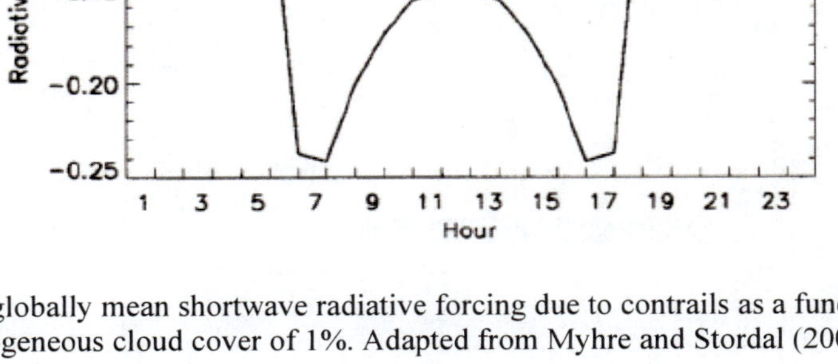

Figure 23. Annual globally mean shortwave radiative forcing due to contrails as a function of local time for a globally homogeneous cloud cover of 1%. Adapted from Myhre and Stordal (2001).

The diurnal variations of contrail coverage and properties are also important for the mean radiative forcing because solar radiative forcing is restricted to the daytime period. The diurnal coverage ratio may be expressed as a ratio of total daily contrail coverage to daytime-only coverage. Diurnal coverage ratios of contrails were found to be about 2 over Europe (Bakan et al., 1994) and 3 over central Europe (Mannstein et al., 1998). The diurnal contrail coverage ratio was consistent with a global mean noon-midnight traffic ratio of 2.8 (Schmitt and Brunner, 1997).

However, the diurnal variation of air traffic is often neglected for contrail radiative forcing (e.g., Marquart et al., 2003; Fichter et al., 2005). The effect of diurnal variations of air traffic on contrail radiative forcing over southeast England was investigated by Stuber et al. (2006). The flights during the nighttime were found to have a disproportionate effect on the annual, diurnal mean contrail radiative forcing because the longwave warming is not offset by any shortwave cooling.

To determine the effects of diurnal variations of air traffic on global mean contrail radiative forcing, Stuber and Forster (2007) calculated a diurnally resolved 3-D distribution of contrail cover. The radiative forcing was calculated for this contrail cover distribution with an assumed constant contrail optical thickness. It was found that less than 40% of the global distance traveled by aircraft is due to flights during local night time and that neglecting diurnal variations in air traiffic/contrail coverage by assuming a diurnal mean contrail coverage can overestimate the global mean radiative forcing by up to 30% (Stuber and Forster, 2007). Figure 24, from Stuber and Forster (2007), shows (a) percentage of flights during local night time and (b) a geographic distribution of the annual mean relative underestimation or overestimation of contrail radiative forcing due to neglect of the diurnal variations of air traffic.

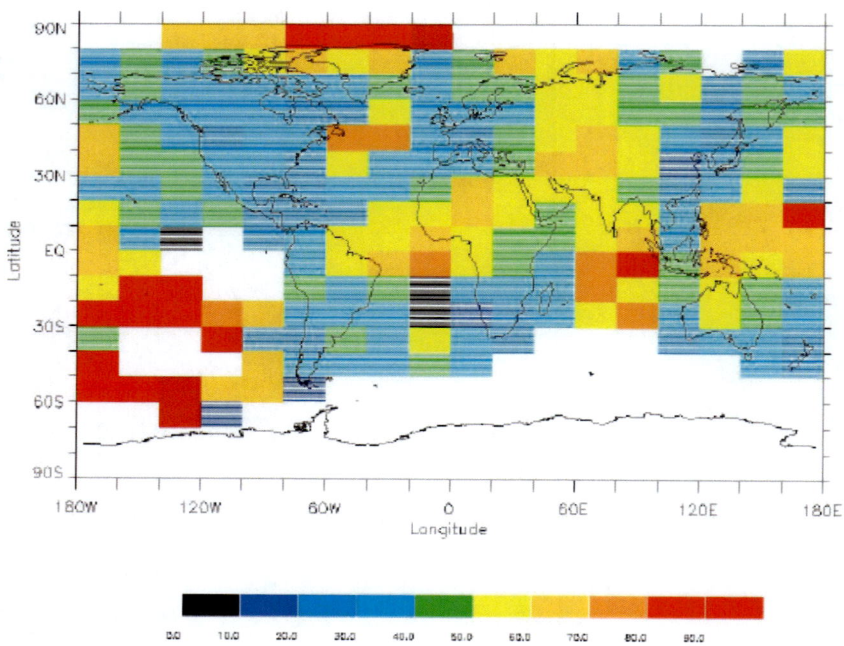

a

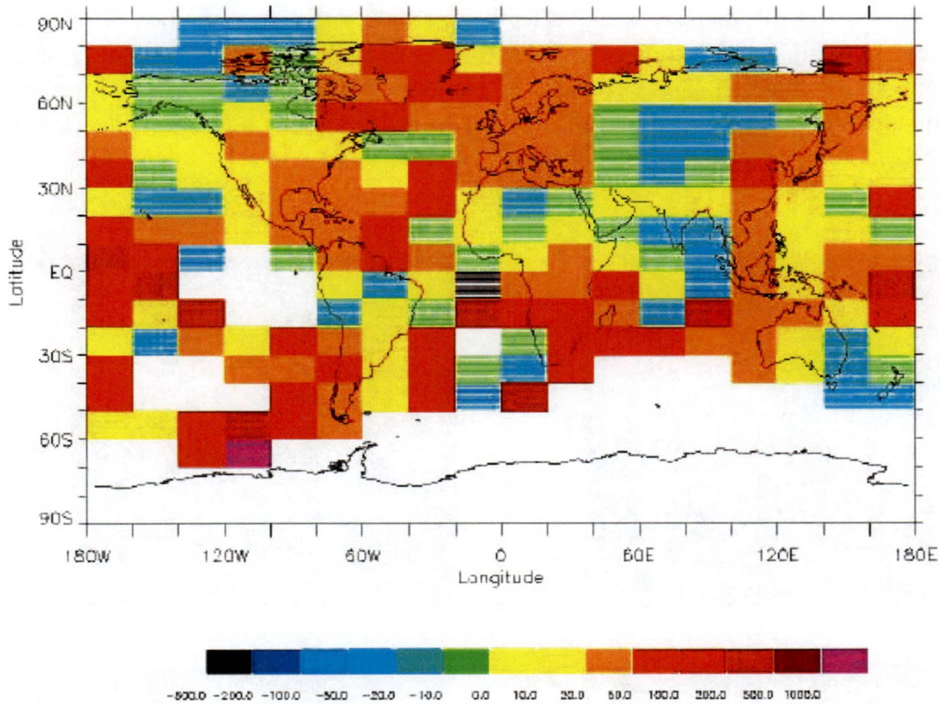

b

Figure 24. (a) Percentage of flights during local night time. (b) Geographical distribution of the percentage overestimation/underestimation of annual mean contrail radiative forcing resulting from neglecting the diurnal cycle of air traffic. Adapted from Stuber and Forster (2007).

For those locations with significant local forcing, the net radiative forcing was overestimated when the diurnal variation of air traffic was neglected. The increase of the net forcing for these regions was due to the increasing number of flights during local nighttime in comparison with a diurnally uniform distribution of flights.

3.2.5. Detection of Contrail and Contrail-Induced Cirrus

Split-window brightness temperature difference techniques (Prabhakara et al., 1988; Parol et al., 1991; Gothe and Grassl, 1993; Duda and Spinhirne, 1996; Duda et al., 1998) have bee n used to estimate optical and microphysical properties of thin cirrus and contrails. Much work has been done to understand the retrieval accuracies of these techniques (Gao et al., 1993; Stubenrauch et al., 1999; Rädel et al., 2003; Hong et al., 2007a). For example, Hong et al. (2007a) investigated the effect of the ice cloud geometrical thickness on the retrieval of optical thickness and effective particle size using split-window bands at 8.5 and 11 i&m (or 12 i&m). The optical thickness in the IR depends strongly on cloud geometrical thickness, and slightly on cloud effective particle size.

Gao et al. (1998) revealed that the narrow channels near the centers of the 1.38 and 1.88 çm water vapor absorption bands are useful for detecting thin cirrus clouds owing to the strong water vapor absorption in the lower atmosphere. The two channels have been used to detect the optical thickness and effective particle size of contrail cirrus by Gao, Meyer, and Yang (2004). Figure 25 shows the observations of 1.38 and 1.88 çm for a contrail cirrus and the retrieved optical thickness and effective particle size. This method is effective for detecting contrail and contrail-cirrus, thereby, investigating the regional or global mean distribution of contrails and cirrus covers and their optical depths, and measuring the variations of properties over different time scales (diurnal, monthly and annual). However, the 1.88 çm is not measured by any existing satellite sensor, although the MODIS band 26 is centered within the 1 .38 -çm water vapor band.

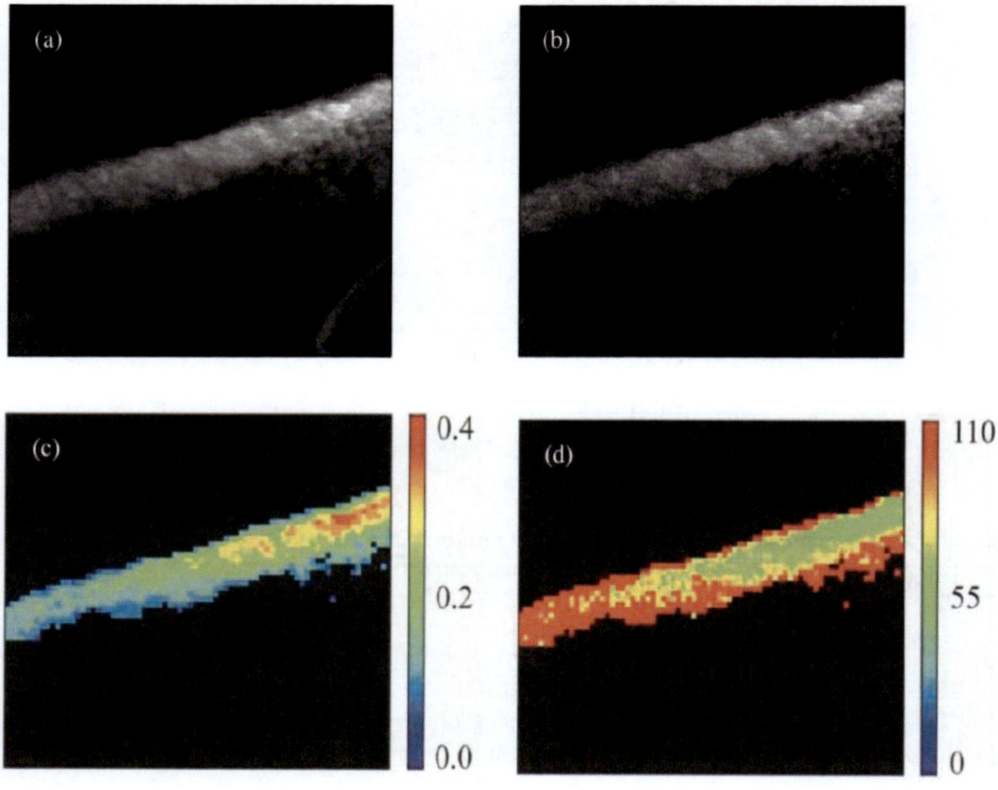

Figure 25. (a) and (b) are for 1.38 and 1.88 çm images for a contrail cirrus, respectively, (c) and (d) are for retrieved optical thickness and effective particle size (in units of çm), respectively (Gaoetal., 1998).

3.2.6. Dependency of Different Types of Aerosols

The ability of aerosols to promote ice formation and the aerosol number density are key factors controlling ice clouds and their radiative forcing. Mangold et al. (2005) investigated the ice water content, ice crystal number concentration, and particle sizes during ice cloud formation experiments for different temperatures and types of aerosol particles. It was found that ice water content generally decreased with decreasing temperature. Figure 26 shows the maximum ice wat er content for different aerosols as ice nuclei for temperatures ranging from 200 K to 225 K (Mangold et al., 2005). No clear influence of the aerosol type on the ice water content was found generally. The ice crystal number increased and their sizes decreased with decreasing temperature for all aerosol types.

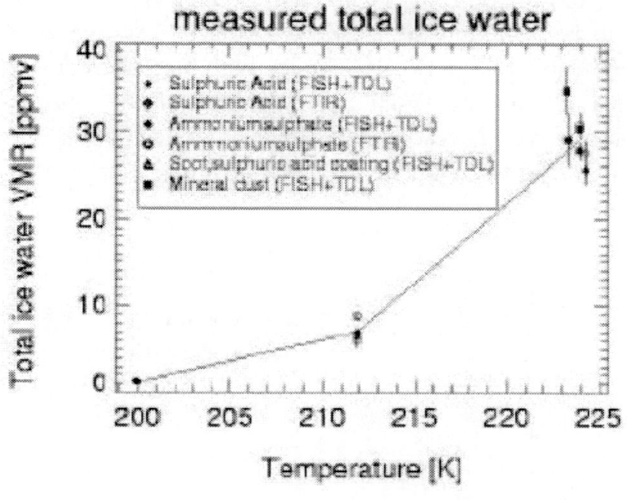

Figure 26. The maximum ice water content for different aerosols as ice nuclei for temperatures ranging from 200 K to 225 K (Mangold et al., 2005).

Contrail particles have been found to contain soot (Kuhn et al., 1998; Petzold et al., 1998; Storm and Ohlsson, 1998a, 1998b). The effect of aircraft soot on ice cloud formation in the absence of contrails is highly complex and depends on a wide variety of aerosol emission and environmental parameters (e.g., Petzold et al., 1997; Kuhn et al., 1998; Storm and Ohlsson, 1998a, 1998b; Kärcher et al., 2007). The difficulty in assessing the role of aircraft soot in ice cloud formation is in separating its influence from those of other aerosols and unambiguously demonstrating that ice nucleates on the exhaust soot particles or those affected by the soot, sulfur, and organic emissions by coagulation and condensation (Kärcher et al., 2007). Kärcher et al. (2007) investigated the role of soot aerosols in ice cloud formation with a combination of process-oriented modeling and evaluation of laboratory results of heterogeneous ice nucleation. They suggested two principal scenarios: "high concentrations of original soot emissions could slightly increase the number of ice crystals; low concentrations of particles originating from coagulation of emitted soot with background aerosols could lead to a significant reduction in ice crystal number." However, a fundamental understanding of the ice nucleation efficiency of soot particles is still lacking. Uncertainties remain in modeling the interaction between direct (contrail cirrus) and indirect (soot -induced cirrus) effects on ice clouds (Kärcher et al., 2007).

Soot within or on an ice particle affects the refractive index of the particle, and hence influences the scattering properties. In turn, this may result in differences in radiative forcing. Chylek and Hallett (1992) documented that the importance of the soot is associated with the type of mixing (external or internal) and the volume fraction of soot enclosures. The question remains whether the black carbon particles are incorporated into contrail ice crystals or deposited on the ice crystal surface, and this is critical for understanding the contrail formation process (Kuhn et al., 1998). Kuhn et al. (1998) found that contrails contain a mixture of pure ice particles, black carbon aerosol, and an internal mixture of these components, using the measurements from the SULFUR-4 experiment in March 1996 over southern Germany.

Figure 27 shows the major components of a young contrail (Kuhn et al., 1998). The contrail was found to be a mixture of 15% ice particles, 32% ice with black carbon, 24% black carbon, and 29% unknown aerosols. The ice with black carbon was found to contain a black carbon volume fraction of 15-20%. Knowledge of the composition is critical for obtaining accurate refractive indices of ice particles and aerosols in the contrails. It is also important to know the background atmospheric composition, such as humidity.

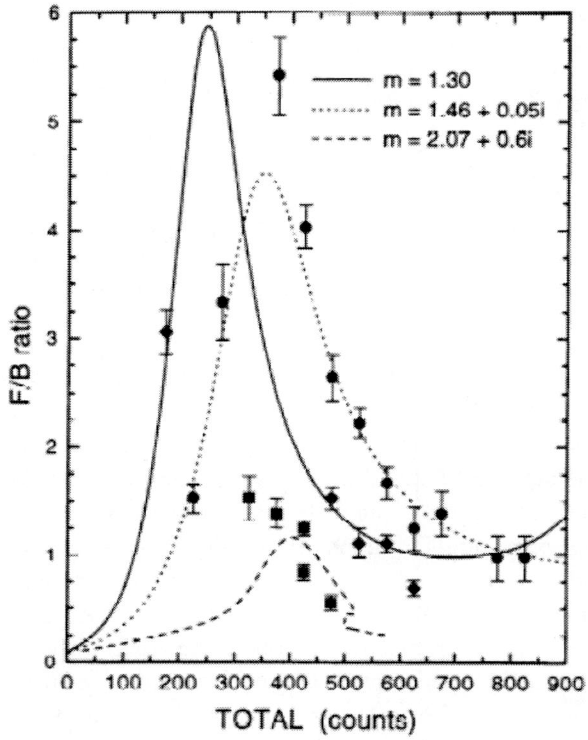

Figure 27. Major components of a young contrail. F/B ratio is the ratio of forward to backward scattered radiation. Ice has a refractive index of 1.30, black carbon has a refractive index of 2.07+0.6i, and ice with black carbon has a refractive index of 1 .46+0.0i. Adapted from Kuhn et al., (1998).

3.2.7. Contrails Occurring in Cirrus Clouds

Since aircrafts generally fly at the same heights at which ice clouds naturally occur, contrails often co-exist with cirrus clouds. Immler et al. (2007) found contrails are often embedded in a partly subvisible cirrostratus. One might expect a difference in microphysical and optical properties for contrails embedded within a cirrus deck from that of contrail-induced cirrus, but no related studies can be found that explore this issue.

The effect of dynamical variability in cirrus cloud formation was investigated by Kärcher and Storm (2003). They found that vertical velocities and mesoscale variability in vertical velocities could significantly modify the proba bility distribution of cirrus ice crystal concentrations. Figure 28 shows the obvious effect of variability in updraft velocity on ice particle number densities.

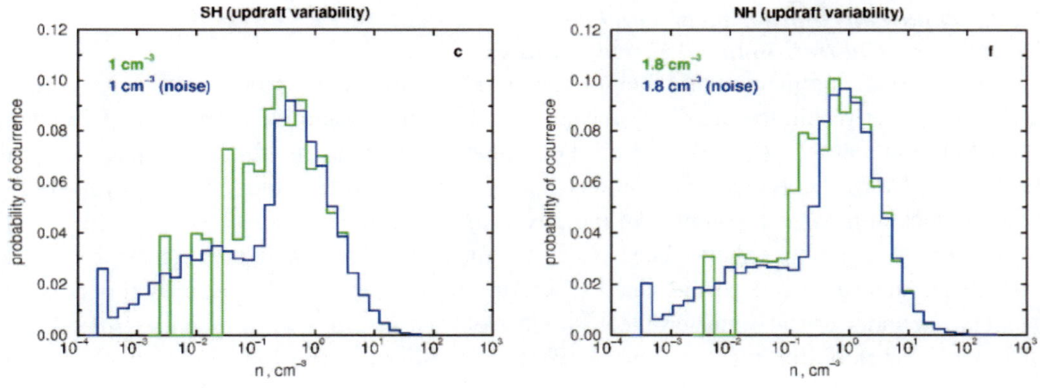

Figure 28. The effect of updraft variability ice particle concentrations in the southern hemisphere (SH) and northern hemisphere (NH) from Kärcher and Storm (2003).

Figure 29 shows a schematic of the dynamic, chemical, and microphysical processes involved in the formation of an internal mixture of soot particles containing black carbon from an aircraft exhaust plume (Kärcher et al., 2007). This scheme was applied to a Lagrangian plume model with interactive entrainment of constituents present in ambient air, gas phase sulfur chemistry and condensation, and coagulation between particles of different size and composition to understand the factors controlling the generation of internally mixed aircraft-soot particles. This scheme was designed for contrails occurring in an otherwise clear sky, not for contrails occurring in already-existing cirrus clouds. It is still unknown what changes would need to be adopted for the scheme for contrails occurring in already existing cirrus clouds. This requires more direct in situ observations, laboratory measurements, and theoretically studies.

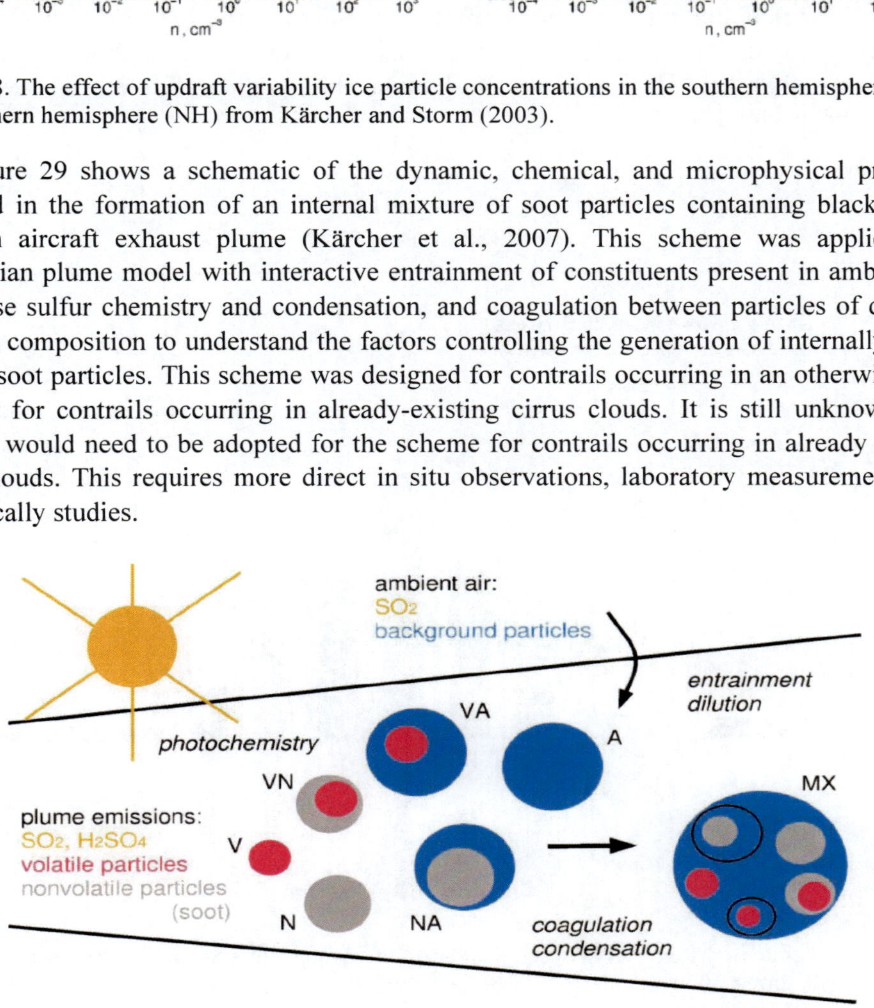

Figure 29. Schematic of the dynamical, chemical, and microphysical processes involved in the formation of internally mixed soot particles containing black carbon from various sources in dispersing aircraft exhaust plumes. V, aircraft-emitted volatile aerosols; N, nonvolatile aerosols; A, entrained background aerosols; VN, coagulated particle resulting from V+N; VA, coagulated particle resulting from V+A; NA, coagulated particle resulting from N+A; MX, coagulated particle resulting from all above particles. Adapted from Kärcher et al. (2007).

3.2.8. Cloud Tops and Physical Thicknesses of Contrail and Contrail-induced Cirrus Clouds

Properties including cloud top height, the cloud vertical distribution, and the properties of each cloud layer within the distribution can also affect cirrus radiative forcing. As pointed out by Sassen and Campbell (2001), "What is currently known of the climatic impact of cirrus has come from theoretical simulations based on idealized cloud macrophysical and microphysical inputs, and a number of radiative transfer approximations."

Different techniques have been used to measure cloud heights, including passive radiometric measurements in the visible, near-infrared, and infrared spectrum (i.e., satellite imagers, sounders, and interferometers) as well as active measurements from lidars. Differences in cloud-top pressure/height results from lidar and the CO_2-slicing method (used for MODIS and HIRS data products) have been shown by Frey et al. (1999) for both single-layered and multilayered clo uds. Figure 30 shows the histogram of differences between lidar and CO_2-slicing cloud heights.

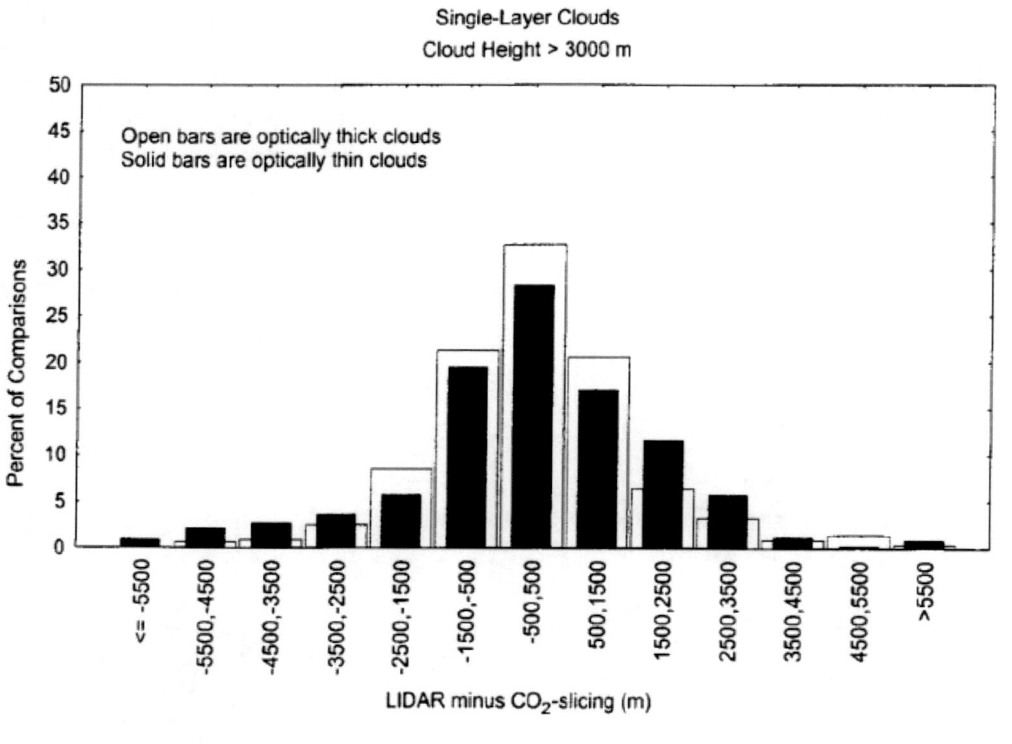

Figure 30. Histogram of differences between lidar and CO_2-slicing cloud heights in meters. Clouds are single-layer with tops > 3000 m. Open bars represent optically thick clouds, while solid bars show optically thin cloud retrievals (Frey et al., 1999).

It was found that the tops of contrails and contrail-induced cirrus clouds are generally at 9-12 km (e.g., Sassen and Hsueh, 1998; Meerkötter et al., 1999; Atlas et al., 2006). The geometrical contrail thickness of contrails was found to be in the range of 50-500 m using high-resolution depolarization lidar studies during SUbsonic aircraft: Contrail and Cloud Effects Special Study (Sassen and Hsueh, 1998). Figure 31 shows a series of spreading contrails between 7.5-11.0 km (Atlas et al., 2006). Different cloud top temperatures and physical thickness result in differences in longwave radiative emission and reflectance of solar radiation (Hong et al., 2007b), and therefore affect the total radiative forcing.

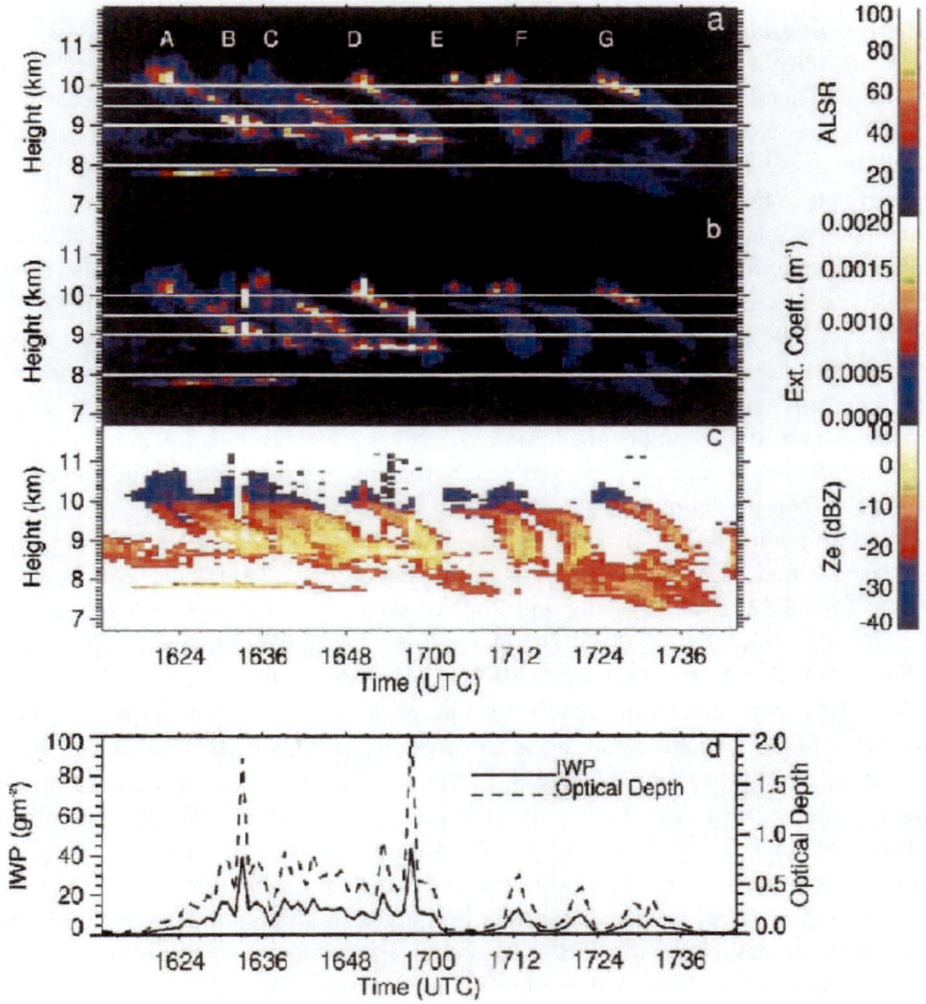

Figure 31. Lidar observations of contrails for (a) attenuated lidar scattering ratio, (b) extinction coefficient, (c) estimated radar reflectivity factor, and (d) corresponding optical thickness. Adapted from Atlas et al. (2006).

Kärcher and Storm (2003) investigated the effects of updraft velocity variability on the formation of ice clouds based on the probability of occurrence of ice crystal number densities. However, their study did not consider the influence of updraft variability on the macrophysical properties of contrails. The processes driving contrail spreading are horizontal diffusion and vertical shear of the horizontal wind (Jensen et al., 1998). Their simulations show that not only the crystal growth, microphysical and optical properties of contrails are strongly dependent on wind shear, but also the vertical distribution and horizontal spreading. However, those studies are hindered by a shortage of in situ observations. The Cloud–Aerosol

Lidar Infrared Pathfinder Satellite Observation (CALIPSO) satellite as a part of A-Train (Winker et al. 2003) can provide global observations of contrails and contrail-induced cirrus clouds. Analysis of the CALIPSO products is relatively recent, and depends on the rapid maturation of officially released cloud products.

3.2.9. Representation Contrails and Contrail-Induced Cirrus Clouds in GCMs

There are only a few studies that have attempted to determine the climate impact of contrails from general circulation model (GCM) simulations (Ponater et al., 1996; Rind et al., 1996, 2000). Those studies estimated the climate impact of a cirrus cloud perturbation introduced by more or less artificial means. Ponater et al. (2002) developed an online parameterization for contrails that was based on a thermodynamic approach (Schmidt-Appleman theory) adapted to the cloud parameterization scheme of the ECHAM4. However, the radiative forcing of the contrails from the ECHAM4 with the contrail parameterization is substantially smaller than the best estimate given by Minnis et al. (1999) and IPCC (1999). Ponater et al. (2002) documented the potential uncertainties of the model results, which include the dependency of ice water content, ice particle size, and the radiative effect of the simulated contrails on ambient parameters such as temperature, humidity, cirrus frequency, surface albedo, and solar zenith angle. They also suggested a classification approach based on the simulated range of contrail properties for various ambient conditions to reduce the uncertainties of their model results.

The contrail parameterization scheme of Poneter et al. (2002) has also been implemented in the IFSHAM model developed at the Danish Meteorological Institute (Guldberg, 2003). The results from the IFSHAM model and Poneter et al. (2002) were compared to radiative forcing estimates of Guldberg (2003). The global mean of the net radiative forcing was found to be almost a factor of ten less in the IFSHAM model with respect to those from the ECHAM4. Also, there are large areas over Europe and USA with negative net radiative forcing where the net radiative forcing is positive in the ECHAM4 (Ponater et al., 2002; Marquart et al., 2003). Figure 32 shows the annual net radiative forcing at the top of atmosphere from the IFSHAM (Guldberg, 2003) and the ECHAM4 (Marquart et al., 2003). Marquart and Mayer et al. (2002) explained the differences as being caused by a problem of the longwave radiation scheme in ECHAM4 which is different from that used in the IFSHAM model. But Guldburg (2003) suggested further investigation, finding that the contrail effective radii and ice water path from the IFSHAM model are smaller than those in Ponater et al. (2002). The influence of those parameters on the contrail radiative forcing is needed to determine by more model experiences.

The GCM parameterization scheme of contrails by Ponater et al. (2002) did not include a contrail-cirrus feedback due to the additional ice water formation caused by the infusion of aerosol particles. This feedback can be added to the direct radiative effect of contrails. This would probably increase the climate impact from contrails. Ponater et al. (2002) proposed an approach to parameterize contrails in GCMs that is to make use of a simulated background distribution of aerosols to include details of the nucleation process; this accounts for the aviation effect on cirrus formation due to an accumulation of aircraft emitted aerosols in the upper troposphere. However, this is dependent on an adequate description of the aerosol-cirrus interaction in GCMs that is currently lacking.

(a)

(b)

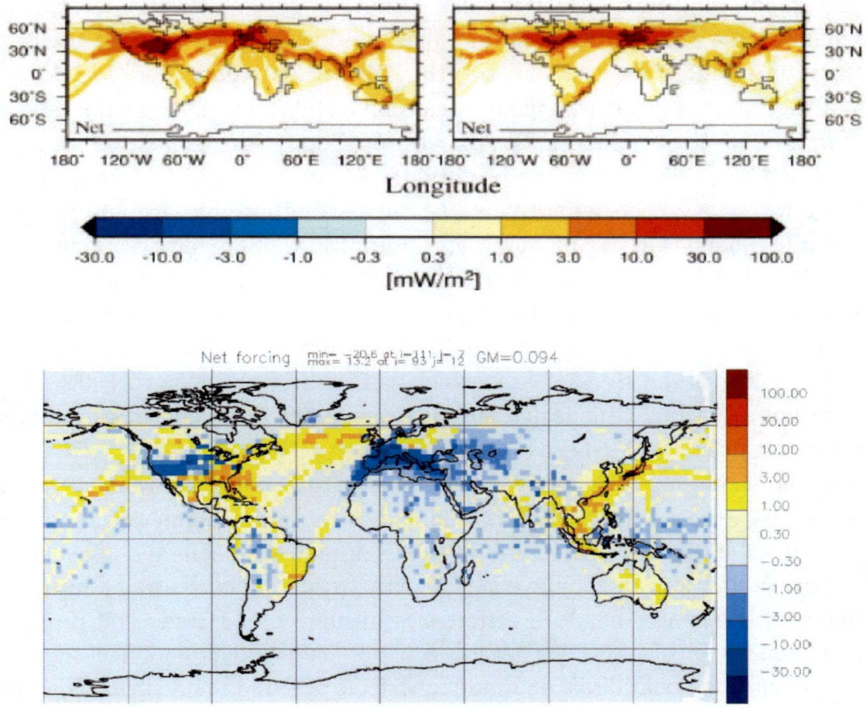

Figure 32. The net radiative forcing of contrails at the top of atmosphere in January (left panel) and July (right panel) from (a) ECHAM4 simulations. Adapted from Marquart et al. (2003) and (b) IFSHAM simulation. Adapated from Guldberg (2003).

3.2.10. Effect of Water Vapor and Temperature on Contrails

Atmospheric water vapor is the most important and efficient greenhouse gas (Held and Soden, 2000). It is transparent to incoming short-wave radiation from the Sun and opaque to long-wave radiation leaving the Earth. In general, an increase in atmospheric temperature is accompanied by an increase in water vapor concentration (e.g, Soden et al., 2002; Minschwaner and Dessler, 2004; Soden et al., 2005; Sherwood and Meyer, 2006), leading to further warming. This positive feedback between atmospheric temperature and water vapor is of crucial importance (IPCC, 2007; Randall et al., 2007). Besides the direct effect of water vapor on the earth radiation balance, water vapor also connects to the cloud feedback through its obvious influence on cloud formation.

Recent work on the water budget over the past decade has shown that the water mixing ratio can be understood almost entirely by advection of vapor by the large-scale atmospheric flow combined with removal of water vapor when the relative humidity exceeds 100% (Sherwood, 1996, Salathe and Hartmann, 1997; Dessler and Sherwood, 2000; Dessler and Minschwaner, 2007; Pierrehumbert et al., 2007).

For cloud formation, relative humidity (RH) is the relevant parameter, rather than water vapor mixing ratio. RH connects to mixing ratio through the temperature, so temperature is therefore also a crucial parameter. An important requirement for long- lived contrails is the ice supersaturation, meaning RH > 100%. Thus, accurate knowledge of ice supersaturation is crucial for quantifying both the direct and indirect effects of aviation on cirrus formation. Measurements have recently come on-line that provide some insight into the distribution of supersaturation (Gettelman et al., 2006). They found that supersaturation occurs in humid regions of the upper tropical tropopause near convection 10%-20% of the time at 200 hPa. Supersaturation is very frequent in the extratropical upper troposphere, occurring 20%-40% of the time, and over 50% of the time in storm track regions below the tropopause.

Some efforts have been initiated to take supersaturation into account (Tompkins et al., 2007) in some models such as ECMWF and European Center/Hamburg General Circulation Model (ECHAM). Tomkins et al. (2007) showed the different schemes of the evolution of relative humidity (RH) in an upper-troposphere parcel of air (figure 33). Figure 33a shows the scheme from Lohmann and Kärcher (2002). It assumes that RH increases with adiabatic cooling until a critical threshold RHcrit which can significantly exceed 100%. At RHcrit, ice crystals are homogeneously nucleated and their uptake of the water vapor reduces towards a value that in most cases just exceeds the 100% level.

Figure 33b shows the scheme assuming any excess humidity is instantaneously converted to ice when RH exceeds 100%. Figure 33c shows the new parameterization developed by Tomkins et al. (2007). This scheme allows supersaturation to occur and converts all humidity exceeding the saturation value to ice instantaneously once the nucleation threshold is attained. The different schemes can result in differences in high- cloud cover and u pper tropopause moisture (Tomkins et al., 2007). Clearly, a better understanding of ice supersaturation is crucial for quantifying both direct and indirect effects of aviation on cirrus cloudiness.

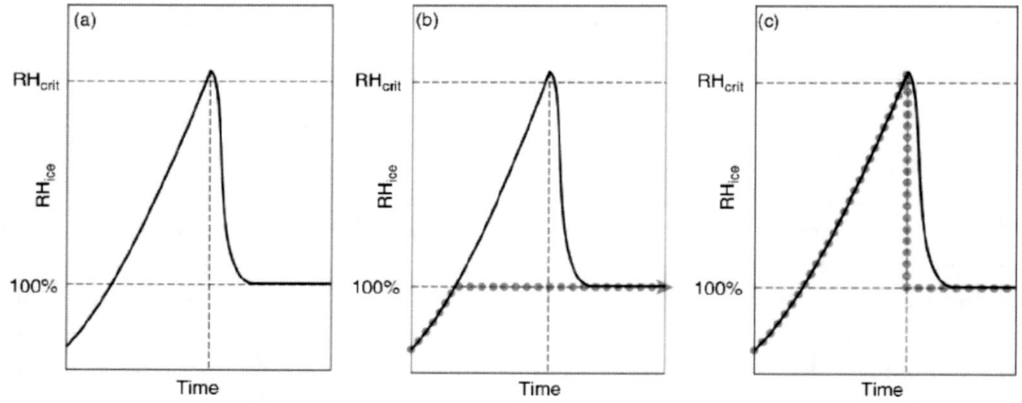

Figure 33. (a) Schematic of evolution of RH in a hypothetical parcel subjext to adiabatic cooling (Lohmann and Kärcher, 2002); (b) the dotted line is the scheme assuming that any excess humidity is instantaneously converted to ice once the RH exceeds 100%; (c) the dotted line is the scheme assuming that supersaturation to occur and convert all humidity exceeding the saturation value to ice instantaneously once the nucleation threshold is attained (Tomkins et al., 2007).

The dependence of contrail formation on temperature has also long been recognized. Appleman (1953) described the basic theory of contrail formation. The critical temperatures for contrail formation and persistence of contrails and contrail- induced cirrus clouds have been extensively studied (e.g., Coleman, 1996; Schumann, 1996; Schrader, 1997; Jensen et al., 1998). Jensen et al. (1998) investigated the threshold temperatures based on Appleman

(1953) theory versus ambient pressure for ambient RH (figure 34). The observed threshold temperatures were compared by Jensen et al. (1998) with theoretical estimates based on simple models of plume evolution. They found that saturation with respect to liquid water must be reached at some point in the plume evolution. The formation of contrails is needed to further investigate when temperature between liquid-saturation threshold temperature and ice-saturation threshold temperature.

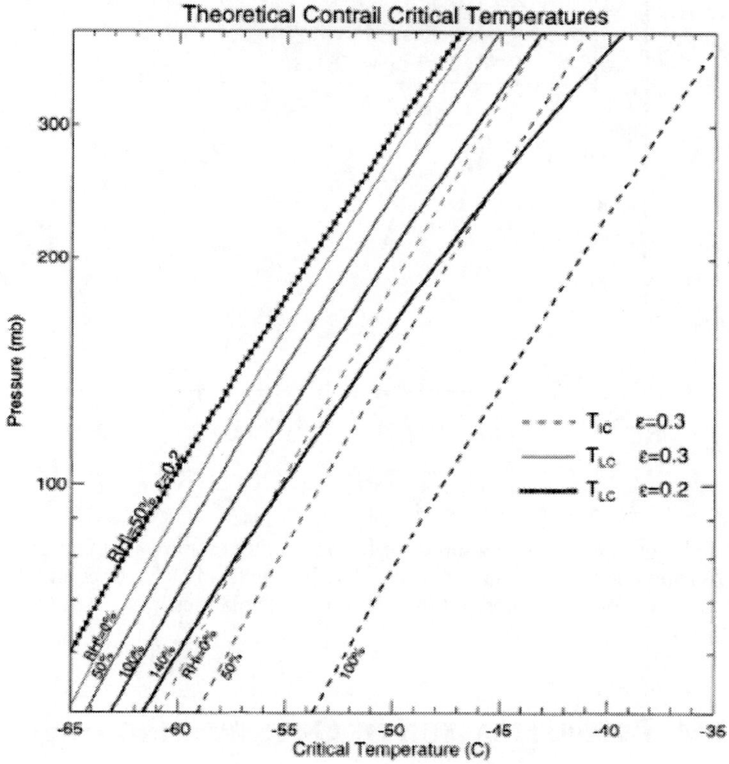

Figure 34. Theoretical threshold temperatures from Appleman (1953) as a function of ambient pressure and RHIs of 0, 50, 100, and 140%. Contrail formation is predicted occur when the ambient temperature is to the left of the solid curves that are for liquid saturation. Dashed curves are for ice saturation. Adapted from Jensen et al. (1998).

The ice-nucleating ability of soot particles in upper troposphere and low stratosphere is also influenced by the environment. The heterogeneous ice nucleation on soot aerosol was investigated at temperatures between 184 and 240 K by Möhler et al. (2003). Figure 35 shows the ice saturation ratios measured for ice nucleation on pure soot and soot coated with sulphuric acid (SA) and ammonium sulphate (AS). The results from DeMoot et al. (1999) are also shown in the figure for comparison to the Aerosol Interaction and Dynamics in the Atmosphere (AIDA) data (Möhler and Kramer, 2003). At temperatures above 235 K, ice nucleation on pure soot particles only occurred close to or slightly above water saturation (dashed line in figure 35). When temperatures become lower, ice is formed below both the liquid water saturation threshold and the threshold for homogeneous freezing nucleation of supercooled liquid solution droplets (solid line in figure 35, Koop et al. 2000). The freezing relative humidity increases 10-20% for the soot particles coated with sulphuric acid (10% mass). These results for the soot particles coated with sulphuric acid from Möhler et al. (2003) is in contrast to those results present by DeMott et al. (1999) showing that increasing amount of sulphuric acid coating decreases freezing relative humidity.

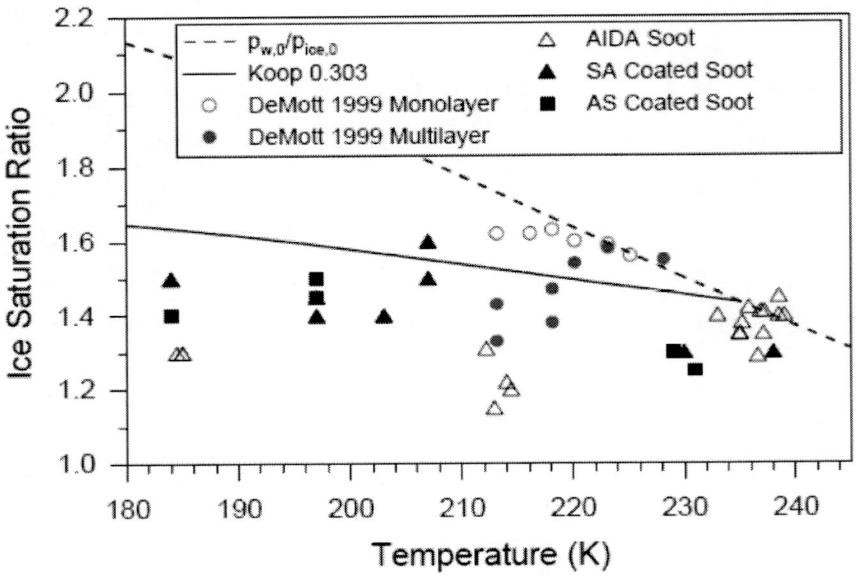

Figure 35. The ice saturation ratios measured for ice nucleation on pure soot and soot coated with sulphuri acid (SA) and ammonium sulphate (AS). The results from DeMoot et al. (1999) are also shown in the figure for comparison to the Aerosol Interaction and Dynamics in the Atmosphere (AIDA) data (Möhler et al. 2003).

4. Prioritization of Outstanding Issues

4.1. Prioritized Areas

4.1.1. Single-scattering Properties of Particles in Contrails and Contrail-Induced Cirrus Clouds

An extensive database of single-scatting properties of ice particles has been built by Yang et al. (2000, 2005) and Baum et al. (2005a, b, c). The habits for droxtals, hollow and solid columns, 3D bullet rosettes (with solid bullets), plates, and aggregates have all been calculated. Further work is ongoing with regards to a new particle: a 3D bullet rosette that has hollow bullets. Inclusions at the ends of the individual bullets are more realistic, and these particles have very different scattering properties than those obtained for the solid bullet rosettes. At present, there is no suitable database for aerosols and ice particles in contrails and contrail-induced cirrus clouds. As a result, the studies on the radiative forcing of contrails and contrail -induced cirrus clouds use scattering properties of either spherical ice particles or of nonspherical ice particles whose optical properties are calculated theoretically. A database of single-scattering properties for specified (more realistic) particle habits in contrails and contrail-induced cirrus clouds will reduce the uncertainties of radiative forcing of contrails and contrail-induced cirrus clouds. These databases are also the basis for the parameterization of radiation from contrails and contrail-induced cirrus clouds for use in GCMs.

As mentioned in the section "present state of modeling capability/best approach for light scattering computation", the most practical approach is a combination of FDTD, DDA, T-matrix methods and the ray-tracing technique. The limitations and gaps of this combination scheme need to be resolved. First, the applicability of geometric optics to light scattering is not clear. Generally, it is believed that geometric optics will produce reasonable results when the particle is very large compared with the wavelength. But in some special cases, the result obtained using geometric optics is wrong. Take an oriented plate as an example; the extinction efficiency is not going to converge to 2 even for a particle with a very large size. This phenomenon is due to missing the edge effect in this method. Because the particle is very large, it is also impossible to use DDA to include the edge effects. Second, there are still no ray-tracing codes to make a "educated guess" of the internal field within an ice crystal. The efficiency of the combination of DDA and the ray-tracing method to include the edge effect is still unknown. The answer to this question needs both theoretical and numerical analyses.

4.1.2. Parameterization of Radiation Forcing of Contrails and Contrail-Induced Cirrus Clouds for Use in GCMs

The radiative forcing of contrails and contrail-induced cirrus clouds in GCMs is based on the shortwave and longwave radiation schemes used for natural cirrus clouds. Building new parameterization schemes suitable for contrails and contrail-induced cirrus clouds will bring more accurate estimates of radiative forcing of these clouds. This also provides an efficient way to understand the differences in the radiative forcing from contrails, contrail -induced cirrus clouds, and natural cirrus clouds.

4.1.3. Detecting Contrails and Contrail-Induced Cirrus Clouds from Multi Satellite Measurements

The detection of cloud properties is sensitive to the satellite sensors and the retrieval algorithms used (Wielicki and Parker, 1992; Hong et al., 2007b). The satellites in the A-Train include passive visible and infrared sensors (MODIS aboard Aqua), active lidar (CALIPSO), and passive polarization measurements (POLDER). These sensors provide a wealth of measurements to study the properties of contrails and contrail-induced cirrus clouds that could reduce the uncertainties in the radiative forcing of contrails and contrail-induced cirrus clouds.

4.1.4. Understanding Formation of Contrails and Contrail-Induced Cirrus Clouds

A better understanding of the format ion mechanisms of contrails and contrail- induced cirrus clouds will improve the representation of these clouds in GCMs. Important factors for contrail formation include both water vapor and temperature, which combine to determine relative humidity. It is also necessary to further investigate the liquid- saturation threshold temperature and ice-saturation threshold temperature for contrail formation. At present, only a handful of climate models allow supersaturation.

4.1.5. Interaction between Aerosols and Cirrus Clouds

Theoretical modeling of the coupling between direct and indirect aviation effects on high clouds is currently very rudimentary (Kärcher et al., 2007). Further modeling, in situ measurements, and laboratory experiments are needed to improve our understanding of this important interaction. This effort will also improve the knowledge of the transition of contrails into cirrus clouds.

4.2. Estimated Costs and Timelines for Prioritized Areas

As stated previously, we have identified five prioritized research topics that are critical to a better understanding the optical and radiative forcing of contrails and contrail-cirrus clouds:

Topic 1. Development of reliable data sets for the optical properties of individual ice crystals with various shapes and sizes, typical in contrails and contrail-cirrus clouds. The data sets will be developed for both the solar and thermal infrared spectra with an adequate spectral resolution. This effort will be based mainly on numerical simulation. However, laboratory and in-situ measurements are required to validate the theoretical simulation.

Topic 2. Parameterization of the radiative properties of contrails and contrail- cirrus clouds for applications to climate models. The single-scattering data sets developed in Topic 1 for individual ice crystals need to be merged with in-situ measured particle size distributions and habit mixtures. The resultant bulk optical properties need to be parameterized as functions of particle effective size. Furthermore, the parameterization needs to be incorporated into the radiative transfer schemes involved in climate models. The validation of the parameterization based on spaceborne and ground-based radiometric measurements is necessary.

Topic3. Detection of contrails and contrail-cirrus clouds from synergetic use of the measurements made by multiple satellite sensors. To understand the radiative forcing of contrails and contrail-cirrus clouds from a global perspective, it is necessary to detect contrails and contrail-cirrus clouds. New techniques using various spectral bands need to be developed to effectively identify contrails and contrail-cirrus in satellite images.

Topic 4. Understanding of the formation of contrails and contrail-induced cirrus clouds. To fully understand the radiative forcing of contrails and contrail- induced cirrus clouds, it is necessary to understand the impact of ambient atmospheric environments on these clouds, which requires an improved knowledge about the formation of contrails and contrail-induced cirrus clouds. The most important factors for contrail formation are water vapor and temperature as discussed in section 3.2.10

Topic 5. Interaction between aerosols and contrails/contrail-cirrus clouds. To understand the indirect radiative effect of contrails/contrail-cirrus clouds, it is necessary to understand the interaction of aerosols and contrails/contrail-cirrus clouds. This is a very broad research area and requires tremendous efforts.

Table 1 lists the estimated costs and timeline for the aforementioned five prioritized research topics.

Table 1. Prioritized areas, estimated costs, and timeline

Prioritized areas	# of FTE	Duration of Efforts
Topic 1	2	3 years
Topic 2	2	2.5 years
Topic 3	3	3 years
Topic 4	3	3 years
Topic 5	4	3 years

5. Recommendations for Best Use of Current Tools for Modeling and Data Analysis

Topic-specific recommendations for the five topics listed in Sec. 4 are as follows:

A combination of the FDTD (or DDA) and geometric optics method (GOM) is recommended. The FDTD and DDA are accurate methods, but their applications are limited to small ice crystals because of tremendous computational demand of these methods. The GOM is approximate method that is valid for large ice crystals. This approach has been proven quite successful for applications to ice crystals in natural cirrus clouds. Because ice crystals in contrails and contrail -cirrus clouds are relatively smaller than naturally occuring cirrus clouds, it is critical to ensure an adequate bridging the FDTD (or DDA) solutions and the GOM solutions.

The parameterization needs to be done for radiative transfer schemes used in several popular climate models such as the NCAR Community Atmosphere Model (CAM). The validation of the parameterization can be carried out by comparing the model simulations of radiation fluxes and their measurement counterparts (e.g., the CERES data and DOE-ARM ground radiometric measurements).

New efforts are recommended to explore the spectral, viewing-geometry, and polarization characteristics of several existing satellite sensors (e.g., MODIS, AIRS, MISR, and POLDER) to effectively detect contrails and contrail-cirrus clouds. Some channels (e.g., the MODIS band 26 centered at 1.37 çm) have proved effective in detecting high clou ds.

More in-situ observations regarding the relationship between contrails and ambient parameters are recommended. Recently, the ECWMF, ECHAM4, and IFSHAM models from Europe have included the capabilities of predicting the supersaturation. Thus, in addition to in situ measurements, modeling is also an efficient way to understand the influence of water vapor and temperature on contrail formation.

The interaction between aerosols and contrails/contrail-cirrus clouds is a quite open research area. In addition to in-situ measurements of the correlation of aerosols and contrails, cloud models may play an important role to accomplish this research task.

REFERENCES

Appleman, H., 1953: The formation of exhaust condensation trails by jet aircraft, *Bull. Amer. Meteor. Soc., 34,* 14–20.

Atlas, D., Z. Wang, D. Duda, 2006: Contrails to Cirrus – Morphology, Microphysics, and Radiative Properties. *J. Appl. Meteor. Climatol., 45,* 5–19.

Bacon, N. J. and B. D. Swanson, 2000: Laboratory Measurements of Light Scattering by Single Levitated Ice Crystals Neil J. Bacon and Brian D. Swanson, *J. Atmos. Sci. 57 2094-2104*

Bacon, N. J., B. D. Swanson, M. B. Baker, and E. J. Davis, 1998: Laboratory Measurements of Light Scattering by Single Ice Particles Neil J. Bacon, Brian D. Swanson, Marcia B. Baker and E. James Davis, *J. Aerosol Sci. Vol. 29, S1317-- S13 18.*

Bakan, S., M. Betancor, V. Gayler, and H. Grassl, 1994: Contrail frequency over Europe from NOAA-satellite images. *Ann. Geophys., 12,* 962–968.

Barkey, B., K.N. Liou, Y. Takano, and W. Gellermann, 2000: Experimental and theoretical spectral reflection measurements of ice clouds generated in a laboratory chamber. *Appl. Opt.*, 39, 3561-3564.

Baum, B. A., A. J. Heymsfield, P. Yang, S. T. Bedka (2005a), Bulk scattering properties for the remote sensing of ice clouds. Part I: Microphysical data and models. *J. Appl. Meteorol., 44,* 1885–1895.

Baum, B. A., P. Yang, A. J. Heymsfield, S. Platnick, M. D. King, Y.-X. Hu, and S. T. Bedka (2005b), Bulk scattering properties for the remote sensing of ice clouds, Part II: Narrowband Models. *J. Appl. Meteorol., 44,* 1896–1911.

Baum, B. A., P. Yang, S. L. Nasiri, A. K. Heidinger, A. J. Heymsfield, and J. Li (2007), Bulk scattering properties for the remote sensing of ice clouds. Part III: High resolution spectral models from 100 to 3250 cm-1, *J. Appl. Meteor. Clim., 46,* 423– 434.

Boucher, O., 1999: Air traffic may increase cirrus cloudiness. *Nature, 397,* 30–31. Broecher, W., Lecture presented AGU 1996 – Baltimore US. 1996, Lamont-Doherty Earth Observatory, Columbia University.

Cai, Q., and K. N. Liou, 1982: Polarized light scattering by hexagonal ice crystals: Theory. *Appl. Opt.* 21, 3569-3580.

Chylek, P. and J. Hallett, 1992: Enhanced absorption of solar radiation by cloud droplets containing soot particles in their surface. *Q. J. R. Meteor. Soc., 118,* 167–172.

Clough, S. A., M. J. Iacono, and J.-L. Moncet, 1992: Line-by-line calculations of atmospheric fluxes and cooling rates: Application to water vapor, *J. Geophys. Res., 97,* 15,761-15, 785.

Coleman, R. F., 1996: a new formulation for the critical temperature for contrail formation, *J. Appl. Meteorol., 35,* 2270-22 82.

DeMott, P. J., Y. Chen, S. M. Kreidenweis, D. C. Rogers, and D. E. Sherman, 1999: Ice formation by black carbon particles. *Geophys. Res. Lett.,* 26, 2429-2432.

Dessler, A. E., and K. Minschwaner, 2007: An analysis of the regulation of tropical tropospheric water vapor, *J. Geophys. Res., 112,* DOT: 10.1029/2006JD007683.

Dessler, A. E., and S. C. Sherwood, 2000: Simulations of tropical upper tropospheric humidity, *J. Geophys. Res., 105,* 20,155–20,163.

Draine, B. T., and P. J. Flatau, 1994: Discrete-dipole approximation for scattering calculations. *J. Opt. Soc. Am. A* 11, 1491-1499.

Duda, D. P., and J. D. Spinhirne,1996: Split-window retrieval of particle size and optical depth in contrails located above horizontal inhomogeneous ice clouds. *Geophys. Res. Lett., 23,* 3711–3714.

Duda, D. P., J. D. Spinhirne, and W. D. Hart, 1998: Retrieval of contrail microphysical properties during SUCCESS by the split-window method. *Geophys. Res. Lett., 25,* 1149–1152.

Duda, D. P., P. Minnis, P. K. Costulis, R. Palikonda, 2003: CONUS Contrail Frequency Estimated from RUC and Flight Track Data. European Conference on Aviation, Atmosphere, and Climate.

Duda, D. P., P. Minnis, and L. Nguyen, 2001: Estimates of cloud radiative forcing in contrail clusters using GOES imagery. *J. Geophys. Res., 106,* 4927–4937.

Duda, D. P., P. Minnis, L. Nguyen, and R. Palikonda, 2004: A case study of the development of contrail clusters over the Great Lakes. *J. Atmos. Sci., 61,* 1132–1146.

Fahey, D. W., U. Schumann, S. Ackerman, P. Artaxo, O. Boucher, M.Y. Danilin, B. Kärcher, P. Minnis, T. Nakajima, and O.B. Toon, 1999: Aviation-produced aerosols and cloudiness. In Aviation and the Global Atmosphere, J. E. Penner, D. H. Lister, D. J. Griggs, D. J. Dokken, and M. McFarland (Eds.), Cambridge University Press, Cambridge, U.K., 65Å120.

Fichter, C., S. Marquar, R. Sausen, and D. S. Lee, 2005: The impact of cruise altitude on contrails and related radiative forcing. *Meteorol. Zeitschrift, 14,* 14,563Å14,572.

Foot, J. S., 1988: Some observations of the optical properties of clouds, II, Cirrus. *Q. J. R. Meteorol. Soc.*, 114, 145-164.

Fu, Q., and K.-N. Liou, 1993: Parameterization of the radiative properties of cirrus clouds. *J. Atmos. Sci., 50,* 2008–2025.

Frey, R., B. Baum, W. Menzel, S. Ackerman, C. Moeller, and J. Spinhirne, 1999: A comparison of cloud top heights computed from airborne lidar and MAS radiance data using CO_2 slicing. *J. Geophy. Res., 104,* 10.1029/1999JD900796.

Gao, B.-C., A. F. H. Goetz, and W. J. Wiscombe, 1993: Cirrus cloud detection from airborne imaging spectrometer data using the 1.38 &m water vapor band. *Geophys. Res. Lett., 20,* 301–304.

Gao, B.-C., Y. J. Kaufman, W. Han, and W. J. Wiscombe, 1998: Correction of thin cirrus path radiance in the 0.4Å1.0 &m spectral region using the sensitive 1.375 &m cirrus detecting channel. *J. Geophys. Res., 103,* 32,169Å32,176.

Gao, B.-C., K. Meyer, and P. Yang, 2004: A new concept on remote sensing of cirrus optical depth and effective ice particle size using strong water vapor absorption channels near 1.38 and 1.88 &m. *IEEE Trans. Geosci. Remote Sens., 42,* 1891–1899.

Garrett, T. J., 2007: Comments on "Effective radius of ice cloud particle populations derived from aircraft probes". *J. Atmos. Oceanic. Technol. 24,* 1492–1503.

Garrett T. J., H. Gerber, D. G. Baumgardner, C. H. Twohy, and E. M. Weinstock, 2003: Small, highly reflective ice crystals in low-latitude cirrus. *Geophys. Res. Lett., 30,* doi: 10. 1029/2003GL01 8153.

Gayet. J.-F., G. Febvre, G, Brogniez, H. Chepfer, W. Renger, and P. Wendling, 1996: Microphysical and optical properties of cirrus and contrails: Cloud field study on 13 October 1989. *J. Atmos. Sci., 53,* 126–138.

Gettelman, A., E. J. Fetzer, A. Eldering, and F. W. Irion, 2006: The global distribution of supersaturation in the upper troposphere from the Atmospheric Infrared Sounder, *J. Climate, 19,* 6089–6103.

Goodman, J., R. F. Pueschel, E. J. Jensen, S. Yerma, G. V. Ferr y, S. D. Howard, S. A. Kinne, and D. Baumgardner, 1998: Shape and size of contrails ice particles. *Geophys. Res. Lett., 25,* 1327–1330.

Gothe, M. B., and H. Grassl, 1993: Satellite remote sensing of the optical depth and mean crystal size of thin cirrus and contrails. *Theor. Appl. Climatol., 48,* 101113. Guldberg, A., 2003: A study of contrails in a General Circulation Model, proceeding of the AAC-Conference, June 30 to July 3, 2003, Friedrichshafen, Germany, 261Å265.

Haywood, J. M., and K.P. Shine, 1997: The effect of anthropogenic sulfate and soot aerosol on the clear sky planetary radiation budget. *Geophy. Res. Lett., 22,* 603–606. Hess, M., and Wiegner, M., 1994: COP: a data library of optical properties of hexagonal ice crystals. *Appl. Opt. 33,* 7740-7746.

Held, I. M., and B. J. Soden, 2000: Water vapor feedback and global warming, *Ann. Rev. Energy Environ., 25,* 441–475.

Heymsfield A. J., C. Schmitt, A. Bansemer, G.-J. Van Zadelhoff, M. J. McGill, C. Twohy, D. Baumgardner, 2006: Effective Radius of Ice Cloud Particle Populations Derived from Aircraft Probes. *J. Atmos. Ocean. Technol., 23,* 361–3 80.

Hong, G., P. Yang, H. -L. Huang, B. A. Baum, Y. X. Hu, and S. Platnick, 2007a: The sensitivity of ice cloud optical and microphysical passive satellite retrievals to cloud geometrical thickness. *IEEE Trans Geosci. Remote Sens., 45,* 13 15–1323.

Hong, G., P. Yang, B.-C. Gao, B. A. Baum, Y. X. Hu, M. D. King, and S. Platnick, 2007b: High cloud properties from three years of MODIS Terra and Aqua data over the Tropics. *J. Appl. Meteor. Climatol., 46,* 1840–1856.

Immler F., and O. Schrems, 2003:Comparison of cirrus cloud properties in the northern and southern hemisphere on the basis of lidar measurements. European Conference on Aviation, Atmosphere, and Climate.

Immler, F., R. Treffeisen, D. Engelbart, K. Kruger, and O. Schrems, 2007: Cirrus, contrails, and ice supersaturated regions in high pressure systems at northern mid latitudes. *Atmos. Chem. Phys. Discuss., 7,* 13,175Å13,201.

Minschwaner, K., and A. E. Dessler, 2004: Water vapor feedback in the tropical upper troposphere: Model results and observations, *J. Climate, 17,* 1272–1282.

Möhler, O, and C. Linke, H. Saathoff, M. Schnaiter, R. Wagner, U. Schurath, A. Mangold, M. Kramer, 2003: Experimental investigation of homogeneous and heterogeneous freezing processes at simulated UTLS conditions. European Conference on Aviation, Atmosphere, and Climate.

Jacobowitz, H., 1971: A method for computing the transfer of solar radiation through clouds of hexagonal ice crystals. J. Quant. Spectrosc. Radiat. Transfer 11, 69 1Å695.

Jensen, E.J., A.S. Ackerman, D.E. Stevens, O.B. Toon, and P. Minnis, 1998: Spreading and growth of contrails in a sheared environment. *J. Geophys. Res., 103,* 3 1,557Å 3 1,568, doi:10.1029/98JD02594.

Jensen, E. J., O. B. Toon, S. Kinne, G. W. Sachse, B. E. Anderson, K. R. Chan, C. H. Twohy, B. Gandrud, A. Heymsfield, and R. C. Miake-Lye, 1998: Environmental conditions required for contrail formation and persistence, *J. Geophys. Res., 103,* 3 929–3 93 6.

Jones, R. L., and J. F. B. Mitchell, 1991, Is water vapour understood? Nature, 353, 210.

Kärcher, B., O. Möhler, P. J. DeMott, S. Pechtl, and F. Yu, 2007: Atmospheric Chemistry and Physics Insights into the role of soot aerosols in cirrus cloud formation. *Atmos. Chem. Phys., 7,* 4203–4227.

Kärcher, B., and J. Strom, 2003: The roles of dynamical variability and aerosols in cirrus cloud formation. *Atmos. Chem. Phys., 3,* 823–838.

Karl T. R., and K. E. Trenberth, 2003: Modern global climate change. Science, 302, 1719–1723.

King, M. D., S. Platnick, P. Yang, G. T. Arnold, M. A. Gray, J. C.Riédi, S. A. Ackerman, and K. N. Liou, 2004: Remote sensing of liquid water and ice cloud optical thickness, and effective radius in the arctic: Application of airborne multispectral MAS data, *J. Atmos. and Ocean. Technol., 21,* 857875.

Koop, T., B. P. Luo, A. Tsias, and T. Peter, 2000: Water activity as the determinant for homogeneous ice nucleation in aqueous solutions. *Nature,* 406, 611-614.

Kuhn, M., A. Petzold, D. Baumgardner, and F. Schröder, 1998: Particle composition of a young condensation trail and of upper tropospheric aerosol. *Geophys. Res. Lett., 25,* 2679–2682.

Liou, K. N., Y. Takano, S. C. Ou, A. Heymsfield, and W. Kreiss, 1990: Infrared transmission through cirrus clouds: a radiative model for target detection. *Appl. Opt., 29,* 1886–1896.

Liou, K.N., Y. Takano, and P. Yang, 2000: Light scattering and radiative transfer in ice crystal clouds: Applications to climate research. In "Light Scattering by Nonspherical Particles: Theory, Measurements and Geophysical Applications." Eds. M.I. Mishchenko, J.W. Hovenier, and L.D. Travis, Academic Press, New York, Chapter 15.

Liou, K.N., P. Yang, Y. Takano, K. Sassen, T.P. Charlock, and W.P. Arnott, 1998: On the radiative properties of contrail cirrus. *Geophys. Res. Lett., 25,* 1161–1164.

Lohmann, U., and B. Kärcher, 2002: First interactive simulations of cirrus clouds formed by homogeneous freezing in the ECHAM GCM. *J. Geophys. Res., 107(D10), 4105,* doi: 10. 1029/2001JD000767.

Macke, A., 1993: Scattering of light by polyhedral ice crystals. *Appl. Opt.* 32, 2780-2788.

Macke, A. J. Mueller, and E. Raschke, 1996: Single scattering properties of atmospheric ice crystal. *J. Atmos. Sci.* 53, 2813-2825.

Mangold, A., R. Wagner, H. Saathoff, U. Schurath, C. Giesemann, V. Ebert, M. Krämer, and O. Möhler, 2005: Experimental investigation of ice nucleation by different types of aerosols in the aerosol chamber AIDA: implications to microphysics of cirrus clouds. *Meteorol. Z., 14,* 485–497.

Mannstein, H., R. Meyer, and P. Wendling, 1999: Operational detection of contrails from NOAAAVHRR-data. *Int. J. Remote Sensing, 20,* 1641–1660.

Marquart, S., M, Ponater, F. Mager, and R. Sausen, 2003: Future development of contrails: Impacts of increasing air traffic and climate change, proceeding of the AAC-Conference, June 30 to July 3, 2003, Friedrichshafen, Germany, 255Å260.

Marquart, S., M. Ponater, F. Mager, and R. Sausen, 2003: Future development of contrail cover, optical depth, and radiative forcing: Impacts of Increasing Air Traffic and Climate Change. *J. Climate, 16,* 2890–2904.

Meerkötter, R., U. Schumann, D. R. Doelling, P. Minnis, T. Nakajima, and Y. Tsushima, 1999: Radiative forcing by contrails. Ann. Geophys., 17, 1070–1084.

Meyer, R., H. Mannstein, R. Meerkötter, U. Schumann, and P. Wendling, 2002: Regional radiative forcing by line-shaped contrails derived from satellite data. *J. Geophys. Res., 1 07(D1 0), 4104,* doi: 10.1029/2001 JD000426.

Minnis, P., J.K. Ayers, R. Palikonda, E. Phan, 2004: Contrails, Cirrus Trends, and Climate. *J. Climate, 17,* 1671–1685.

Minnis, P., U. Schumann, D. R. Doelling, K. Gierens, and D. Fahey, 1999: Global distribution of contrail radiative forcing. *Geophys. Res. Lett., 26,* 1853–1856.

Minnis, P., L. Nguyen, D. P. Duda, and R. Palikonda, 2002: Spreading of isolated controls during the 2001 air traffic shutdown. Preprints, 10th Conf. on Aviation, Range, and Aerospace Meteorology, Portland, OR, Amer. Meteor. Soc., J9–J12.

Minnis, P., D. F. Young, L. Nguyen, D. P. Garber, W. L. Smith, Jr., and R. Palikonda, 1998: Transformation of contrails into cirrus during SUCCESS. *Geophys. Res. Lett., 25,* 1157–1160.

Mishchenko, M. I., and L. D. Travis, 1994: Light scattering by polydispersions of randomly oriented spheroids with sizes comparable to wavelengths of observation, *Appl. Opt., 33,* 7206-7225.

Mishchenko, M. I., and K. Sassen, 1998: Depolarization of lidar returns by small ice crystals: an application to contrails, *Geophys. Res. Lett., 25,* 309-3 12.

Mishchenko, M. I., L. D. Travis, and A. Macke, 2000: *T*-matrix method and its applications. In *Light Scattering by Nonspherical Particles: Theory, Measurements, and Applications* (M. I. Mishchenko, L. D. Travis, and J. W. Hovenier, Eds.), Academic Press, San Diego, pp. 147-172.

Myhre, G., and F. Stordal, 2001: Global sensitivity experiments of the radiative forcing due to mineral aerosols. *J. Geophys. Res., 106,* 18,193–18,204.

Muinonen, K., 1989: Scattering of light by crystals: a modified Kirchhoff approximation. *Appl. Opt., 28,* 3044-3050.

Noel, V., D. M. Winker, T. J. Garrett, and M. McGill, 2007: Extinction coefficients retrieved in deep tropical ice clouds from lidar observations using a CALIPSO-like algorithm compared to in-situ measurements from the cloud integrating nephelometer during CRYSTAL-FACE. Atmos. Chem. Phys., 7, 1415–1422.

Palikonda, R., D. N. Phan, V. Chakrapani, and P. Minnis, 2004: Contrail coverage over the USA from MODIS and AVHRR data. Preprints, European Conf. on Aviation, Atmosphere, and Climate, Friedrichshafen at Lake Constance, Germany, Institut fu" r Physik der Atmospha"re, DLR, in press. [Available online at http://www-pm.larc.nasa.gov/sass/pub/conference/rabi.AAC.03 .pdf.]

Parol, F., J. C. Buriez, G. Brogniez, and Y. Fouquart, 1991: Information content of AVHRR channels 4 and 5 with respect to the effective radius of cirrus cloud particles. *J. Appl. Meteor., 24,* 973–984.

Petzold, A., and Coauthors, 1997: Near field measurements on contrail properties from fuels with different sulfur content. *J. Geophys. Res., 102,* 29,867–29,880.

Petzold, A., A. Döpelheuer, C. A. Brock, and F. Schröder, 1999: In situ observations and model calculations of black carbon emission by aircraft at cruise altitude. *J. Geophys. Res., 104,* 22 171–22 181.

Pierrehumbert, R. T., H. Brogniez, and R. Roca, 2007: On the relative humidity of the Earth's atmosphere, in The Global Circulation of the Atmosphere, Schneider, T., and A. H. Sobel (eds.), Princeton Univ. Press.

Ponater, M., S. Brinkop, R. Sausen, U. Schumann, 1996: Simulating the global atmospheric response to aircraft water vapor emission and contrails: a first approach using a GCM. *Annales De Geophysique, 14,* 941–960.

Ponater, M., S. Marquart, and R. Sausen, 2002: Contrails in a comprehensive global climate model: parameterisation and radiative forcing results. *J. Geophys. Res., 107,* 10. 1029/2001JD000429.

Prabhakara, C., R. S. Fraser, G. Dalu, M. L. C. Wu, R. J. Curran, and T. Styles, 1988: Thin cirrus clouds: Seasonal distribution over oceans deduced from Nimbus-4 IRIS. *J. Appl. Meteor., 27,* 379–399.

Purcell, E. M. and C. R. Pennypacker, 1973: Scattering and absorption of light by nonspherical dielectric grains. *Astrophys. J. 186,* 705-7 14.

Ramanathan, V., R. D. Cess, E. F. Harrison, P. Minnis, B. R. Barkstrom, E. Ahmad, and D. Hartmann, 1989: Cloud radiative forcing and climate: Results from the ERBE, *Science, 243,* 57–63.

Randall et al., 2007: Cilmate Models and Their Evaluation. In: Climate Change 2007: The Physical Science Basis. Contribution of Working Group Ito the Fourth Assessment Report of the Intergovernmental Panel on Climate Change, Solomon, S., D. Qin, M. Manning, Z. Chen, M. Marquis, K. B. Averyt, M. Tignor and H. L. Miller (eds.), Cambridge University Press, Cambridge, United Kingdom and New York, NY, USA.

Rädel, G., C. J. Stubenrauch, R. Holz, and D. L. Mitchell, 2003: Retrieval of effective ice crystal size in the infrared: Sensitivity study and global measurements from the TIROS-N Operational Vertical Sounder. *J. Geophys. Res., 108,* 428 14292.

Rind, D., P. Lonergan, and K. Shah, 1996: Climatic Effect of water vapor release in the upper troposphere. *J. Geophy. Res., 101,* 29,395Å29,405.

Rind, D., P. Lonergan, and K. Shah, 2000: Modeled impact of cirrus cloud increases along aircraft flight paths. *J. Geophy. Res., 105,* 19,297Å19,940.

Rossow, W. B., and R. A. Schiffer, "Advances in understanding clouds from ISCCP," *Bull. Am. Meteorol. Soc., 80,* 2261-2287 (1999).

Salathe, E. P., and D. L. Hartmann, 1997: A trajectory analysis of tropical upper- tropospheric moisture and convection, *J. Climate, 10,* 2533–2547.

Sassen, K., 1997: Contrail-cirrus and their potential for regional climate change. *Bull. Amer. Meteor. Soc., 78,* 1885–1903.

Sassen, K., and J. R. Campbell, 2001: A midlatitude cirrus cloud climatology from the Facility for Atmospheric Remote Sensing: I. Macrophysical and synoptic properties. *J. Atmos. Sci., 58,* 481–496.

Sassen, K., and C. Hsueh, 1998: Contrail properties derived from high-resolution polarization lidar studies during SUCCESS. *Geophys. Res. Lett., 25,* 1165–1168. Sassen, K., and K.N. Liou, 1979a: Scattering of polarized laser light by water droplet, mixed phase, and ice crystal clouds: I. Angular Scattering patterns. *J. Atmos. Sci.*, 36, 838-851.

Sassen, K., and K.N. Liou, 1979b: Scattering of polarized laser light by water droplet, mixed phase and ice crystal clouds: II. Angular depolarizing and multiple scattering behavior. *J. Atmos. Sci.*, 36, 852-861.

Sausen, R., K. Gierens, M. Ponater, and U. Schumann, 1998: A diagnostic study of the global distribution of contrails. Part I: Present day climate. *Theor. Appl. Climatol., 61,* 127–141.

Sausen, R. I. Isaksen, V. Grewe, D. Hauglustaine, D. Lee, G. Myhre, M. Köhler, G. Pitari, U. Schumann, F. Stordal, C. Zerefos, 2005: Aviation radiative forcing in 2000: An update on IPCC (1999). *Meteorologische Zeitschrift, 14-4,* 555–561.

Schmitt, A., Brunner, B., 1997: Emissions from aviation and their development over time. In: Schumann, U., Chlond, A., Ebel, A., KaÕrcher, B., Pak, H., Schlager, H., Schmitt, A., Wendling, P. (eds.) Pollutants from Air Traffic Results of Atmospheric Research 1992-1997. DLR Mitteilung 97-04, 37Å52.

Schörder, F.; B. Kãrcher, C. Duroure, J. Strm, A. Petzold, J.-F. Gayet, B. Strauss, P. Wendling, S. Borrmann, 2000: On the transition of contrails into cirrus clouds. *J. Atmos. Sci., 57,* 464–480.

Schneider, E. K., B. P. Kirtman, and R. S. Lindzen, 1999: Tropospheric water vapor and climate sensitivity, *J. Atmos. Sci., 36,* 1649–1658.

Schrader, M. L., W. D. Meyer, and C. L. Weaver, 1997: Comments on "A reexamination of the formation of exhaust condensation trails by jet aircraft." *J. Appl. Meteor., 36,* 623–626.

Schumann, U., 1996: On conditions for contrail formation from aircraft exhausts, *Meteor. Z., N.F. 5,* 3Å22.

Schumann, U. and P. Wendling, 1990: Determination of contrails from satellite data and observational results. Air traffic and the environment–background, tendencies, and potential global atmospheric effects, SpringerÅVerlag, Berlin, Heidelberg, New York, USA, 13 8Å153.

Sherwood, S. C., 1996: Maintenance of the free-tropospheric tropical water vapor distribution. Part II: Simulation by large-scale advection, *J. Climate, 9,* 29 19–2934.

Sherwood, S. C., and C. L. Meyer, 2006: The general circulation and robust relative humidity, *J. Climate, 19,* 6278–6290.

Shine K. P., and A. Sinha, 1991: Sensitivity of the earth's climate to height dependent changes in the water vapor mixing ratio. Nature, 354, 382Å384.

Sinha, A., and J. E. Harris, 1995: Water vapour and greenhouse trapping: the role of far infrared absorption, *Geophys. Res. Lett., 22,* 2147–2150.

Soden, B. J., D. L. Jackson, V. Ramaswamy, M. D. Schwarzkopf, and X. Huang, 2005: The radiative signature of upper tropospheric moistening, *Science, 310,* 84 1–844.

Soden, B. J., R. T. Wetherald, G. L. Stenchikov, and A. Robok, 2002: Global cooling after the eruption of Mount Pinatubo: A test of climate feedback by water vapor, *Science, 296,* 727–730.

Strauss, B., 1994: Uber den Einfluss naturlicher und anthropogener Eiswolken auf das regionale Klima mit besonderer Berucksichtigung des mikrophysikalischen Einflusses. Dtsch. Luft Raumfahrt Forschungsber. 94-23, 97.

Stordal, F., G. Myhre, E. J. G. Stordal, W. B. Rossow, D. S. Lee, D. W. Arlander, and T. Svendby, 2005: Is there a trend in cirrus cloud cover due to aircraft traffic? *Atmos. Chem. Phys., 5,* 2155–2162.

Strom J., and S. Ohlsson, 1998, In situ measurements of enhanced crystal number densities in cirrus clouds caused by aircraft exhaust. *J. Geophys. Res. 103 (D10):* 11355–11361.

Stubenrauch, C. J., R. Holz, A. Chédin, D. L. Mitchell, and A. J. Baran, 1999: Retrieval of cirrus ice crystal sizes from 8.3 and 11.1 &m emissivities determined by the improved initialization inversion of TIROS-N Operational Vertical Sounder. *J. Geophys. Res., 104,* 31,793–31,808.

Stuber, N., P. Forster, G. Rädel, and K. Shine, 2006: The importance of the diurnal and annual cycle of air traffic for contrail radiative forcing. *Nature, 441,* 864–867.

Stuber, N., and P. M. Forster, 2007: The impact of diurnal variations of air traffic on contrail radiative forcing. *Atmospheric Chemistry and Physics, 7,* 3153–3162. Sussmann R., 1997: Optical properties of contrail-induced cirrus: Discussion of unusual halo phenomena. *Appl. Opt, 36,* 4195–4201.

Takano, Y., and K. Jayaweera, 1985: Scattering phase matrix for hexagonal ice crystals computed from ray optics. *Appl. Opt., 24,* 3254-3263.

Takano, Y., and K.N. Liou, 1989a: Solar radiative transfer in cirrus clouds. Part I. Single-scattering and optical properties of hexagonal ice crystals. *J. Atmos. Sci., 46,* 3- 19.

Takano, Y., and Liou, K.N., 1989b: Radiative transfer in cirrus clouds. II. Theory and computation of multiple scattering in an anisotropic medium. *J. Atmos. Sci., 46,* 20-36.

Tompkins, A. M., K. Gierens, and G. Radel, 2007: Ice supersaturation in the ECMWF integrated forecast system. *Q. J. R. Meteorol. Soc., 133,* 53–63.

Travis, D. J., A. M. Carleton, and R. Lauritsen, 2002: Jet contrails and climate: anomalous increases in U.S. diurnal temperature range for September 11-14, 2001. *Nature, 418,* 601.

Waterman, P. C., 1971: Symmetry, unitarity, and geometry in electromagnetic scattering, *Phys. Rev., D3,* 825-839.

Weickmannn, H., 1945: Formen und Bildung atmosph ̃arischer Eiskristalle. *Beitr. Phys. Atmos. 28,* 12–52.

Weickmann, H., 1949: Die Eisphase in der Atmosphäre. Ber. Dtsch. Wetterdienstes U.S.-Zone, 6, 3–54.

Wielicki, B. A., and L. Parker, 1992: On the determination of cloud cover from satellite sensors: The effect of sensor spatial resolution. *J. Geophys. Res., 97,* 12 799–12 823.

Wendling, P., R. Wendling, and H. K. Weickmann, 1979: Scattering of solar radiation by hexagonal ice crystals. Appl. Opt. 18, 2663–2671.

Yang, P., and K. N. Liou, 1995: Light scattering by hexagonal ice crystals: comparison of finite-difference time domain and geometric optics methods. *J. Opt. Soc. Amer. A,, 12,* 162-176.

Yang, P., and K. N. Liou, 1996a: Finite-difference time domain method for light scattering by small ice crystals in three-dimensional space. *J. Opt. Soc. Amer., A13,* 2072-2085

Yang, P., and K. N. Liou, 1996b: Geometric-optics-integral-equation method for light scattering by nonspherical ice crystals. *Appl. Opt., 35,* 6568-6584.

Yang, P., and K. N. Liou, 1997: Light scattering by hexagonal ice crystals: solution by a ray-by-ray integration algorithm. *J. Opt. Soc. Am. A., 14,* 2278-2288.

Yang, P., and K. N. Liou, 1998: Single-scattering properties of complex ice crystals in terrestrial atmosphere. *Contr. Atmos. Phys., 71,* 223-248.

Yang, P., K. N. Liou, K. Wyser, and D. Mitchell, 2000: Parameterization of the scattering and absorption properties of individual ice crystals. *J. Geophys. Res., 105,* 4699– 4718.

Yang, P., H. Wei, H.-L. Huang, B. A. Baum, Y. X. Hu, G. W. Kattawar, M. I. Mishchenko, and Q. Fu, 2005: Scattering and absorption property database for nonspherical ice particles in the near- through far-infrared spectral region. *Appl. Opt., 44,* 5512–5523.

Yang, P. and K. N. Liou, 2006: Light Scattering and Absorption by Nonspherical Ice Crystals, in *Light Scattering Reviews: Single and Multiple Light Scattering,* Ed. A. Kokhanovsky, Springer-Praxis Publishing, Chichester, UK, 31-71.

Yee, S. K., 1966: Numerical solution of initial boundary value problems involving Maxwell's equations in isotropic media. *IEEE Trans. Antennas Propag., 14,* 302- 307.

Yurkin, M. A., and A. G. Hoekstra, 2007: The discrete dipole approximation: an overview and recent developments, *J. Quant. Spectrosc. Radiat. Transfer,* doi: 10. 1016/j.jqsrt.2007.01 .034.

In: Aviation and the Environment
Editor: Jon C. Goodman
ISBN: 978-1-60692-320-7

Chapter 7

METRICS FOR COMPARISON OF CLIMATE IMPACTS FROM WELL MIXED GREENHOUSE GASES AND INHOMOGENEOUS FORCING SUCH AS THOSE FROM UT/LS OZONE, CONTRAILS AND CONTRAIL-CIRRUS

Piers Forster and Helen Rogers

EXECUTIVE SUMMARY

Issues of Contention

The United Nations Framework Convention on Climate Change (UNFCC) entered into force in 1994 with the objective for 'stabilization of greenhouse gas concentrations in the atmosphere at a level that would prevent dangerous anthropogenic interference with the climate system'. The Kyoto Protocol (1997) set out to reduce emissions of most long-lived greenhouse gases in developed countries to below their 1990 levels. Probably as a result of convenience and simplicity, the chosen metric to compare the climate impact of these greenhouse gases was the 100-year Global Warming Potential (GWP), as calculated by the Intergovernmental Panel of Climate Change Second Assessment Report (IPCC, 1995).

As an integral and growing part of the global economy and transportation sector, aviation has the potential to significantly contribute to changes in the Earth's climate. However, the impact of short-lived species (e.g. nitrogen oxides (NOx), an ozone precursor which in turn impacts on methane) and effects (e.g. aviation induced contrails) on the climate system depends upon geographical and altitudinal location, season, time of the day and the background meteorology and chemistry during their release (Rogers et al., 2000; Sausen et al., 2005). Such short-lived species therefore require an appropriate metric which takes into consideration these dependencies (Rogers et al., 2002a). For the aviation sector the potential climate impact is dependent upon both long-lived and short-lived emissions and effects, making the choice of a suitable metric that integrates over all effects more difficult.

Gaps

In 1999, the Intergovernmental Panel on Climate Change published a landmark report, 'Aviation and the Global Atmosphere' (IPCC, 1999) which saw the first sectoral examination by the IPCC and estimates of the potential impact resulting from aircraft emissions and their effects. The IPCC (1999) report identified the factors that influence climate. Using radiative forcing as the chosen metric, it found that aviation gives a small but significant climate forcing that is somewhat uncertain in overall magnitude. However, the IPCC (1999) report came out strongly against the use of GWPs in the context of aircraft emissions. In contrast, the most recent IPCC (2007) report presented a range of possible GWPs for aviation NOx emissions, although not for other aviation effects (Forster et al., 2007).

Due to a pressing need to provide policy-relevant answers to regulatory bodies and industry, many researchers have developed their own metrics to assess the impact of these short-lived species. Unfortunately, these approaches are often scientifically flawed. The strong statements of IPCC (1999) have certainly affected the landscape of metric design not only for aviation but also for other sectors. With climate change very much on the agenda of international policy and with a need to quantify the climate impact of human emissions, metric evaluation and metric design literature has flourished. Metric design is no longer solely undertaken by physical scientists, but social scientists, economists and industry are developing a plethora of metrics to suit individual needs.

Limitations

There is considerable controversy about the application of emission metrics to assess the effect of aviation non-CO_2 emissions. IPCC (1999) stated that the global warming potential "has flaws that make its use questionable for aviation emissions" and that "there is a basic impossibility of defining a GWP for aircraft NO_x". Wit et al. (2005) echo these sentiments, concluding that "GWPs are not a useful tool for calculating the complete suite of aircraft effects". An undesirable side effect of the negative stance is that it has led some policymakers and other groups to apply a Radiative Forcing Index (RFI) as if it is some kind of alternative to the GWP (see Forster et al., 2006).

It is certainly true that major caveats are required in the presentation and application of any currently proposed emissions metric. However, it needs to be clearly recognised that some difficulties are not a function of the metric design but are due to more fundamental limitations of our understanding of atmospheric processes. One example is the impact of persistent contrails on cirrus clouds; these certainly do preclude confident evaluation of values of GWPs, but the problem is much deeper than the evaluation of metrics – any attempt to quantify their impact, using even the most sophisticated climate models, would face similar limitations. Other limitations are more structural, such as the problem in using global-mean values for NO_x emissions, when compensation between negative forcings at a global level may not apply at the hemispheric level.

Priorities

A list of recommended priorities for tackling the outstanding issues related to the development and implementation of an appropriate metric for determining aviation's climate impact are given below: All of the tasks listed are achievable and will significantly improve our understanding of climate impacts whilst reducing scientific uncertainty

Understand that metric choice is not solely a science issue –policy comes into play. Therefore a range of people from different disciplines, including policy makers and scientists need to be involved in metric choice.

Assessment of the literature on alternative approaches to the use of GWPs as a suitable metric of climate change.

Diagnosis of the variation of the climate sensitivity parameter with forcing agent.

A study of climate impacts and their robust beyond global mean temperature change, with particular emphasis on the local response

Assessment of the potential range of impacts diagnosed using a spectrum of metrics and timescales.

Appropriateness of cancelling negative and positive climate effects – improved understanding as to whether multiple climate effects can be combined and how global cancellation affects local responses.

Appropriateness of pulsed or sustained emissions of realistic scenarios – improved understanding of scenario choice leads to different implications of aviation impact.

Improved understanding of how background climate change and atmospheric conditions affect forcing, climate impact and metric choice.

Recommendations for Research Needs

Improved description of NO_x and NO_y chemistry, sources and sinks particularly related to the chemistry of the UTLS region and potential anthropogenic impacts.

Improved model prediction of dynamical climate feedback processes throughout the lower atmosphere.

Investigations of how regional localised emissions affect climate both locally and globally

Study of the processes and radiative effects of contrails and aircraft induced cirrus.

Development of methods for ascertaining and forecasting supersaturation for use in cloud and contrail prediction

Model-model intercomparison and model-measurement intercomparison - understanding of the interaction between ozone and methane.

Impact of a pulse emission of NO_x emitted under different atmospheric conditions and seasons.

Quantification of the full effect of aviation under potential operational and technical procedures.

Long-term observational capability for integrated monitoring of climate gases and clouds.

Coniuted development of social and economic metric approach , with an acknowledgement of their limitations

'Practical' Application of Current Knowledge and Capability

In general, we recommend continued science studies to reduce uncertainties where achievable, and the use of simple metrics. We recommend quoting ranges for a number of metrics, as different metrics give different indications of importance. This also prevents metrics being deliberately chosen to advocate particular policy choices. Development of our understanding of the atmosphere and computational power should eventually enable sophisticated coupled climate models to be used to explore metrics of aviations impact.

Specifically, our recommended approaches involve simple metrics only (GWP and GTP) and includes all forcing factors that are relatively well quantified (currently excluding the role of aviation induced cirrus). Since likely future policy will be directed towards reductions by a particular target date, we recommend the adoption of ASGTP(H), limited probably to a target date around 2060. Further, with present knowledge we recommend only applying these metrics at the globally-averaged emission level, i.e. not applying different GWPs to emissions from different regions/heights/seasons etc.

1. Introduction and Background

The Earth's climate is warming and human activity is *very likely* (90% certain) to be responsible for the warming observed over recent decades (IPCC WG1, 2007). The largest contribution to both past climate change and expected future climate results from emissions of long-lived greenhouse gases. Due to their long life-time in the atmosphere (greater than 10 years) the climate effects of these emissions are not location specific and are readily comparable using simple metrics (Forster et al., 2007).

The United Nations Framework Convention on Climate Change (UNFCC) entered into force in 1994 with the objective for 'stabilization of greenhouse gas concentrations in the atmosphere at a level that would prevent dangerous anthropogenic interference with the climate system'. The Kyoto Protocol (1997) set out to reduce emissions of most long- lived greenhouse gases in developed countries to below their 1990 levels. As a clear climate-change target was never defined, the Kyoto protocol aimed simply to limit emissions of several greenhouse gases: carbon dioxide (CO_2); methane (CH4); nitrous oxide (N2O); hydrofluorocarbons (HFCs); perfluorocarbons (PFCs) and sulphur hexafluoride (SF6). Probably as a result of convenience and simplicity, the chosen metric to compare the climate impact of these greenhouse gases was the 100-year Global Warming Potential (GWP), as calculated by the Intergovernmental Panel of Climate Change Second Assessment Report (IPCC, 1995). In recent years a more targeted approach has been developed to directly address the issue of 'dangerous climate change'. A 2005 UK initiative (Avoiding Dangerous climate Change, 2005) suggested that a globally average temperature rise of 2K or more from pre-industrial times would be 'dangerous' - largely because of the possibility of destabilising high latitude ice caps (especially Greenland) and permafrost melt. This would cause rapid sea-level rise and other positive feedbacks. A similar description of temperature thresholds beyond which climate change becomes 'dangerous' has recently become internationally recognised in European Union climate change policy. The IPCC (2007) WGIII Fourth Assessment report (AR4) also analysed mitigation polices to keep global mean temperatures below certain target thresholds and such an approach is likely to feature in any agreement made at the UN Climate Change conference in Bali at the beginning of December 2007.

Predicting future warming depends both on climate model behaviours (such as climate sensitivity) and future emission scenarios – both are uncertain. Nevertheless, based on standard future emission scenarios we expect 'dangerous' warming (a globally averaged temperature rise of 2K or more from pre-industrial times) to be reached before the end of this century (figure 1). Potential impacts of these target thresholds are shown in figure 1.

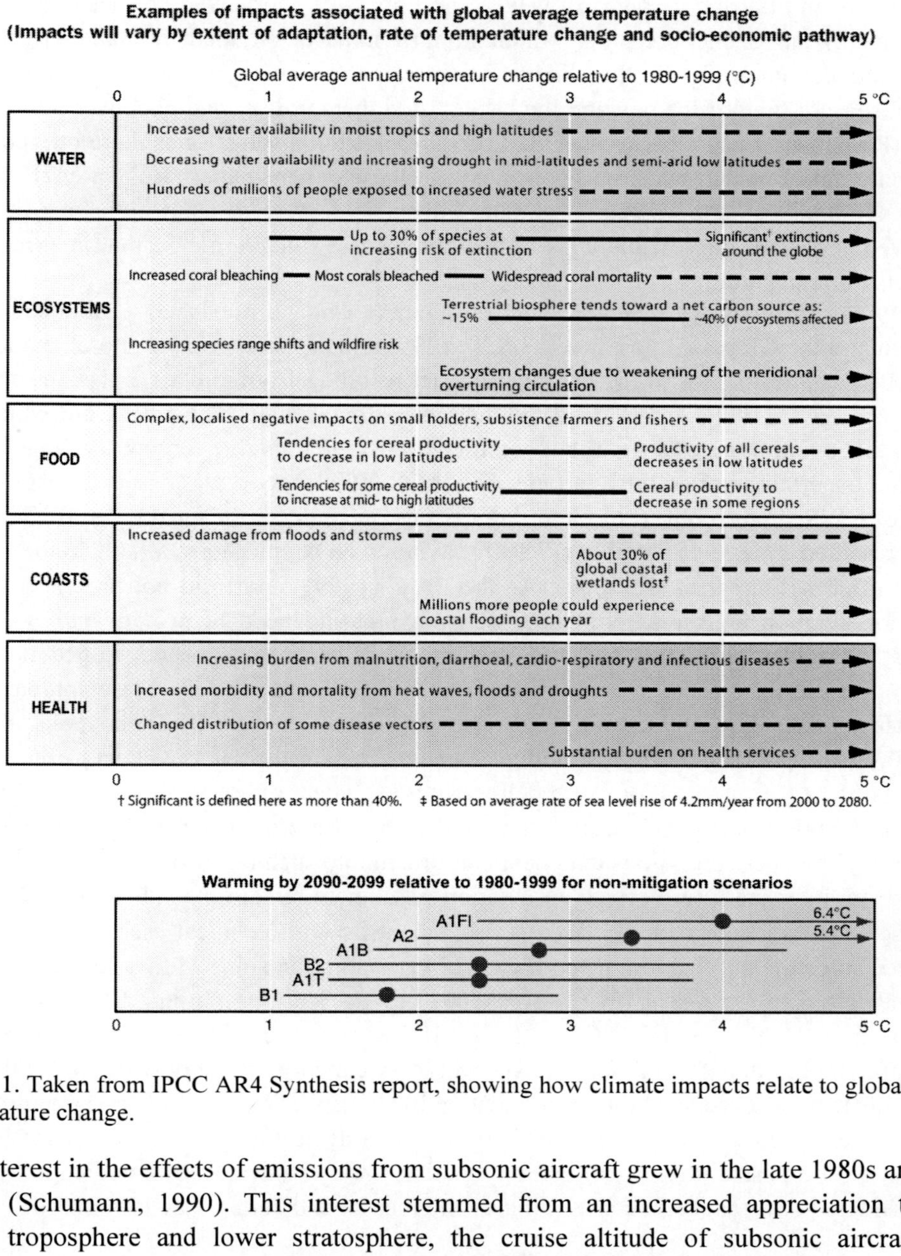

Figure 1. Taken from IPCC AR4 Synthesis report, showing how climate impacts relate to global mean temperature change.

Interest in the effects of emissions from subsonic aircraft grew in the late 1980s and early 1990s (Schumann, 1990). This interest stemmed from an increased appreciation that the upper troposphere and lower stratosphere, the cruise altitude of subsonic aircraft, is a sensitive region of the atmosphere for both chemistry and climate changes. Initially the attention was placed upon the effects of NOx emission from aviation on tropospheric O_3 production (e.g. the EU AERONOX and the US SASS projects, Schumann 1997; Friedl *et al.*, 1997). More recently the potential climate impact of other effects such as those of condensation clouds (contrails) and cirrus have been the focus of intensive investigation (e.g. Sausen et al., 2005).

The aviation sector has continued to grow strongly over the 1990s and early 2000s, despite events such as the Gulf War, 9-11 and SARS. As an integral and growing part of the global economy and transportation sector, aviation has the potential to significantly contribute to changes in the Earth's climate. However, the impact of short-lived species (e.g. nitrogen oxides (NOx), an ozone precursor which in turn impacts on methane) and effects (e.g. aviation induced contrails) on the climate system depends upon geographical and altitudinal location, season, time of the day and the background meteorology and chemistry during their release (Rogers et al., 2000; Sausen et al., 2005). Such short-lived species therefore require an appropriate metric which takes into consideration these dependencies (Rogers et al., 2002a). For the aviation sector the potential climate impact is dependent upon both long-lived and short-lived emissions and effects, making the choice of a suitable metric that integrates over all effects more difficult.

In 1999, the Intergovernmental Panel on Climate Change published a landmark report, 'Aviation and the Global Atmosphere' (IPCC, 1999) which saw the first sectoral examination by the IPCC and estimates of the potential impact resulting from aircraft emissions and their effects. The IPCC (1999) report identified the factors that influence climate. Combining these it found that aviation gives a small but significant positive radiative forcing of climate that is somewhat uncertain in overall magnitude. The IPCC (1999) report was however dismissive in the use of GWPs in the context of aircraft emissions. In contrast, the most recent IPCC (2007) report presented a range of possible GWPs for aviation NOx emissions, although not for other aviation effects (Forster et al., 2007). As the IPCC (1999) report did not present a suitable metric for aviation emissions, and because of a pressing need to provide policy-relevant answers to regulatory bodies and industry, many researchers have developed their own metrics to assess the impact of these short-lived species. Unfortunately, these approaches are often scientifically flawed. Currently only domestic emissions of CO_2 are covered under the Kyoto Protocol (i.e. departure and landing locations within the same country). International emissions of CO_2 from aviation were deliberately excluded, although the International Civil Aviation Organisation (ICAO) Committee on Aviation Environmental Protection (CAEP) is considering how these emissions may be incorporated into such protocols.

Concern over the future effects of aviation on climate remain the subject of debate both in the science and policy arena. As a result, scientific and technical assessment work has continued since the publication of the IPCC (1999) report and some of this has been reported and synthesized in the recent IPCC AR4 *2007) by its Working Groups I (science) and III (adaptation and mitigation). WGI and WGIII addressed disparate aspects of aviation, although there are important linkages, especially associated with metrics. In the WGI report, the aspects that have received the most attention in atmospheric science, namely contrails and aviation-induced cloudiness were considered in some detail. The WGIII report focussed its attention on the possibilities of mitigating aviation impacts from a technological standpoint, and considered other aspects such as policies and measures that might be introduced.

This SSWP relies heavily on published literature, together with state-of-the-art research from appropriate academic initiatives (e.g. UK-OMEGA, EU-QUANTIFY, EU- ATTICA, USA-PARTNER) in order discuss the metric problem in detail, assessing current levels of understanding, gaps in our knowledge and future possibilities.

2. Review

Before reviewing the literature on metrics it is important to briefly assess our overall understanding of aviation's role in climate change. It is also important to introduce past and future predicted trends in aviation traffic and discuss flight locations. As all of these features influence metric discussion.

2.1. Current state of science

2.1.1. Air Travel – Its Emissions and Its Trends

Aviation is a fundamental part of business and commerce, and as the globalisation of industry and commerce has increased so aviation has undergone spectacular growth, outstripping GDP. There are many forecasts available for the future growth of civil aviation traffic. Aerospace companies, aircraft manufacturers and airlines provide forecasts for business projections. The UK Department for Business Enterprise and Regulatory Reform provides its own market forecasts in order to inform UK government policy. Most aviation growth forecasts rely upon assessments of global economic trends, due to the close linkage between global GDP growth and aviation traffic growth. Passenger traffic is expected to average around 5.3% annual growth over the coming years (see figure 2). The increased global capacity in aviation will be provided by around 14,000 new aircraft between 1999 and 2018. Approximately half of this demand is expected to be derived from the replacement of existing aircraft retired from the fleet, with the other half generated by anticipated traffic growth. The environmental performance of civil aviation maintains a growing profile in social awareness and imposes pressures on the aviation industry to which it will need to respond.

Members of the European Regions Airline Association (ERA) have recorded significant growth for the first six months of 2007. Scheduled passenger traffic increased by 7.7% compared to the first half of 2006 with scheduled passenger kilometers increasing by 9.7% on the same period last year. Capacity levels for ERA member airlines have also been growing with seat numbers up 5.3% and available seat kilometers up 7.8% in the first six months of 2007 when compared to the same period in 2006.

For reasons of economy of operation, range and market demand, there has been a constant drive towards more fuel-efficient aircraft. Following the introduction of jet aircraft into the civil aviation fleet, approximately 40 years ago, fuel consumption per passenger-km has been reduced by approximately 70%. The most significant gains have been achieved through engine improvements and further improvements in efficiency are forecast to continue into the future.

Early research on aircraft emissions was focused primarily on improvements in the combustor technology required to meet the emerging landing/takeoff regulations. Today, the focus has widened beyond the locality of the airport to include emissions at higher altitude. Improvements to all aircraft components are required to meet the environmental concerns.

Gas turbine exhausts contain concentrations of CO_2, water vapour (H_2O), NO_x, sulphur compounds (SO_x, originating from sulphur in the fuel) and trace amounts of numerous other chemical species. In general, emissions of NO_x, CO, HCs and particles are relevant to local air quality issues whilst CO_2, H_2O, NO_x, SO_x and particles are of particular interest for climate change. Table 2 outlines the distance flown, fuel usage and emission products from civil and military aviation for 2002, as provided by the AERO2K database.

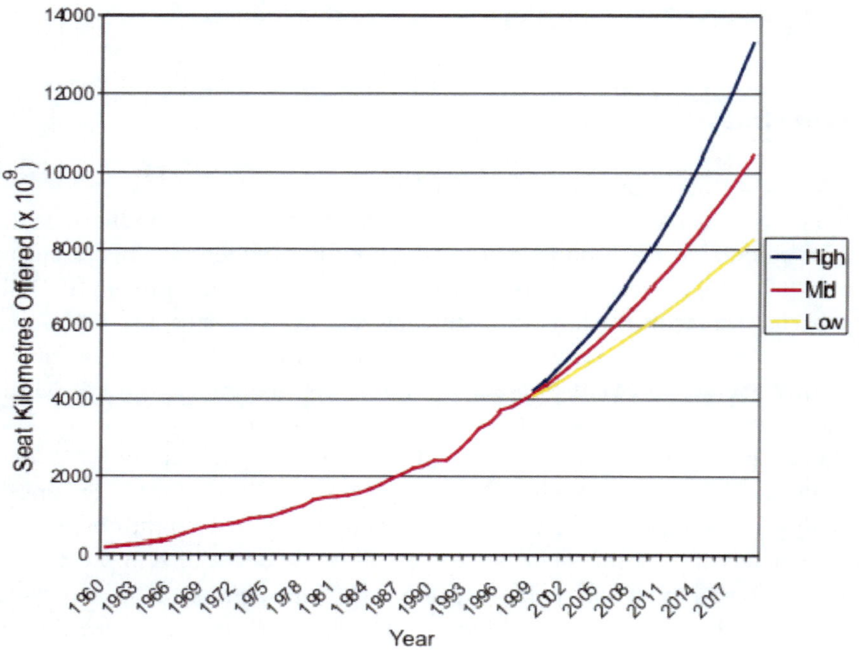

Figure 2. Aviation growth in terms of global SKO (seat kilometres offered) between 1960 and 2020 (source: UK. DTI data) – as in Rogers et al., 2002a.

Past and future aviation growth significantly influences the metric discussion. For example past rapid growth in aviation is responsible for the currently large non-CO_2 forcings from aviation, compared to the CO_2 forcing, which rises more slowly. Growth in the future will also affect choice of metric

2.1.2. Aviation's Climate Impact

This assessment largely draws on the IPCC AR4 assessment report (Forster et al., 2007) which in turn was largely based on Sausen et al. (2005). Together these works provide a valuable overview of the significant developments achieved following the IPCC (1999) report.

Table 1. Emission for AERO2K dataset in 2002 (Eyers et al,. 2004).

	Distance Flown	Fuel Used	CO_2 Produced	H_2O Produced	CO Produced	NOx Produced	HC Produced	Soot Produced	Particles Produced
	Nautical miles x 10^{-9})	(Tg)	(Tg)	(Tg)	(Tg)	(Tg)	(Tg)	(Tg)	(X 10-)
Civil Aviation	17.9	156	492	193	.507	2.06	.063	.0039	25)
Military Aviation	n/a	19.5	61.5	24.1	.627	.178	.064	n/a	4.03
AERO2K Total	n/a	176	553	217	1.13	2.24	0.127	n/a	n/a

Aviation emits gases and particles that in turn affect the climate by changing the atmospheric abundance of constituents and/or cloudiness. These effects are typically assessed by calculating the radiative forcing (RF, with units of Wm^{-2}) imbalance at the tropopause (see Forster et al., 2007 for details). These effects arise from:

emission of CO_2, which has a warming effect (positive RF);
emission of NOx, which results in the production of tropospheric O_3 (positive RF) and the reduction of ambient CH4, a cooling effect (negative RF);
direct emissions of H_2O (positive RF);
the formation of line-shaped contrails (positive RF);
the increase of cirrus clouds by spreading contrails (positive RF);
the emission of sulphate particles (negative RF) and;
the emission of soot particles (positive RF).
the indirect effects of aviation aerosols on background cloudiness (unknown RF)

and are typically quantified in terms of a global average RF -see figure 3. Each mechanism can be given a level of scientific understanding which incorporates both the evidence for the mechanism's existence and the consensus on the degree to which individual studies agree. It is important to note however that these mechanisms may each have different geographical distributions and timescales, and that, with the exception of CO_2, the impact is determined using the steady state change in concentrations resulting from 2005 emissions. Another necessary consideration when designing metrics is how radiative forcing translates into surface temperature change and/or other impacts. For example, studies have indicated that contrails may have a direct local impact on surface temperatures over the US including the diurnal temperature range (Travis et al., 2002). Another example, Ponater et al. (2005), found that in an ECHAM modelling study the equilibrium surface temperature response due to a Wm^{-2} forcing from contrails only produced around 60% of the response due to a Wm^{-2} forcing from CO_2. The ratio of a mechanisms response to the CO_2 response is called efficacy and, in fact, all aircraft forcings could have different efficacies compared to carbon dioxide. Table 2 presents a range of efficacies from an example model study that it relevant to aviation. The differences between the climate impact of the various aviation emissions and the trends in aviation itself need to be bourn in mind for the metric discussion which follows.

Table 2. Efficacies for aviation and idealized ozone changes from the ECHAM model. Taken from Grewe et al. (2007) – Table 7

	CO_2	CH_4	O_3 Lower strat	O_3 Upper trop	O_3 subsonic	H_2O subsonic	contrails
Efficacy	1	1.18	1.8	0.75	1.2-1.56	0.14	0.59

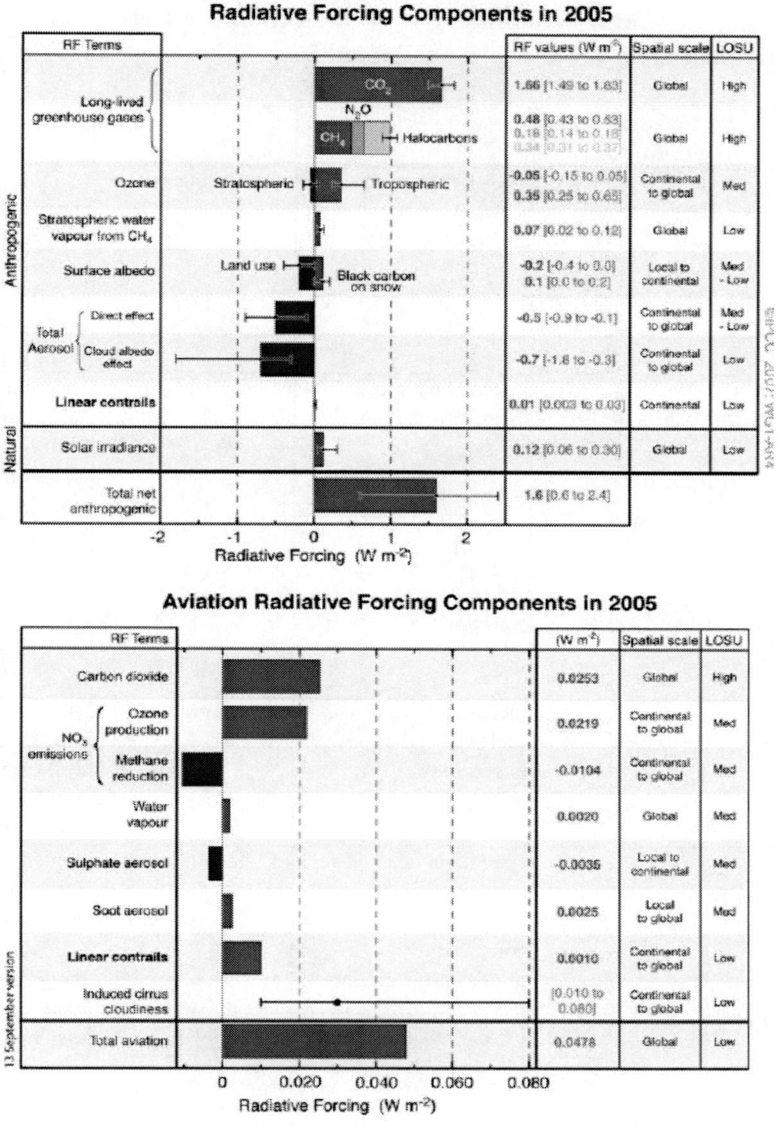

Figure 3. a) Radiative forcings from Forster et al. (2007). Showing aggregated forcing terms (implicitly including aviation effects) and b) RFs from aviation emissions, based on Sausen et al. (2005). Note that linear contrails are equivalent on the two plots. Columns represent spatial scale and level of scientific understanding. (Dave Fahey, Pers. Comm.).

2.1.3. Review of the RF Characteristics and Uncertainties of Mechanisms

2.1.3.1.Chemistry of Importance to Aviation

Aviation impacts on the atmosphere by perturbing the composition and microphysics of the system. A summary of the effects together with notes on the uncertainty of our understanding and/or modelling ability is provided in table 3.

Table 3. Summary of aviation climate effects and their quantification (adapted from Faber et al. 2006)

Effect	Emission quantification	Notes	Effect calculation	Notes
CO2	Yes	Relatively easy – scales with fuel; low uncertainty	Concentration, RF	Requires historical emissions data; moderate uncertainty. Can validate by sales of aviation fuel
O_3	No	Secondary species formed from NOx emissions	Concentration, RF	Secondary species formed from NOx emissions: modeldependent, large uncertainty
CH_4	No	Secondary species affected by NOx emissions:	Concentration (reduction), RF	Secondary species affected by NOx emissions: modeldependent, large uncertainty
H_2O	Yes	Relatively easy – scales with fuel; low uncertainty	Concentration, RF	Water vapour concentrations not well characterized in UTLS; moderate uncertainty
Sulphate	Yes	Relatively easy if S content of fuel is known; consequently moderate uncertainty	Concentration, RF	S content of fuel not well characterized. Calculation of RF model dependent, requires assumptions on size distribution; moderate uncertainty for direct effect, large uncertainty for impact on cloud properties

Table 3. (Continued).

Effect	Emission quantification	Notes	Effect calculation	Notes
Soot	Yes	Engine/combustor dependent, poorly characterized from measurements; large uncertainty	Concentration, RF	Concentrations and size poorly characterized; large uncertainty for both direct effect and impact on cloud properties
Contrails	No	Occurrence of contrails relatively easy to calculate if suitable atmospheric and engine data available	Coverage, RF	Coverage is modeldependent, RF model requires assumptions (size/shape of ice crystals); large uncertainty
Contrailinduced Cirrus	No	No current methodology for measurement/ modelling	Enhancement or coverage, RF	Coverage model/data dependent, poorly characterized optical properties; very large uncertainty

The main impacts of aviation on ozone, methane and contrails/cirrus are briefly discussed below. Full details can be found in SSWPs 2,4,5 and 6.

Ozone is produced in the troposphere and lower stratosphere by photochemical oxidation of CO and HCs, catalysed by NOx and HOx radicals. The production rate of O_3 is mainly dependent upon the abundance of NO and HO2, with increases in the ozone production rate with NO at low NO concentrations (Brasseur *et al.*, 1998). For NOx concentrations between 0.1 and 0.4 nmol/mol the production rate is however predicted to reach a maximum. Above this concentration, high levels of NOx cause a reduction of OH and hence a reduction in the ozone production rate (see figure 2-1, IPCC, 1999). As a result the change in ozone production rate due to the inclusion of aircraft emissions is highly dependent upon the background atmospheric conditions.

Methane (CH4) is emitted from both anthropogenic and natural sources, and is a greenhouse gas. Stevenson et al. (1997) and Isaksen et al. (2001) have shown that NOx emissions from aviation are very efficient within the upper troposphere in producing O_3 and thereby a positive impact on radiative forcing. As a result of the enhancement in NOx and O_3 due to aviation the hydroxyl radical (OH) concentration also increases. It is this hydroxyl radical that is primarily responsible for the oxidizing capacity of the troposphere. The increase in OH significantly reduces the lifetime of CH4 in the atmosphere and as such results in a negative radiative forcing signal due to CH_4.

Line-shaped clouds due to aviation (contrails) are formed when a mixture of hot and humid exhaust gases becomes mixed with cold ambient air in an environment saturated with respect to liquid water. This mechanism can be represented by the Schmidt-Appleman criterion (Schmidt, 1941; Appleman, 1953; Schumann, 2002) which predicts, to better than 1K, the threshold temperature for contrail formation based on the ambient pressure and relative humidity, the combustion temperature and overall propulsion efficiency, and the emission index of the water vapour from aviation. As well as the radiative importance of contrails, Borrmann et al. (1996) & (1997), Solomon et al. (1997) and Lelieveld et al. (1999) have suggested a potential role for cirrus particles in the heterogeneous chemistry of the atmosphere although further research on this topic is still required.

Radiative Effects: Emissions of NOx result in an enhancement of O_3 concentrations with an almost global reduction in CH_4 concentrations. The enhancement of O_3 results in a positive globally averaged radiative forcing, whilst the reduced CH4 concentrations result in a reduction in radiative forcing. As with thin cirrus clouds, contrails act to reduce the amount of both incoming short wave radiation (which acts to cool the climate system) and long-wave radiation (which acts to warm the climate system). The consensus (e.g. IPCC, 1999; Minnis et al., 2004) is that the impact on the longwave dominates such that contrails act to warm the climate.

2.1.3.2.Modelling the Impact Of Aviation

Global chemistry transport models (CTMs) and chemistry general circulation models (CGCMs) have become paramount to our understanding of aviation's impact on the atmosphere and the possible implications for our future climate. These models are frequently used for estimating the contributions due to individual pollutant sources on regional and global scales. Of particular importance for the climate system are changes to greenhouse gases occurring in the upper troposphere/lower stratosphere (Ramaswamy et al., 2001). Ozone chemistry in the upper troposphere and lower stratosphere is particularly sensitive to NO_x and is therefore dependent upon the transport of NO_x to and from this region. The ability of a model to correctly predict the atmospheric lifetime of ozone is necessary if the impact on the hydroxyl radical, and in turn methane, is to be determined. Accurately representing these processes relies on the skill of the atmospheric model involved and as such experiments are necessary, with a variety of atmospheric models, to provide confidence in the impact of aviation on the atmosphere under varying meteorological and chemical conditions.

It is important to note that modelling the various chemical and dynamical processes occurring within this region is a particularly challenging task. For example, the correct representation of lightning activity, which in the upper troposphere/lower stratosphere (UTLS) is an important source of NO_x, is poorly quantified (Hauglustaine et al., 2001). Another important consideration for the photochemistry of the upper troposphere, is the transport, both large scale vertical ascent and rapid convective activity, of pollutants from the surface into the UTLS (Berntsen and Isaksen, 1999; Jaeglé et al., 2001). Finally, the downward transport of stratospheric ozone into the troposphere is particularly sensitive the model's dynamical formulation and together with the other mechanisms discussed briefly above can result in a large uncertainty in the ozone budget of the UTLS and therefore any perturbation to it resulting from the aviation emissions.

Models involved in the prediction of aviation's impact on the atmosphere have often shown significantly differing results both in terms of their background concentrations of key species such as NO_x and in their calculation of the perturbation to atmospheric composition due to aircraft emissions. Brunner et al. (2003) & (2005) provided a rigorous evaluation of several European CTMs and CGCMs. Comparisons were made with trace gas observations from a number of research aircraft measurement campaigns during the period 1995-1998 inclusively. Their results revealed individual model deficits and suggested areas for further improvement. In general the models exhibited a weakness in their ability to represent both trace gas mean concentrations and vertical gradients (for example, O_3, CO and NO_x) in the tropopause region. Enhanced mixing across the tropopause accounted for large-scale differences between modelled and observed CO and O_3 concentrations, with deficiencies in the biomass burning emissions having a significant impact on CO concentrations. Poor correlations between modelled and observed NO_x concentrations suggested weakness in current parameterisations of convection and lightning. In contrast, however, modelled OH concentrations showed good agreement with observations. Overall, Brunner et al. (2003) & (2005) highlighted that a better description of NO_x and NO_y chemistry, sources and sinks was probably the key to any future model improvements with regard to accurately representing the chemistry of the UTLS region and potential anthropogenic impacts.

Following the IPCC (1999) report, Rogers et al. (2002b) provided a model intercomparison of the transport of aircraft-like emissions from both sub- and supersonic aircraft. Whilst the IPCC (1999) report highlighted the variability between model calculations, the results of Rogers et al. (2002b) emphasised the importance of correctly modelling the transport processes within the lower atmosphere when determining the impact of aviation on atmospheric composition and climate. The tracer transport experiments of Rogers et al. (2002b) revealed that the transport of aircraft-like tracers across dynamical 'barriers' was particularly important. For example, in the case of supersonic aircraft-like tracers, the correct reproduction of the 'tropical pipe' was critical in isolating any sub-tropical aircraft emissions from the mid and high latitudes. By isolating emissions within the tropics, these emissions can be effectively transported up into the middle stratosphere where effective NO_x chemistry can act to reduce O_3 at altitudes of ~30-35km. Of particular importance for subsonic aircraft, the degree of stratosphere-troposphere exchange of the prescribed aircraft-like tracers revealed further differences in the transport diagnosed between the various models compared in the study. The results suggest that the variability in stratosphere-troposphere exchange may be a possible cause of the discrepancies between IPCC (1999) model values of upper tropospheric ozone resulting from subsonic aircraft emissions. Rogers et al. (2002b) state that if aircraft emissions are considered to be inactive then within the course of only two years model calculations predict that emissions from the mid-latitude upper troposphere can be transported into the polar middle stratosphere. This result highlights the importance of atmospheric models to correctly predict transport processes throughout the lower atmosphere when determining the impact of both sub- and supersonic aircraft.

Prather (2002) suggested that to quantify the full impact of a trace gas emission on the climate system it is necessary to integrate the radiative forcing effects over the lifetime of the impact. For the troposphere, Prather (1994) showed that the adjustment time of methane (estimated at 12 years by IPCC, 2001) was the critical step in determining the longest lifetime. Whilst Prather (2002) demonstrated that the cumulative impacts of an emission can be evaluated by taking the steady-state response and scaling by the steady-state lifetime of the source gas, Stevenson et al. (2004) never-the-less adopted the approach of introducing a pulse emission from aviation within a climate-chemistry model and examining the resultant change in atmospheric composition after a sufficiently long time period (100 years). Stevenson et al. (2004) showed that the size of the initial positive ozone anomaly, resulting from a pulse emission of NO_x, determines the sign and magnitude of the overall net forcing. Further work however is clearly required (for example a range of pulse sizes needs to be considered) in order to test the robustness of this result. Additional research is also required to examine the impact of a pulse emission of NO_x emitted under different atmospheric conditions and seasons (Stevenson et al., 2004 only considered emissions during the months of January and July). This is particularly important as both ozone and the hydroxyl radical exhibit strong meteorological and seasonal dependencies.

Sausen et al. (2005) summarised some of the main conclusions of the EC funded TRADEOFF project, thereby providing an update to the aviation-induced radiative forcings for the year 2000. The largest difference with those presented in IPCC (1999) resulted from the reduction, by a factor of ~3-4, of the RF resulting from (linear) contrails. The impacts due to CO_2, O_3 and CH_4 were also reduced but to a far lesser extent. Overall the total radiative forcing impact due to aviation in 2000 (not including aviation induced cirrus) was calculated at 48 mWm^{-2}, similar to the total calculated in IPCC (1999) for 1992. It is important however to note that the radiative forcing due to aviation induced cirrus is not included in either the Sausen et al. (2005) or IPCC (1999) final estimates of the total impact of aviation due to uncertainties in the magnitude of such an impact. Hartmann et al. (1992) have shown that optically thin cirrus clouds on average warm the climate system however there are examples where the radiative forcing from aviation induced cirrus can be negative (Meerkotter et al., 1999; Myhre and Stordal, 2001). Sausen et al. (2005) suggest that the total aviation RF could be significantly larger than that given in the IPCC (1999) estimate, but that further research is required not only to correctly quantify the full effect but to examine potential operational and technical procedures which could be adopted by the aviation community if the impact were to be considered as significant.

2.1.4. Regional and Timescale Issues

Different forcing agents have different spatial patterns (see figure 2 and figure 6.7 of Ramaswamy et al. 2001). These are broadly associated with timescale – the shorter a timescale of a forcing agent the more localised the pattern of radiative forcing. CO_2 and CH_4 are long-lived and have global forcing patterns, whilst contrail and O_3 forcings are shorter lived and remain fairly localized to the Northern Hemisphere and flight corridors.

Each emission can affect atmospheric concentrations and the resulting RF on different timescales. These timescales are crucial in determining the climate impact of a given emission. As outlined in Section 2.1.3, aircraft emissions are associated with multiple lifetimes. Carbon dioxide lifetime ranges from years to millennia (a tiny fraction remaining permanently in the atmosphere). As CO_2 is long-lived (having an average lifetime longer than the atmospheric circulation), a tonne of CO_2 from aviation emitted into the upper troposphere is no different than that emitted by any other surface-based industry and its concentration, and hence RF, can easily be estimated using simplified but established methods based on carbon-cycle modelling. In contrast, timescales associated with aviation NOx emissions are different than those associated with NOx emissions at the surface. Stevenson et al. (2004) presents a useful discussion of the various timescales. Initially NOx produces ozone on short timescales (weeks-months), but it also decreases CH_4, which has an associated timescale of roughly 12 years. As CH_4 in turn also affects ozone, there is also a component of ozone change that occurs on this longer timescale. Contrails, in contrast, only last for a few hours. It is important to consider than forcings which may last no more than a few hours still influence climate for many years after, due to the time-lag of the Earth system (for example, the Earth's ocean takes decades to respond). Therefore forcings such as contrails still have a significant climate role.

Global average forcing has been a useful measure of global average equilibrium temperature response – climate models show a robust temperature response, especially when efficacy is accounted for (Forster et al., 2007). However, less work has been done on assessing how forcing relates to regional impacts. The surface temperature response certainly covers a wider area than the radiative forcing. Minnis et al. (2004) suggested a local response to aviation effects warming over the US, but this has been disputed by several studies that point to systematic flaws in the Minnis analysis. (Shine et al., 2005a, Ponater et al., 2005; Hansen et al., 2005). These modelling studies all support the view that the response to local forcing spreads over much of the globe. For example, high latitudes, generally warm more than low latitudes, even when the forcing is confined to low-latitudes (Forster et al., 2000).

Importantly, global cancellations between the responses of different forcings do not necessarily represent regional cancellation between their responses. In the metric context this is particularly important for NOx, where the O_3 warming effect remains confined to the hemisphere of emissions and the CH_4 cooling effect occurs globally. The net effect, given the regional pattern of airline flights, is therefore a Northern Hemisphere warming and Southern Hemisphere cooling (see figure 4).

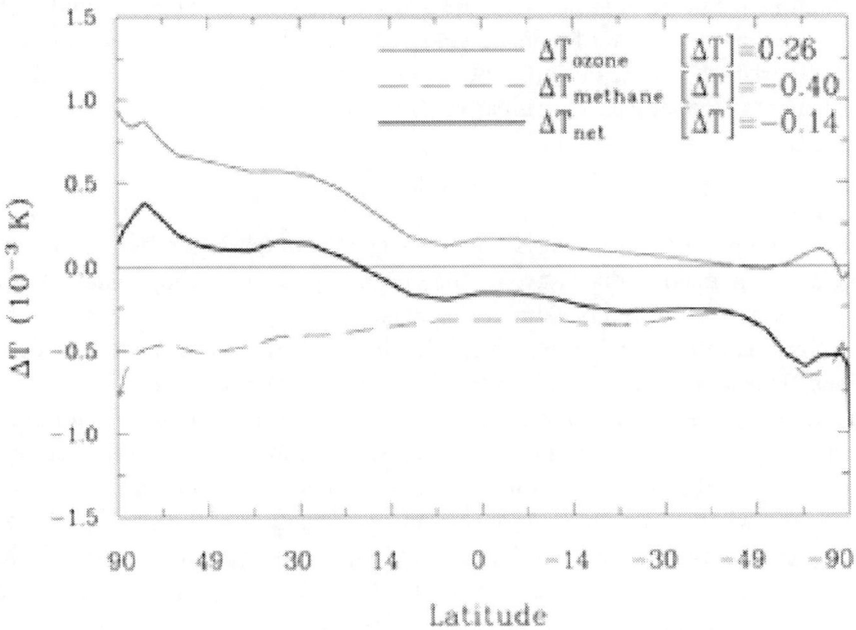

Figure 4. Surface temperature changes from calculations where an idealised emission of NOx from the surface in Europe is traced through its impacts on ozone, methane, radiative forcing and temperature

change. The surface temperature changes are shown for ozone changes only (thin solid line), methane changes only (dashed line) and the net effect (thick solid line). It shows that the strong global-mean cancellation between the two impacts (see [ ] values in legend) are made up of a northern hemisphere warming, where the ozone impact dominates over methane, and a southern hemisphere cooling where methane dominates over ozone. (From Shine et al, 2005b).

Regional climate change prediction has improved since the IPCC TAR report. However, it is still far less certain than prediction of global climate change (IPCC, 2007, Chapter 11). Regional surface temperature changes are still not adequately evaluated for aviation.

Observational studies have suggested that aviation plays a role in local diurnal temperature range change (Travis et al., 2002; 2004) and the possibility of an aviation induced weekend effect in diurnal temperature range has been mooted (Forster and Solomon, 2003). Other effects, such as surface energy budget changes, hydrological cycle effects and other climate impacts have not currently been evaluated for aviation. For future climate impact analysis these impacts are often simply associated with global mean temperature response irrespective of the cause of the temperature change itself (see Section 1).

2.2. Critical Role of the Specific Theme

2.2.1. Advancements Since the IPCC 1999 Report

Section 2 and other SSWPs discuss the development of RF understanding for aviation emissions. Here we focus on metric development only. As stated in the introduction, IPCC (1999) was somewhat dismissive of aviation GWPs as a metric. Their strong statements have certainly affected the landscape of metric design not only for aviation but also for other sectors. With climate change very much on the agenda of international policy and with a need to quantify the climate impact of human emissions, metric evaluation and metric design literature has flourished. Metric design is no longer solely undertaken by physical scientists, but social scientists, economists and industry are developing a plethora of metrics to suit individual needs.

2.2.2. What Is a Metric?

A metric, within this context, is simply a way of comparing differing influences on climate change in a quantifiable way so that users (typically policy makers) can make informed choices about the likely climate impacts of different future scenarios. They can explicitly be used as mitigation instruments, allowing tradeoffs to be made between various policy options. The design of a suitable metric is dependent upon an explicit set of choices made by the user. These may include a knowledge of the desired end-effect for comparison (e.g. economic cost of climate impact, surface temperature change, sea-level rise); the timeframe over which the *end-effect* is to considered; whether the emissions are sustained or act as a pulse; and whether the metric provides an accumulation of the effects throughout the timeframe. Figure 5 shows the cause and effect chain for climate emissions. The further down the chain you can evaluate a metric, the more directly relevant a policy choice can be made for its direct impact on climate and human welfare. However, uncertainty also increases, making metrics less quantifiable and transparent.

The assumption here is that a relatively transparent and simple methodology is required for quantifying the climate impact of non-CO_2 aviation effects. Several such measures exist and have been applied to aviation specifically or more generally. Each metric has disadvantages and advantages, and within each, several parameter choices have to be made. First we discuss non-emission based metrics and then we discuss emission based metrics.

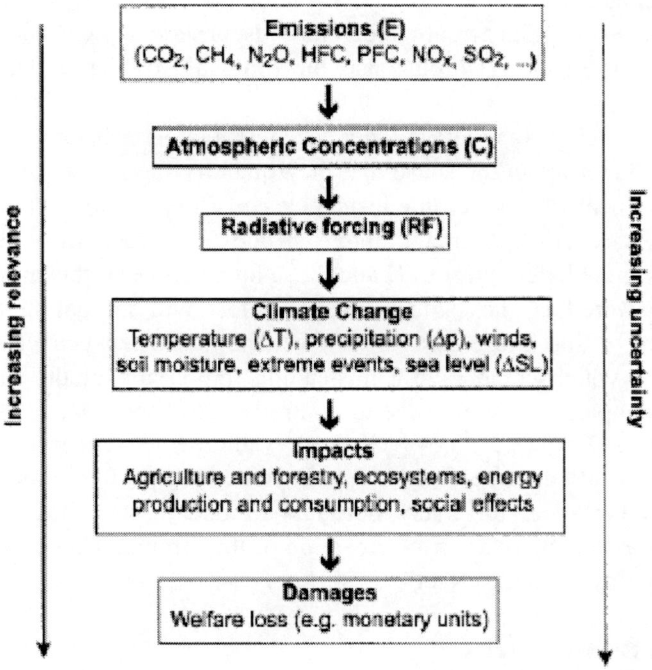

Figure 5. Cause and effect chain of the potential climate effect of emissions (from Fuglestvedt et al., 2003).

2.2.2.1.Non-Emission Based Metrics

Non-emission based metrics with do not specifically involve emissions but have been used to quantify and understand climate change effects.

Radiative forcing: Radiative forcing can be used as a metric, it quantifies, at a given time H, the perturbation to the Earth's radiation balance over some given time period (e.g. from pre-industrial times to the present day). At H, the total forcing is due to the remaining concentrations of all radiatively-active species in the atmosphere as a result of all emissions during the given time period. In the case of aviation, emissions of CO_2 from decades before H contribute to the CO_2 concentration at time H. By contrast, for short-lived species, it is emissions near H that contribute – in the case of contrails, it will be the effect of emissions only in the few hours before H.

Radiative Forcing Index (RFI): IPCC (1999) introduced the RFI as one way of characterising the importance of non-CO_2 forcings from aviation. It is simply the ratio of the total forcing to the CO_2-only forcing. Regrettably, the concept has been mis-applied as a measure of the relative impact of non-CO_2 species of *emissions* at a given time (see Forster et al., 2006 and 2007 corrigendum, also Section 2.2.4).

Temperature response: Given a time-history of radiative forcing, the resulting global averaged surface temperature response at a time H can be calculated; often this is done using quite simple models of the climate system (e.g. Sausen and Schumann 2000, Lim et al. 2007). The thermal inertia of the climate system means that the temperature change at H is less dependent on the emissions at times near H, as the climate system will have had less time to respond to these emissions. The actual temperature response to any emission will then depend on the lifetime of the resulting forcing and the timescale of the response of the climate system.

The radiative forcing (and RFI) and the temperature change can be considered "backward-looking" metrics in the sense that they quantify the impact of all emissions prior to H and are thus dependent on the time history of emissions (or for future times, the choice of future emission scenarios). As noted above, it does not necessarily distinguish between emissions at times immediately prior to H and those long before H; this may be an issue if the question to be answered is "how much climate effect will mitigating today's emissions have?" And related to this, these metrics do not distinguish between the timescales of the different emissions, which could give a misleading impression of the impact of emission controls. As an example, the forcing due to contrails may appear to be as important as the forcing due to CO_2 (see figure 3); however, if all aviation emissions were suddenly to cease, the contrail forcing would disappear within hours, while the CO_2 forcing would remain, albeit with decreasing importance, for many decades. In both cases, though, the temperature response remains for some time after the cessation of the forcing. Thus it is very important to define what is meant by "climate effect".

2.2.2.2. Emission Based Metrics

An alternative framework to the metrics above is to consider *emission* metrics, which attempt to quantify some measure of the climate impact on, for example, a per kg, or per kilometre flown, basis. Various possibilities are presented here, which are shown schematically on figure 6.

A very general formulation of an emission metric can be given by (e.g. Kandlikar,1996):

$$AM_i = \int_0^{\infty} \left[\left(I(\Delta C_{(r+i)}(t)) - I(\Delta C_r(t)) \right) \times g(t) \right] dt$$

where $I(\Delta Ci(t))$ is a function describing the impact (damage and benefit) of change in climate (ΔC) at time t. The expression $g(t)$ is a weighting function over time (e.g., $g(t) = e^{-kt}$ as a simple discounting giving short-term impacts more weight) (Heal, 1997; Nordhaus, 1997; IPCC WGIII 4AR Section 3.6.1.2). The subscript r refers to a baseline emission path. For two emission perturbations i and j the absolute metric values AMi and AMj can be calculated to provide a quantitative comparison of the two emission scenarios. In the special case where the emission scenarios consist of only one component (as for the assumed pulse emissions in the definition of GWP), the ratio between AMi and AMj can be interpreted as a relative emission index for component i versus a reference component j (as CO_2 in the case of GWP).

For a gas x, if A_x is the radiative forcing per kg, 6_x is the lifetime, and H is the time horizon then

$$APGWP^x(H) = \int_0^H A_x \exp(-t/\alpha_x)dt = A_x\alpha_x[1-\exp(-H/\alpha_x)] \qquad (2.1.)$$

The APGWP for CO_2 is more complicated, because its atmospheric lifetime cannot be represented by a simple exponential decay. All GWPs depends on the APGWP for CO_2. The APGWP of CO_2 again depends on the radiative efficiency for a small perturbation of CO_2 from the current level of about 378 ppm. The radiative efficiency per kilogram CO_2 has been calculated using the same expressions as in IPCC (2001), but with an updated background CO_2 mixing ratio of 378 ppm. For a small perturbation from 378 ppm the RF is 0.01413 W m^{-2} ppm^{-1}. The CO_2 response function is based on an updated version of the Bern carbon-cycle model, using a background CO_2 concentration of 378 ppm. The increased background concentrations of CO_2 means that the airborne fraction of emitted CO_2 is enhanced, contributing to an increase in the APGWP for CO_2. The APGWP values for CO_2 for 20, 100, and 500 years time horizons are $2.47X10^{-14}$, $8.69X10^{-14}$, and $28.6X10^{-14}$ W m^{-2} yr $(kg(CO_2))^{-1}$.

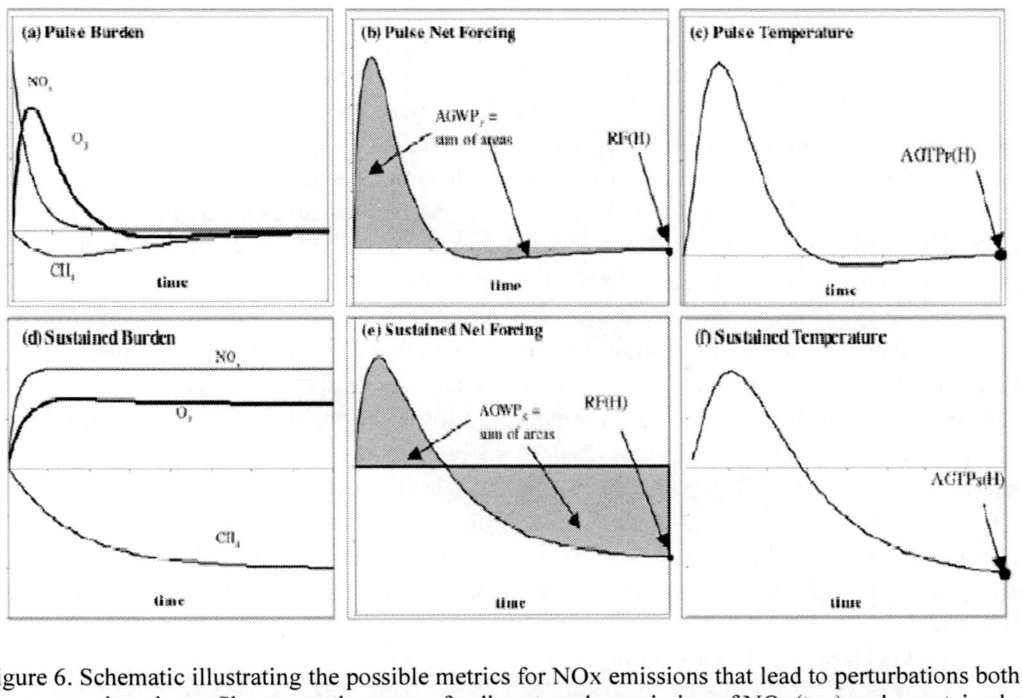

Figure 6. Schematic illustrating the possible metrics for NOx emissions that lead to perturbations both in ozone and methane. Shown are the cases of a discrete pulse emission of NOx (top) and a sustained emission change (bottom). (a) and (d): The evolution of the concentrations of NOx, ozone and methane. (b) and (e): The net (ozone plus methane) RF (the individual ozone and methane RFs follow the curves for the burden in (a) and (d) and the parameters that can be used for climate metrics. The absolute GWP (AGWP) is the time-integrated RF over some time horizon (H). The RF at some time H could also be used in a metric. (c) and (f): The global-mean surface- temperature change in response to the RF from (b) and (e). The absolute global temperature potential (AGTP) at some time H is another possible metric. (From Shine et al., 2005b). Note that when considering the integral of all impacts, independent of the number and atmospheric residence times of the secondary effects, Prather (2002)

demonstrated that this is equal to the steady-state pattern of impacts (caused by the specified emissions) multiplied by the steady-state lifetime of the source gas for that emission pattern.

The Sustained Global Warming Potential: A related metric is the version of the GWP for a sustained (rather than pulse) emission (or SGWP) which gives the time-integrated radiative forcing for a sustained step change in emissions. The SGWP has been in use for a number of years, but its formulation is clearly spelt out in the appendices of Berntsen *et al.* (2005).

The change in concentration, ΔC, as a function of time for a unit mass emission is given by

$$\Delta C(t) = \alpha_x(1 - \exp(-{}^{t}/_{\alpha_x})) \tag{2.2.}$$

and so the ASGWP is given by

$$ASGWP^x(H) = \int_0^H A_x\alpha_x(1 - \exp(-{}^{t}/_{\alpha_x}))dt = A_x\alpha_x[H - \alpha_x(1 - \exp(-{}^{H}/_{\alpha_x}))] \tag{2.3.}$$

Again, the formulation of the ASGWP for CO_2 is more complex, and is given in Appendix A of Berntsen et al. (2005), using the same carbon cycle model as used for the GWP (and hence consistent with IPCC, 2001).

The Global Temperature Change Potentials: A more recently proposed group of metrics (Shine et al., 2005a) are the pulse and sustained Global Temperature Change Potential (PGTP and SGTP) which have rather different characteristics (they are "end-point" metrics i.e. the temperature change at a particular time in the future, rather than a time integrated one). Arguably the GTPs are more relevant, as they address an actual climate impact (temperature change), rather than the more abstract integrated radiative forcing. Note that although not an integrated quantity they still rely on integrating the radiative forcing over time. A disadvantage of these is that they are not accepted for widespread use. To allow a transparent formulation of the GTPs, Shine et al. (2005a) adopted a simple climate model which allowed analytical forms of the GTPs to be derived, although this is by no means a requirement. The inclusion of this climate model means that additional parameters are required to be defined – the timescale of the climate response, τ, and the heat capacity of the climate system, C (or equivalently, C and the climate sensitivity parameter, λ – the three parameters are related since $\tau = C\lambda$).

The APGTP for gas x is given by

$$APGTP^x(H) = \frac{A_x}{C(\tau^{-1} - \alpha_x^{-1})}[\exp(-{}^{H}/_{\alpha_x}) - \exp(-\frac{H}{\tau})] \tag{2.4.}$$

Again, a more complex relationship is required for CO_2 and (2.4) is only applicable provided τ is not equal to б. Details are given in Shine et al. (2005a).

Shine et al. (2005a) point that although the pulse form of the GTP has some appeal, it appears that the simple climate model does not well represent the response of the climate system to a pulse emission; it will be retained here for illustrative purposes only. Also, for any case where $H >> \alpha_x$ (which is often the case for aviation emissions), the PGTP will be very small, as the climate system will have "forgotten" about the pulse emission. However, Shine et al. (2007) have proposed an alternative use of the PGTP, consistent with EU policy of restricting warming below some target amount at some future time.

This application shows clearly that as the target is approached, it becomes more "valuable" to reduce short-lived emissions. At times well before the target time, it is the long-lived species that exert more influence on the temperature at the target time.

The ASGTP for gas x is given by

$$ASGTP^x(H) = \frac{\alpha_x A_x}{C}\left\{\tau[1-\exp(-H/\tau)] - \frac{1}{(\tau^{-1}-\alpha_x^{-1})}[\exp(-H/\alpha_x) - \exp(-\frac{H}{\tau})]\right. \tag{2.5.}$$

Shine et al. (2005a) provide details of the CO_2 and τ=6 cases. As detailed by Shine et al (2005a), and, for long time horizons, the PGWP and SGTP asymptote to the same result, which allows an alternative interpretation of the GWP, and makes the distinction between the choice of pulse and sustained emissions arguably less important. It would be straightforward to develop metrics which are analogous to the PGTP and the SGTP, but which consider the forcing at time H.

2.2.3. Uncertainties of Metric Approaches

There is considerable controversy about the application of emission metrics to assess the effect of aviation non-CO_2 emissions. IPCC (1999) stated that the global warming potential "has flaws that make its use questionable for aviation emissions" and that "there is a basic impossibility of defining a GWP for aircraft NO_x". Wit et al. (2005) echo these sentiments, concluding that "GWPs are not a useful tool for calculating the complete suite of aircraft effects". An undesirable side effect of the negative stance is that it has led some policymakers and other groups to apply the RFI as if it is some kind of alternative to the GWP (see Forster et al., 2006).

Others have taken a more pragmatic stance than IPCC, and attempted to develop GWPs for aviation emissions, whilst recognising the caveats. The first attempt appears to be by Klug and colleagues in a series of unpublished reports as part of the EC Framework 5 Cryoplane project. More recently Svennson et al. (2004) has provided GWP values for aviation, based partly on the Klug approach. Wild et al. (2001) and Stevenson et al. (2004) have generated GWP values (although they did not label them as such) for aviation NO_x emissions. These are presented in the AR4 IPCC report. Forster et al. (2006) have also quoted GWP values for a range of aviation emissions, based on the Stevenson and Wild numbers.

It is certainly true that major caveats are required in the presentation and application of any currently proposed emissions metric. However, it needs to be clearly recognised that some difficulties are not a function of the metric design but more fundamental limitations of our understanding of atmospheric processes. One example is the impact of persistent contrails on cirrus clouds; these certainly do preclude confident evaluation of values of GWPs, but the problem is much deeper than the evaluation of metrics – *any* attempt to quantify their impact, using even the most sophisticated climate models, would face similar limitations. Other limitations are more structural, such as the problem in using global-mean values for NO_x emissions, as discussed in Section 2.1.4, when compensation between negative forcings at a global level may not apply at the hemispheric level.

One other cited difficulty with emissions metrics in the context of aviation is that some effects, particularly persistent contrail production, are not clearly related to emissions by the engine. Contrails are more a function of the background atmosphere, than they are of the emissions, with the water vapour (and particulate) emissions providing a trigger. Forster et al. (2006) propose that the contrail forcing is related to CO_2 emissions, which it is argued is valid provided that a fleet-wide approach is taken, and that the height and latitude distribution of emissions remains similar to the present day fleet. Indeed this approach of using fuel use as a proxy is embedded in calculations of global mean contrail cover (e.g. Sausen et al. 1998). It has been argued that flight km is a better way of doing this, but either approach can only be applied at some time or space aggregated basis, rather than for an individual flight.

Quantification uncertainties also need to be assessed when evaluating metrics. In particular more uncertain effects should not necessarily be given an equal weight to the role of carbon dioxide emissions in which we have a good level of confidence. These uncertainties are indicated by error-bars for NO_x and contrails in Section 2.4. Efficacy (see Section 2.1.2) can also influence this judgement.

Each metric and timescale chosen essentially gives a different viewpoint on the importance of various effects. Failing to show error bars for non-CO_2 effects may not give an accurate measure of understanding. Also different metrics address different policy concerns and apply different weightings to these. They therefore factor in policy decisions (e.g. about the relative importance of temperature change in the next 20 or 100 years). These metric choices and the effects of making them need to be carefully considered. We recommend that a range of metrics covering different time periods are given.

There are uncertainties associated with GWPs. The 95% uncertainty in the AGWP for CO_2 was estimated by Forster et al. (2007) to be ±15%, with equal contribution from the CO_2 response function and the RF calculation. The uncertainties of other long lived greenhouse gas GWPs were taken to be ±20%. The simplifications made to derive the standard GWP index include, set $g(t) = 1$ (i.e., no discounting) up until the time-horizon (TH), and then $g(t)=0$ thereafter, the choice of a 1 kg pulse emission, the definition of the impact function, $I(\Delta C)$ as the global mean RF, the assumption that the climate response is equal for all RF mechanisms, and the evaluation of the impact relative to a baseline equal to current concentrations (i.e., setting $I(\Delta C_r(t)) = 0$). The criticism of the GWP metric have focused on all of these simplifications (e.g. Smith and Wigley, 2000, O'Neill, 2000; Bradford, 2001; Godal, 2003). However, as long as there is no consensus on what is the relevant impact function ($I(\Delta C)$) and temporal weighting function to use (both involve value judgements), it is difficult to assess the implications of the simplifications objectively (O'Neill, 2000; Fuglestvedt *et al.*, 2003).

Berntsen et al. (2005) have examined the climate response due to ozone perturbations resulting from regional emissions of NO_x or CO. Using a combination of chemical transport models and general circulation models they have studied the response in O_3 and OH concentrations from emission perturbations in Europe and southeast Asia. The results for radiative forcing and climate sensitivities have been incorporated to examine the potential for improving the concept of GWPs in order to represent more fully the forcings due to short-lived species. They propose a modified GWP for a sustained-step emission change which includes variations in the climate sensitivity parameter under different climate change mechanisms. Their results indicate a higher latitudinal gradient in O_3 due to NO_x emissions than calculated with CO emissions. Although they state that they are unable to conclude whether real O_3 perturbations will in general result in a different climate sensitivity from CO_2, they are able to conclude that for O_3 high-latitude emissions of NO_x lead to climate perturbations with ~10-30% higher climate sensitivities. Their results for CO however showed little regional dependency. Berntsen et al. (2005) therefore support the idea that regionally different weighting factors for the climate sensitivity parameter are necessary for emissions of NO_x whilst for CO a single global number may suffice. They note however that calculating metrics for short-lived species by necessity requires the use of atmospheric models and that the derived metrics will be more model dependent than those calculated for long-lived species.

The adequacy of the GWP concept has been widely debated since its introduction (O'Neill, 2000; Fuglestvedt et al., 2003). By its definition, two sets of emissions that are equal in terms of their total GWP weighted emissions, will not give equivalence in terms of temporal evolution of the climate response (Smith and Wigley, 2000; Fuglestvedt *et al.*, 2000). Using a 100 year time horizon as in the Kyoto Protocol, the effect of current emissions reductions (e.g. during the first commitment period under the Kyoto Protocol) that contain a significant fraction of short-lived species (e.g. methane) will give less temperature reductions towards the end of the time horizon compared to reductions of CO_2 emissions only. GWPs can really only be expected to produce identical changes in one measure of climate change – integrated temperature change following emissions impulses – and only under a particular set of assumptions (O'Neill, 2000). The GTP metric (section 2.2.2.2) provides an alternative approach by comparing global mean temperature change at the end of a given time horizon. Compared to the GW P, the GTP gives equivalent climate response at a chosen time, whilst placing much less emphasis on near term climate fluctuations caused by emissions of short-lived species (e.g. methane). However, as long as it has not been determined, neither scientifically, economically nor politically, what is the proper time horizon for evaluating "dangerous climate change", the lack of temporal equivalence does not invalidate the GWP concept or provide a guidance to replace it. O'Neill (2003) have argued that the disadvantages of GWPs are likely to be out-weighed by the advantages. This can be done by showing that the cost difference between a multi-gas strategy and a CO_2-only strategy is likely to be much larger than the difference between a GWP-based multi-gas strategy and a cost-optimal strategy (accounting for damage and mitigations costs). Thus although it has several known short comings, the GWP remains the recommended metric to compare future climate impact of emissions of long lived climate gases. although it is possible to calculate the GWP for short-lived species, these have not been adopted by policy makers for a variety of reasons (IPCC, 2001; Berntsen et al., 2005 and Shine et al., 2005b). These include for example the robustness of model simulations used to predict the response in ozone (and methane) due to an emission of NO_x, and the ability to determine the global impact resulting from regional perturbations to short-lived species.

Shine et al. (2007) have examined the dependence of the climate sensitivity parameter, λ, on a pulse emitted Global Temperature Potential (GTP). The climate sensitivity parameter was varied from 0.4 $K(Wm^{-2})^{-1}$ to 1.2 $K(Wm^{-2})^{-1}$ (as suggested by IPCC, 2001) and the impact on the time for the climate response to reach an increase of 2°C above pre- industrial times was recorded. Their results showed a marked shift in the time for the climate response from 2067 with λ =0.4 $K(Wm^{-2})^{-1}$ to 2035 with λ =1.2 $K(Wm^{-2})^{-1}$. This result clearly emphasises that any uncertainty in the climate sensitivity parameter can have a significant impact on the appropriate metric. Any application of such a metric will therefore have to include a time dependency as our knowledge of the climate system increases and we move towards the target date.

For any purely physical metric it is important to note the difficulties when attempting to maintain climate stabilisation close to and after the target time. Irrespective of these difficulties the GTP has distinct advantages over GWP not least because it is further down the cause-and-effect chain. It maintains a level of transparency similar to the GWP metric and could provide valuable information to policymakers in determining appropriate new technological and economic options.

2.2.4. Incorrect Application of Metrics – Radiative Forcing Index, an Example

In the context of aviation, a common metric approach is to use an uplift factor of 2-3 to account for non-CO_2 effects of aviation. For example the recent inclusion of aviation within the EU emissions trading scheme has suggested an RFI value of 2 be used to compensate for the additional impacts of emissions from aircraft at altitude (see Section 3.5). The use of an uplift factor originates from a mis-application of the radiative forcing index (RFI). It is worth spending some time discussing its specific flaws here. An RFI of 2.7, calculated from the IPCC-1999 Special Report is often used as an uplift factor to weight the impact of CO_2 emissions from aviation in order to account for the non-CO_2 effects. Such an approach is scientifically flawed for a number of reasons.

1) Most importantly RFI is an instantaneous evaluation that does not account for the lifetime of emission and thereby overestimating the role of short-lived effects. This is highlighted by Forster et al. (2006) which illustrates how, with constant emissions for the year 2000, the forcings and RFI would vary with time (see figure 7). It is important to note that due to the long lifetime of carbon dioxide, CO_2 concentrations and the associate RF increases gradually with its emission. Aviation has grown rapidly over recent decades and as a result other non-CO_2 forcings have outgrown the RF for CO_2 alone, thereby culminating in a relatively high value for the RFI.

Using such a metric may not bring climate-benefit. For example the aviation industry could argue for a reduction in an uplift factor, by flying lower to produce less contrails at the expense of increased CO_2 emissions. Although in the long-term the increased CO_2 would warm climate, using an RFI metric would incorrectly predict climate benefit, where none existed.

2) The current RFI depends on past emissions, using it to evaluate future emissions is flawed. The current high value results from rapid past growth in aviation traffic, where non-CO_2 forcing effects have grown considerably faster than the CO_2 forcing. Therefore using such a metric effectively penalises the aviation industry's past rapid growth, which may be unfair. Although, if aviation continues to grow rapidly its use may be more justifiable.

3) Uncertainties are not taken into account. As discussed earlier in this section, uncertainties in the non- CO_2 effects of aviation preclude an accurate evaluation of the non-CO_2 forcing terms. Using latest RF estimates for aviation from Sausen et al. (2005) would reduce the RFI to around 1.9. However, if aviation induced cirrus effects were included RFI could be much bigger (~4, taking RF estimates from Sausen et al., 2005).

4) Similar uncertainties also exist for RFI as they do for the other metrics. RFI does not account for regional variation in forcing or response and it sums over very different effects, happening on different spatial scales and different timescales.

5) A similar RFI-type metric may need to be applied to other sectors for consistency (see Section 3.5). An RFI for shipping would likely be negative, due to SO_2 emissions leading to sulphate aerosol formation and an indirect effect on clouds. These effects have a larger negative instantaneous forcing than their positive forcing resulting from CO_2 emissions.

However, in the long-term ships will still produce climate warming because the long-lived CO_2 warming outlasts the sulphate cooling, yet applying such an RFI metric would suggest incorrectly that ships are actually beneficial for climate change (see Section 3.5 for further discussion).

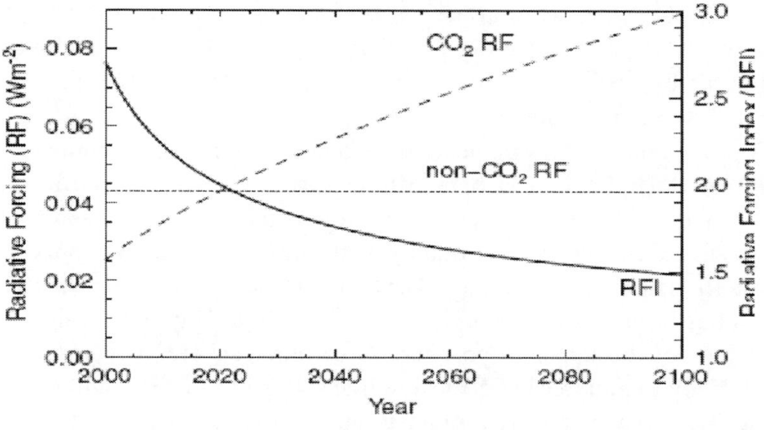

Figure 7. A scenario for sustained present-day emissions illustrating how CO_2 and its RF (dashed line) will continue to increase, whereas the non-CO2 effects (dotted line) have roughly stabilised with the emissions and are not expected to change. As a consequence of this the RFI (solid line) does not remain constant, but decreases over time (from Forster et al. 2006).

2.3. Present State of Measurements and Data Analysis

International assessments by WMO/UNEP, IPCC, IGAC, SPARC and EUROTRAC have all indicated that the largest uncertainties when assessing air quality and climate change result from:

the transport of aerosols, ozone and gases that control the concentration, over long distances;

possible changes in the oxidising capacity of the troposphere, with direct consequences for the removal of pollutants from the atmosphere;

the potential influence of water vapour, aerosol and clouds on the climate, including trends and the indirect effect of aerosols on cloud formation;

and variations in stratosphere-troposphere exchange as a result of climate change.

As emphasised in the WMO (2007) report, 'changes to the temperature and circulation of the stratosphere affect climate and weather in the troposphere', highlighting the importance of indirect perturbations to the highly-coupled atmospheric system. The impact of aviation on the global environment occurs through the emission of gases and particles directly into the atmosphere, which contribute to global change by altering the concentration of atmospheric greenhouse gases and triggering the formation of contrails and aviation induced cirrus. Localised air pollution, in the vicinity of airports, results from the emission of gases and particles from aircraft and associated ground transport and infrastructure. It is evident that not only could the aviation industry benefit from the provision of a long term monitoring network, but that it could substantially contribute through the use of commercial in-service aircraft as observational platforms of atmospheric composition.

In the early 1970s NASA's Global Atmospheric Sampling Programme (GASP) attempted to make regular atmospheric observations using commercial aircraft. This philosophy was again adopted in the early 1990s with research projects both in Europe and Japan. Whilst the European (MOZAIC, NOXAR) approach was to provide routine observations, Japan (JAL) opted for a biweekly 'grab' sampling technique. By the late 1990s this later approach was also utilised in the European CARIBIC project with an instrumented freight container for use primarily on short-haul destinations.

The EC programmes Measurement of Ozone and Water Vapour on Airbus Inservice Aircraft (MOZAIC I, II and III) demonstrated the enormous scientific value of regular observations made on board commercial aircraft in the monitoring and assessment of the causes for observed changes in air quality and climate. MOZAIC ended in 2004 having collected over 10 years worth of O_3 and H_2O vapour data, and 2 years of CO and NO_y data. This approach has been shown to provide an invaluable facility with which to maintain long term observations of the upper troposphere lower stratosphere, a region of the atmosphere notoriously difficult to monitor but critical to improving our understanding of climate change. Measurements from space and the ground in this region are difficult to perform and do not achieve the necessary spatial resolution attainable with in situ observations. Not only this, but with over 40,000 vertical profiles (obtained during landing and take-off) from more than 100 airports world-wide, a large database of observations have been made in developing countries where such data would otherwise have been difficult to obtain.

The scientific and technological expertise gained through the MOZAIC process is now being used in the design of a sustainable infrastructure suitable for routine global observations onboard a fleet of commercial aircraft. IAGOS differs from MOZAIC in many of its aims, including the design of instrument packages specifically aimed at measuring aerosol and cloud parameters, which, as stated by the IPCC, are the most uncertain contributors to climate change. IAGOS will also measure the important trace gases thereby providing information crucial to our understanding of climate change (including aviation's contribution) and the intercontinental transport of air pollution.

2.4. Present State of Modeling Capability/Best Approaches

Minimising the impact of aviation on the environment depends crucially upon the robust understanding of our atmosphere and aviation's contribution to its change. Potential areas of research cut across the disciplines of atmospheric science, economics and engineering and require a holistic view of the potential gains to be made from improved technologies (including alternative fuels) and operations. Mitigation options need to be carefully considered in order to provide accountability within all transportation sectors without inadvertently encouraging the misuse of resources which may result in environmental damage. Ongoing scientific research aims to improve our understanding of the atmosphere and the role of natural and anthropogenic emissions. A description of the major activities currently focussed on aviation's contribution to atmospheric change are described below.

In the USA, the PARTNER Center of Excellence is closely aligned with national and international needs by providing a world-class research organization with leverage from a broad range of stakeholder capabilities PARTNER fosters technological, operational, policy and workforce advances for the benefit of mobility, economy, national security and the environment, with involvement from 10 research institutes and more than 100 students. Particular emphasis is given to providing quantitative predictions and qualitative assessments of aviation noise, emissions and their impacts. A key objective of PARTNER is the improved communication and decision-making in addressing the interdependent environmental effects of aviation.

To assist in the development and communication of future strategies for a sustainable UK aviation industry, HEFCE provided financial support for a UK activity which combines academic capability with knowledge transfer to the stakeholder community. Opportunities for Meeting the Environmental Challenge of Growth in Aviation (OMEGA) is a 2 year programme of activities which started in January 2007, and aims to develop a consolidated knowledge basis within the UK; an overview of where the 'gaps' in our understanding remain, together with potential solutions; and a 'neutral space' for dialogue between academia and the stakeholder community.

The EC funded Integrated Project QUANTIFY aims to determine the climate impact of both present and future transport systems, including aviation, shipping and land-surface. The project, which began in March 2005 with funding for 5 years, uses improved emission inventories and more reliable models. The project provides forecasts and other policy-relevant advice with the assessment of several transport scenarios, and incorporates the exploitation of existing data with new field measurement, state-of-the-art numerical models and focused policy-relevant metrics for climate change. The project has already provided initial transport emission inventories, which have been incorporated into the appropriate modelling tools, and a variety of climate change metrics are under consideration. Through a European 'specific support action', ATTICA, also aims to provide a coherent set of assessments of the impact of transport emissions on ozone depletion and climate change.

2.5. Current Estimates of Climate Impacts and Uncertainties

In this section we present specific case studies in order to perform a quantitative comparison with which to evaluate different metrics on different timescales. For reasons previously discussed we only consider emission metrics here. We use 2002 emission data from AERO2K (see section 2.1.1) and associate each forcing agent with a particular emission (see table 4). Table 4 also provides information on how each forcing agent is evaluated within these example metric frameworks.

As examples of variation between metric choices, three metrics are evaluated in table 5 (pulse GWPs, pulse GTPs, and sustained GTPs), for three time horizons (20 years, 50 years and 100 years).Table 5 presents the "per kg emitted" metrics. To evaluate the actual impact of a fleet, these values must be multiplied by the actual mass emissions. Figures 8 and 9 do this for the AERO2K fleet (table 1).

Table 4. Mechanism characteristics for metrics

Mechanism	Time-scale (alpha)	Associated emission source	Notes
Carbon dioxide	Multiple	CO_2	Metric evaluated with 4 term approximation to Bern carbon cycle model (Shine et al. 2005a)
Short-lived ozone production from NOx	Weeks-month	NOx	100 yr GWPs taken from Stevenson et al. (2004) or corrected Wild et al. (2001). For other time horizons
Methane reduction from NOx	~12 Years	NOx	
Ozone reduction from methane loss	~12 Years	NOx	
Contrails	Hours	Distance travelled by aircraft fleet, assumed to relate to CO_2 emissions	No associated emission, but assumed to be CO_2 for simplicity. Using AERO2K and IPCC (2007) numbers the associated metrics are calculated assuming that 550 Tg CO_2 corresponds to an RF of 10 mW/m^2, with a factor of three uncertainty
Water vapour	Days (troposphere); few years (stratosphere)	Water vapour	Not evaluated here as only thought significant for supersonic fleet
Aerosols	Days	SO2, soot	Not evaluated – believed to be small effect
Aviation induced cirrus	Hours	N/A	Very uncertain for evaluate of metrics. However, as an example, A range of AIC values is used based on an RF between 10 mWm^{-2} and 80 mWm-2 with a best estimate of 30 mWm^{-2}. These rough values are taken from Forster et al. (2007), Table 2.9. These RFs are assumed to correspond to 550 Tg CO_2

Uncertainties also need to be assessed when evaluating metrics. In particular more uncertain effects should not necessarily be given an equal weight to the role of carbon dioxide emissions in which we have a good level of confidence. These uncertainties are indicated by error-bars for NOx and contrails. Efficacy (see Section 2.1) can also influence this judgement. Ponater et al. (2005) suggest that the efficacy for contrails is roughly 0.6, which would mean that the contrail numbers in table 5f could be weighted by this factor, reducing their overall contribution.

Figure 8 shows that at the 20-year time horizon, the short lived emissions are competitive with CO_2 for all three metrics. The net NO_x effect varies between the cases but all three metrics tell a generally similar story. At longer time horizons, CO_2 becomes increasingly dominant, especially using the PGTP. The values using PGWP and SGTP become increasingly similar at long time horizons.

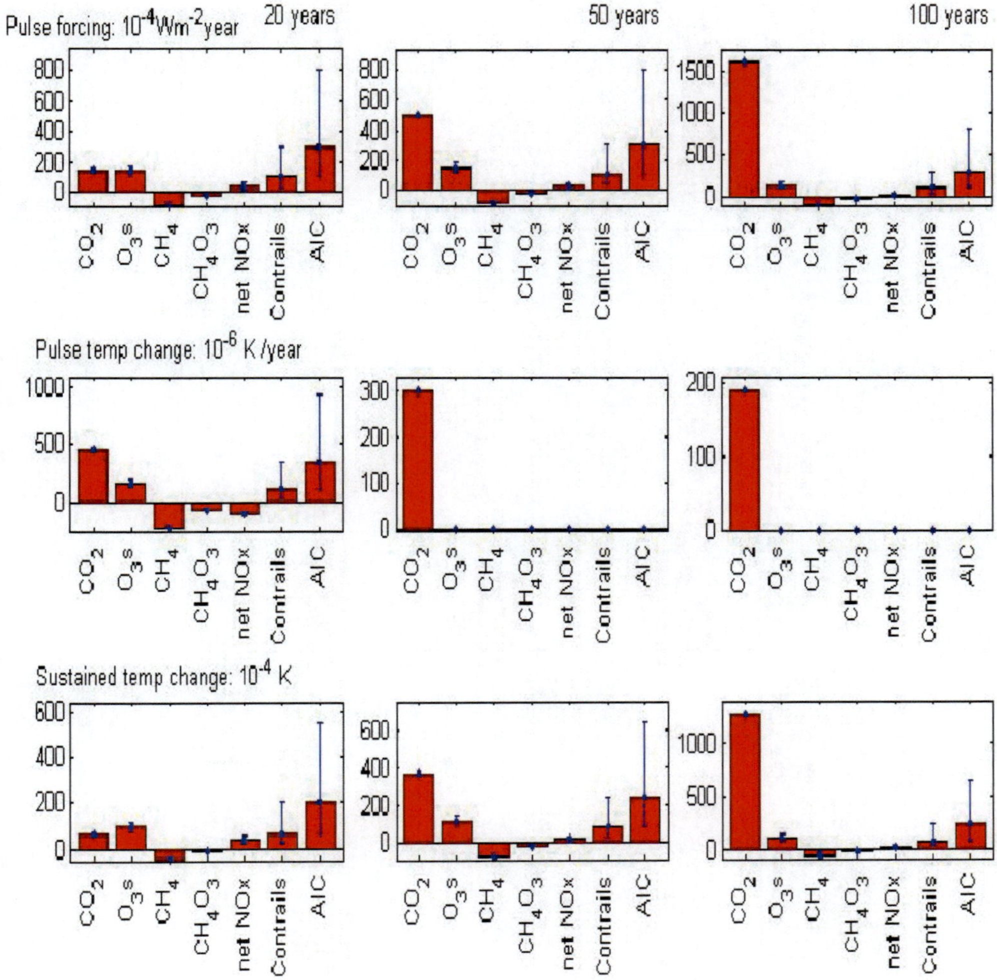

Figure 8. Examples of the use of three metrics using AERO2K emissions (rows: using PGWP, PGTP, SGTP to evaluate climate effect) evaluated at three time horizons (columns: 20, 50, 100 years). Units are 10^{-4} Wm^{-2}year (row 1); 10^{-6} K year^{-1} (row 2); 10^{-4} K (row 3). NO_x evaluations are based on averages of Stevenson et al. 2004 and Wild et al. 2001 numbers. AIC is aviation induced cirrus. Note that the scale on the y-axes varies between frames. Note that no uncertainty is given for CO_2 as there are none which are specific to their evaluation in the context of aviation. Typically quoted uncertainties for CO_2 are ±10%.

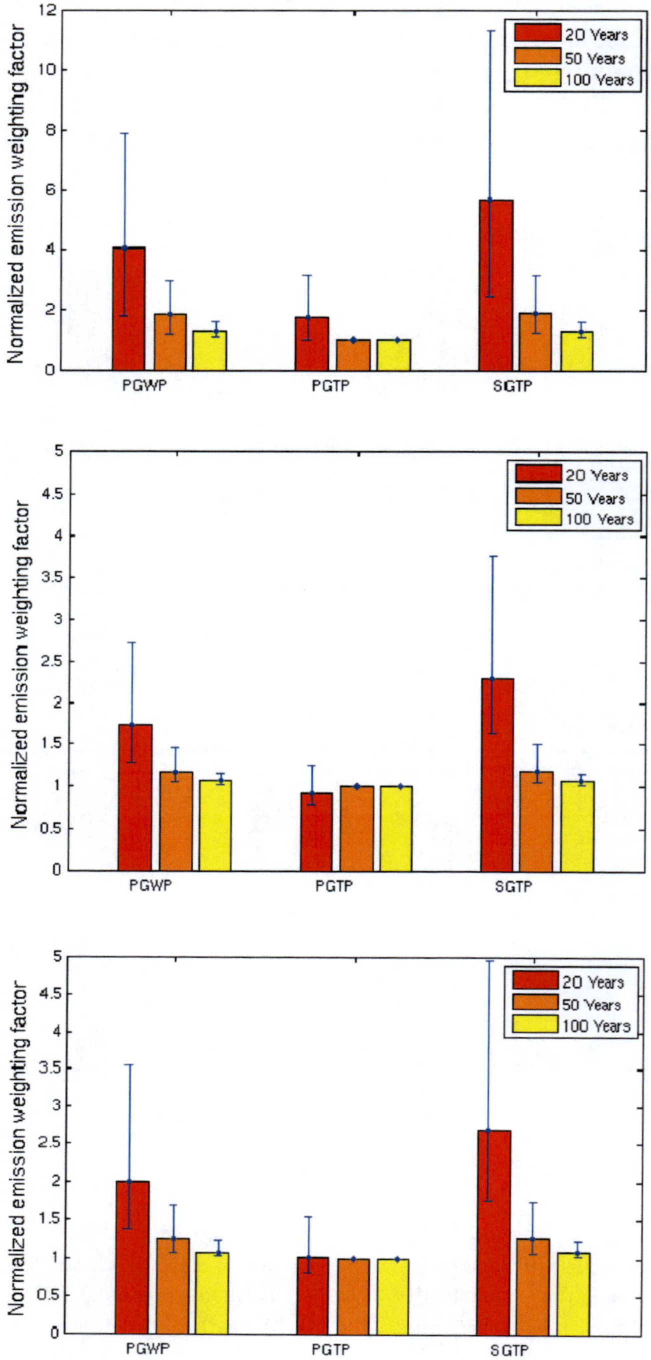

Figure 9. Summary of figure 8, where total aviation impact has been normalized to CO_2 impact creating an emission weighting factor appropriate to the current fleet. Error bars present uncertainties arising from NOx and contrail forcings. Top: excluding the highly uncertain aviation induced cirrus (AIC). The uncertainties are the range of values presented in table 5. Middle: Including AIC. Bottom: Excluding AIC, and assuming an efficacy of 0.6 for contrail forcing. Note that the scale on the y-axes varies between frames.

Figure 9 presents an emissions form of an RFI where the total impact is divided by the CO_2 only effect. Figure 9a neglects the highly uncertain aviation induced cirrus (AIC). It illustrates that the emissions index tends to 1 (i.e. CO_2 dominance) as the time scale increases, especially when using the PGTP. However, for the 20 year time horizon, the non-CO_2 effects are clearly important when using the PGWP and SGTP, a characteristic that could become even more marked if a shorter time horizon was chosen.

Figure 9b shows the impact of including the AIC, which has a particularly marked impact at shorter time horizons. Figure 9c excludes the AIC but, for illustration, assumes that the efficacy for contrails is 0.6, following Ponater et al. (2005), this acts to reduce the effect of the short lived emissions, enhancing the dominance of CO_2.

As emphasized in Section 2.2, the choice of metric and time horizon depends on the application to which the metrics are put, and there appears some merit in presenting multiple indices/horizons, to illustrate these dependencies.

Table 5. Absolute values of the metrics for 3 different time horizons. a) for carbon dioxide emissions. b) Short lived ozone production from NO_x emissions. c) CH_4 reduction from NO_x emissions. d) The longer timescale ozone change associated with the CH_4 reduction. e) the net effect of NO_x emissions (i.e. the sum of (b), (c) and (d)). f) contrails, based on CO_2 emissions. Contrails metrics are given in terms of CO_2 and have an associated uncertainty that is estimated to be a factor of three. g) Aviation-induced cirrus (AIC) based on CO_2 emissions. A range of AIC values is used based on an RF between 10 mWm^{-2} and 80 mWm^{-2}. These ranges are taken from Forster et al. (2007). Metrics in Tables b)-e) are quoted in terms of NO_x emission. Uncertainties are evaluated by quoting numbers from the two available studies (Stevenson et al., 2004 and Wild et al., 2001)

a) Carbon dioxide (using Shine et al., 2005 parameterization)

	Time Horizon (years)		
Metric	20	100	500
APGWP ($x10^{-14}$ $Wm^{-2}kg(CO2)^{-1}year$)	2.7	9.1	29
APGTP ($x1 0^{-16}$ $Kkg(CO2)^{-1}$)	8.3	5.5	3.5
ASGTP ($x10^{-14}$ $K(kg(CO2)\ year^{-1})^{-1}$)	1.2	6.7	23

b) NO_x ozone production on short timescales. Stevenson (Wild)

	Time Horizon (years)		
Metric	20	100	500
APGWP	510	510	510
($x10^{-14}$ $Wm^{-2}kg(NO2)^{-1}year$)	(790)	(790)	(790)
APGTP	590	0.33	0.0
($x10^{-16}$ $Kkg(NO2)^{-1}$)	(920)	(0.52)	(0.0)
ASGTP	340	410	410
($x10^{-14}$ $K(kg\ year(NO2)^{-1})^{-1}$)	(530)	(630)	(630)

c) NO_x induced CH_4 reduction. Stevenson (Wild)

	Time Horizon (years)		
Metric	20	100	500
APGWP	-350	-420	-420
(x10^{-14} $Wm^{-2}kg(NO_2)^{-1}year$)	(-380)	(-460)	(-460)
APGTP	-900	-3.4	0.0
(x10^{-16} $Kkg(NO_2)^{-1}$)	(-990)	(-3.7)	(0.0)
ASGTP	-180	-340	-340
(x10^{-14} $K(kg\ year(NO_2)^{-1})^{-1}$)	(-200)	(-370)	(-120)

d) Long-term ozone loss from CH_4 changes. Stevenson (Wild)

	Time Horizon (years)		
Metric	20	100	500
APGWP	-78	-95	-95
(x1 0^{-14} $Wm^{-2}kg(NO_2)^{-1}year$)	(-130)	(-150)	(-150)
APGTP	-200	-0.77	0.0
(x10^{-16} $Kkg(NO_2)^{-1}$)	(-330)	(-1.2)	(0.0)
ASGTP	-41	-76	-76
(x10^{-14} $K(kg\ year(NO_2)^{-1})^{-1}$)	(-65)	(-120)	(-120)

e) Net NO_x Changes associated with all methane and Nox effects. Stevenson (Wild)

	Time Horizon (years)		
Metric	20	100	500
APGWP	82	-8.8	-8.9
(x10^{-14} $Wm^{-2}kg(NO_2)^{-1}year$)	(286)	(178)	(178)
APGTP	-510	-3.8	0.0
(x10^{-16} $Kkg(NO_2)^{-1}$)	(-390)	(-4.4)	(0.0)
ASGTP	120	-6.7	-7.1
(x10^{-14} $K(kg\ year(NO_2)^{-1})^{-1}$)	(270)	(140)	(140)

f) Contrails, assuming 10 m Wm^{-2} for 550 Tg CO_2, factor of three uncertain

	Time Horizon (years)		
Metric	20	100	500
APGWP (x1 0^{-14} $Wm^{-2}kg(CO2)^{-1}year$)	1.8	1.8	1.8
APGTP (x10^{-16} $Kkg(CO2)^{-1}$)	2.1	0.0	0.0
ASGTP (x10^{-14} $K(kg(CO2)\ year^{-1})^{-1}$)	1.2	1.5	1.5

g) AIC, assuming 30 mWm-2 for 550 Tg CO_2, range based on an RF between 10 mWm-2 and 80 m Wm-2. These ranges are taken from Forster et al. (2007), table 2.9.

	Time Horizon (years)		
Metric	20	100	500
APGWP ($x10^{-14}$ $Wm^{-2}kg(CO_2)^{-1}year$)	5.5	5.5	5.5
APGTP ($x10^{-16}$ $Kkg(CO_2)^{-1}$)	6.3	0.0	0.0
ASGTP ($x10^{-14}$ K(kg(CO_2) year-1)-1)	3.7	4.4	4.4

Interconnectivity with other SSWP theme areas

The magnitude of any climate response due to aviation will rely heavily on our understanding of the background atmosphere (composition and meteorology) as well as our ability to accurately represent any perturbations to the atmosphere due to aviation. This SSWP will inevitably draw upon the conclusions and recommendations found in all other SSWPs. It is important however to note that other SSWPs may not be dependent upon the outcomes of this SSWP which is aimed at providing an overview of the metrics available for comparison of the climate impacts due to aviation.

3. Outstanding Limitations, Gaps and Issues that Need Improvement

3.1. Science

Assessment

It is now over 7 years since the publication of the IPCC Special Report on Aviation and the Environment and during this time substantial advances to our understanding have been made. It is therefore timely to consider whether a new IPCC report, again focusing on aviation and/or the transportation sector as a whole, should be instigated. The specific support action ATTICA started in June 2006 and will provide 3 assessment reports covering the impact of emissions from the individual transport sectors: land traffic; shipping and aviation. A further assessment will consider the metrics that describe, quantify and compare the atmospheric impacts of transport emissions. It is important to note however that focus within ATTICA will be given to European research. Godal (2003) also suggested that an assessment of the literature on alternative approaches to the use of GWPs as a suitable metric of climate change is necessary, and that this would not only represent a major step forward in improving our understanding of these issues, but that it is necessary if a different metric is to be implemented in the future. An assessment of this kind may in turn generate further studies on the political feasibility of various metrics, a critical issue when it comes to their implementation. A further discussion of these policy-related issues is given in Section 3.4.[*Priority Task As1a & b, Section 4*]

Efficacy

Joshi et al. (2003) found that, in a study of three GCMs, the climate sensitivities (λ), defined as the ratio of the globally averaged surface temperature change to radiative forcing, revealed generic deviations from a base case with global CO_2 perturbations. In general, upper tropospheric O_3 increases produced lower values of λ whilst lower stratospheric O_3 perturbations lead to higher values of л. Extratropical forcings also indicated higher л values than found for tropical forcings. Forster et al. (2007) also found that the efficacies were within about 50% of 1.0 for a range of mechanisms and models. The efficacy for contrails was considerably smaller than 1.0 in one model (Ponater et al., 2005). Further examination of the efficacy for contrails and ozone especially are needed in a variety of different models to understand this further. [*Priority Task A1, Section 4*]

Impact of Local Effects on Regional/Global Change – Variations with Metrics

A modelling intercomparison is required to examine the impacts of local radiative effects (e.g. contrails, ozone) on global climate change. Historically the radiative responses due to all effects were added together irrespective of either their sign or geographical extent. It is this addition of the effects that has led to the formulation of a radiative forcing index (RFI) for aviation of 2.7 (IPCC, 1999) in order to account for the non-CO_2 effects of aviation. The true impact of all radiative effects (positive and negative, local and global) on the climate system therefore needs to be addressed in order to confirm whether an additive approach is appropriate. [*Priority Task A2, Section 4*]

Timescales

Probably as a result of convenience and simplicity, the chosen metric to compare the climate impact of these greenhouse gases was the 100-year Global Warming Potential (GWP) as calculated by the Intergovernmental Panel of Climate Change Second Assessment Report (IPCC, 1995). The 100 year timescale may have been chosen arbitrarily as this was the middle value of 20, 100 and 500 year GWPs presented in the report. A full assessment of the range of impacts, using a spectrum of metrics and timescales, should be conducted with a variety of models on a single future climate scenario. Note the decision of timescale has a large socio-political element involved and also impacts discount rates – do we care as much about our grandchildren as our children, and what about our great, great grand children? (see Section 3.4). [*Priority Task A3, Section 4*]

Cancelling Negative and Positive Effects

Metrics could be adopted which consider local inputs (averaged globally) rather than global mean inputs. One difficulty with this approach however is the degree to which the local impact on the climate system remains local and whether the amount of 'spread' varies depending upon the mechanism (species) responsible for the initial climate change. [*Priority Task A4, Section 4*]

Pulse Emissions, Sustained Emissions or Realistic Scenario

Using pulse or sustained emissions can give very different interpretations of climate impact (See Section 2.4). Advantageously, pulse and sustained emissions lead to simple often analytic reproducible metrics that are not prejudicing the future scenario of aviation emissions and would be more or less invariant with time. However, choosing a realistic growth scenario (e.g. Lim et al 2007; Wit et al. 2005) can give a more relevant metric. For example, if aviation continues to grow at an exponential rate, aviation's non-CO_2 effects on climate change would remain proportionally similar to CO_2 as that expected using the current radiative forcing index of around 2, whereas using a GWP metric would underestimate the role of non-CO_2 effects. [*Priority Task A5, Section 4*]

Background Scenario

The background scenario choice affects metric evaluation. Further, as background atmospheric composition and temperature changes into the future metric values will change. The most obvious and predictable change is that as concentrations of CO_2 rise its radiative effect saturates, therefore non-CO_2 effects become more significant. A question leads from this as to whether metrics, when is use, should be revaluated from time to time depending on the current background atmosphere. Also development of knowledge and understanding could lead to future metric re- evaluation. [*Priority Task A6, Section 4*]

3.2. Measurements, Analysis and Modelling Capability

IPCC (2001) highlighted that 'further action is required to address remaining gaps in information and understanding'. Focus should therefore be given to the necessary research needed in order to improve the ability to detect, attribute and understand climate change, with a reduction in the uncertainties, and an aim to forecast future perturbations. Special emphasis should also be given to the need for additional long term observations following the decline in monitoring networks, an effort encouraged by the IPCC report. Together with improved observational capacity however is the need for appropriate modelling and process studies. Of relevance to the aviation industry, the IPCC report notes:

Systematic observations and reconstructions:

Reverse the decline of observational networks in many parts of the world

Sustain and expand the observational foundation for climate studies by providing accurate, long term, consistent data including implementation of a strategy for integrated global observations

Improve the observations of the spatial distribution of greenhouse gases and aerosols

Modelling and process studies:

Improve understanding of the mechanisms and factors leading to changes in radiative forcing

Improve methods to quantify uncertainties of climate projections and scenarios, including long-term ensemble simulations using complex models

Improve the integrated hierarchy of global and regional climate models with a focus on the simulation of climate variability, regional climate changes and extreme events.'

As stated in IPCC (2007) one of the largest uncertainties in predicting future climate change is still related to the potential impact of aerosols and clouds on the global radiation budget. These uncertainties are critical to determining the full contribution of aviation to total anthropogenic climate change. Additional research on contrails and aviation induced cirrus (including their occurrence and radiative properties), together with the provision of data on aerosols, clouds and radiatively active gases and precursors, is paramount to the construction of appropriate mitigation options.

An initial report of findings and recommendations by the PARTNER and the USA Joint Planning and Development Office, based on a workshop on The impacts of aviation on climate change, June 2006, (recently published in summary form by Wuebbles et al. 2007) highlighted the need for focused research efforts to 'address uncertainties and gaps in our understanding of current and projected impacts of aviation on the climate and to develop metrics to characterise these impacts'. They also went further to suggest that this could be achieved through the coordination and/or expansion of existing and planned climate research programmes together with new activities. The short term research needs identified, included:

A model and measurement intercomparison.
In-situ probing and remote sensing (including space-borne sensors) of aging contrail-cirrus and aircraft plumes.
Regional modelling studies of supersaturation and contrail formation, including evaluation of satellite observational capability.
Calculation of radiative forcing from cirrus and contrails including studies of efficacy.
Exploration of alternative metrics including their reliability.

In the long term the following were suggested:

Field campaigns to examine HO_x-NO_x chemistry in the upper troposphere.
Forecasting methods for supersaturation (possibly based on commercial aircraft measurements).
Development of prognostic methods for the calculation of cloud fraction within atmospheric models.

3.3. Interconnectivity with Other SSWP Theme Areas

See Section 2.8

3.4. Interconnectivity with Comprehensive Transport Policy

3.4.1. Policy Interface Issues

Lee & Sausen, 2000 concluded that if aviation participated in an open regime of CO_2 emissions trading (i.e. intersector with capped global CO_2 emissions), where overall aviation was a purchaser of CO_2 permits from other sectors, the result would be a larger radiative forcing from aviation emissions (including NO_x) than if the emissions had originated from sectors operational at the Earth's surface. Alternatively, if aviation participated in a closed regime of CO_2 emissions trading (i.e. intrasector with capped global CO_2 emissions) the total radiative forcing from aviation emissions could be greater or lesser depending on the temporal and geographical location of emissions. It is therefore possible to envisage a scenario where the effects of emissions trading with capped global CO_2 emissions could increase the radiative forcing from aviation.

This section is provided to give a short perspective of the way metric use may depend on the policy question being asked. It is emphasized that the authors of this report are climate scientists, and are not experts in policy issues. It presents one, perhaps rather limited, perspective on this issue.

The overall stated aim of the UN Framework Convention on Climate Change (UNFCCC) (http://unfccc.it) is to stabilise greenhouse gas *concentrations* at a level that will avoid dangerous climate change; the required level has not been defined and is the subject of intense debate. The Kyoto Protocol, which incorporates the UNFCCC set emission targets, relative to 1990 levels, for signatories to the treaty. These emission targets do not appear to have stabilisation, let alone a defined stabilisation target, in mind. The targets are set in terms of CO_2 equivalent emissions for 6 groups of greenhouse gases (CO_2, N_2O, CH_4, the HFCs, the PFCs and SF_6), where CO_2 equivalence is determined using the 100 year (pulse) GWP. The Kyoto Protocol covers the period up until 2012 with the negotiations for the period beyond 2012 currently active. It is not clear whether any new protocol would include emissions beyond the group of six gases mentioned above. It could be argued that for consistency with the operation of the Kyoto Protocol, the 100-year GWPs, despite all the caveats in their derivation, are the most appropriate metric to use in assessing non-CO_2 emissions from aviation.

The 100 year timescale may have been chosen arbitrarily as this was the middle value of 20, 100 and 500 year GWPs presented in the report. It is also interesting to note that since the Second Assessment Report (IPCC, 1995) there has been considerable revision to many of the 100-year GWPs (e.g. the methane GWP has increased by over 25%), yet all accounting under the Kyoto Protocol retains values from the original IPCC (1995) report. Cost effectiveness of mitigation policy would likely improve with more accurate metrics. Yet there is also an argument for a consistent policy landscape, allowing businesses and sectors to make longer-term plans. These issues need to be considered when developing new metrics.

More recently, the European Union has adopted a more specific target stating that the global annual mean surface temperature increase should not exceed 2°C above pre-industrial levels. (www.europa.eu/bulletin/en/200503/i1010.htm). It has been argued (see for example Shine et al. 2007 for discussion and references), that metrics like the GWP are ill-suited to such targets. The argument is that the GWP places equal emphasis on emissions of long and short-lived gases, irrespective of when they are emitted. The argument then follows that at times distant from when the target will be achieved, the emphasis should be on the longer-lived gases; emissions of short-lived gases will have only a small impact on climate change at the target time. However, as the time of the target is approached, increasing emphasis should be placed on the short-lived gases, as their influence on temperatures becomes greater. Hence, in this view, the value of metrics, relative to CO_2 changes as the target is approached. Results indicate that it is only at times less than 20 years before the target is reached that aviation's non-CO_2 emissions become important. Before that time CO_2 emissions are the dominant effect. Such arguments assume that the rate of change of climate is much less important than the total change at some distant point.

Multi-component abatement strategies to limit anthropogenic climate change need a framework and numerical values for the trade-off between emissions of different forcing agents (gases and aerosols). GWPs or other emission metrics provides the necessary tool to operationalize comprehensive and cost-effective policies (Article 3 of the UNFCCC) in a decentralised manner so that multi-gas emitters (nations, industries) can compose cost-effective mitigation measures according to a specified target by allowing for substitution between different climate agents. The metric formulation depends on whether a long-term target to comply with the UNFCCC goal of avoiding dangerous climate change is set (either by a cost-benefit analysis or by a more political judgement), or if we are concerned about reducing the impacts of climate change, but so far have not agreed on any specific long-term target (as in the Kyoto Protocol). In both cases the metric formulation requires knowledge of the contribution to climate change from emissions of various components over time, i.e. their radiative efficiency and atmospheric residence time. In addition, both formulations also involve input from economics. Economists have argued that, ideally, the metric should be the outcome of an analysis that minimizes the discounted present value of damages and mitigation costs (e.g. Manne and Richels, 2001). If a climate forcing reduction trajectory is formulated to achieve a long-term target the proper trade-off between gases is then their relative contribution to that trajectory, that is, the ratio of the shadow prices[1]. Otherwise, if a long-term target is not set, the proper trade-off is the relative contribution of various gases to the impacts, that is, the ratio of the marginal damage costs[2]. Substitution of gases within an international climate policy with a long-term target and including economic factors is discussed in Sections 3.3.2 and 3.6 of IPCC WG III AR4.

The UNFCCC has requested that the International Civil Aviation Organisations (ICAO) takes action on aviation emissions in recognition that a global approach is crucial to the success of any action. In response ICAO has formed a Committee on Aviation Environmental Protection (CAEP) with current tasks including the development of guidance for states wishing to participate in emissions trading schemes and an improved understanding of the potential tradeoffs between improvements in emissions of CO_2 and the effect on other environmental effects. It is important however to note that the current tasks within ICAO-CAEP do not themselves constitute the regulation of emissions. The international co-ordination of taxes is difficult to implement since it is contrary to the ICAO rules to levy the tax on fuel carried on international flights. The majority of bilateral air service agreements responsible for regulating international air travel also forbid air fuel taxation. It is manly for this reason that the level of taxation experienced by the aviation industry is currently low relative to road fuel taxes.

ICAO has recently endorsed the concept of emissions trading schemes for the aviation industry and the European Union (EU) has now released a Directive to include aviation within the EU's emission trading scheme with a view that the guiding principles can be replicated in a workable worldwide model. For example, the EU suggest that the coverage must be clear (e.g. including domestic, intra-European Union and all flights landing or leaving the EU), trading entities should be all aircraft operators and carriers, and the allocation of permits should occur at the EU level. Importantly they have voted for a multiplier, of at least two, to be used to compensate for the additional impacts of emissions from aircraft at altitude. The Stern Review (2007), chapter 15, suggested that the auctioning

[1] The shadow price of gas g is the reduced cost of meeting the desired policy if we were allowed to emit one extra unit of gas *i* at time *t*. This shadow price therefore tells you the cost benefit of slightly relaxing the emission constraint.

[2] The marginal damage cost is the economic cost of climate impact per unit increase in an emission (e.g. impact measured in dollars per tonne of CO_2 emitted or dollars per tonne of NO_x emitted)

of permits would raise valuable revenue and increase the speed of adjustment to a carbon market. Not only this, but combining emissions trading with taxation could provide additional revenue with strong incentives towards innovative approaches to reduce aviation emissions. The EU emissions trading scheme states that for aviation only 25 percent of emissions permits are to be auctioned (with an option to increase this at a later date).

The Stern Review (2007) stated that the ultimate choice in taxation, trading or alternative economic instruments is likely to be driven as much by political viability as by economics. It was also suggested that a lack of international co-ordination could lead to serious carbon leakage as the aviation sector would be incentivized to fuel-up in countries where carbon pricing was not included. The Stern Review (2007) went further however to recommend that any carbon price faced by aviation should reflect the full climate change contribution due to emissions from aviation and noted that non-CO_2 effects should be included, through the design of an appropriate tax or trading scheme, and that a form of discounting could be used analogous to GWPs. Uncertainties in the conversion of CO_2 emissions into the full CO_2 equivalent quantity were however highlighted.

Voluntary approaches to a reduction in the climate impact of aviation are also important. Existing international co-operation through, for example, the Advisory Council for Aeronautics Research in Europe (ACARE) requires that all new aircraft produced after 2020 be 50% more fuel efficient per passenger seat kilometre relative to an equivalent aircraft in 2000. Currently these targets, though technically challenging, are broadly on track. Similar goals have also been set in the USA through the National Aeronautics and Space Administration (NASA).

3.4.2. Interface with Air-Quality

Global averaged GWPs can be calculated for short-lived species (e.g. ozone precursors and aerosols). On a global level the mean metric values can be used to give an indication of the total potential of mitigating climate change by including a certain forcing agent in climate policy. As discussed by Hansen and Sato (2004) and Rypdal et al. (2005) there might be a potential for more effective climate mitigation strategies if climate mitigation and air quality issues are viewed together. Assessing the climate impact of key species affecting air quality is therefore needed. However, the metric values for short-lived compounds vary significantly by region and time so that for operationalization on a decentralized level, robust regionally varying GWPs must be established and agreed upon. Improved scientific understanding of O_3 chemistry and the climate effects of aerosols are needed before this can be established, with the possible exception of carbon monoxide (Berntsen et al., 2005). A more fundamental question related to the application of GWPs for short lived species is whether the more short-term climate fluctuations caused by pulse emissions of these components should be weighted equally to long-term climate warming by long lived gases, as is implicitly assumed through application of the GWP concept. However, as long as there is no consensus on what constitutes 'dangerous anthropogenic interference with the climate system' there is no clear conclusion to this question. A more long term perspective, e.g. by calculating the contribution from current emissions to climate change at a time (or time interval) when global warming is predicted to reach a given threshold value would lead to reduced emphasis on the short lived species.

3.4.3. Comparison to Other Sectors

During the 1990s global CO_2 emissions increased by 13%. Of these emissions road and aviation each experienced a growth in CO_2 emissions of 25%. In Eastern Asia road transport emissions of NO_x and CO_2 doubled during this period (Olivier & Berdowski, 2001). In the European Union, whilst the majority of sectors reduced their greenhouse gas emissions during this period, emissions from the transportation sector increased by ~21% (EEA, 2003). Nakicenovic et al., (2000) has predicted that the growth in greenhouse gas emission from the global transportation sector will continue and that by 2050 between 30 and 50% of total CO_2 emissions will originate from the transportation sector compared to 2000 levels of 20-25%.

The first comprehensive analysis of the radiative forcing impact due to road, rail, shipping and aviation, using both a historical and futuristic perspective, has been performed by Fuglestvedt et al. (2008). They have found that since pre-industrial times the transportation sector has contributed to more than 20% of the total man-made CO_2 emissions (figure 10) equating to 15% of the total man-made CO_2 forcing and 30% of the total man-made O_3 forcing. Furthermore their research indicates that the current emissions from the transportation sector are responsible for 17% of the net integrated forcing (100 years) of all current man-made emissions. The dominating effects are from CO_2 and tropospheric O_3 and it is important to note therefore that much of the forcing from the transport sector originates from emissions not included within the Kyoto Protocol (e.g. SO_2, organic carbon and O_3 changes due to precursors such as NO_x, CO and VOCs). As shown in figure 11 the dominant subsector is road, followed by aviation. In contrast to the other subsectors, shipping emissions result in a negative radiative forcing primarily due to sulphate emissions.

Fuglestvedt et al. (2008) argues that the adoption of 100 years as a time horizon for examining the climate forcing from the transportation sector has implications involving value judgements and that other time horizons should also be considered. For example, figure 12, from Fuglestvedt et al. (2008), shows the global mean net radiative forcing per sector due to 2000 transport emissions. The results are normalised to the values for road transport for time horizons of 20, 100 and 500 years. The importance of the time horizon is shown in the critical role that short-lived sulphate has on the impact of shipping. In the short to medium timescales the impact of shipping is negative (due to the negative impact of sulphate emissions) whilst over longer timescales the impact becomes positive. A similar argument is applicable to rail. In general the largest scientific uncertainties in calculating the climate impact due to the transportation sector results from the quantification of the indirect effects of aerosols, together with contrails and aviation- induced cirrus. Uncertainties are however also apparent in the estimates of the emissions themselves.

As shown by Fuglestvedt et al. (2008) by only including well mixed mixed greenhouse gases in the Kyoto Protocol the full climate impacts of the transportation sector will not be captured. This is particularly apparent when determining the climate response due to emissions from shipping.

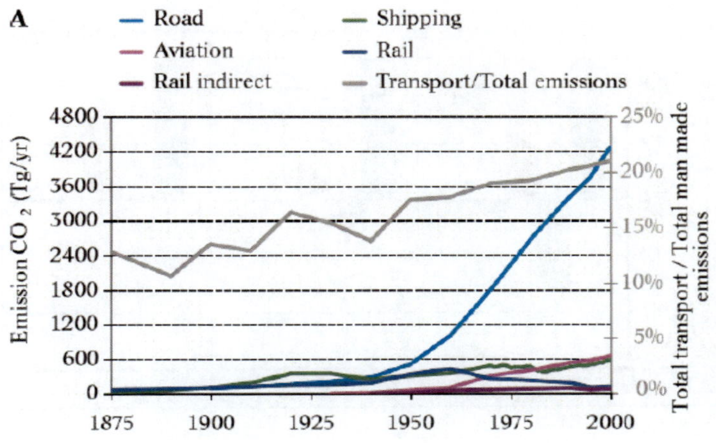

Figure 10. Development of CO_2 emissions from various transport subsectors and the fraction of the total man-made fossil fuel CO_2 emissions – Fuglestvedt et al. (2008).

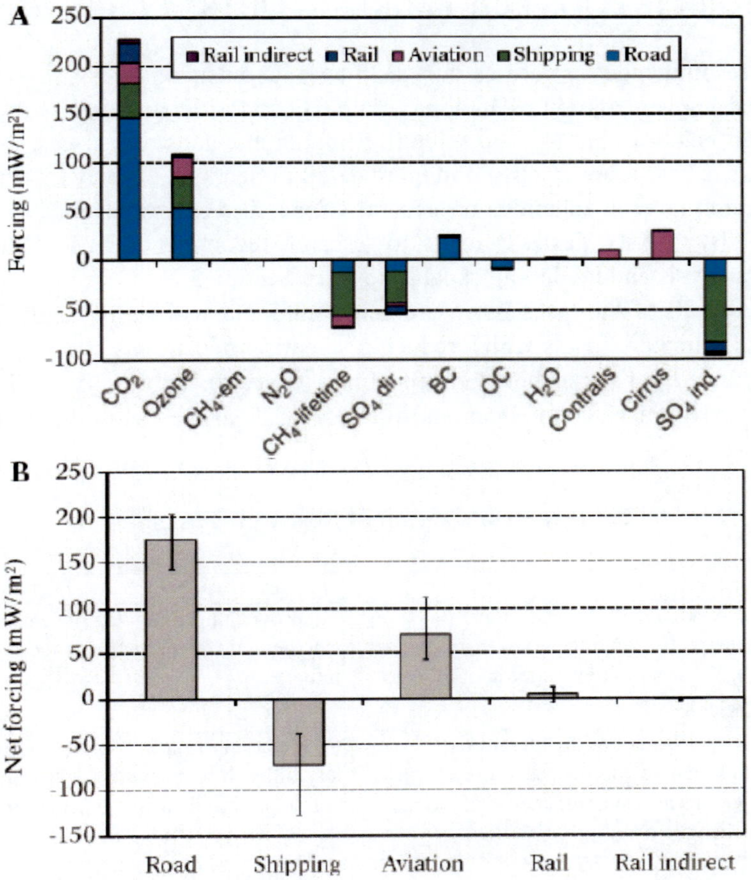

Figure 11. A: Global mean radiative forcing for 2000 due to transport relative to preindustrial times; B: Global mean net radiative forcing – Fuglestvedt et al. (2008).

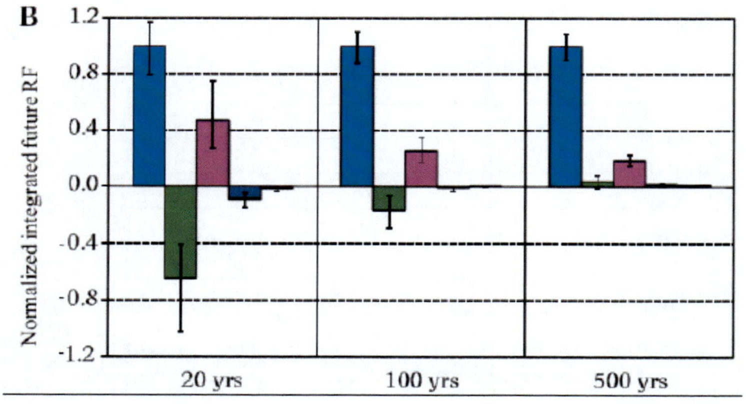

Figure 12. Integrated global mean net radiative forcing per sector due to 2000 transport emissions, normalised to the values for road transport for various time horizons (20, 100 and 500 years) – Fuglestvedt et al. (2008).

4. PRIORITIZATION FOR TACKLING OUTSTANDING ISSUES

A list of recommended priorities for tackling the outstanding issues related to the development and implementation of an appropriate metric for determining aviation's climate impact are given below (table 6). The scientific limitations, gaps and issues, on which this selection of tasks is based, are discussed in more detail in Sections 3.1 and 3.2. Priority Tasks A relate to research recommendations on general science issues of relevance to metrics (see Section 3.1) whilst Priority Tasks B relate to research recommendations of importance to measurements, analysis and modelling capabilities (see Section 3.2).

In our opinion all of the tasks listed are achievable and will significantly improve our understanding of climate impacts whilst reducing scientific uncertainty. Priority Tasks listed under A are predicted to have a short-term timeline (<5 years). Priority Tasks listed under B are predicted to have varying timelines and practical uses and as such these are explicitly given.

Table 6. Prioirization of Research Tasks

Priority Task	Task	Impact
As1a	Assessment of the literature on alternative approaches to the use of GWPs as a suitable metric of climate change	Improved understanding of issues and whether a different metric is necessary in the future.
As1b	Assessment of the literature on alternative approaches to the use of GWPs as a suitable metric of climate change	Generation of further studies on the political feasibility of various metrics, a critical issue with regard to their implementation.
A1	Efficacy	Diagnosis of the variability in the climate sensitivity parameter.
A2	Confirmation as to the importance of local impacts on global climate	Impact of local effects on regional/global change –variations with metrics

	change	
A3	Assessment of the potential range of impacts diagnosed using a spectrum of metrics and timescales	Improved understanding of the potential impact of aviation under various metrics and timescales
A4	Appropriateness of cancelling negative and positive climate effects	Improved understanding as to whether multiple climate effects can be combined
A5	Appropriateness of pulsed or sustained emissions of realistic scenarios	Improved understanding of how scenario choice leads to different implications of aviation impact
A6	Choice of background scenario	Improved understanding of how background climate change and atmospheric conditions affect metric choice

Priority Task	Task		Impact	
B1	Improved description of Nox and Noy chemistry, sources and sinks	Accurately represent the Hox – Nox chemistry of the UTLS region and potential anthropogenic impacts	Model improvement requiring additional observations, laboratory measurements and observations	Long-term (>10 years)
B2	Improved prediction of transport processes throughout the lower atmosphere	Correct determination of the impact of both sub- and supersonic aircraft	Model improvement requiring additional computational resources and longterm observations	Long-term (>10 years)
B3	Model-model intercomparison and model-measurement intercomparison	Improved understanding of the interaction between ozone and methane	Model improvement through comparison and validation	Mediumterm (5-10 years)
B4	Impact of a pulse emission of Nox emitted under different atmospheric conditions and seasons	Improved understanding of climate impact of Nox emissions under different atmospheric conditions and seasons	Sensitivity analysis	Short-term (<5 years)
B5	Impact of a range of Nox pulse sizes	Confirmation as to whether the size of the initial positive ozone anomaly, resulting from a pulse emission of Nox, determines the sign and magnitude of the overall net forcing	Sensitivity analysis	Short-term (<5 years)
B6	Study of impact of cirrus particles on atmospheric composition	The potential role for cirrus particles in the heterogeneous chemistry of the atmosphere	Model investigation requiring additional laboratory studies and in situ observations	Long-term (>10 years)

Table 6. (Continued).

B7	Study of the processes and radiative effects of contrails and aircraft induced cirrus	Quantification of contrail/cirrus effects	Model investigations with laboratory studies and observations (including in situ and satellite)	Long-term (>10 years)
B8	Forecasting of regions of supersaturation	Development of methods for forecasting supersaturation for use in cloud and contrail prediction	Model investigations with observations	Long-term (>10 years)
B9	Quantification of the full effect of aviation under potential operational and technical procedures	Alternative operational and technical procedures could be adopted by the aviation community if the impact were to be considered as significant	Sensitivity Analysis	Short-term (<5 years)
B10	Long-term observational networks	Long-term observational capability for integrated monitoring of climate gases	Observations	Long-term (>10 years)

5. Recommendations for Best Use of Current Tools for Modeling and Data Analysis

5.1. Options

Currently, when determining any climate impact, a choice exists between:

simple analytical models such as GWPs and GTPs;
models of intermediate complexity that calculate induced temperature change for various scenarios (in the case of aviation those given by Lim et al., 2007; Sausen and Schumann, 2000; Wit et al., 2005); and
the option of running integrations in complex coupled climate models.

The range of possible metric options are shown in table 7, and provide a basis for the best available options and approaches with which to quantify the climate impact under varying scenarios.

It should be noted that it is important to consider aviation climate issues within the wider context of the political landscape, air quality concerns and other transport sectors (Section 3.4). There remains however issues about which emissions and factors should be included in policy decisions and whether to have separate policies for different emissions (CO_2 and NO_x) or one unified metric, such as the GWP. A multiple-agent metric will likely have more cost-effective benefit when applied, provided it is scientifically robust (see Section 3.4). These aspects we feel are still very much an open question. The inclusion of short- lived climate gases in any climate policy will require scientific robustness and therefore a substantial degree of model independence. The results of Berntsen et al. (2005) indicate that short-lived species could be included in future climate policies however their level of credibility will remain less than that of the long-lived species.

Our recommended approach for the best use of current tools involves simple metrics only (GWP and GTP) and including in these all forcing factors that are relatively well quantified (currently excluding the role of aviation induced cirrus). Since likely future policy will be directed towards reductions by a particular target date, we recommend the adoption of ASGTP(H), limited probably to a target date around 2060, as this time horizon features in draft European union policy and UNFCC- Bali discussions. The reasons for this selection are given in the following subsection (5.2).

Table 7. Metric options

Metric	Usage and advantages	Disadvantages
All	Combining climate impact of more than one emission source in a quantifiable way	Difficulty in quantifying many effects, given current scientific understanding Conceptual difficulty in handling the compensation between opposing forcings on a global level when they do not compensate locally
RF(present), ΔT(present)	Gives impact of all current and past emissions on RF and AT at the present. Includes "responsibility" for past emissions	Temperature metrics add complexity and uncertainty to calculations, as the climate sensitivity parameter is poorly quantified. Nothing can be done now about past emissions
RF(future), ΔT(future)	Gives impact of all current and past emissions on RF and AT at some future date. Could also include scenarios of emissions between present day and future date	As above, but with additional uncertainty due to scenario
RF or ΔT due to emissions in one year	Use of PGTP(H) (or similar metric for forcing) to give impact of current emissions on temperature at some time in the future	Choice of time horizon has much stronger effect on results than is the case for GWPs, and could be manipulated to suit "world view"
RF(target), ΔT(target)	Similar to above, but could be used if the policy were to aim to restrict the contribution to RF or AT at some future target date, it would say how much current emissions are contributing to that target. Impact of short-lived emissions would grow as target time is approached	As above. Additional difficulties in choosing the target date. Some argue that the rate of change of temperature is as important as the actual change in temperature

Table 7. (Continued)

Metric	Usage and advantages	Disadvantages
Time integrated RF due to emissions in one year	Use of Standard GWP(H) would characterise the impact of current emissions in a manner that is consistent with the Kyoto Protocol and the accepted method of achieving carbon equivalence by most other sectors. Choice of H=100 years would be fully consistent with Kyoto, but could be presented for range of H (e.g. 20, 100, 500 years)	Strong negative comments made about use of GWP for aviation, in high profile places, notably IPCC (1999). These would need countering when presented
Sustained GWP(H) and GTP(H)	Sustained versions of the pulse GWP and GTP, in which the effect at time H is considered if the current emissions are sustained between now and H	Difficulty in explaining usage and the assumption of constant future emissions
Economic impact metrics	Monetary unit based on global temperature or local climate effects, (precip, storms etc.); also could be based on impacts (flooding, drought etc) or livelihood change indices. Has advantage of being closer to real world effects.	Hugely uncertain. Combines uncertainties. In regional climate modelling and socioeconomic modelling.

Specific modeling integrations should be performed on an individual basis dependent upon the scientific and/or political question that is to be addressed. If, for example, we are interested in the global impact of a tripling in the aviation system capacity (and as such a related doubling in aviation emissions) then we recommend that, with input from a range of global atmospheric models, the metric ASGTP(2060) be applied for comparison with other scenarios (including alternative transportation options and future climates). We refer to other SSWPs theme areas for recommendations on the choice of atmospheric models, emissions and background conditions.

5.2. Supporting Rationale

Considering aviation's effects within complex climate models is firstly problematic because aviation is only a minor perturbation within the context of natural variability. The advantage of using these models is that they are able to capture physical interactions. However, physical processes such as aviation induced cirrus are not understood and to include simple empirical parameterizations within climate models would be unnecessarily complicated (we would be building in interactions we didn't understand). Therefore we conclude that their use in a metric context brings no clear benefit.

Intermediate models give global temperature evolution and allow the user to explore mitigation options and give a suggestion of climate impact. However, we argue against them for giving a misleading confidence to the user. Because they show temperature evolution over the next 100 years, people may interpret these as reality when in fact they are have many uncertainties: quantification of forcing and efficacy, uncertainty in background scenario and uncertainty in climate response, such as ocean heat up take. We therefore do not endorse them. This is especially true when making such models publicly available for end users to experiment with, as end users may not understand their limitations or valid ranges of applicability.

The choice of a simple analytical model to determine the sustained emission GTP is based on its transparency and ease of use (only a small number of input parameters are required in the calculation). The derivation of GTP is robust to simplifications and key uncertainties, and the unambiguous interpretation and increased relevance, due to its progression down the cause-effect chain of climate impacts, makes it a valuable metric for policy makers.

We recommend that all metrics be applied at a globally integrated level as there is too much uncertainty to distinguish either global differences in response from similar emissions in different regions or to determine the local response to global emissions. Therefore even if Asian NOx emissions are worse than European NOx emissions in terms of their climate impact, we believe that uncertainties are too large to be able to quantify these differences adequately within a policy framework

Our recommendation that aviation induced cirrus should be excluded from both GWP and GTP metrics is due to the current lack of knowledge regarding the quantification of the full (both direct and indirect) impact due to this effect. Line shaped contrails, although not related to a particular emission can be easily associated with distance flown or emissions for CO_2. As in this report, associating their emissions with that of CO_2 enables simple comparison with the effects of other factors. Note that such an association is only valid on a globally-integrated sense due to the dependence of contrail formation on background conditions – this again reinforces the use of global metrics, compared to local ones. We particularly emphasize, both for contrail and for other factors, that uncertainties should be quoted whenever a metric is deployed.

The choice of time horizon is not just a science issue. Although the Kyoto protocol adopts 100 a 100 year time horizon, current policy discussion centres on shorter time scales. A 50 year timescale seems appropriate as it is still primarily concerned with addressing long-term climate change, but within a typical human lifetime. At this timescale shorter lived emissions still play a significant role

5.3. How to Best Integrate Best Available Options?

We recommend continued science studies to reduce uncertainties where achievable, and the use of simple metrics. We recommend quoting ranges for a number of metrics, as different metrics give different indications of importance. This also prevents metrics being deliberately chosen to advocate particular policy choices. Development of our understanding of the atmosphere and computational power should eventually enable sophisticated coupled climate models to be used to explore metrics of aviations impact. Approaches of integration of air quality and climate change requires incorporation into economic models of climate impact (as in the Stern, 2007 review). Assessing the available options here is beyond the scope of our expertise and would require input from economists with knowledge of costing climate mitigation options who would also ideally have a knowledge of the aviation industry.

We finally note that metric choice is very much a policy issue and people from a range of disciplines including policy makers should ultimately decide on the most appropriate metric choice. Time horizon etc. cannot be chosen on purely physical science grounds.

ACKNOWLEDGMENT

The numbers in table 5 and a many of the ideas are derived from a unpublished manuscript led by Keith Shine that Piers Forster and Helen Rogers were co-author of.

REFERENCES

Appleman, H. (1953) The formation of exhaust contrails by jet aircraft, *Bull. Am. Meteorol. Soc.,* 34, 14-20.

Berntsen, T.K. and Isaksen, I.S.A. (1999) Effects of lightning and convection on changes in tropospheric ozone due to NOx emissions from aircraft. *Tellus Series B-Chemical and Physical Meteorology,* 51 (4), 766-788.

Berntsen, T., Fulestvedt, J., Joshi, M., Shine, K., Stuber, N., Ponater, M., Sausen, R., Hauglustaine, D and Li, L. (2005) Response of climate to regional emissions of ozone precursors: sensitivities and warming potentials. *Tellus B Chem. Phys. Meteorology,* 57, 283-304.

Borrmann, S., Solomon, S., Dye, J. and Luo, B. (1996) The potential of cirrus clouds for heterogeneous chlorine activation. *Geophys. Res. Lett.,* 23 , 2133-2136.

Borrmann, S., Solomon, S., Avallone, L., Toohey, D. and Baumgardner, D. (1997) On the occurrence of ClO in cirrus clouds and volcanic aerosol in the tropopause region. *Geophys. Res. Lett.,* 24 , 2011-2014.

Bradford, D. (2001) Time, money and tradeoffs, *Nature,* 410, 649-650

Brasseur, G.P., Cox, R.A., Hauglustaine, D., Isaksen, I., Lelieveld, J., Lister, D.H., Sausen, R., Schumann, U., Wahner, A. and Wiesen, P. (1998) European scientific assessment of the atmospheric effects of aircraft emissions. *Atmospheric Environment.* 32, 2329–2418.

Brunner, D., Staehelin, J., Rogers, H., Kohler, M., Pyle, J., Hauglustaine, D., Jourdain, L., Bernsten, T., Gauss, M., Isaksen, I., Meijer, E., van Velthoven, P., Pitari, G., Mancini, E., Grewe, V. and Sausen, R. (2003) An evaluation of the performance of chemistry transport models by comparison with research aircraft observations. Part 1: Concepts and overall model performance. *Atmospheric Chemistry and Physics,* Vol. 3, pgs. 1609-1631.

Brunner, D., Staehelin, J., Rogers, H., Kohler, M., Pyle, J., Hauglustaine, D., Jourdain, L., Bernsten, T., Gauss, M., Isaksen, I., Meijer, E., van Velthoven, P., Pitari, G., Mancini, E., Grewe, V. and Sausen, R. (2005) An evaluation of the performance of chemistry transport models – Part 2: Detailed comparison with two selected campaigns. *Atmospheric Chemistry and Physics*, Vol. 5, pgs. 107-129.

EEA (2003) Greenhouse gas emission trends and projections in Europe 2003 (EEA, Copenhagen).

Eyers C. J., Addleton, D., Atkinson K., Broomhead M. J., Christou R., Elliff T., Falk R., Gee I., Lee D. S., Marizy C., Michot S., Middel J., Newton P., Norman P., Plohr M., Raper D. and Stanciou N. 2004: AERO2K global aviation emissions inventories for 2002 and 2025. QinetiQ/04/01113, Farnborough. Available from http://www.cate.mmu.ac.uk/reports_aero2k.asp

Faber, J., Boon, B., Berk, M., den Elzen, M., Lee, D.S. (2006) Aviation and maritime transport in a post 2012 climate policy regime. *CE-Delft,* The Netherlands.

Fluglestvedt , J., Bernsten, T., Godal, O. and Skodvin, T. (2000) Climate implications of GWP based reductions in greenhouse gas emissions. *Geophysical Research Letters,* 27, 409-412. doi:10.1029/1999GL010939.

Fuglestvedt, J.S., Berntsen, T.K., Godal, O., Sausen, R., Shine, K.P. and Skodvin, T. (2003) Metrics of climate change: Assessing radiative forcing and emission indices. *Climatic Change.* 58:267–331

Fuglestvedt, J.S., Berntsen, T.K., Myhre, G., Rypdal, K. and Skeie, R.. (2008) Climate forcing from the transport sectors. *Proc. Nat. Acad. Sci. Vol.* 105. doi:10.1073/pnas.0702958104.

Forster, P.M.F., Blackburn, M., Glover, R. and Shine, K.P. (2000) An examination of climate sensitivity for idealised climate change experiments in an intermediate general circulation model. *Clim. Dyn.,* 16, 833-849.

Forster, P. and Solomon, S. (2003) Observations of a 'weekend effect' in diurnal temperature range. *Proc. Nat. Acad. Sci. USA,* 100, 20, 11225-11230.

Forster, P.M., K. Shine and N. Stuber, N. (2006) It is premature to include non- CO_2 effects of aviation in emission trading schemes, *Atmospheric Environment,* 40, pp.1117-1121. doi:10.1016/j.atmosenv.2005.11.005

Forster, P., Ramaswamy, V., Artaxo, P., Berntsen, T., Betts, R., Fahey, D.W., Haywood, J., Lean, J., Lowe, D.C., Myhre, G., Nganga, J., Prinn, R., Raga, G., Schulz, M. and Van Dorland, R. (2007) Changes in Atmospheric Constituents and in Radiative Forcing. In: Climate Change 2007: The Physical Science Basis. Contribution of Working Group I to the Fourth Assessment Report of the Intergovernmental Panel on Climate Change. Cambridge University Press, Cambridge, United Kingdom and New York, NY, USA.

Friedl, R. et al. (1997) Atmospheric Effects of Subsonic Aircraft: Interim Assessment Report of the Advanced Subsonic Technology Program, Reference Publication 1400, National Aeronautics and Space Administration, Washington, D.C.

Godal, O. (2003) The IPCC's assessment of multidisciplinary issues: The case of greenhouse gas indices. *Climate Change,* 58, 243-249.

Grewe, V., Stenke, A., Ponater, M., Sausen, R., Pitari, G., Iachetti, D., Rogers, H., Dessens, O., Pyle, J., Isaksen, I.S.A., Gulstad, L., Søvde, O.A., Marizy, C., and Pascuillo, E. (2007): Climate impact of supersonic air traffic: an approach to optimize a potential future supersonic fleet – results from the EU-project SCENIC, *Atmos. Chem. Phys.,* 7, 5129-5145.

Hammitt, J.K., A.K. Jain, J.L. Adams, and D.J. Wuebbles, 1996: A welfare- based index for assessing environmental effects of greenhouse-gas emissions. *Nature,* 381, 301-303.

Hansen, J., et al., (2005) Efficacy of climate forcings. *Journal of Geophysical Research*, 110, D18104, doi:10.1029/2005JD005776

Hansen, J. and Sato, M. (2004) Greenhouse gas growth rates. Proceedings of the National Academy of Sciences of the United States of America. 101 (46): 16109-16114.

Hartmann, D., Ockertbell, M. and Michelsen, M. (1992) The effect of cloud type on Earth's energy-balance – Global Analysis. *Journal of Climate,* 5 (11), 1281-1304.

Hauglustaine, D,. Emmons, L., Newchurch, M., Brasseur, G., Takao, T., Matsubara, K., Johnson, J., Ridley, B., Stith, J., Dye, J. (2001) On the role of lightning NOx in the formation of tropospheric ozone plumes: A global model perspective. *Journal of Atmospheric Chemistry,* 38 (3): 277-294.

IPCC (1995) Climate Change 1995. Second Assessment Report of the Intergovernmental Panel on Climate Change, Cambridge University Press, UK.

IPCC (1999) Aviation and the Global Atmosphere. A Special Report of IPCC Working Groups I and III in collaboration with the Scientific Assessment panel to the Montreal Protocol on Substances that Deplete the Ozone Layer, Cambridge University Press, UK.

IPCC (2001) Climate Change 2001. Third Assessment Report of the Intergovernmental Panel on Climate Change, Cambridge University Press, UK.

IPCC (2007) Climate Change 2007. Fourth Assessment Report of the Intergovernmental Panel on Climate Change, Cambridge University Press, UK.

Isaksen, I.S.A., Berntsen, T.K., Wang, W.C. (2001) NOx emissions from aircraft: Its impact on the global distribution of CH4 and O-3 and on radiative forcing. Terrestrial Atmospheric and Oceanic *Sciences,* 12 (1): 63-78.

Jaegle, L., Jacob, D.J., Brune, W.H. and Wennberg, P. (2001) Chemistry of Hox radicals in the upper troposphere. *Atmospheric Environment,* 35 (3), 469-489.

Joshi, M., Shine, K., Ponater, M., Stuber, N., Sausen, R. and Li, L. (2003) A comparison of climate response to different radiative forcings in three general circulation models: towards an improved metric of climate change. *Climate Dynamics,* 20 (7-8): 843-854.

Lacis, A., Wuebbles, D. and Logan, J. (1990) Radiative forcing of climate by changes in the vertical distribution of ozone. *J. Geophys. Res.,* 95, 9971-9981.

Lee, D. and Sausen, R. (2000) New Directions: Assessing the real impact of CO_2 emissions trading by the aviation industry. *Atmospheric Environment,* 34, 5337-5338.

Lelieveld, J., Bregman, A., Scheeren, H., Ström, J., Carslaw, K., Fischer, H., Siegmund, P. and Arnold, F. (1999) Chlorine activation and ozone destruction in the northern lowermost stratosphere. *J. Geophys. Res.,* 104, 8201-8213.

Lim, L.L., Lee, D.S., Sausen, R., Ponater, M. (2007) Modelling the climate impacts of aviation with a climate response model. Manuscript in draft.

Meerkotter, R., Schumann, U., Minnis, P., Doelling, D., Nakajima, T. and Tsushima, Y. (1999) Radiative forcing by contrails. *Ann. Geophysics,* 17, 1080-1094.

Minnis, P., Ayers, J., Palikonda, R. and Phan, D. (2004) Contrails, cirrus trends, and climate *J. Climate,* 17,1671-1685.

Myhre, G. and Stordal, F. (2001) On the tradeoff of the solar and thermal infrared radiative impact of contrails. *Geophysical Research Letters,* 28, 3119-3122.

Nakicenovic, N., Alcamo, J., Davis, G., Vries, B., Fenhann, J., Gaffin, S., Gregory, K., Grübler, A., Jung, T. Kram, T. et al. (2000) Special Report on Emissions Scenarios. Cambridge University Press, UK.

Olivier, J. and Berdowski, J (2001) The Climate System, eds. Berdowski, Guicherit, Heij, 33-78.

O'Neill, B.C. (2000) The jury is still out on global warming potentials. *Clim. Change*, 44, 427–443.

O'Neill, B. (2003) Economics, natural science, and the costs of global warming potentials. *Clim. Change,* 58, 251–260.

Ponater, M., Marquart, S., Sausen, R., Schumann, U. (2005) On contrail climate sensitivity. *Geophysical Research Letters,* 32.

Prather, M. (1994) Lifetimes and eigenstates in atmospheric chemistry. *Geophysical Research Letters,* 21, 801-804.

Prather, M. (2002) Lifetimes of atmospheric species Integrating environmental impacts. *Geophysical Research Letters,* 29, 2063, doi:10.1029/2002GL016299.

Ramaswamy, V., et al., 2001: Radiative forcing of climate change. In: Climate Change 2001: The Scientific Basis. Contribution of Working Group I to the Third Assessment Report of the Intergovernmental Panel on Climate Change [Houghton, J.T., et al. (eds.)]. Cambridge University Press, Cambridge, United Kingdom and New York, NY, USA, pp. 349-416.

Rogers, H.L., Chipperfield, M.P., Bekki, S. and Pyle, J.A. (2000) The effects of future supersonic aircraft on stratospheric chemistry modeled with varying meteorology, *Journal of Geophysical Research.* 105, 29359–29367

Rogers, H.L., Lee, D.S., Raper, D.W., Forster, P.M., Wilson, C.W. and Newton, P.J. (2002a) The impacts of aviation on the atmosphere. *The Aeronautical Journal,* 521-546, October 2002.

Rogers, H., Teyssedre, H., Pitari, G., Grewe, V., van Velthoven, P. and Sundet, J. (2002b) Model intercomparison of the transport transport of aircraft-like emissions from sub- and supersonic aircraft. *Meteorol. Z.,* 11, 151-159.

Rypdal, K., Berntsen, T., Fuglestvedt, J.S., et al. (2005) Tropospheric ozone and aerosols in climate agreements: scientific and political challenges. *Environmental Science and Policy,* 8 (1), 29-43.

Sausen, R., Gierens, K., Ponater, M. and Schumann, U. (1998) A diagnostic study of the global distribution of contrails Part 1: Present day climate. *Theor. Appl. Climatol.* 61:127-141

Sausen, R. and Schumann, U. (2000) Estimates of the climate response to aircraft CO_2 and NOx emissions scenarios. *Clim. Change,* 44, 27-58.

Sausen, R., Isaksen, I., Grewe, V., Hauglustaine, D., Lee, D., Myhre, G., Kohler, M., Pitari, G., Schumann, U., Stordal, F. and Zerefos, C. (2005) Aviation radiative forcing in 2000: An update on IPCC (1999). *Meteorologische Zeitschrift,* 14, 4, 555-561. doi: 10.1127/0941-2948/2005/0049.

Schmidt, E. (1941) Die Entstehung von Eisnebel aus den Auspuffgasen von Flugmotoren. Schriften der Dt. Akad. *der Luftfahrtforschung,* 44, 1-15.

Schumann, U. (1990) Air traffic and the environment–background, tendencies and potential global atmospheric effects. Proceedings of a DLR International Colloquium, Bonn, November 15–16, 1990, Lecture Notes in Engineering Vol. 60, Springer, Berlin, pp. 170.

Schumann, U. (1997) The impact of nitrogen oxide emissions from aircraft upon the atmosphere at flight altitudes—Results from the AERONOX project, *Atmospheric Environment.* 31 (1997), pp. 1723–1733.

Schumann, U. (2002) Contrail Cirrus, IN: Cirrus, Eds. D. Lynch et al., Oxford University Press, 231-255.

Shine, K., Berntsen, T., Fuglestvedt, J., and Sausen, R. (2005a) Scientific issues in the design of metrics for inclusion of oxides of nitrogen in global climate agreements. *Proc. Nat. Acad. Sci. USA,* 102, 15768-15773.

Shine, K., Fuglestvedt, J., Hailemariam, K. and Stuber, N. (2005b) Alternatives to the global warming potential for comparing climate impacts of emission of greenhouse gases. *Climate Change,* 68, 281-302.

Shine, K., Berntsen, T., Fluglestvedt, J., Skeie, R. and Stuber, N., (2007) Comparing the climate effect of emissions of short- and long-lived climate agents. *Phil. Trans. R. Soc.* A., 365, 1903-1914, doi: 10.1098/rsta.2007.2050.

Smith, S. and Wigley, T. (2000) Global warming potentials: 1. Climatic implications of emissions reductions'. *Climate change,* 44, 445-457.

Solomon, S., Borrmann, S., Garcia, R., Portmann, R., Thomason, L., Poole, L., Winker, D. and McCormick, M. (1997) Heterogeneous chlorine chemistry in the tropopause region. *J. Geophys. Res.,* 102, 21411-21429.

Stern, N. (2007) The Economics of Climate Change: The Stern Review. Cambridge University Press, UK.

Stevenson, D.S., Collins, W.J., Johnson, C.E. and Derwent, R. (1997) The impact of aircraft nitrogen oxide emissions on tropospheric ozone studied with a 3D Lagrangian model including fully diurnal chemistry. *Atmospheric Environment,* 31 (12), 1837-1850.

Stevenson, D., Doherty, R., Sanderson, M., Collins, W., Johnson, C. and Derwent, R. (2004) *Journal of Geophysical Research,* 109, D17307.

Svensson, F., Hasselrot, A. and Moldanova, J. (2004) Reduced environmental impact by lowered cruise altitude for liquid hydrogen-fuelled aircraft. *Aerospace Science and Technology,* 8:307-320.

Tol, R.S.J. (2002) Estimates of the damage costs of climate change, Part II. Dynamic estimates. *Environ. Resour. Econ.,* 21, 135–160.

Travis, D.J., Carleton, A.M., and Lauritsen, R.G. (2002) Contrails reduce daily temperature range. *Nature,* 418, 601-602.

Travis, D.J., Carleton, A.M., and Lauritsen, R.G. (2004) Regional variations in U.S. diurnal temperature range for the 11-14 September 2001 aircraft groundings: evidence of jet contrail influence on climate. *J. Clim.,* 17, 1123-1134.

Wild, O., Prather, M.J. and Akimoto, H. (2001) Indirect long-term global radiative cooling from NOx emissions. *Geophysical Research Letters.* 28:1719-1722.

Wit, R.C.N., Boon, B.H., van Velzen, A., Cames, M., Deuber, O., and Lee, D.S. (2005) Giving wings to emission trading. Inclusion of aviation under the European emission trading system (ETS): design and impacts. CE-Delft, No. ENC.C.2/ETU/2004/0074r, the Netherlands.

WMO. (2007). Scientific Assessment of Ozone Depletion: 2006. Global Ozone Research and Monitoring Project Report No. 50. World Meteorological Organization, Geneva.

Wuebbles D, Gupta M and Ko M. (2007) Evaluating the impacts of aviation on climate change. EOS Transactions American Geophysical Union, 88:157-168.

In: Aviation and the Environment
Editor: Jon C. Goodman
ISBN: 978-1-60692-320-7

Chapter 8

AVIATION-CLIMATE CHANGE RESEARCH INITIATIVE (ACCRI) SUBJECT SPECIFIC WHITE PAPER (SSWP) ON METRICS FOR CLIMATE IMPACTS; CLIMATE METRICS AND AVIATION: ANALYSIS OF CURRENT UNDERSTANDING AND UNCERTAINTIES, SSWP # VIII, JANUARY 22, 2008

Donald J. Wuebbles[1], Huiguang Yang[1] and Redina Herman[2]
[1] Department of Atmospheric Sciences,
University of Illinois USA
at Urbana-Champaign, Urbana, IL, USA
[2] Department of Geography,
Western Illinois University, Macomb, IL, USA

EXECUTIVE SUMMARY

The impact of climate-altering agents on the atmospheric system is a result of a complex system of interactions and feedbacks within the atmosphere, and with the oceans, the land surface, the biosphere and the cryosphere. Climate metrics are used as a proxy to simplify interpretation of the complex science and associated feedbacks to indicate the ultimate effect of constituent changes in the atmosphere. Aviation is just one contributor to these constituent changes in the atmosphere but the potential impact of aviation on climate is expected to grow over the coming decades as demand for air travel increases. It is necessary to quantify the impact of aviation so that appropriate policy actions may be defined. The objective of this report is to examine the capabilities and limitations of current climate metrics in the context of the aviation impact on climate change, to analyze key uncertainties associated with these metrics and, to the extent possible, to make recommendations on future research and about how best to use metrics currently to gauge aviation-induced climate change.

Climate change not only involves changes in temperature, but also changes in precipitations and changes in extreme events. Nonetheless, globally averaged surface temperature is generally used as a proxy for climate change because temperature changes are easier to predict and the effect of temperature changes are better understood than other atmospheric variables. When deciding which metric to use for aviation considerations, some general questions must first be answered, such as: What is the function or purpose of the metric? Can the metric be applied to various scenarios and forcings? What is the effectiveness of the metric for the user, whether it is for technology or policy considerations? Is the metric flexible enough to incorporate advances in scientific understanding? A useful metric should also be applicable to other transportation and / or energy sectors as well.

A useful metric must be easy to use and understand, as well as firmly supported by the science. When developing a metric or choosing between existing metrics one must balance the applicability of the metric to a wide range of climate altering scenarios with ease of use of the metric within the limits of scientific understanding. Aviation presents a very specific situation where emissions are deposited largely in the upper troposphere/lower stratosphere (UT/LS) region rather than at the Earth's surface like other transportation or energy related emissions. Some emissions from aviation are both long-lived (e.g., a century or longer for carbon dioxide) while others are very short-lived (e.g., minutes to a few days for contrail and cirrus effects). Also, the total amount of emissions and corresponding changes in climate resulting from the existing aviation fleet is currently relatively small compared to the total human-induced emissions that are leading to climate change. Some specific questions that must be answered with regard to aviation-induced climate change are: What are the climate effects of aviation relative to other transportation sectors? What technology choices will minimize the impacts on climate? Which forcing agent in aviation should be the highest priority for policy considerations? What are the trade-offs between reductions of different forcing agents? What are the trade-offs between different policy considerations? How can the industry maximize the benefit while minimizing the cost of abatement? What metric or metrics would be most useful for analyses of the potential climate impacts from aviation emissions? Or from other transportation and energy sectors? The "best" metric probably depends on which question(s) are being addressed and no metric should be used blindly.

The most widely used metric for climate change has been radiative forcing (RF). It is also an integral part of many of the existing climate metrics. In fact, there is no single "radiative forcing" metric; there are several "flavors" of radiative forcing based metrics. Although the use of the stratospheric adjusted radiative forcing metric is often used for aviation studies (as well as many other climate analyses) and has been proposed by some policymakers for use in possible policy development relative to aircraft emissions, the classic evaluation of this metric has limited suitability for that purpose and it is clear that it only provides part of the story regarding aircraft effects on climate.

Of all the problems associated with RF (in all its flavors), the most serious limitation may come from the fact that not all forcing agents cause the same climate impact (for the discussion here, change in globally averaged surface temperature) for a given change in radiative flux. This means that RF from one cause cannot be compared to RF from another cause easily. One way to get around this problem is to define an "equivalent" RF where the forcing is weighted by its climate sensitivity. This additional multiplier term is called "efficacy".

Existing metrics can be grouped into one of three categories: (1) concentration-based metrics which use constituent concentrations to gauge the change in radiative forcing; (2) emissions-based metrics that aim to control emissions and examine trade-offs; and (3) economics- and damage-based metrics which attempt to account for damages and abatement costs. The discussion in the report largely centers on the first two groups, the science-based metrics.

The most widely used metrics in climate assessments and policy considerations are stratospheric adjusted Radiative Forcing and Global Warming Potentials, but many other metrics have been proposed. At this point, the most promising metrics for future climate analyses including aviation are: Equivalent Radiative Forcing (Radiative Forcing with efficacies applied), Global Warming Potentials (GWPs), Global Temperature Potentials (GTPs) and Linearized Temperature Response (LTR) metrics. Efficacy factors should be applied to these metrics to account for the fact that not all constituents have the same impact on climate change. All of these metrics have strengths and some limitations towards addressing key policy questions related to the potential impacts of aviation on climate. However, all of these climate metrics should be further evaluated for their applicability to aviation-induced climate change because so far it is unclear which metric is most suitable to address the needs of policymakers.

In order to determine which metric is most applicable for which question, the applicability and robustness of individual metrics must be tested. These metrics must be tested both for global and regional applicability. A Metrics Working Group should be formed to evaluate the different metrics and their value for addressing policy questions using a variety of climate and chemistry- climate models. The Metrics Working Group will meet with policy makers to establish priorities because a metric preference particularly depends on the choice of questions to be addressed. One of the initial tasks of this working group will be to establish criteria for evaluating metrics, and then existing metrics will be compared in the context of the priorities established by policy makers. Efficacy factors will also need to be evaluated to determine if efficacies can adequately correct for differences in climate sensitivity to various aviation scenarios. The possible effects of changes in the background atmospheric conditions (effects of composition and climate changes) on derived aviation impacts need to be evaluated. Finally, metrics will be evaluated based on applicability to other transportation and energy sectors. These priorities will greatly enhance the understanding of climate metrics within the next five years using the current suite of tools, which include state-of-the-art chemical-transport and chemical-climate models, as well as the set of existing metrics.

The GWP concept cannot be ignored because it still is the most accepted metric in the international climate assessments and corresponding policy considerations. However, the GTP concept and the linearized temperature response (LTR) approach also have many advantages and may be the preferred approaches for technological and policy analyses relative to aviation. GTP has the advantage of being relatively simple, transparent, and flexible, but, like GWPs, they have not been adequately tested for application to aviation impacts on climate.

The latest LTR approaches, namely the APMT and AirClim assessment tools, appear to be quite promising for future studies of aviation. The AirClim approach may even provide a capability for analyzing regional impacts not considered otherwise. However, these tools are dependent on the validity of much more complex representations and understanding of the science, including the carbon cycle, chemistry interactions, aerosol direct and indirect effects, contrail formation and evolution, and the resulting impacts on climate. Current tools need much further development and evaluation before they will be applicable to policy considerations.

It will be important to take a systems point of view in any new study using existing metrics to evaluate the climate impacts from aviation. As such, it will be important to consider all of the uncertainties associated with current understanding of the effects of aviation emissions on climate, including the fact that with the exception of carbon dioxide, the effects of other emissions on climate are still not very well understood. In particular, it would be very difficult to provide a meaningful evaluation of the effects of contrails or the effects of contrails and aerosols on cirrus. However, metrics may be able to better consider the effects NOx emissions from aviation. To provide a perspective relative to prior assessments of aircraft effects, any new study done at this time should start with the use of stratospheric adjusted radiative forcing, but also include consideration of efficacies to the degree possible. The effects of uncertainties in the evaluation of the climate effects and in the metric itself will need to be clearly stated. The radiative forcing could be evaluated for the current time period but it can also be worthwhile to consider projections of effects on aviation based on reasonable scenarios for future emissions. Such scenarios, however, need to be carefully considered, and should be based on best available projections from ICAO and the FAA (or associated organizations like JPDO). Emissions-based metrics should also be considered, but interpretation is currently limited by the lack of a community-consensus on which metrics should be adopted and the by the limited application currently of the GWP and GTP approaches to evaluation of aviation impacts. The LTR approaches are promising as assessment tools but have not been evaluated by the science community and need further development to reduce existing uncertainties.

1. Introduction

Metrics have long been used in studies of climate change to simplify interpretation of the complex science and associated feedbacks and interactions that determine the ultimate effect of gaseous or particulate emissions on the atmosphere. Several different types of metrics have been developed, each with its advantages and disadvantages. Several of these metrics have been applied in various ways to study the effects of aviation on climate. However, there has been little attempt to assess what is known about climate metrics in order to evaluate the relevance and applicability of these metrics to aviation.

Climate is defined as the typical behavior of the atmosphere, the aggregation of the weather, and is generally expressed in terms of averages and variances of temperature, precipitation and other physical properties. A climate metric, in general, is a variable (or a set of variables) designed to parameterize a set of known or deduced influences on the climate system that may result in climate change. The climate metric variable is then used as a proxy to indicate the impact of forcing on the climate system resulting in a change in the energy balance of the earth-atmosphere system. This forcing results in a change in both the instantaneous and long-term equilibrium conditions of the Earth's atmosphere, and a shift in the long-term average conditions of the Earth's atmosphere. Climate change may be manifested by a variety of important parameters, including temperature, precipitation, humidity, cloudiness, soil moisture, sea surface temperature, and sea ice location and thickness.

Whereas comprehensive models of the climate system can be used to study the much larger climate effects of fossil fuel use and other human-related emissions at the Earth's surface, the climate effects from current aircraft emissions are only a small fraction of the total impacts of human activities on climate (e.g., emissions of carbon dioxide from aviation are currently approximately two percent of the total emissions from fossil fuel burning and changes in land use). As a result, it is very difficult to use a climate model to directly evaluate the climate effects resulting from aviation. Metrics thus provide the primary means for evaluating the relative effects of different emissions, including policy or tradeoff options, from aviation on climate and for comparing the effects of aviation on climate relative to other human factors affecting climate.

However, the potential importance of aviation on climate is expected to grow over the coming decades, further increasing the need for well-defined metrics to study and understand the role of aviation on climate. For example, the U.S. projects demand for air transportation services to grow three fold by 2025 (e.g., Next Generation Air Transportation System, 2004). It is a daunting challenge for both the scientific and technological communities to satisfy this increasing demand, while still protecting our environment, including potential impacts on the Earth's climate. With extensive growth demand expected in aviation over the next few decades, it is imperative that vigorous action be taken to understand the potential impacts of aviation emissions to help policymakers address climate and other potential environmental impacts associated with aviation. To meet the challenges presented by this growth, the President of the United States signed 'Vision 100 – Century of Aviation Reauthorization Act" in 2003 and created a multi-agency integrated plan for the development of a Next Generation Air Transportation system (NGATS). The vision of the NGATS is "A transformed aviation system that allows all communities to participate in the global market-place, provides services tailored to individual customer needs, and accommodates seamless civil and military operations." One of the challenges posed by the vision is achieving growth while reducing environmental impacts. At the same time, other countries (e.g., the European Union) and the United Nations' International Civil Aviation Organization (ICAO) face similar concerns and issues.

As stated in the 2006 Workshop on the Impacts of Aviation on Climate Change (Wuebbles et al., 2006; available from http://web.mit.edu/aeroastro/partner/reports/climatewrksp-rpt-0806.pdf), the integrated national plan for implementation of the NGATS initiative in the U.S. is carried out by a Joint Planning and Development Office (JPDO). The JPDO is comprised of a number of U.S. agencies: National Aeronautics and Space Administration (NASA), Federal Aviation Administration (FAA), Department of Transportation (DOT), Department of Homeland Security (DHS), Department of Commerce (DOC) and the Whitehouse Office of Science and Technology Policy (OSTP). The Environmental Integrated Product Team (EIPT) of JPDO has been tasked with incorporating environmental impact planning into the NGATS. To fulfill this strategy, it is necessary to quantify the climatic impacts of aviation emissions to enable appropriate policy considerations and actions. Understanding aviation's climate impact is also critical to informing the United States in the best considerations and trade-offs for setting standards in engine emissions, special flight operations, or other potential policy actions through the International Civil Aviation Organization. This cannot be adequately done until the policymakers can correctly capture the environmental effects of aviation emissions, including climate impacts. The extensive investment of new aircraft in the marketplace, with their long service lifetime (25-30 years or longer), emphasizes the urgent need for improving our current understanding of the effects of aviation on climate.

The vast majority of the emissions from aviation occur at cruise altitudes in the upper troposphere and lower stratosphere (UT/LS). The chemical species released during the fuel combustion process in aircraft engines include carbon dioxide (CO_2), water (H_2O), nitrogen oxides (NO and NO_2 or NOx collectively) and sulfur oxides (SO_x) along with small amounts of soot carbon (C_{soot}), hydrocarbons (HC) and carbon monoxide (CO). Once released at cruise altitudes within the UT/LS, these species interact with the background atmosphere and undergo complex processes, resulting in climate impacts and related damages. However, one also needs to bear in mind that the background atmosphere is also changing over time as a result of both natural and human drivers. As the background atmosphere changes, the response of atmospheric chemistry and the climate system to emissions from aviation may also change.

The schematic in figure 1 illustrates how emissions from aviation can cause resulting climate impacts and subsequent damages. The impact of climate-altering agents leads to the following chain of events: Emissions lead to changes in atmospheric concentrations of gases and particles; these in turn lead to changes in the radiative transfer affecting the climate system, referred to as the radiative forcing on climate; changes in radiative forcing alters key climate parameters like temperature and precipitation (e.g., IPCC, 1999; IPCC, 2007a). These changes in the climate system can have resulting social and ecosystem impacts and can result in a variety of societal and economic impacts (IPCC, 2007b; O'Neill, 2000; Smith and Wigley, 2000; Fuglestvedt *et al.*, 2003). As one moves down the diagram, there is increasing policy relevance (in terms of observed changes that are likely to produce measurable economic or other types of social welfare damages) but there is also increasing uncertainty regarding the exact magnitude of the change as it depends not only on the forcing of the climate system by emissions but also on the vulnerability of individual natural and human systems.

Climate metrics are often used as an indicator of these climate impacts. Some climate metrics go further and indicate the influence of climate change on human-related factors, like economic damage or cost of abatement. This report is largely restricted to physical metrics, which do not consider costs or other economic factors. However, because there is a lot of potential interest in the use of metrics containing economic factors, we do provide a cursory discussion on the potential use of and the current issues associated with using economics in climate metrics.

The specific ways that aircraft emissions can alter the radiative budget of the Earth and contribute to human-induced climate change are:

Aircraft engines emit CO_2 and water vapor, important greenhouse gases, that directly affect climate through their absorption and reemission of infrared radiation;

Aircraft emitted NO_x (and hydrogen oxides (HOx) produced from water vapor emissions into the stratosphere) can modify atmospheric ozone concentrations through chemical interactions. Ozone affects the radiative balance of the climate system through both its shortwave and infrared (greenhouse effect) absorption;

Through its resulting net production of upper tropospheric and lower stratospheric ozone, NOx emissions from subsonic aircraft reduce the atmospheric abundance of CH_4, another important greenhouse gas, through enhancing the concentrations of tropospheric hydroxyl radicals (OH), the primary reactant for destruction of methane;

Aircraft emit aerosols in the form of liquid particles containing sulfate and organics, and soot particles. Emissions of sulfur dioxide also increase the aerosol mass in aging plumes. These aerosols can be radiatively active themselves, either by scattering (sulfates) or absorbing (soot) solar radiation or can indirectly affect climate by triggering the formation of persistent condensation trails or altering natural cloudiness; Under the right meteorological conditions, aircraft emissions of water vapor (and aerosols) can lead to formation of contrails and possibly result in effects on upper tropospheric cirrus clouds – these effects may exert spatially inhomogeneous radiative impacts on climate.

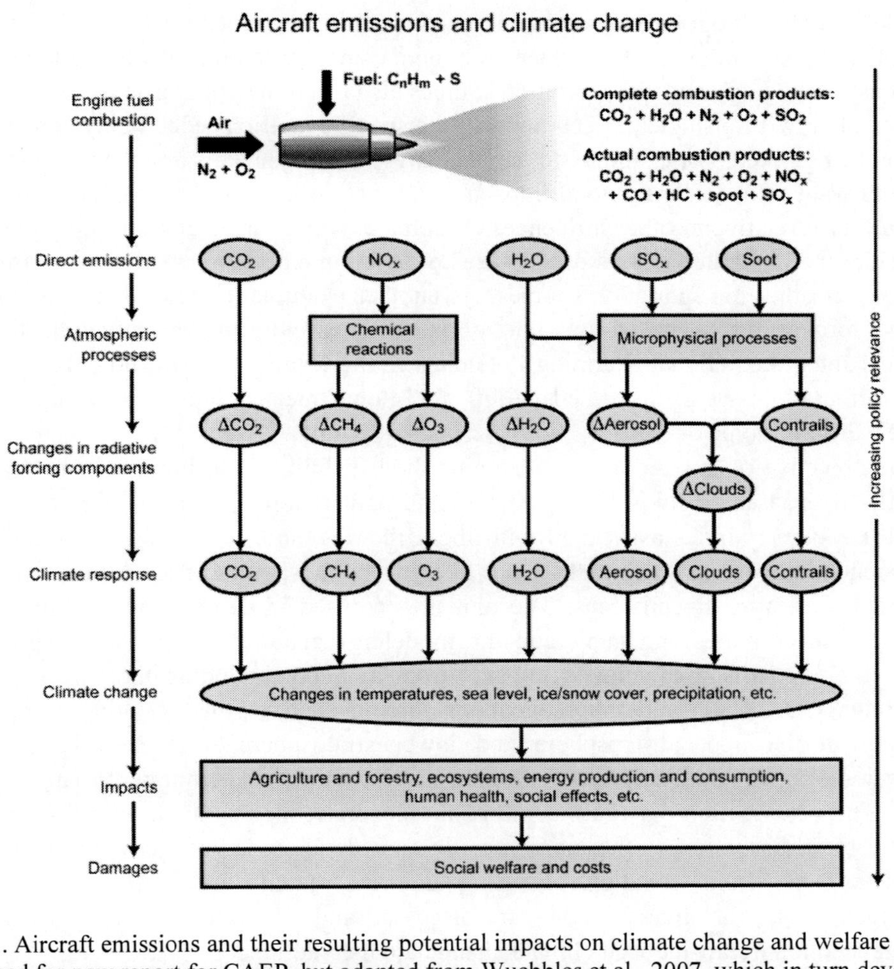

Figure 1. Aircraft emissions and their resulting potential impacts on climate change and welfare loss (developed for new report for CAEP, but adapted from Wuebbles et al., 2007, which in turn developed this figure based on IPCC, 1999 and Fuglestvedt et al., 2003).

As will be discussed further in subsequent sections of this report, the current scientific understanding of the potential effects on climate from aviation emissions range from good for the carbon dioxide emissions to fair for the NOx, water vapor, direct particle, and contrail effects to poor for the effects on cirrus clouds (also see the report from the 2006 Workshop on the Impacts of Aviation on Climate Change).

Although current fuel use from aviation is only a few percent of all combustion sources of CO_2, one of the dominant radiatively important gases currently affecting the climate system as a result of human activities, the expectation is that this percentage will increase in the future. On a multi- decadal time scale, aircraft emissions could become a more significant factor in climate change because of the projected increase in passenger demand and associated flights, and because of the likely decrease in other combustion sources as the world moves away from fossil fuels towards alternative and renewable energy sources. Although the long atmospheric lifetime of CO_2 implies little dependence on where emissions occur, the effects on climate from the other emissions from aviation are strongly affected by emissions primarily occurring at cruise altitudes in the upper troposphere and lower stratosphere. For example, aircraft nitrogen oxides released at these altitudes generally have a larger climate impact than those emitted at the surface, although a small fraction of the much larger surface emissions from energy and transportation sources also reach the upper troposphere.

There likely is no single perfect metric -- the specific metric needed likely depends on the question being asked. For example, for some analyses, policymakers and the aviation industry may both want to consider the total impacts that aviation is having on climate currently and into the future relative to other influences on climate, while for other studies, they may want to consider the integrated effects of a "pulse" of aviation emissions on climate relative to the emissions of other transportation sources. As another example, a metric based on integrated radiative forcing over a chosen time horizon is consistent with the current application of the 100- year integrated Global Warming Potentials in the Kyoto Protocol; however, a different target formulation – e.g. a defined ceiling for global mean temperature change – would require a different type of metric.

The objective of this report is to examine the capabilities and limitations of the metrics currently being used to study human-related and natural forcings on the climate system, to analyze key uncertainties associated with these metrics, and, to the degree possible, make recommendations about which metrics are likely to be most suitable for various applications associated with aircraft emissions. The aim is a focused in-depth review of the scientific principles, uncertainties and gaps, and the modeling capabilities, for determining suitable metrics for comparison of climate impacts from aviation, including those for well-mixed gases (e.g., CO_2, CH_4) and inhomogeneous forcing such as that resulting from changes occurring in the upper troposphere and lower stratosphere from perturbations to the distribution of ozone and particles, the formation of contrails and from perturbations to cirrus clouds. The next section discusses some of the general concerns about metrics for climate, followed by a discussion of the more specific considerations associated with analyzing the climate effects from aviation. Existing metrics being used are then discussed. Recommendations for aircraft studies are discussed and research needs to address specific issues related to aircraft-induced climate change are then defined.

2. A Review of Metrics for Climate Impacts

2a. General Comments About Climate Metrics

There are a number of general concerns that must be considered when trying to find a metric that is the most useful for analyses of aviation and other human-related impacts on climate.

First, it should be recognized that there is now overwhelming scientific consensus regarding the role of human activities in causing the changes in climate that have occurred over the last few decades. The science community has become increasingly convinced that the changes in climate being seen are primarily due to burning of fossil fuels and other human-related activities (IPCC, 2001, 2007a). Nonetheless, there are some significant uncertainties remaining in our understanding of the feedbacks on climate and the resulting impacts. Quantifying the role of aviation is further complicated by uncertainties in understanding the specific mechanisms whereby aviation can affect climate - for example, determining the effects of emissions of nitrogen oxides from aircraft on tropospheric and stratospheric ozone and the resulting effects on hydroxyl and methane concentrations. Since ozone and methane are radiatively important "greenhouse" gases that can affect climate, these effects need to be well understood. An even larger uncertainty is the extent of persistent contrails from aviation and the resulting effects on climate, the role of these contrails and the aerosol (particle) emissions from aviation on cirrus cloud production in the upper troposphere, and the role of cirrus in climate change.

Second, projections of regional changes in climate are, at this point, still less well understood than the global effects on climate. Regional impacts are driven by regional feedback mechanisms and the local distribution of forcing agents. Regional feedback mechanisms can be driven by such things as proximity to a large body of water, local climate and elevation. Local distribution of forcing agents is particularly important for short-lived species. These issues are particularly important for aircraft emissions because aircraft emissions contain both long- and short-lived constituents and span a wide range of geographic regions.

Also, the effects of temperature changes are also better understood than precipitation changes. It is for this reason that globally-averaged surface temperature is generally used as the primary model-derived output variable for climate change. As our ability to model other variables, such as precipitation, cloud cover, etc., improves, the climate change variable of choice may change as well.

Emissions-based metrics (e.g., Global warming Potentials) are often defined based on emissions put into the current atmosphere. However, the atmosphere is not at a steady state. The atmospheric composition, plus temperature and other physical variables, are changing, largely as a result of human-related activities. As a result of nonlinear relationships in atmospheric chemistry and in radiative and other physical processes, a metric calculated assuming the background corresponds to 2050 may result in very different values than if the metric is calculated relative to the background corresponding to the current atmosphere.

Some additional difficulties in developing metrics for climate change include the choice of an appropriate structure for the metric (which may depend on its intended use), the quantification of input values (due to underlying uncertainties) and the need for value judgements in the choice of parameters within these metrics (e.g., the evaluation of long term impacts versus short term impacts). Such value judgements go beyond natural sciences. In the choice of impact parameter there is also a trade-off between relevance and uncertainty.

The scientific limitations in our understanding of climate change and the impact of aircraft emissions will be discussed in more detail as we look at specific metrics and their usefulness. There are some general questions that must be answered in order to evaluate a metric. These questions include: What is the function or purpose of the metric? Can the metric be applied to various scenarios and forcings? What is the effectiveness of the metric for the user, whether it is for technology or policy considerations? Is the metric flexible enough to incorporate advances in scientific understanding?

2b. The Characteristics of a Climate Metric

Development of meaningful metrics for climate change requires a reasonably accurate capability for the evaluation of the effects of human-related and natural factors affecting climate. Such capabilities require complex state-of-the-art models that include representations of, and interactions among, the atmosphere, its chemical composition, the oceans, biosphere, cryosphere, etc. These models encapsulate our understanding of physical, chemical and biological processes. However, they are not useful in directly providing metrics for, for example, policymaking for several reasons. They require very large computer resources and considerable expertise to perform calculations and to diagnose results from the large amount of output that they produce. Hence, there is a limit on the number of different cases (e.g., emission scenarios) that can be considered. Alternatively, simplified models or metrics (that build on the results of the complex models) can be used.

Climate metrics have a number of potential uses, including:

Providing flexible, rapidly-available input regarding the relative ability of various approaches to minimize the potential impact of human activities on the climate system;

Assessing the relative contributions of emissions from different human activities to climate change;

Comparing (and ranking) climate effects from competing technologies, energy uses – or the different emissions in a given sector like aviation;

Ranking the emissions from various countries;

Establishing a basis for comparing reductions in climate effects in various countries;

Functioning as a signal for policy considerations to encourage some activities and discourage others;

As an analysis tool for industries and countries to determine the best approaches for meeting commitments to reduce climate impacts

In general, a metric must be *scientifically well grounded*, but also *simple to use and easy to understand*. It must be an effective tool for communication between scientists, industry, and policymakers. Users, whether it is industry, policymakers, or others, should be able to make use of the metric without further input from the scientific community, so the metric should be *transparent* enough to convey a meaning all on its own. One main concern with developing new metrics is the need to weight *applicability of the metric* versus *ease in understanding the results*. So, the metric needs to be simple, yet users must be confident enough in the scientific quality of the metric to trust it and use it; therefore it should be *subject to a minimum of uncertainties* or have the effects of scientific uncertainties reduced (or at least represented) as much as possible. In the choice of impact parameter, there is also a trade-off between relevance and uncertainty. As stated before, the metric has to be *applicable to the questions or policy concerns of interest*.

Making the right choices is an important part of formulating a metric for climate change. The spatial and temporal scales of interest need to be considered. Are globally- and annually-averaged effects and impacts of climate change adequate or is it necessary to consider regional impacts. Generally metrics have been used at the global scale because of the uncertainties in representing regional impacts.

A choice also must be made as to what are the key parameters to use in representing climate change in the metric. While one could consider parameters like change in precipitation or change in sea level, the most commonly considered parameters are change in radiative forcing, change in temperature, or some sort of economic impact, such as change in damages and abatement costs. The first two (radiative forcing and temperature changes) have wide acceptance in the science community. While economists often argue that damages and abatement costs must be included and that this may be the only way to really compare climate change impacts across different emissions sources and at different geographic locations, there is no general consensus on what the best approaches are for doing so.

A choice must also be made as to how to consider temporal changes in the climate parameter and/or the emissions of interest, e.g., whether to consider the absolute change in the climate parameter over a given time period, the integrated change over a given time period, and/or to consider the effects of pulsed or sustained emissions. Such choices can affect decisions using the metric, e.g., whether it is best to reduce emissions of long-lived gases or short-lived gases or particles.

In considering a metric, it is important to recognize the current state of scientific understanding. It would be very difficult, for example, to define an accurate metric based on regional (or even global average) precipitation because current regional and global climate models have significant uncertainties in representing precipitation processes and their interaction with the global climate system well enough. Essentially all of the climate metrics being used in analyses of human- related emissions to date are based in some way on the change in globally-averaged (and annually-averaged) surface temperature as the measure of climate change since that is the projection in which we have the most confidence from a scientific perspective. As our scientific understanding improves, the metrics of choice might change.

Other considerations in metrics choice include also the choice of an appropriate structure (e.g., to be applicable to temperature targets) for the metric (this choice will likely depend on the design of any climate policy it is intended to serve), the quantification of input values (due to underlying uncertainties) and the need for value judgements in the choice of parameters within these metrics (e.g., the evaluation of long term impacts versus short term impacts). Such value judgements go beyond natural sciences.

2c. Special Considerations for Aviation Analyses

Emissions from aviation present some special problems for climate metrics. First, these emissions are deposited largely into the upper troposphere and lower stratosphere while other human-related emissions are mostly at the Earth's surface. Second, the total emissions from aviation are relatively small when compared to the total emissions from other anthropogenic sources of radiatively active (either direct or indirect) constituents. Third, aircraft emissions contain both long- and short-lived constituents, meaning that both direct radiative effects and the indirect radiative effects via complex chemical and physical processes, such as impacts on ozone, methane and cloudiness, all need to be considered. Aircraft emissions also contain aerosols, which are difficult for climate metrics to accurately depict because of the non-linear effects of indirect forcings (Lohmann and Feichter, 2005).

Emission Region

A number of past studies have examined the relationship between radiative forcing and temperature change. Typically these have examined the effects resulting from long-lived gases or well distributed changes in forcing, such as changes in the solar flux. For example, Hansen et al. (2005) examined the climate sensitivity to CO_2 and solar irradiance changes. They found that the climate sensitivity does depend on the magnitude of the forcing, but for forcings close to the current state the sensitivity is nearly constant. As the forcing from CO_2 or solar irradiance in the model was changed, the climate sensitivity changed as well.

Aircraft emissions are deposited locally, both geographically and in altitude. Aircraft emissions are deposited predominately in the upper troposphere and lower stratosphere in the Northern Hemisphere mid-latitudes. Part of the difficulty in understanding the chemical and physical impacts on climate from aviation emissions is because the upper troposphere / lower stratosphere (UT/LS) is a highly coupled region where dynamics, chemistry, microphysics and radiative processes are fundamentally interconnected. Water vapor and ozone, perhaps the two most important greenhouse gases in the UT/LS, are controlled by both transport processes, such as stratosphere-troposphere exchange, and chemical processes including multiphase chemistry, and cloud microphysics, which in turn are influenced by the temperature and aerosol distributions. The UT/LS is a region of much scientific scrutiny (e.g., Pan et al., 2007) because of the uncertainties surrounding these complex interactions.

Since aircraft emissions have such a unique region of influence, one might think that they would have an equally unique forcing signature. Unfortunately, Boer and Yu (2003b) and other studies suggest that this is not the case for different geographic distributions. Rather, they found that the geographic distribution of temperature change is predominately determined by the geographic distribution of the feedback mechanisms and only secondarily determined by the geographic distribution of the forcing agent.

Hansen et al. (2005) also determined that it was difficult to use the geographic pattern of the temperature response to determine the climate forcing agent responsible. They tested the climate response to different geographic patterns of CO_2, CH_4, O_3, BC (black carbon, soot) aerosols, N_2O and CFCs, as well as land use, volcanic emission and solar irradiance change, and found that the temperature response preferentially occurred in certain places, particularly high latitudes. In fact, Hansen et al. (2005) examined the geographic distribution of the temperature response normalized by the magnitude of the forcing (assuming constant sea surface temperature) so that the global average radiative forcing is the same for all runs and found that for well-mixed greenhouse gases "changes evoke nearly identical normalized response" patterns. This pattern also held for the all-forcings-at-once scenarios, but broke down somewhat for scattering aerosols and more so for absorbing aerosols.

On the other hand, Hansen et al. (2005) found that the vertical distribution of temperature change could be used to indicate a vertical distribution of forcing agent. Aircraft have a very distinct vertical influence, so it is possible that the vertical distribution of forcing can be linked to a change in environment lapse rate. Further studies are needed to determine if this is a reliable way to detect aircraft impacts. *This also raises the question of whether the normal surface temperature-based metric is capable of adequately capturing the climate impacts of aviation.*

Total Emission Size

Aircraft emissions are not large when compared to other anthropogenic sources of radiatively active constituents. It is not possible to evaluate emission signatures of the non-CO_2 short-lived emissions from aviation in climate models because the signal does not rise above the natural climate variability and model noise. In order to detect an aircraft signature in a climate model relative to natural climate variability, aircraft emissions have to be scaled to a larger size. Scaling presents its own set of problems because if the scaling factor is too large then the model is no longer in the linear regime of the emission-response function. As an example, scaling the NOx emissions from aviation to be able to detect effects on climate may be affected by nonlinearities in the chemistry and physical processes leading to the resulting changes in ozone and methane. For aircraft emissions, as with other anthropogenic emissions of short-lived it is unclear just how important such non-linear effects are in determining the climate response.

Short-Lived Species

In addition to long-lived atmospheric constituents like CO_2, aircraft also emit short-lived pollutants that are either themselves radiatively active (e.g., aerosols) or can affect radiatively important gases, particles, or clouds. Short-lived emissions, which last from minutes to days, can affect the geographic region where they are emitted and the effect will likely be different for different geographic regions, even for the same emissions. In addition, the lifetime of gases like CH_4 depend on the chemical composition of the background atmosphere. In order for a climate metric to work effectively for aircraft emissions, the metric must take into consideration short- lived species.

Concentration-based metrics like radiative forcing are often being used to examine the change in climate forcing over a period of time and ignore the transient effects. Because it is unlikely that a transportation source like aviation is suddenly going to have no emissions tomorrow or even in a few years, it can be worthwhile to use a concentration-based metric like radiative forcing to consider what effects emissions are having on climate over a given time period. However, there is also significant value in considering the transient effects. The very different atmospheric lifetime of the emission effect associated with CO_2, NOx/O_3, CH_4, and contrails suggest that technology or policy changes could lead to vastly different short-term versus long-term effects on climate. Metrics that consider these transient effects thus can provide useful insights.

Contrails present a problem that is unique to aircraft emissions. Current models do not adequately simulate the ice-supersaturation environment necessary for persistent contrails, nor do they have the spatial resolution to represent individual contrails, so it is difficult to adequately model contrails. In addition, contrails typically have very short lifetimes as compared to other radiatively important aircraft effects. As a result, the climate effect from contrails is still poorly understood. Hansen et al. (2005) did climate simulations using "observed" contrail coverage multiplied by a factor of 10. Nonetheless, the climate effect may be large enough locally to be important to climate analyses. The problem is how to account for such uncertain effects in metrics being used for studying the climate effects associated with aviation.

Aerosols emitted by aircraft have a relatively small direct effect on climate but may be important as condensation nuclei for cirrus formation. The direct radiative effect of aerosols is reasonably well understood compared to the indirect effects on cloudiness. The indirect effects are harder to understand than the direct effect because of the poorly understood interactions between aerosols, cloud condensation nuclei and cloud properties. In addition to the indirect effects there is also a semi-direct effect caused by soot. Black carbon warms the air in the immediate vicinity and leads to cloud evaporation (Hansen et al., 1997). Chylek et al. (1996) also points out that the location of soot relative to the cloud is very important to radiative transfer. If soot is above the cloud layer, it behaves very differently than if it is below the cloud layer. Aerosols also change the optical properties of clouds and cause an increase in the ice nucleation efficiency of mixed-phase clouds (Lohmann, 2002). Smaller liquid droplets from aerosol-influenced clouds would decrease the freezing efficiency and allow supercooled droplets to penetrate higher into the cloud.

For subsonic aircraft, NOx emitted from aircraft are short-lived (lifetime of days) but the NOx emissions in the UTLS generally lead to O_3 formation and CH_4 destruction, depending on the background environment. Regional dependence of O_3 production depends on solar flux (varies by latitude), background NOx concentration, and local chemistry and emissions (IPCC, 2001; Prather et al., 1999; Collins et al., 2006; Jacob et al., 2005). As a result, the impact of NOx emissions depends on where the emission occurs.

Current global-averaged analyses imply that cooling effects of CH_4 decreases and warming effects of O_3 increases from aviation are roughly of the same magnitude. CH_4 is well distributed globally because of its longer lifetime (~8 years, but recovery time after a CH_4 perturbation is closer to 12 years because of the resulting interactions with atmospheric hydroxyl), but aviation effects on O_3 not globally distributed because of the relatively short atmospheric lifetime of tropospheric (and lower stratospheric) ozone. As a result, the distribution of warming/cooling effects from ozone and methane perturbations from aviation will not be equally distributed across the globe. In addition, it has been shown than the regional climate response is not the same for all regions of the Earth. Equatorial latitudes show a stronger response to emissions than mid-latitudes (Bernsten et al., 2005; Fuglestvedt et al., 2003; Derwent et al., 2001).

Metric Considerations

There are a variety of potential questions that a user may want to address in terms of aviation applications using climate metrics. Depending on the question, more than one type of metric may be needed to fully address all aspects to be evaluated. Some examples of potential questions include:

What are the climate effects of aviation relative to other transportation sectors?
What technology choices will minimize the impacts on climate?
Which forcing agent in aviation should be the highest priority for policy considerations?
What are the trade-offs between reductions of different forcing agents?
What are the trade-offs between different policy considerations?
How can the industry maximize the benefit while minimizing the cost of abatement?

In order to answer such questions, a climate metric (or metrics) should be able to weight the different forcing agents and put them all on the same scale for comparison. While there has not been universal agreement, many studies of climate forcings compare the impact of various climate forcings with the forcing from changes in CO_2, the gas currently having the largest human-related impact on climate. Forcing agents are often considered in terms of their "CO_2 equivalent" forcing effect. Of course, then one has to decide what is meant by equivalence. Are forcings equivalent in terms of their radiative forcing, integrated radiative forcing, change in global average surface temperature, integrated change in global average surface temperature, etc.?

There may be metrics that would be particularly suitable for aviation emission, e.g., a metric that applies best to the climate effects associated with changes occurring in the upper troposphere and lower stratosphere. However, even if such a metric exists, another factor is just how useful the metric is for other climate policy considerations because metrics for aircraft emissions must also fit into the framework being used by policymakers and others for sectors analyzing human- related emissions effects on climate.

2d. Development of Radiative Forcing as a Metric

The most widely used metric for climate change has been radiative forcing. Since it is used in many of the concentration-based and emissions-based metrics, it is worthwhile to first look at the definition and historical development of radiation forcing. In fact, as seen in later sections, there is no single "radiative forcing" metric; there are several "flavors" of radiative forcing based metrics. Although the use of the stratospheric adjusted radiative forcing metric is often used for aviation studies (e.g., IPCC, 1999; Sausen et al., 2005) and has been proposed by some policymakers for use in possible policy development relative to aircraft emissions, the classic evaluation of this metric has limited suitability for that purpose and it is clear that it only provides part of the story regarding aircraft effects on climate. Other metrics will need to be considered – for example, emissions-based metrics provide important information not provided by the traditional use of radiative forcing as a concentration-based metric.

The term 'radiative forcing' as a metric applied to climate change has been used since the 1980s. It has been a central tool in all of the international assessments of climate change. The IPCC Assessment (2001) describes radiative forcing as "a useful concept, providing a convenient first- order measure of the relative climatic importance of different agents" without the need to actually conduct time consuming and computationally expensive climate model simulations. However, as discussed later, this concept has significant limitations for spatially inhomogeneous perturbations to the climate system and can be a poor predictor of the global mean climate response. As a result, alternative definitions have been developed.

Essentially, radiative forcing for a given greenhouse gas or other forcing agent requires two primary factors, its three-dimensional distribution and how this has changed over time, and its interactions with solar and thermal infrared radiation (Shine and Forster, 1999; Myhre et al., 2001).

Over time, the radiative forcing concept has been broadened to not only include changes in solar flux and changes in relatively long-lived greenhouse gases like CO_2, O_3, CH_4 and various halocarbons, but also to include the climate effects resulting from changing emissions and concentrations of short-lived gases and particles. Short-lived gases generally have little direct effect on climate but can have indirect climate effects through chemical interactions affecting radiatively important constituents like O_3 and CH_4. Emissions of and secondary production of atmospheric particles can have both direct effects on climate and indirect impacts on climate resulting from their effects on cloudiness.

The concept of radiative forcing arose directly from the assumption that the Earth-atmosphere system is always approximately in radiative convective equilibrium. Assuming radiative-convective equilibrium, the heating rate of the atmosphere can be derived as:

$$\frac{dH}{dt} = F - \frac{\Delta T}{\lambda},$$

where

$$H = \int_{z_b}^{\infty} \rho C_p T dz$$

is the heat content of the atmosphere, F is the forcing on the system, T is temperature change in the system, is the climate sensitivity parameter that accounts for the effects of climate feedbacks, is the density of the atmosphere, C_p is the specific heat, and z_b is the depth that heat penetrates into the atmosphere. For analyses of changing solar flux and changes in the concentration of carbon dioxide, climate model calculations found an approximately linear relationship between global-mean radiative forcing at the tropopause and the change in equilibrium global mean surface air temperature. Because of the close linking of the troposphere to the surface through convection, climate models have typically found that the land surface, ocean mixed layer, and troposphere together respond to a radiative forcing for such perturbations with a relatively uniform increase in globally-averaged temperature.

As a result, the steady state form of the heat change equation is:

$$\Delta T = \lambda F.$$

This equation has traditionally been used to estimate surface temperature change given the radiative forcing, with an estimated value or uncertainty range in the climate sensitivity parameter (generally is taken to be the value corresponding to that expected for a doubling of the atmospheric concentration of CO_2 from pre-industrial levels, namely a 1.5 to 4.5 degree C change in surface temperature for a 4 Wm-2 increase in radiative forcing). The first applications of a radiative-convective model to predict radiative forcing effects of greenhouse gases and clouds in the Earth's atmosphere were done by Manabe and Strickler (1964) and Manabe and Wetherald (1967). These early studies demonstrated that the climate of the Earth can be affected by the influences (or forcings) of changes in solar irradiance and albedo and changes in the atmospheric distribution of certain radiatively active gases and aerosols.

A number of studies of examined the sensitivity factor , but without much success in reducing the uncertainty range (NRC, 2003; Meehl et al., 2004a; Schwartz, 2004; Andronova et al., 2007; Kiehl, 2007; Roe and Baker, 2007; plus discussion and references in the various IPCC assessments). The primary factors affecting the range of sensitivity factors founds in existing climate models appear to be uncertainties associated with the treatment of aerosols and cloud processes. However, Stuber et al. (2005) suggest that the two largest factors in the variability of λ are the varying strength of stratospheric water vapor feedback and the sea ice-albedo feedback.

Ramanathan et al. (1985) found that the climate sensitivity or climate feedback parameter, , was almost invariant to the type of forcing used in a one-dimensional radiative convective model. Many other climate modeling studies have shown an approximately linear relationship between the global mean change in radiative forcing at the top of the atmosphere resulting in a change in the equilibrium global mean temperature at the surface. Models have shown a large difference in λ between different climate models (thus the range of values mentioned above), but an approximately constant value for within a particular model for changes in solar flux and atmospheric concentrations of long-lived gases like $_{CO2}$, CH_4, and N_2O.

Ramanathan et al. (1987), as well as a number of later studies (e.g., Wang et al., 1986; Hansen et al., 1997; Jain et al., 2000; Naik et al., 2000; Forster et al., 2001; Gauss et al., 2003; Gohar et al., 2004; Huang and Ramaswamy, 2006; Meehl et al., 2004b; Tett et al., 2002), examined the effects of various trace gases on climate. Many trace gases absorb infrared radiation and can have a significant surface warming effect. Some gases can also affect climate indirectly by chemically altering the composition of the atmosphere. Wang et al. (1991) noted that global climate models had either neglected trace gases altogether in model simulations or did not study the differences in climate responses between trace gases and CO_2. Wang et al. (1991) recognized that the behavior of CO_2 is very different from that of other trace gases, because different gases absorb at different wavelengths and have different atmospheric lifetimes.

A number of studies have since examined the definition of radiative forcing. As stated in Chapter 15 (Ramanathan et al., 1985) of the WMO (1985) global atmospheric ozone assessment, "Radiative forcing due to trace gases can be considered either in terms of the changes in the fluxes of radiative energy into and out of the entire system (i.e., surface-troposphere system) or in terms of the change in the vertical distribution of the radiative heating rates. The choice between the two quantities depends on the region of interest. Within the troposphere, the vertical mixing of sensible and latent heat by convection and large scale motions is considered to be quite rapid compared to the time scales associated with radiative adjustment. As a result, the vertical distribution of the tropospheric temperature change is largely governed by the radiative forcing of the column. Hence, as a first approximation, we can ignore details of the vertical distribution of the tropospheric radiative forcing and focus, instead, on the radiative forcing of the entire surface-troposphere system."

Using this knowledge, column radiative transfer models were developed. Column models are much less computationally intensive than global climate models (GCMs). Column models can compute the globally averaged radiative forcing in a small fraction of the time it takes to run a full GCM and at a fraction of the cost. In addition to saving both time and money, the model noise level in column models is much lower than it is for global models so the impact of relatively small perturbations like those for the current aviation fleet is much easier to detect.

Later uses of radiative forcing built upon the fact that the climate responses differed for different substances in the atmosphere. The concept of radiative forcing was originally implemented for the global climate system, but during the 1990s, its use was extended to determine regional mean radiative forcing for various seasons in order to account for the effects of short-lived gases and aerosols that occur over certain regions (Wang et al., 1992; Haywood and Ramaswamy, 2006). Wang et al. found that the use of "effective CO_2" in climate models (as often used still) as a proxy for other gases such as methane and N_2O was generally fine for determining global average surface temperature (as long as the forcing was dominated by well-mixed gases), but it is not sufficient to assess future climate changes on a regional scale. Wang et al. (1992) emphasized the need for trace gases to be included in regional calculations. Cox et al. (1995) brought attention to the fact that the cooling effects of regional anthropogenic aerosols were "offsetting a substantial fraction of the global mean response to forcing due to greenhouse gases." Cox et al. (1995) found that the hemispheric temperature response was considerably less than expected, and the regional forcing also demonstrated substantial differences between forcing and temperature response. These differences are an indication that there is a need to represent the spatial and seasonal distribution of aerosol forcing when examining climate responses more detailed than the global and annual mean (Cox et al., 1995).

The generally accepted definition of radiative forcing, as adopted initially by IPCC (1990) is the change in net irradiance (in Wm-2) at the tropopause after allowing stratospheric temperatures to readjust to radiative equilibrium, but with surface and tropospheric temperatures held fixed at the unperturbed values. Comparisons of radiative forcing from different forcing agents relied on the assumption that the climate sensitivity factor was constant, therefore a particular radiative forcing led to the same change in globally-averaged surface temperature. Recent studies have shown that the climate sensitivity parameter, , is not constant within a particular model for all climate forcings. For example, Hansen et al. (1997) found that there is sensitivity in the climate response to the altitude and latitude of the forcing. In particular, forcings that are inhomogeneously distributed, like aircraft-induced changes in ozone and the effects of contrails, can have very different (even negative) climate sensitivities (IPCC, 1999). Indirect effects due to unevenly distributed aerosols also may have different climate sensitivities.

Radiative forcing is a particularly attractive concept for well-mixed gases because it can be calculated either within a comprehensive climate or Earth system model, or it can be calculated, almost as accurately in a simple column radiative transfer model (RTM) (or more accurately since the column model can use a higher wavelength resolution form of solution). Though column models do not have grid-to-grid interactions, they are less noisy than full climate models so it is easier to pick the small aircraft signal out of the model noise. Column models are also much cheaper and much faster than larger, more comprehensive models.

2e. Existing Metrics

There are basically two families of science-based metrics that are currently being used in studies of and policy considerations relative to climate change. The first, referred to as concentration-based metrics do not directly account for emissions, but instead are based on the forcing or temperature change over a given time period. The other family of climate metrics are emissions- based, either assuming pulse, sustained or an emissions scenario over time. The following discussion is aimed at examining the advantages and limitations for each of the major metrics currently used. Other less used metrics are also discussed, along with the limitations that have kept them from being widely used and/or accepted. Some early climate metrics (e.g., Rogers and Stephens, 1988; Fisher et al., 1990) aimed at comparing chlorofluorocarbons and other halogenated gases are not discussed here.

Concentration-Based Climate Change Metrics

The concentration-based metrics are largely different "flavors" of the radiative forcing concept and its application. Some new approaches (e.g., fixed surface temperature forcing; use of efficacies) may improve upon the traditional definition but have not yet gained wide acceptance and also appear at this point to have their own limitations.

Instantaneous Radiative Forcing

Instantaneous radiative forcing at the top of the atmosphere and/or at the tropopause is the most straightforward form of radiative forcing to derive because it involves the least amount of effort and does not account for feedbacks within the climate system. However, it was recognized early on that when forcings occur in the stratosphere, the temperature responds rapidly locally in order to restore the radiative balance in the stratosphere (IPCC, 1990; Hansen et al., 1997). This change in stratospheric climate in turn affects the tropospheric temperature. As a result, stratospheric adjustment has been adopted universally in the calculation of radiative forcing. IPCC has adopted the stratospheric-adjusted radiative forcing as the preferred climate metric. While instantaneous radiative forcing is often reported (e.g., in some cases in the IPCC, 1999, assessment of aviation), it is not generally used in assessing the potential impacts on climate.

Stratospheric Adjusted Radiative Forcing

The most widely used metric as a proxy for climate change has been globally-averaged annual mean stratospheric adjusted radiative forcing (RF) at the tropopause (which is the same as the RF at the top of the atmosphere after stratospheric adjustment.) For this metric, as discussed in the previous section, globally-averaged annual mean surface temperature is assumed to be equal to the RF multiplied by a climate sensitivity factor. This method works well for well-mixed greenhouse gases, solar irradiance, surface albedo, and homogeneously distributed non-absorbing aerosols (IPCC, 2001). However, the linear relationship between

RF at the tropopause and global mean surface temperature may not hold for forcing agents that have a strong response near the surface but very little response at the top of the atmosphere. This relationship also breaks down if the forcing agent is not homogeneously distributed. The classical definition of RF also applies best for global-mean climate response and does not account for regional climate change. In addition to the RF, we must also consider the efficiency of a particular forcing agent in causing climate change. This "efficacy" is not considered in current RF calculations using the traditional definition of RF. The effects of including efficacies in a revised definition of RF are provided later.

As a key example of the application of RF to aviation, an update of the IPCC (1999) globally averaged annual mean RF from aviation for the "current" time period (relative to no aircraft) has been presented by Sausen et al. (2005). Specifically, the forcing from CO_2 was calculated from the cumulative change in concentration of CO_2 from historical operation of the aircraft fleet. The other forcings were calculated from the steady state change in concentrations of O_3, CH_4, and H_2O to the 1992 emissions. The forcing from sulphate, soot, contrails and contrail-cirrus also correspond to steady responses. Figure 2 summarizes their results as well as the findings from IPCC (1999). In view of the large error bars of IPCC (1999), the RF from CO_2, H_2O and direct effect of sulfate aerosols have not changed significantly, apart from the increase in air traffic from 1992 to 2000.

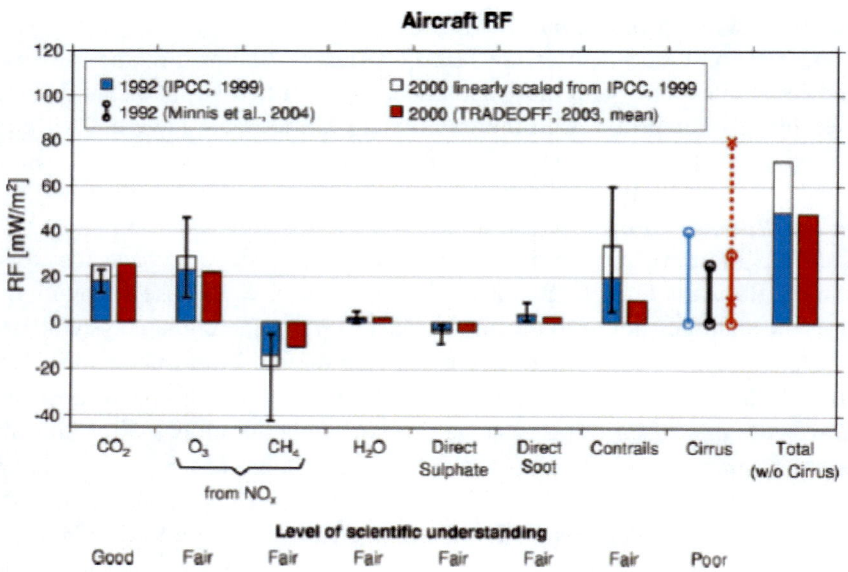

Figure 2. Global radiative forcing (RF) [mW/m2] from aviation estimated for the years 1992 and 2000, based on IPCC (1999) and the European Union's TRADEOFF program results. The whiskers denote the 2/3 confidence intervals of the IPCC (1999) values. The lines with the circles at the end display different estimates for the possible range of RF from aviation induced cirrus clouds. In addition the dashed line with the crosses at the end denotes an estimate of the range for RF from aviation-induced cirrus. The total does not include the contribution from cirrus clouds (Sausen et al., 2005).

The O_3 and CH_4 effects are changed due to more recent analyses from European chemical-transport models. The other major change is found for the direct global RF from (linear) contrails; the new value is roughly a factor of 3 smaller than IPCC (1999) based on results from Marquart et al. (2003) and Myhre and Stordal (2001), which were scaled (by fuel burn) to the year 2000 resulting in 6 mW/m2 and 15 mW/m2, respectively. As indicated in the bottom part of figure 2, the overall conclusion from these analyses is that significant uncertainties still remain in quantifying the impacts of aviation emissions on climate. Except for carbon dioxide, the understanding of the climate effects from other aviation emissions range from fair to poor. Note that the RF for direct soot in figure 2 are based on the atmospheric soot concentrations, and does not include the soot incorporated into clouds or long-term deposition to the ground.

Below is a list of strengths and weaknesses associated with the globally averaged annual mean RF calculations:

Strengths

Widely used in many climate assessments, including aviation studies (e.g., IPCC, 1999; Sausen et al., 2005).

Forms the basis for evaluation of the emissions-based metric Global Warming Potentials, which is widely used in climate policy considerations, particular for emissions trading between different transportation and energy systems.

Global mean surface temperature change is linearly related to the top of the atmosphere RF for many forcing agents, especially well-mixed greenhouse gases (Boer and Yu, 2003a; Hansen et al., 1997; IPCC, 1995; Joshi et al., 2003; Rotstayn and Penner, 2001).

Easy to search parameter space.

Fast and inexpensive to run using a radiative transfer model (RTM), so a number of detailed studies can be done and many factors can be considered.

Much less concern about climate variability and model noise in RTMs than the complex global climate models, so smaller forcings can be considered.

Easy to compare effects of different forcing agents, assuming the climate sensitivity is the same.

Relatively easy to compare different models.

Benchmarks relative to highly accurate line-by-line RF values exist for many gases.

Observation-based estimates of radiative balance provide constraints to the RF values.

Limitations

Does not account for the lifetime expected for the forcing agent or the temporal response after the perturbation is initiated. Generally based on a "snapshot" atmospheric perturbation over a given time period.

Difficult to determine RF from indirect changes using simple models.

Difficult to interpret relative RFs for direct and indirect effects from gases and particles having short atmospheric lifetimes and inhomogeneous distributions.

No hydrological response information is included.

Light-absorbing aerosols are not fully treated (indirect aerosol effect and semi-direct effect).

Does not characterize the regional responses.

Non-linear response from large perturbations or perturbations that are not well mixed may not be accurate.

RF comparisons depend on climate sensitivity, which is not well understood.

Models show that climate sensitivity is not the same for aerosols and ozone as it is for CO_2 (Cook and Highwood, 2004; Hansen et al., 1997; Hansen et al., 2005).
Models show that changes in ozone in th e upper troposphere and lower stratosphere don't have the same climate sensitivity and that they are also different from the climate sensitivity for CO_2 (Joshi et al., 2003; Stuber et al., 2001).
Does not consider dynamic feedback.
Does not characterize non-RFs on climate (e.g., land use changes).
Assumption of a constant, linear relationship between RF at the top of the atmosphere and global mean surface temperature.
Requires a tropopause height.
RF is sensitive to the choice of tropopause height (Forster et al., 1997; Myhre and Stordal, 1997; Freckleton et al., 1998).

Radiative Forcing Index (RFI)

The Radiative Forcing Index (RFI) was introduced in IPCC (1999) -- it is defined as the ratio of total RF to that from CO_2 emissions alone. In FRI, total RF induced by aircraft is the sum of all forcings, including direct emissions (e.g., CO_2, soot) and indirect atmospheric responses (e.g., CH_4, O_3, sulfate, contrails). RFI is intended to be a measure of the importance of aircraft-induced climate change other than that from the release of fossil carbon alone. However, it does not take into account the relative time scales of the climate effects or the atmospheric lifetimes of the direct and indirect effects on climate resulting from emissions of the gases and particles (Forster et al., 2006). Because of this, the simple sum of individual forcings used in deriving the total RF can lead to misinterpretation in policy considerations using the single value of the RFI as the basis for policy.

RFI as a climate metric has undergone much criticism since it was proposed. One major concern is that RFI is actually not an intrinsically fixed number (Wit et al., 2005). It is entirely dependent upon either the actual history of the emission or the assumed future scenario, or alternatively, background concentration of CO_2. Wit et al. (2005) and Lee and Wit (2006) show that the RFI will decrease over time even though the aviation emissions were held constant from year 2000 onwards. This is because CO_2 would assume a more and more important role as the time growing due to its long lifetime.

Global-Mean Radiative Forcing at the Surface

For forcing agents that change the vertical distribution of heat in the atmosphere, the RF at the tropopause may not be directly related to surface temperature change. One example of this is forcing due to absorbing aerosols, which have a large impact on RF near the surface but very little effect on the tropopause-level RF. Global-mean RF can also be calculated at the surface. Ramaswamy et al. (2001) and Menon et al. (2002a) suggest that this may be a more appropriate metric. If the RF at the tropopause and the surface are compared then we have an idea of how the lapse rate has changed and we may be able to account for some indirect changes like cloud response, precipitation and vertical mixing changes. This approach still does not account for regional climate change, nor does it consider the lifetime of forcing agents. This approach also does not account for dynamic and thermodynamic feedback, but by comparing the tropopause and surface RF values, we may get a sense of how strongly the dynamic and thermodynamic feedbacks could influence climate change. This may lead to an estimate of how much confidence we have in the resulting RF and whether we need to go to a more inclusive climate change, like a full GCM output. Sokolov (2006) suggests calculating a surface climate sensitivity and an atmospheric climate sensitivity, then using these values to modify the stratospheric adjusted RF.

Some of the strengths and limitations of the global mean RF at the surface are: Strengths

Gives surface energy budget information.

By comparing surface RF with tropopause RF, we may get an idea of how strongly dynamic and thermodynamic feedback will influence climate change.

Accounts for forcing agents that strongly influence the surface temperature, but minimally affect the RF at the tropopause.

Easy and fast.

Limitations

Has most of the same limitations as the traditional stratospheric adjusted RF definition.

No dynamic or thermodynamic feedback

Surface RF values have not been tested adequately in climate models to determine the climate sensitivity, or even if the surface RF can be directly related to surface temperature change

Fixed Sea Surface Temperature Forcing / Fixed Surface Temperature Forcing

Hansen et al. (2002) developed the concept of fixed sea surface temperature (SST) forcing. This metric measures the RF at the top of the atmosphere as computed in a global climate model by holding the sea surface temperature (SST) constant and allowing tropospheric and stratospheric temperatures to reach a new equilibrium. This method has many of the same limitations as the stratospheric adjusted RF metric, but allows the inclusion of the direct and semi-direct aerosol effects within a GCM. This method still does not quantify the regional climate impacts, but it seems to have a more constant climate sensitivity parameter than stratospheric RF (Hansen et al., 2005). Because it depends on the use of a complete climate model, it is much more computationally intensive than the use of a RTM to calculate the traditional RF.

Shine et al. (2003) extended this idea by setting both the land and ocean temperatures constant and allowing the atmosphere to adjust. Their new forcing is called the "(global-mean) adjusted troposphere and stratosphere forcing". The Reading Intermediate GCM (IGCM) is used to illustrate the performance of this forcing. The calculations presented are based mainly on model integrations from a study of the semi-direct aerosol forcing by Cook and Highwood (2004) which used 2 m mixed layer ocean to speed the approach to equilibrium. Two additional calculations examining the impact of ozone changes are presented in Joshi et al. (2003), using a 25 m mixed layer ocean. The results presented were rescaled so the two sets of results have the same climate sensitivity parameter for increases in carbon dioxide concentration. RF is calculated using a 5-year integration of the model with spatially varying sea and land surface temperatures taken from a monthly mean, annually-repeating observed climatology. The global- mean equilibrium surface temperature response is calculated from the temperature change using the mixed-layer ocean after 30 years. Shine et al. (2003) shows an intercomparison of RF results and "fixed sea surface temperature forcing" (Hansen et al., 2002) for several forcing agents, as well as "stratospheric adjusted

RF". The results show that the new forcing is a good predictor of the IGCM's surface temperature change for all of the forcing agents considered.

Hansen et al. (2005) further tested these metrics and determined that the fixed surface temperature metric yields a climate sensitivity factor that is closer to 1.0 than stratospheric adjusted RF for aircraft-related scenarios, such as: stratospheric water vapor, tropospheric and stratospheric ozone, and indirect aerosol effects. The "fixed sea surface temperature" and "fixed surface temperature" metrics require the use of a GCM. As discussed earlier, GCMs typically cannot differentiate the aircraft forcing signature from model noise (Hansen et al., 2005 tested 10 times present day contrail coverage). The results from aircraft studies still need to be tested further. One way to do this is to scale the aircraft forcing effect so that it is larger than model noise, but then the question is whether such studies would distort the actual effect of aviation on climate. Studies need to be done to determine if these scaled forcings still lie within the linear forcing-response regime.

Some of the strengths and limitations of the Hansen et al (2002) and Shine et al. (2003) approaches are:

Strengths

Although this metric does require the use of a GCM, relatively short integrations are needed because the sea surface temperature is not allowed to vary. Nonetheless, this metric is much more computationally intensive than RTM-based metric calculations.

Existing studies suggest these metrics are more accurate than other RF approaches.

Includes the direct and semi-direct aerosol effects.

RF can be calculated at any altitude.

Fast atmospheric feedback is used to simulate climate change.

Allows some dynamic and thermodynamic feedback as the atmosphere "relaxes" to a new equilibrium.

Does not require the tropopause height to be explicitly declared. Limitations

Computationally more intensive than RTM-based metric calculations.

Requires the use of a GCM, and thus is subject to uncertainties inherent in climate models, e.g., treatment of clouds.

Use of a GCM makes it difficult to determine the aviation signature on climate relative to the model noise.

Still subject to most of the limitations of the stratospheric adjusted RF approach.

Much more difficult to compare between models.

Does not consider non-radiative forcings.

Does not fully account for lifetime of forcing agents because the results are still steady-state.

Climate sensitivity parameter is not constant, though it is less variable than the climate sensitivity parameter for stratospheric adjusted RF.

Not simple or fast.

Time-Varying Radiative Forcing

Time–varying radiative forcing or radiative forcing time series has been used for natural forcing like solar flux variations for some time. Time-varying radiative forcing could be either a concentration-based or an emissions-based metric. As a concentration-based metric, it could be derived for a given scenario of changing concentrations and other forcing agents over time. As an emissions-based metric, it could be based on a pulse of emissions, sustained emissions, or a scenario of emissions over a given time period.

Although it is much more difficult to determine time-varying RF for ozone and aerosols because of the necessity to account for the past emissions, transport, chemistry and other processes affecting the concentration of constituents, there have been several attempts at this. For example, IPCC (2001), Myhre et al. (2001), and Hansen et al. (2002) provide time histories for RF. Time- varying RF has also been applied to aviation, for example, in IPCC (1999) and more recently at a presentation by MIT's Ian Waitz at the AIAA/AAAF Aircraft Noise and Emissions Reduction Symposium.

As applied by Waitz, this metric would calculate RF due to aircraft emissions as the emissions are emitted. RF is calculated for a time period, X, based on the emissions during that time period. The RF is then calculated at time X+dX using the emissions in time dX plus the emissions remaining in the atmosphere that were emitted at time X. This process would continue to yield a time-varying RF based on the time-varying emissions and the removal rate of previously emitted constituents. This approach has not been applied to specific scenarios for aviation emissions at this point. Essentially, this approach involves derivation of a time-dependent snapshot of RF that depends on the given assumptions of emissions.

In order to do this correctly, the adjustment time of the ocean-atmosphere system needs to be taken into account. The RF that will determine temperature for any given time would be a weighted average of the RFs during the previous years. It is not clear that this time-varying RF metric would yield different results than the stratospheric adjusted RF calculations using steady- state species concentrations, but it does have the benefit of explicitly considering short-lived species.

Some of the strengths and limitations of the time-varying RF approach are:

Strengths

Easy to understand concept, but not necessarily easy to calculate.

RF can be calculated at any time.

Lifetime of the species can be explicitly considered in the calculations. As such, it could be considered to be an emissions-based metric. However, applications to this point have basically used observed changes in the forcing agents. The Waitz approach, if applied, would be an emissions-based metric.

Limitations

Depending on how derived (RTM vs. climate model), it still subject to many of the limitations of the previously discussed RF approaches.

As applied using observed changes in forcing agents, this metric really has not caught on and remains little used.

Indirect effects require special consideration before can be considered.

More computationally intensive than stratospheric adjusted RF calculation using a column model.

No dynamic or thermodynamic feedback.

Computationally more intensive than stratospheric adjusted RF.

If column model RFs are used then this method still requires a declared tropopause height.

Much more difficult to compare between models.

Does not consider non-radiative forcings.
Climate sensitivity parameter is unclear. Climate model studies would have to be done to determine how the RF calculated in this way are related to surface temperature change.

Equivalent (or Efficacy-Corrected) Radiative Forcing

Of all the problems associated with RF (in all its flavors), the most serious limitation may come from the fact that not all forcing agents cause the same climate impact (for the discussion here, change in globally averaged surface temperature) for a given change in radiative flux. This means that RF from one cause cannot be compared to RF from another cause easily. One way to get around this problem is to define an "equivalent" RF where the forcing is weighted by its climate sensitivity. This additional multiplier term is called "efficacy".

The equivalent RF metric appears to be becoming the new standard as a concentration-based metric for climate change. The equivalent RF is defined as the efficacy (climate sensitivity of the particular forcing agent divided by the climate sensitivity of CO_2) multiplied by the RF. The stratospheric adjusted RF is the most logical RF parameter to use because it does not require a GCM to calculate it.

Since aircraft forcing signals get lost in GCM noise, a metric that does not require the continual use of a GCM is highly desirable. As a result, for analyses of the effects of changes in aviation effects on the atmosphere over a given time, when a concentration-based approach is useful, the equivalent RF metric is likely the best choice.

However, while this approach is certainly a significant improvement over the standard RF definitions, it still has a major problem, namely the accurate determination of the efficacy factors. Determining the climate sensitivity to various forcing agents is the hard part and requires the use of a GCM. As the spatial distribution of emissions change over time or the background atmosphere changes, there is also the question of whether the efficacy has to calculated all over again. So far, the literature has not really addressed this question. For aviation, there remains the problem of signal to noise ratio, adding further to the potential uncertainties associated with using efficacies. All we can really say at this point is the use of efficacies are likely to be more meaningful than the traditional RF approaches.

Appendix A provides a discussion of currently available evaluations of efficacy factors. Existing efficacies, in general, have limited usefulness for application to aviation even though some scientists are adapting results from Hansen et al. (2005) for that purpose. The problem is that either of the efficacies have been based on the idealized changes in the distribution of a constituent or that they have been based on only a single model that may or may not have wide spread applicability. Some of the strengths and limitations of equivalent RF approach are:

Strengths

Easy to understand concept.
If efficacy factors can be accurately determined, then it is easy to calculate.
Indirect effects can be considered through efficacy values, but not explicitly.
Equivalence is determined in a way that is widely accepted.

Limitations

Lifetime of forcing agents is not directly considered. Perhaps an efficiency factor could be used to scale a response depending on its lifetime, but at this point there has been no attempt to do so.

Most of the limitations of stratospheric adjusted RF also apply to equivalent RF

Requires a spatially-varying tropopause height location.

Emissions-Based Climate Change Metrics

These metrics all begin with emissions as their starting point. Many policy analyses are aimed at controlling emissions or examining tradeoffs relative to emissions – as a result, those types of analyses require emissions-based metrics.

Time-Dependent Radiative Forcing

When applied in terms of the emissions instead of just observed or modeled concentration changes, the time-dependent RF metric can be an emissions-based metric. The analysis can assume either a pulse, sustained, or a time-dependent scenario of emissions.

Time-dependent RF can account for the atmospheric lifetime of the emissions and can evaluate indirect effects as well as the direct effects of the emissions being considered. As with some of the other metrics, because of nonlinearities in atmospheric chemical and climate processes, RF can also depends on the initial conditions assumed and on the history of all emissions. Like other metrics, this metric is strongly dependent on the model of chemical and physical processes used for analyzing short-lived gases, particles, contrails and cirrus. It is also less simple and less transparent than other metrics. Efficacies can be used with metric (as they can with any metric using RF) towards creating an improved equivalence across different types of emissions.

Stevenson et al. (2004) uses pulse emissions and resulting RF to examine the effects of aviation NOx emissions on ozone and methane. With this approach, they are able to clearly show the effects of atmospheric lifetimes on the resulting RF with time. In general however, time-dependent RF is not commonly used. One of the difficulties with it as a metric is how to interpret time-dependent RF relative to the time-dependence of the resulting climate response. As pointed out by Shindell et al. (2005), the resulting climate effects of using emissions rather than concentration perturbations are quite different.

Global Warming Potentials (GWPs)

The concept of GWPs as generally used was developed for the first IPCC assessment (IPCC, 1990) by Wuebbles, Rodhe and Derwent (growing out of previous development of the Ozone Depletion Potential concept and alternative concepts for GWP-like metrics proposed by Lashof and Ahuja (1990), Rodhe (1990), Wuebbles (1989), and others). This concept has been extensively utilized, discussed, and criticized ever since (e.g., see discussions in other IPCC assessments). Despite all of the criticisms of its limitations (e.g., Wuebbles, 1995; Wuebbles et al., 1995; Smith and Wigley, 2000a, b; Fuglestvedt et al., 2000; Godal and Fuglestvedt, 2002), it remains the most popular emissions-based metric and it is likely that it will be used into the foreseeable future. GWPs have been adopted as an instrument for the Kyoto Protocol of the United Nations Framework Convention on Climate Change (UNFCCC). Lashof and Ahuja (1990) developed a similar, but somewhat different concept that uses steady-state calculations (which unfortunately do not apply readily to CO_2 because of its complex decay function).

Global Warming Potentials (GWPs) provide a means of quantifying relative potential integrated forcing on climate from emissions of various greenhouse gases. In the international assessments, GWPs have been defined as the time-integrated RF from the instantaneous release of a unit mass of a gas expressed relative to that of the same mass of the reference gas, generally taken as carbon dioxide, the gas of most current concern to climate change. Thus, the concept of GWPs is an index to estimate the relative impact of emission of a fixed amount of one greenhouse gas compared to another for the globally averaged RF over a specified time scale. GWPs provide a better measure of the relative greenhouse impacts than RF alone as they help differentiate between gases that would reside in the atmosphere for vastly different amount of time, from days to, in some case, many centuries. The GWP concept is based on the science of greenhouse gas effects, but does not include climatic or biospheric feedbacks nor consider resulting impacts on the environment. GWPs have generally been applied to gases that are well mixed in the atmosphere, but they can be applied to short-lived gas emissions as well. Although it has not been done at this time, efficacies could be applied in the radiative forcing values used. GWPs are calculated from the RF as follows:

$$GWP(H)_i = \frac{\int_0^H RF_i(t)c_i dt}{\int_0^H RF_{CO2}(t)c_{CO2} dt} = \frac{AGWP_i}{AGWP_{CO2}},$$

where H is the time horizon over which a forcing is integrated, RF is the RF for a particular forcing agent (i) or CO_2, and c is the remaining abundance of a particular forcing agent (i) or CO_2 after a time-decaying pulse emission. AGWP (discussed as a separate metric below) is the Absolute Global Warming Potential for a particular forcing agent (i) or CO_2. The climate sensitivity is assumed to be equal for both the numerator and denominator and therefore cancels out. (This assumption can easily be modified to account for different climate sensitivities of different forcings, but the traditional GWP definition assumes the same sensitivity factor.) Uncertainties in GWPs depend on uncertainties in RF per unit molecule and the lifetime of a particular forcing agent. Efficacies can be also incorporated as a multiplier on the RF – this modified approach is likely better for emissions (e.g., aviation) that are short-lived enough so as to not result in well-mixed forcings on climate.

GWPs allow the direct comparison of integrated forcing for any forcing agent and the forcing due to CO_2. The basis for this is that CO_2 is the greenhouse gas of primary concern to climate change. While GWPs are relatively simple to derive for long-lived well-mixed gases, they are more difficult to derive for short-lived gases with indirect effects, e.g., like NOx emissions on ozone and methane. GWPs have a high degree of transparency in the methodology compared to other emissions-based metrics, which allows other scientists to easily verify calculations and policy makers to easily compare different forcing agents.

Unlike Ozone Depletion Potentials (ODPs), the metric used in the Montreal Protocol and other stratospheric ozone policy that can be calculated to steady-state it is not possible to integrate the AGWP for CO_2 to steady-state. Because of the complexity of the carbon cycle, the decay of atmospheric carbon dioxide is a complex function that generally is represented as the sum of a series of exponential removal terms. For this reason, GWPs are usually determined for select integration times. However, these integration times are arbitrary.

IPCC assessments have adopted multiple time horizons for the integration, generally 20, 100, and 500 years, reflecting that specific questions being addressed might need to consider different time horizons (e.g., what has the largest impact in the near term? in the long term?). Of these time horizons, the most discussed in policy considerations has been a time horizon of 100 years.

For example, the U.S. EPA has adopted the 100-year time horizon in its uses of GWPs for emissions trading. Policymakers tend to prefer having one value of a metric per forcing, not the range of values for different integration periods.

O'Neill (2000) uses a short time horizon and keeps track of the impact of current and future emissions on future RF and assigns responsibility for that forcing to a particular species. This method accounts for different lifetimes of different species, but it is computationally much more intensive. Smith and Wigley (2000a) found that GWPs used for short-time horizons were reasonably accurate, but accuracy declined as time horizon increased. Smith and Wigley (2000b) determined that the impulse-response function did not accurately capture the relationship between emissions and climate response due to RF (perhaps correctable by the use of efficacies).

Manne and Richels (2001) criticize the use of 100-year GWPs because it is not a time variant metric and therefore cannot account for fixed targets, like a given temperatures or amount of damages. However, time-dependent GWPs without a fixed time horizon would satisfy the objectives they present. The GTP concept would also satisfy their analyses (Shine et al., 2007).

Like the ODP concept for gases affecting ozone, the original GWP concept developed for IPCC was primarily aimed at comparing the relative potential effects of different gases. The GWP metric represents the accumulated RF over a certain period of time and was never intended to represent equivalent climate impacts and is not a very useful tool for evaluating future climate development.

For aviation, IPCC (1999) suggests that the flaws in the basic definition of GWPs may make it questionable to use them in addressing aviation emissions. For example, the formation of contrails is not only dependent on emissions of water vapor but also on atmospheric conditions being suitable for ice formation. IPCC (1999) also based their statement on the NOx effect on ozone not only depending on the amount of NOx emitted but also when and where it is emitted. It is possible that including efficacies into the RF analyses may be able to correct for this problem for a given fleet and assumed operations.

Although they are traditionally based on pulse emissions, GWPs can also be defined in terms of sustained emissions (e.g., Harvey, 1993; Shine et al., 2005b; Berntsen et al., 2005). Berntsen et al. (2005) also allow for the climate sensitivity factor to depend on the type of perturbation thus allowing for the use of efficacies. For surface NOx emissions, Shine et al. (2005b) find little difference in the resulting GWPs, but Berntsen et al. (2005) find a significant effect when efficacies are included.

Some of the important strengths and limitations of the GWP approach are:

Strengths

Easy to understand concept and easy to calculate.

Successful at transforming various gases to a common unit (CO_2 equivalent).

Performs a time integration of the RF to project climate change to some future time.

Can possibly be modified to include equivalent forcing using efficacies.

Widely used in existing policy.

Limitations

Only considers effects for which RFs are calculated.

Does not evaluate the temperature change or the time evolution of temperature change.

Not clear what time integration of radiative forcing means.

Comparison of short-lived or inhomogeneous forcings is difficult (like all existing metrics).

All of the limitations inherent in RF are also limitations for GWPs except that atmospheric lifetime is fully accounted for.

Characterization of the impact of a gas is not robust with respect to the climate impact. For example, difficult to account for contrail formation using GWP approach.

Primarily because of rapid improvements in the understanding of the carbon cycle, GWP values have changed essentially each IPCC assessment, leading to criticism from users who want stable metrics.

Difficult to know what an appropriate time horizon should be, although the 100-year horizon has become the standard.

Not applicable in traditional configuration (fixed integration period integration) for fixed target policy analyses.

Absolute Global Warming Potentials (AGWPs)

Absolute GWPs (AGWPs) as defined under the GWPs section (the numerator and denominator terms in GWPs) can have advantages for certain applications because they are not dependent on comparisons with CO_2. Comparison with CO_2 may not always be desired, e.g., comparisons of NOx emissions effects from aviation relative to NOx emissions from ground-based transportation systems.

AGWPs may have more associated uncertainties than GWPs because it is generally assumed that GWPs cancel out uncertainties about the climate sensitivity between the numerator and denominator. AGWPs have been determined for various greenhouse gases, but this metric is not commonly used.

Global Temperature Potentials (GTPs)

Global Temperature Potentials (GTPs) was proposed by Shine et al. (2005a) as an alternative to the GWP climate metric. Similar integrated temperature approaches had previously been proposed (e.g., Rotmans and Elzen, 1992) but did not gain wide acceptance.

GTP gives the global temperature change as a function of time rather than that integrated over a certain time. GTP starts out in much the same way as RF, but instead of assuming a steady-state solution, GTP looks at the time evolution of the solution. Following Shine et al., GTP can be defined either for pulse (GTP_P) or for sustained (GTP_S) emissions. GTPs may also be applicable to emission scenarios but have not been evaluated.

GTP assumes that the global mean surface temperature is given by:

$$C\frac{d\Delta T(t)}{dt} = \Delta F(t) - \frac{\Delta T(t)}{\lambda},$$

which has the general solution:

$$\Delta T(t) = \frac{1}{C}\int_0^t \Delta F(t')\exp\left(\frac{t'-t}{\lambda C}\right)dt',$$

where the exponential is an impulse response function to a forcing at some initial time t', t is some time in the future, T is the change in temperature as a function of time, F is the change in RF, C is the heat capacity of the mixed-layer ocean and is the (assumed) climate sensitivity. Thermal inertia is represented by an ocean mixed-layer heat capacity, so the climate system has a single time constant, rather than a slow time constant (ocean) and a fast time constant (land). The concentration change over time, given a known time-independent increase (or decrease) in concentration (S) of forcing agent, is given by:

$$\Delta X(t) = \alpha \Delta S\left[1 - \exp\left(-\frac{t}{\alpha}\right)\right].$$

Assuming the forcing (F) is given by X(t), AGTPs (absolute GTP for a sustained emission change) at a particular time for a forcing x is given by:

$$AGTP_S^x(t) = \frac{\alpha_x A_x}{C}\left\{\tau\left[1 - \exp\left(-\frac{t}{\tau}\right)\right] - \frac{1}{\tau^{-1} - \alpha^{-1}}\left[\exp\left(-\frac{t}{\alpha_x}\right) - \exp\left(-\frac{t}{\tau}\right)\right]\right\} \text{ for } \tau \neq \alpha_x,$$

where is the time constant for removal of the gas x, A is the RF for a 1 kg change in concentration of gas x, C is the heat capacity of the mixed-layer ocean, and is the time constant (C) for the climate system. The AGTPs for CO_2 is more complicated because it has a more complex response function. Finally, time changing GTP for a forcing agent, x, is the ratio of AGTP for x divided by AGTP for CO_2 and given by:

$$GTP_S^x(t) = \frac{AGTP_S^x(t)}{AGTP_S^{CO2}(t)}.$$

Like GWPs, GTP is a relative change as compared to a known forcing due to CO_2. GTP moves one more step down the chain of events from forcing to temperature change caused by the forcing. AGWPs give the integral of a decaying pulse, while AGTPs give an exponential approach to an asymptotic temperature change due to either a decaying pulse or a sustained emission. GTP could be considered to be better than GWP because it calculates a temperature change over time, which is a clearer physical meaning. However, Shine at al. (2005a) found that the pulse emission effects compared poorly with an energy balance model and therefore may not be the metric of choice (more analysis needed however). The sustained emissions approach gives much better results, but then one has to assume sustained emissions. GTP still requires a climate sensitivity parameter, but this climate sensitivity is in the numerator and denominator so the effect of unknown sensitivity cancels out assuming the sensitivity is the same for the perturbation and reference forcing agent. (This assumption has come into question in recent studies, so GTP has the same problem in its traditional conception as GWP and RF.). One major benefit of GTP is that it can be used for short-lived gases because it better accounts for variations is forcing strength and lifetime of the gas.

Major strengths and limitations of the GTP approach include:

Strengths

Relatively simple and transparent.

Requires few input variables.

Allows calculation of time-dependent change in temperature (not RF), which GWP does not.

Limitations

May be limited to sustained emissions applications, but more studies of pulse emission effects are needed.

Depends on the numerical value of climate sensitivity, which is not well known.

No clear choice for how to define equivalence (could inclusion of efficacies help this?).

Like GWPs and other emissions-based metrics, difficult to include non-emission related effects, like those occurring with the formation of contrails.

Global Temperature Index (GTI)

Akin to RFI but using pulse-based GTPs as the basis, this index was proposed by Lee and Wit (2006) as perhaps being a better approach for trading schemes. However, this index is totally untested and requires much more evaluation.

Linearized Temperature Response (LTR)

Using carbon cycle and climate models, linearized response functions have been developed in various research studies (e.g., Hasselmann et al., 1993, 1997; Hooss et al., 2001; Joos et al., 2001) as a way of deriving CO_2 from emissions and temperature changes without using a full climate model in further studies, mostly for examining effects of projections of future CO_2 emissions.

Studies to determine these response functions have typically included a year of emissions of CO_2 treated as a pulse emission. In the past, such studies typically have not included emissions of short-lived emissions.

Sausen and Schumann (2000) use a combination of linearized response models in analyses of the effects of carbon dioxide and ozone (from NO_x) emissions from current aircraft on surface temperature and on sea level. For the carbon cycle, they use linearized functions determined from the analyses of Hasselmann et al. (1997). RF is then derived using simple expressions from the literature (a logarithm function for CO_2). Finally, temperature change is derived using the response functions from Hasselmann et al. (1993, 1997) (with a climate sensitivity factor based on studies by Ponater and colleagues). The study by Sausen and Schumann (2000) found that, even though the RFs from CO_2 and from NOx were comparable, the aircraft-induced ozone increase causes a larger temperature change than the CO_2 forcing. Although regional climate effects are not considered, they note that regional effects may be larger than the global mean responses.

Lee and Sausen (2004) use the climate response model of Sausen and Schumann (2000) for a similar study except that they base the climate sensitivity factor on IPCC (2001). Like Sausen and Schumann (2000), they found a larger temperature response from ozone relative to CO_2 than would have been expected based on the RFs. However, they also recognize that this conclusion is highly dependent on the equilibrium response temperature function used and recommend that analyses from coupled climate (GCMs) and chemistry-transport models (CTMs) are needed to better understand the ozone temperature response.

Marais et al. (2007) and the companion report by Mahashabde et al. (2007) have adapted the concept of linearized temperature response (LTR) functions to the evaluation of the climate impacts from aviation. This APMT (Aviation environmental Portfolio Management Tool) modeling system has been developed for the U.S. Federal Aviation Administration. They likewise borrow from the approach of Sausen and Schumann (2000), but then build upon it.

Like earlier studies, the APMT model conceptualizes a year of aviation emissions as a pulse emission. They use published linearized response functions of the carbon cycle for CO_2 (Hasselmann et al., 1993, 1997; Hooss et al., 2001) and the response functions from the very simple Bern carbon cycle model (Joos et al., 2001). It should be noted that all of these response functions, including the Bern model, are all based on earlier versions of the ECHAM model, versions of this model that are generally recognized as being well out of date of the current state- of-the-art. For determining the CO_2 climate impact, they follow the approach of Hasselmann et al. (1997) and base the linearized temperature response functions on the earlier versions of the ECHAM model (Hasselmann et al., 1993, 1997; Hooss et al., 2001; Cubasch et al., 1992). They also use the simple energy balance model of Shine et al. (2005) with a fixed climate sensitivity value. Although they recognize this approach has "lower fidelity than the impulse response functions derived from the more complex (climate) models", they also recognize that the other functions were based on papers from out-of-date climate models.

The RF (normalized to RF for the doubling of CO_2 relative to the preindustrial atmosphere, as generally used in deriving the linearized temperature response functions) times the resulting concentrations using these functions are then integrated with a given linearized temperature response function to determine the change in globally averaged temperature. Uncertainties in the climate sensitivity are accounted for via a scaling of the sensitivity of the model used for the linearized temperature response function derivation through the use of a simple energy balance model.

For short-lived emissions, they scale the normalized RF for different climate responses relative to CO_2 (much like Sausen and Schumann, 2000). Except for the methane and resulting ozone effect, all effects are assumed to only last for a period no more than the one year of the emissions. Efficacies are used in this scaling (based on either a value of one or values from Hansen et al., 2005). For ozone and methane effects, the emissions index is proportional to the NOx inventory. For all other impacts, the emissions index is proportional to the fuel burn.

Another new model, mentioned in Wit et al. (2005) uses a very similar approach developed by L.L. Lim, D. Lee, and R. Sausen (unpublished except for Wit et al. and one page on the Manchester Metropolitan University website under the Centre for Air Transport and the Environment). There are some other, more minor differences in the two approaches, but not enough is known to discuss this model in detail at this time.

All of the LTR metrics discussed so far represent the climate system through global-mean surface temperature, which may be misleading for the effects resulting from emissions of NOx (e.g., due to similar responses in each hemisphere for the methane effects but different hemispheric responses in ozone) and perhaps for the resulting effects from aerosols and contrails. However, other simple metrics generally have not addressed this issue either.

A related but somewhat different approach is proposed by Grewe and Stenke (2007). Although their temperature response is based exactly on that used by Sausen and Schumann (2000), the rest of their model is very different. Their assessment tool is called AirClim. For CO_2, they assume a constant 100-year lifetime, an overly simplified representation of the complex decay function for CO_2. On the other hand, their treatment of the RF for CO_2 and the other emissions from aircraft, as well as their residence times, includes representation of altitude and regional effects not considered as fully, if at all, in other metrics. Basically, they use a coupled climate- chemistry model (based on a recent version of ECHAM), to derive factors for 4 latitude regions and for 6 pressure (altitude) levels. This paper focuses on determining the effects from an assumed fleet of supersonic aircraft but the approach used should be expandable to subsonic aircraft. At this point, the modeling approach developed by Grewe and Stenke (2007) appears promising, but largely untested. More evaluation is required. In addition the treatment of the temperature response function needs to be upgraded (based on state-of-the-art climate model or models) and the carbon cycle complexity needs to be better accounted for.

While it could be argued that the simplified LTR models are not classic metrics in the way that radiative forcing or GWPs are metrics, the ability to greatly simplify the complexity of determining climate impacts from aviation or emissions from other transportation sectors could be a very useful tool to policy analysis and, as such, are a metric. By developing parametric models based on the results from much more sophisticated climate, carbon and chemistry models, the LTR approaches discussed here represent a pathway towards a potentially powerful capability that allows for extensive analyses of aviation and other climate forcings and evaluation of uncertainties. This new approach to a metric has not been adequately tested at this time, but the approach is certainly promising. A key problem with the existing models though is that they are all largely dependent on out-of-date linearized response functions developed from older versions of carbon cycle and climate models. The one exception may be APMT, which also uses a simplified energy balance climate model (from Shine et al., 2005a). However, such simplified models are only as good as the science and more sophisticated models they are based on. Thus, the choice of such simple models needs further evaluation.

As discussed earlier, GTP, whether for pulse emissions, GTP_P, or for sustained emissions, GTP_S, is defined as the ratio of the Absolute GTP (AGTP) for X relative to the AGTP for CO_2; in this way, it follows the ration approach developed for GWPs. On the other hand, the LTR approach derives the change in temperature with time akin to the AGTP. As such, LTR and AGTP are similar except that the goal in LTR is to use the results from complex climate models as the basis for the carbon cycle and temperature derivations rather than the simpler treatments used in GTP. However, use of the simplified energy balance model in AMPT may produce results very similar to those derived for AGTP using the same energy balance model.

Strengths

Allows determinations of time dependent changes in globally-averaged temperature.

Thus, readily understood response compared to using RF.

Has a methodology for accounting for short-lived emission.

Allows some sense of uncertainties to be included, by using different derived response functions for CO_2, temperature change and efficacies.

Could be a very useful approach for addressing some technological and policy question, but may not be so useful for other questions (e.g., changing the flight altitude or a change in routing).

Limitations

Methods have not been adequately tested and evaluated at this time.

Limited by uncertainties in determined linearized temperature response functions.

Requires knowledge that requires a GCM to calculate.

Could potentially be applied to other sectors but this has not been done at this time.

Requires more input parameters and is more difficult to determine than GWPs. Requires more complex input from scientists than other metrics.

Not clear yet whether this approach really has much advantage over GWPs or GTPs.

Global Temperature Index

Wit et al. (2005) present another metric (developed by David Lee) in their report that combines the GTP concept with the use of linearized impulse response functions. This metric is called Global Temperature Index (GTI) and is supposedly analogous to using RFI. Like GTP, GTI assumes sustained emissions integrated over a certain time period (100 years). Efficacies are included. However, the overall methodology is not fully developed or tested (or even explained very well at this point). It is difficult to tell at this time just how useful this metric will be in future aviation and other sector studies.

Economics- and Damages-Based Metrics

Following figure 1, it has long been recognized that development of climate policy would benefit from analyses of welfare and damages (Eckaus, 1992; Schmalensee, 1993; Kandlikar, 1995). A number of economists and policy experts have criticized existing physical-based metrics like GWPs because they do not account for damages and abatement costs (e.g., Manne and Richels, 2001).

A number of different studies have used economic approaches to assess impacts associated with future scenarios of climate change (e.g., Mendelsohn et al., 2000; Nordhaus and Bauer, 2000; Tol et al., 2002a, b; Manne and Richels, 2001; Bradford et al., 2001; Sygna et al., 2002; O'Neill, 2003; Hammond et al., 1990; Kandlikar, 1996). Especially designed for analyses of aviation impacts on climate, Marais et al., (2007) (and the corresponding report on the AMPT system for the FAA, Mahashabde et al., 2007) assume either a linear damage function or the damage function developed by Nordhaus and Boyer (2000), which assumes a quadratic relationship with the change in temperature. They also include discounting (e.g., see Nordhaus, 1997) to express future value in terms of present monetary terms. There have also been a number of attempts to develop alternative metrics that are welfare-based. For example, Hammitt et al. (1996) proposed the Economic-Damage Index (EDI).

While there is a large body of existing studies considering damages and their assessment through various indices, there is no widely accepted approach. There is no straightforward way to aggregate spatially and temporally diverse impacts into a single damages estimate. Such an index or metric would only be useful for policy considerations if it can successfully enumerate all of the relevant potential impacts on society and the environment resulting from climate change. This holds for studies of aviation-induced climate change as well. Part of the problem is that it is difficult to determine what "successfully" means in this regard. As a result, unlike the generally accepted metrics within the science community, RF and GWPs, even with recognition of their flaws, there are no community-wide accepted approaches for damages and abatement costs being used in policy considerations.

3. Uncertainties, Limitations, Gaps, and Needed Improvements

A variety of different metrics have been discussed in the previous sections. Some are physical science-based metrics like Radiative Forcing (Stratospheric Adjusted RF has been the standard, but Equivalent RF should likely be considered to be the new standard) and GWPs that have become the currently "accepted" approaches for evaluating climate policies and legislation related to reducing emissions of multiple greenhouse gases. Others, like LTR modeling, are relatively new and untested in climate assessments. Still others attempt to incorporate the human dimension of change through estimating the relative impact of emissions on economic or social damages.

Several different designations of climate metrics have been considered, and strengths and limitations of these metrics have been discussed in the previous section. This section is aimed at further understanding of the uncertainties, limitations, and gaps in knowledge and capability of these metrics (or at least those that seem most relevant to future use). In addition, this section examines issues that need improvement before these metrics can be used to fully address policy-related questions relating to the effects of aviation on climate. The first designation of metrics is based on concentration-based analyses using some form of radiative forcing. Table 1 provides further insight into some of the key uncertainties, gaps and issues needing improvement for these metrics. The second designation of metrics is emission-based analyses. The key uncertainties, gaps and issues needing improvement for emissions-based metrics are further discussed in table 2. The third designation of metrics discussed earlier were those associated with economics or social damages, but there is no generally accepted treatment of these impacts at this time and there is no attempt here to further discuss these metrics.

The question is, what metric or metrics would be most useful for analyses of the potential climate impacts from aviation emissions? Or from other transportation and energy sectors? There is no simple answer to this question; in fact, there is no one answer.

Table 1. Uncertainties, gaps and issues needing improvement for application of selected Concentration-based Climate Change Metrics to aviation

Metric	Uncertainties	Gaps	Improvement issues
Stratospheric adjusted radiative forcing (RF)	Except for CO_2, RFs for other aviation climate impacts are not well known. Traditional definition does not account for nonlinear climate response due to location and timing of the forcing. Depends on model used in the derivation and time period evaluated.	RF has not been defined for regional emissions. Effect of atmospheric lifetime on resulting climate response is not accounted for. Unknown whether RF could be applied for regional analyses.	The basic science for determining the climate effects from non- CO_2 aviation emissions needs significant improvement. Effects of contrails and changes in cirrus are particularly uncertain.
Global-mean surface RF	The basic concept has not been tested adequately, but may provide useful info on dynamic and thermodynamic feedbacks relative to tropopause based RF. Large uncertainties about value of this approach until it is further evaluated.	Not clear at this point if it will really add to better understanding of climate effects relative to traditional tropopause based RF. Likely not applicable to regional analyses.	The value of this approach needs to be tested in climate models. This approach has not been applied to aviation.
Fixed land/ocean surface temperature RF			Need to determine how dependent values will be to different climate models.
Equivalent RF	Efficacies for aviation effects on climate are still poorly known.	This could be applied to any of the above approaches but this still needs to be done. Not clear if applicable to regional analyses.	Need systematic model intercomparison for efficacy evaluation. Test use of efficacies relative to the above RF approaches compared to climate models (for nonaviation forcing and then for aviation (bearing in mind possible scaling problems when multiplying aviation emissions to get sufficient climate signal).

Table 2. Uncertainties, gaps and issues needing improvement for application of selected Emissions-based Climate Change Metrics to aviation

Metric	Uncertainties	Gaps	Improvement issues
Time-Dependent Radiative Forcing	Value not clearly known even though it has had some application to aviation.	Interpretation of this approach relative to resulting climate impacts is not understood.	Requires much further testing. Relative usefulness of pulse, sustained, and scenario emissions needs to be evaluated. Needs to be tested using efficacies.
GWPs -- Global Warming Potentials	Although commonly used in climate studies and policy considerations, it is not known how well this metric could be applied to aviation. Difficult to know what an appropriate time horizon should be, although the 100-year horizon has become the standard. Not clear if GWPs could be applied to regional analyses.	Not clear what time integration of radiative forcing means. Characterization of the impact of a gas is not robust with respect to the climate impact. Difficult to account for contrail formation and other non-emission related effects using GWPs.	Applicability for aviation needs to be evaluated. Applicability for comparing aviation with other transportation /energy sectors needs to be tested. Testing needed using efficacies.
		Not applicable in traditional configuration (fixed integration period integration) for fixed target policy analyses.	
GTPs – Global emperature Potentials	The advantages and disadvantages of applying GTPs to pulse or sustained emissions are still poorly known. Similarly whether GTPs could be applicable to emissions scenarios. Not clear if GTPs could be applied to regional analyses	Like GWPs, difficult to include non-emission related effects, like those occurring with the formation of contrails.	Overall method needs further testing. Also, need to include efficacies. Applicability for aviation needs to be evaluated. Applicability for comparing aviation with other transportation /energy sectors needs to be tested.

Table 2. (Continued).

Metric	Uncertainties	Gaps	Improvement issues
Linearized Temperature Response	A major advantage of LTR is the ability to couple to the capabilities of global climate models, but existing linear response functions are not based on state-of-the-art GCMs. Same concerns apply to the carbon cycle applications.	Like GWPs, difficult to include non-emission related effects, like those occurring with the formation of contrails.	LTR has not been adequately tested and evaluated at this time for either aviation or other sectors. One study suggests that LTR may be applicable to regional analyses, but this needs much further evaluation.

The best metric to use for a given situation depends on the question that is being asked. In order to make some generalized recommendations, it is instructive to first look at several other studies that address at least parts of this question. We can then make recommendations regarding additional research that is required to further address this question.

First, users of climate metrics need to bear in mind that simplified climate metrics should not be used in isolation without considering more fully the literature and assessments that take into account the many complexities affecting climate change. At the same time, it is not sufficient to only use emissions as the basis for policy – it is important to go further down the chain of figure 1 towards evaluating the resulting climate impacts.

As mentioned in Forster et al. (2006), there have already been attempts to use simple multipliers (2-4, with a value of 2.5 used in some UK policy discussions) on the climate effect (radiative forcing) due to CO_2 effects from aviation by itself. The use of such a multiplier, e.g., based on RFI, has been used extensively in climate model calculations, but primarily in accounting for the effects of other long-lived greenhouse gases. While, as mentioned earlier, the total RF does have value in considering the climate effect of aviation over a given period of time, it not only does not present the whole story needing to be considered in developing policy, and bears little relationship to the metric being applied in most current policy considerations from non-aviation emissions, namely GWPs. The GWP concept not only considers the lifetime of the emissions, but also provides a time-integrated RF from a pulse emission, a very different metric than RF.

If the total sum of RF were applied to other sectors, it would lead to a very misleading interpretation of the climate effects. For example, emissions from coal burning power plants without extensive scrubbing capabilities emit a significant amount of sulfur gases that rapidly transform to sulfate aerosols in addition to their emissions of CO_2 and NO_x and some less important gases. The RF due to the cooling effect from the sulfate aerosols would counteract a large amount of the warming due to the CO_2 emissions and effects from the NO_x emissions on tropospheric ozone, and the "total" RF would suggest that coal burning power plants are beneficial to climate. Similarly, using total RF as the only metric for aviation, one might conclude that reducing the cruise altitude to prevent contrails (e.g., Williams et al., 2003) would be beneficial to climate. However, the decreased energy efficiency would lead to more CO_2 emissions and in fact, the reduced flight altitude may be more harmful to climate. If the

RF metric is to remain useful, then perhaps hemispheric or even regional "equivalent" RF could be derived using efficacy factors.

Forster et al. (2006) suggest that much more extensive evaluation of the impacts of short-lived aviation emissions be done before they are applied to any emission trading scheme. They conclude that RFI should not be used as an emissions index without giving due consideration to the timescales of the climate effects. RFI exaggerates the climate impact of aviation emissions, potentially putting too much weight on very short lived climate forcings. They also conclude that a number of other issues need to be considered in any emissions scheme used for emissions trading. First, any emissions-based weighting of non- CO_2 climate effects should be applicable to all sectors – not just aviation. Secondly, it is important to choose an index that is emissions-based. Uncertainties need to be considered in the analyses. Third, a suitable time horizon needs to be chosen, e.g., say 100 years (but to what degree is this choice arbitrary?).

Other studies have compared several different climate metrics. Shine et al. (2005b) compares several different emissions-based metrics, both RF (e.g., GWPs) and temperature based (e.g., GTPs), for surface NOx emissions and finds little difference in the results. Shine et al. (2005b) also examines two more regionally-based metrics, based on the absolute value of the local change temperature relative to the same for a reference gas, called Linear Damage Potential (LDP) and the square of the local temperature change, called the Square Damage Potential (SDP). Such regional metrics may be useful, but their limited testing done for NOx emissions in Asia versus Europe is insufficient.

Wit et al. (2005) discuss different metrics for examining emissions trading relative to aviation impacts on climate. They conclude that RF and RFI are not useful for emissions trading because they do not account for effects occurring in the future. They also criticize GWPs as not being useful for emissions trading because (1) it is difficult to account for particles or their indirect effects; (2) the O_3 effects from NOx emissions is subject to large uncertainties; (3) the GWP concept is based on a per unit mass of emissions which does not apply readily to contrails; and (4) GWPs do not account for the climate sensitivity parameter. However, there is a response to all of these issues, the GWP concept can be appropriately modified to include these, e.g., GWP analyses are already being applied to NOx effects on O_3 from surface sources and one can include efficacies to account for the effects of the climate sensitivity parameter. Most of the remaining GWP issues raised would equally apply to any existing metric. The Wit et al. (2005) analysis does not account for current adaptations to the GWP concept. The one criticism of GWPs that cannot be readily addressed is that it lacks an equivalence to a climate response at some given point in time.

Wit et al. (2005) suggest that the GTP concept eliminates some of the key concerns about GWPs. The GTP concept does indeed have a number of key advantages. However, Shine et al. (2005a) suggests that GTPs don't work very well for pulse emissions, only for sustained emissions. This may not be a serious concern for most applications – long term integrations of 100 years or more tend to give similar results with GTPs and GWPs – but the particular use of a metric needs to carefully consider whether a pulse or sustained emission is desirable.

Despite the many criticisms, GWPs at this point are still the metric of choice for climate analyses by policymakers. This is largely because they are seen as simple (a table of values are published in the international climate assessments), transparent (easily reproduced), and flexible (new knowledge can be incorporated). While each of these points could be argued (and rightly so), the controversies in the science community about GWPs are not readily perceived by policymakers.

The GWP concept cannot be ignored because it still is the most accepted metric in climate analyses. However, the GTP concept and the linearized temperature response (LTR) approach also have many advantages and may be the preferred approaches for technological and policy analyses relative to aviation. GTP has the advantage of being relatively simple, transparent, and flexible, but, in the long run, it could be argued that a well tested and evaluated version of the LTR approach will better represent changes in the scientific understanding. However, LTR is largely untested at this point and it relies on more scientific input from complex numerical climate, carbon cycle, and chemistry models. Some of the same information is needed from such models for other metrics, so this may not be a real issue.

4. Prioritization for Tackling Outstanding Issues

Further evaluation of climate metrics is required before the right choices can be made for application to aviation policy studies. In particular, the individual questions of interest – e.g., whether requiring comparison of one species of aviation emissions with another or of aviation emissions with emissions from other sources - will determine the most appropriate metric to use. Input from policymakers as to what questions they see as priorities will be important to determining where efforts should go into further development of climate metrics for aviation.

At this time, it is not at all clear which metrics will be most suitable for addressing the questions related to aviation impacts on climate, or for possible considerations of tradeoffs relating to aviation emissions and climate. Even more difficult would be to consider tradeoffs of aviation climate concerns relative to air quality or noise issues associated with aviation (the difficulty in doing such tradeoffs is discussed in the 2006 workshop report, Wuebbles et al., 2006). As a result, at this time, the suite of metrics discussed in sections 2 and 3 should be tested, evaluated and prodded in every possible way in order to get to the point over the next few years where specific recommendations can be made regarding appropriate choices for the possible sets of questions related to aviation. Each of the uncertainties and issues discussed in section 3 will need to be considered. New metrics should also be considered. Input from policymakers regarding what they actually see as the key questions for metrics to address will be an important element of this evaluation. Also, the interest of policymakers in global (entire fleet) versus regional (as little as a single flight) evaluation of aviation impacts on climate needs to be known, so that priorities can be determined for global versus regional analyses. If the gaps listed in section 3 limit the metrics applicable to a given set of policy questions, effort may need to go into development of new metrics.

This section discusses priorities for research to greatly enhance the understanding of climate metrics for aviation studies so that within a five year time period policymakers will have a much enhanced set of tools for addressing key questions related to the impacts of aviation emissions on climate. Table 3 then summarizes the discussion in this section into a series of potential projects along with a rough estimate of the required effort (in full time equivalents) required and an associated estimate of cost. Within table 3, there is also an attempt to provide a rough timeline for such studies.

In addition to assessing appropriate applications for individual metrics, the robustness of the existing metrics all need further evaluation. The most effort should likely go into testing and further developing the Equivalent Radiative Forcing, Global Warming Potentials, Global Temperature Potentials, and Linearized Temperature Response metrics. The usefulness of efficacies needs to be evaluated for all of these metrics. The Radiative Forcing and GWP metrics are already well-accepted approaches with well-known limitations, but the use of efficacies in these is relatively new and not fully tested. The GTP and LTR metrics and their various forms are not yet as accepted in the science and policy communities, but may be very useful. The capabilities of the various metrics should be further examined in comparison with each other and relative to their ability to address a range of policy questions. Such studies may also lead to the development of new metrics.

Table 3. Research priorities over next 5 years towards enhanced capabilities of climate metrics for addressing the impacts of aviation on climate

Project	Effort required
Near term (0-1 year)	Cost of meetings.
Establish Metrics Working Group (MWG) that will interact on evaluating and testing metrics for application to aviation impacts on climate. Develop criteria for evaluating aviation impacts in climate metrics.	Cost of meetings.
Meeting of Metrics Working Group with policymakers interested in aviation impacts to establish priorities for key questions to be addressed with climate metrics.	Cost of meeting.
Mid term (1-3 years) The Bakeoff: Evaluation, testing, and further development of existing metrics (different forms of RF; GWPs; GTPs; LTR metrics) first for aviation NOx emissions using chemistry-transport models and climate models (or coupled chemistry-climate models) first for ozone effect and then ozone and methane. Global models necessary for evaluating capabilities of metrics. Incorporate improved efficacies and improved understanding of science effects for various emissions as they become available. Determine capabilities for including contrails and cirrus effects in metrics. Determine needs for regional studies and test metrics relative to such needs as appropriate. Evaluate effects of background atmosphere.	MWG members: ~3-5 FTE*, roughly \$500K per year for 2 to 3 years; quarterly meetings of MWG.
Development of improved efficacies for aviation emissions, starting with NOx emissions.	MWG members: 1-2 FTE, roughly \$250 K per year for 2 years.
Development of scenarios for future growth of aviation and resulting emissions. Initial studies with metrics (after initial phases of Bakeoff).	Emissions scenario developers: 1- 2 FTE, roughly \$200K for 1 year; MWG: 1-2 FTE, roughly \$250K for 1 year.
Studies with 2-3 existing state-of-the-art climate models (e.g., NCAR, GFDL; NASA Goddard) to develop new linearized functions for temperature and carbon cycle. These will be used in future LTR studies.	2-4 FTE, roughly \$400K for 1 year.
Initial meetings (of MWG) with economics and others	Cost of meetings.

communities to determine best way forward for incorporating damages into metrics.	
Long term (3-5 years)	
After first stage of Bakeoff completed, test metrics for aviation sector relative to other transportation / energy sectors.	MWG: 2-4 FTE, roughly $400K per year for 2 years.
If determine that 2nd stage of Bakeoff is needed, then proceed with further evaluation and testing of metrics. At this point, we should know whether additional metrics are needed as well.	Not known; could be as much as 2-3 FTE, $400K per year for 2 years).
Initial studies using metrics in addressing climate tradeoffs. Update as science knowledge of climate impacts improves. Include damages if there is community agreed upon approach.	MWG: 2-3 FTE, $400K per year for 2 years.

* FTE = Full-time equivalent (assumes mixture of PhD scientists, post-docs, and graduate students).

A combination of modeling tools will be needed for assessing the different metrics, including global and regional climate models, atmospheric chemistry-transport models (either coupled or decoupled from the climate models), and radiative transfer models. *Since different scientists have different experiences with different metrics, it may be worthwhile to develop a working group that together would evaluate the different metrics and their value for addressing different policy questions.* Detailed comparison with results from state-of-the-art climate models will be a necessary part of the evaluation of metrics (as well as in the development of better treatments of efficacies). As mentioned earlier, there is a possible issue with scaling of aviation effects within climate models to be able to fully detect the climate signal; this uncertainty will need to be considered within the evaluation of the different metrics. It is important to also recognize that the evaluation of climate metrics can only be as good as our understanding of the scientific understanding of the processes affecting climate impacts from the different aviation emissions.

Efficacies will likely become a norm for most of the future studies using metrics but they have not been adequately evaluated for aviation-based emissions. The sensitivity of efficacies to the background atmosphere and to a range of possible aviation emissions scenarios need to be evaluated for each of the separate climate concerns associated with aviation (including NOx effects on ozone and methane, aerosols, contrails, cirrus). These analyses will of course have to go hand in hand with improved understanding of the emissions effects themselves.

As stated in Fuglestvedt et al. (2003), there are no unambiguously agreed upon criteria for evaluating metrics. In examining potential uses of metrics for aviation, it would be useful to have a special meeting to establish these criteria, to set the stage for the studies to be done. Feedback from those involved in aviation policy will be a necessary part of this – the lack of clear goals currently for combating climate change from aviation affects the choice of metrics and the criteria to be evaluated. The scientists involved in evaluating and developing climate metrics also need to understand what tradeoffs are likely to be most important to the considerations of the aviation policy community.

Fuglestvedt et al. (2003) do suggest that different climate and/or coupled chemistry-climate models evaluate the robustness of radiative forcing for consistency across a variety of issues, e.g., to what degree are high latitude forcings more effective at affecting climate than low latitude forcings or shortwave forcings are more effective than infrared ones. Can efficacies adequately correct for such differences?

Climate modeling and coupled chemistry-climate modeling studies will play an important role in further evaluating metrics, but these modeling tools are computationally intensive, so the tests using these models need to be carefully considered.

Both of the latest LTR approaches, namely the APMT and AirClim assessment tools, appear to be quite promising for future studies of aviation. The AirClim approach may even provide a capability for analyzing regional impacts not considered otherwise. However, these tools are dependent on the validity of much more complex representations and understanding of the science, including the carbon cycle, chemistry interactions, aerosol direct and indirect effects, contrail formation and evolution, and the resulting impacts on climate. Current tools need much further development and evaluation before they will be applicable to policy considerations. In particular, both models need to have a much more carefully-considered representation of the carbon cycle and temperature response functions in order to better represent the state-of-the-art of the science.

Any metric being considered for aviation should also be applicable to other transportation sectors to enable comparisons between sectors. At this point, the GWP concept has been applied in a limited manner to such sectors, but, there has been no attempt at applying the GTP or the LTR concepts to such sectors. Further research is needed to test these capabilities.

One of the next step needs to be testing and comparison of the Equivalent RF, GWP, GTP and LTR metrics for NOx-O_3-CH_4 effects from aviation. These effects are known better than the effects from contrails and changes in cirrus and there is a real possibility that the effects, as well as remaining uncertainties, of NOx emissions can be better quantized within the next few years. Three-dimensional steady-state modeling studies could be done of these effects, but the applications of the concepts and interpretation of the results as used in metrics will require much analysis and thought. These analyses will be crucial in determining which metric or metrics) should be the primary focus for future aviation applications. One could also attempt to do rough analyses for contrails (using an approach akin to Hansen et al., 2005) although current science understanding of the contrail and cirrus effects may make it difficult to fully include these effects at this time.

5. Recommendations for Best Use of Current Tools

It will be important to take a systems point of view in any new study using existing metrics to evaluate the climate impacts from aviation. As such, it will be important to consider all of the uncertainties associated with current understanding of the effects of aviation emissions on climate, including the fact that with the exception of carbon dioxide, the effects of other emissions on climate are still not very well understood. In particular, it would be very difficult to provide a meaningful evaluation of the effects of contrails or the effects of contrails and aerosols on cirrus. However, metrics may be able to better consider the effects NOx emissions from aviation. Modeling capabilities for understanding the UT/LS region have improved greatly in the last few years (although there are definitely remaining uncertainties), such that determining the effects of NOx emissions from aviation on ozone and methane should be more possible than previously; it may be possible to get a stronger understanding of those effects and remaining uncertainties using analyses from current state-of-the-art chemistry-transport and chemistry-climate models.

To provide a perspective relative to prior assessments of aircraft effects, any new study done at this time should start with the use of stratospheric adjusted radiative forcing, but also include consideration of efficacies to the degree possible. The effects of uncertainties in the evaluation of the climate effects and in the metric itself will need to be clearly stated. The radiative forcing could be evaluated for the current time period but it can also be worthwhile to consider projections of effects on aviation based on reasonable scenarios for future emissions. Such scenarios, however, need to be carefully considered, and should be based on best available projections from ICAO and the FAA (or associated organizations like JPDO).

Emissions-based metrics should also be considered, but interpretation will be limited by the lack of a community-consensus on which metrics should be adopted and the lack of current application of the GWP and GTP approaches to evaluation of aviation. The LTR approaches are promising as assessment tools but have not been evaluated by the science community and need further development to reduce existing uncertainties.

It will be difficult to make useful policy decisions involving tradeoffs within the climate sector at this time.

6. Summary

A number of the existing metrics for climate have been considered. Advantages and limitations of the various metrics have been discussed. To some degree, we arrive at more questions than answers. Ultimately, the specific metric of choice in a given situation will always depend on the question being addressed. For aviation, there is no single metric currently in existence that does not have well-recognized shortcomings in either its application to this sector or in evaluation of its capabilities and limitations.

This said, there are still some metrics that demonstrate clear advantages over others, and may be appropriate for use in specific situations and/or after further research and testing, as recommended below.

Beginning with the well-accepted metrics of radiative forcing and GWPs, we find that they have major limitations that affect their interpretation when used to address many of the policy questions of interest to climate.

For example, the equivalent RF concept can be useful to address questions related to changes in climate for the atmospheric agents that have been emitted over a specific period of time. However, equivalent radiative forcing is not an emissions-based metric. Emissions-based metrics are likely the primary choice for addressing most questions of interest for technological or policy considerations and/or trade-offs.

GWPs (and AGWPs) are well established but may be difficult to apply to aviation emissions. We recommend that the existing concept be modified to include efficacies, and tests done to see if all effects can be conceptually included. While there have been many criticisms about this, no one has really attempted to see if the concept could be readily modified to include contrails and other cloud effects, e.g., by basing these effects in a more general sense on the emissions associated with fuel burn. Despite its limitations, the GWP concept is so well engrained in current international climate policy considerations that it might actually impede the progress of negotiations to promote use of an alternative metric. As a result, decision-makers are faced with weighing scientific precision relative to practical applicability (Fuglestvedt et al., 2000).

The answer may lie in using similar metrics that address some of the scientific concerns raised by GWPs. Specifically, the GTP and the LTR approaches have some major advantages, but neither has been adequately tested. GTPs assume either pulse or sustained emissions while LTR generally uses a pulse of one year of emissions. Both may also be applicable to emissions scenarios.

Additional research needs to be done to identify appropriate metrics for evaluating emissions from aviation and from other transportation and energy sectors. The application of existing metrics to aviation emissions needs to be evaluated individually and relative to each other. Some metrics such as the LTR approaches need further development to be scientifically robust. New metrics should also be considered.

Any new assessment of aviation impacts on climate done at this time, before the research outlined above has been done, will have to be limited in scope and subject to large uncertainties. A systems approach will be necessary so that the resulting metric studies are considered relative to remaining uncertainties in the scientific understanding of the processes affecting atmospheric composition and climate from aviation emissions.

ACKNOWLEDGMENT

This report was supported in part by the U.S. Federal Aviation Administration through the DOT/RITA/Volpe National Transportation Systems Center under contract DTRT57-07-C-10059.

APPENDIX A: DISCUSSION ON EFFICACY FACTORS

Efficacy is the factor relating surface temperature change from a particular forcing agent to that from equivalent CO_2 radiative forcing. It is defined as the ratio of the climate sensitivity parameter for a given forcing agent to the climate sensitivity parameter for CO_2 changes (Joshi et al., 2003). Joshi et al. (2003) tested the climate sensitivity to idealized forcing agents (mainly ozone) in three very different GCMs. They found that the climate sensitivity to any given forcing type was varied greatly between the models, but once the sensitivities were normalized by the climate sensitivity of CO_2 within the same model the efficacies were within 30% of one another.

The effective radiative forcing for a given forcing agent would then be the radiative forcing (for this work, radiative forcing refers to the stratospheric adjusted radiative forcing discussed earlier) for a particular forcing type multiplied by the efficacy factor. The effective radiative forcing is then independent of forcing type and can be compared directly to CO_2 RF. Global mean surface temperature can then be calculated as:

$$\Delta T_s = \lambda_{CO2} E * \Delta F$$

where T_s is the global mean surface temperature change, CO_2 is the climate sensitivity for CO_2, E is the efficacy for a particular forcing type, and F is the radiative forcing associated with a particular forcing type. Using an efficacy factor with RF is likely to give a much closer approximation to global surface temperature change than using RF alone (Sausen and Schumann, 2000; Hansen et al., 2005; Lohmann and Feichter, 2005). The difficult part is determining the efficacy for the many forcing types that are currently considered.

According to Boer and Yu (2003b), the efficacy associated with a particular forcing type depends on the spatial distribution of the forcing and how the forcing projects onto the climate feedback mechanisms. Numerous studies have shown that different patterns (both geographic and vertical) of forcings and any non-linearities associated with the forcing will affect the efficacy. It is generally found that higher latitude forcings (regardless of source) have a higher efficacy than tropical forcings (Boer and Yu, 2003b; Joshi et al., 2003; Hansen et al., 2005; Sokolov, 2006; Stuber et al., 2005; Sausen et al., 2002). Most of this effect is thought to be from the change in snow and ice albedo (Stuber et al., 2001; Joshi et al., 2003; Stuber et al., 2005).

Regional efficacies and efficacy for regionally distributed forcing agents have also been examined (Forster et al., 2000; Boer and Yu, 2003b; Joshi et al., 2003). Forster et al. (2000) examined efficacy for regional increases in CO_2 and solar irradiance. Joshi et al. (2003) extended this study to include O_3 and ran experiments using three different GCMs. Each of the GCMs treats feedback mechanisms in different ways. In both Joshi et al. (2003) and Forster et al. (2000) it was found that while climate sensitivity for a particular forcing varied greatly from model to model, the climate sensitivity normalized by the climate sensitivity of CO_2, were similar. This normalized climate sensitivity is the efficacy. Efficacies were generally within 30% of each other across models for a given forcing scenario. Efficacy was found to be lower for upper tropospheric O_3 changes and higher for lower stratospheric O_3 changes; lower for tropical changes and higher for extratropical changes. This systematic error in the stratospheric adjusted RF implies that an effective RF would be a better predictor of globally averaged surface temperature change. This work also seems to suggest that more regionally (upper troposphere, lower stratosphere, tropical, extratropical) appropriate efficacies be used in calculating the effective (globally averaged) RF.

Boer and Yu (2003b) looked in more detail at the spatial distribution of the forcing response. They determined that the geographic location of temperature change is strongly influenced by the feedback mechanisms that dominate that region. In fact they determined that the geographic location of the feedback mechanisms were more important than the geographic location of the forcing agent in determining the temperature distribution. Joshi et al. (2003), on the other hand, noticed that when a forcing maximum was located in the tropics/extratropics then the tropics/extratropics showed the greatest response. Boer and Yu (2003b) also noted that there was a tendency for certain areas (like the Northern Hemisphere high latitude region) to show a strong temperature response for all of the forcing scenarios tested, except those with sharp gradients. Some regions were preferentially changed even if the forcing was remote. Since GCMs treat climate feedback mechanisms in many different ways, it is not currently possible to determine efficacies for small geographic regions until we have a better understanding of climate feedback mechanisms.

Vertical distribution of the forcing and its effect on efficacy has also been examined in some detail (Hansen et al., 1997; Christiansen, 1999; Joshi et al., 2003; Cook and Highwood, 2004; Roberts and Jones, 2004; Forster and Joshi, 2005; Sokolov, 2006; Stuber et al., 2005).

It is generally found that upper-troposphere forcings have smaller efficacy than forcings that affect the surface. However, climate feedback considerations, such as cloud cover and water vapour content, make it difficult to generalize this finding with confidence (Govindasamy et al., 2001b; Joshi et al., 2003; Sokolov, 2006).

Efficacies Reported in the Literature

We now examine the efficacies that are currently available in the literature (also see Table A). Efficacies that may be relevant for aircraft issues include: long-lived GHGs, stratospheric ozone, upper tropospheric ozone, scattering aerosols, absorbing aerosols, contrails and stratospheric water vapor. Efficacies are also given in the literature for total solar irradiance change (Gregory et al., 2004; Joshi et al., 2003; Cook and Highwood, 2004; Sokolov, 2006; Forster et al., 2000; Hansen et al., 2005) and for tropospheric ozone change near the surface (Hansen et al., 2005; Lohmann and Feichter, 2005; Mickley et al., 2004), but these are not directly relevant to aircraft studies and will not be discussed in this report.

Existing derived efficacies for gas and particle emissions or concentration perturbations to atmospheric concentrations have been adopted recently by various authors to aviation application – however, these efficacies were not sp ecifically based on aviation emissions studies and may not be appropriate for the spatial and temporal emissions associated with aviation. At this point, there are no reliable efficacies for aviation impacts on climate.

For the forcing types relevant to aircraft issues, in looking at the existing analyses of efficacies, the exact experiment done to calculate the efficacy will determine whether the value may be of use for aircraft studies because aircraft forcings tend to have very specific characteristics (for example, geographic location and altitude.) Long-lived GHGs, contrails, stratospheric ozone, upper tropospheric ozone and stratospheric water vapor are directly relevant to aircraft issues regardless of the experiment, but we still examine the experiments used to determine efficacies for these forcing types. Scattering aerosols and absorbing aerosols efficacies reported in the literature may or may not be relevant to aircraft studies, depending on how they were determined. The efficacy value, as with RF, depends strongly on the definition of tropopause height (Ramaswamy et al., 2001; Chipperfield et al., 2003; Hansen et al., 2005). Efficacies for each aircraft-related forcing agent are given, along with an overview of efficacy values in the literature, a description of how the efficacy was calculated and a statement of how relevant this efficacy value is likely to be for aircraft studies.

Long-Lived Greenhouse Gases

IPCC (1995; and references therein) determined that the climate sensitivity for a wide range of forcing agents is invariant. Most of the climate forcings examined were long-lived greenhouse gases that are approximately spatially homogeneous. Hansen et al. (2005) suggests around 1.04 as an average efficacy for all well-mixed GHGs. Generally, long-lived GHG efficacies are thought to be around 1.0 (with an error of 10%). Long-lived GHGs include CO_2, N_2O and CFCs. CH4 is also a long-lived gas, but is considered in more detail because of its chemistry importance in the atmosphere

Very few studies have examined the efficacy for individual GHGs. Hansen et al. (2005) suggest slightly higher efficacies for individual GHGs with N_2O having an efficacy of 1.04 and CFC-11 and CFC-12 having a value of 1.32. On the other hand, some studies suggest that efficacies for CFCs should be slightly smaller than 1.0, such as Forster and Joshi (2005) who report 0.94. This suggests that the efficacies being derived are also dependent on the model used and the specific experiment.

Hansen et al. (2005) found that CH_4 had an average efficacy of 1.1.Two separate CH_4 experiments were done with concentrations of 2 and 6 times the current concentration. Efficacies were 1.10 and 1.13, respectively. This illustrates the potential nonlinearity associated with climate sensitivity. Indirect effects, such as the effect of CH_4 and CFCs on O3 and the effect of CH4 on water vapor are not included in these efficacies. Bernsten et al. (2005) determined that efficacies for methane were 1.08 and 0.95 for the ECHAM4 and UREAD models, respectively.

In summary, there is very little model consensus on the efficacies for individual long-lived greenhouse gases. The general consensus among journal articles that do not directly test the efficacy of long-lived GHGs remains that long-lived well-mixed GHGs have efficacies around 1.0 and most model experiments support this consensus for CH_4 and N_2O within about 10%. These efficacies should apply to aircraft studies without qualification because the species tend to be well mixed in the atmosphere.

UT/LS Ozone

Stratospheric ozone efficacies have been examined by Stuber et al. (2001), Joshi et al. (2003), Hansen et al. (2005) and Stuber et al. (2005) using idealized ozone changes. Hansen et al. (2005) used realistic stratospheric ozone changes. Ozone changes throughout the atmosphere and in the troposphere only were examined. It was found that both of these cases led to the same efficacy, implying that a stratospheric ozone change would have the same efficacy if the effects are linearly additive. This linearity was not tested but it would be a relatively easy experiment.

Stuber et al. (2005) examined the radiative forcing temperature response for ozone in the upper troposphere and lower stratosphere separately. They also examined homogeneous and inhomogeneous distributions for ozone for both UT and LS experiments. The inhomogeneously distributed O_3 had a maximum concentration at about 60 N. The Northern Hemisphere upper tropopause experiment matched the ozone distribution from aircraft emissions. The GCM used did not have a chemistry model, so the production of stratospheric water vapor from oxidation of CH4 is not included. Efficacies were found to be: 1.8 for a homogeneous distribution in the LS; 0.72 for homogeneous distribution in the UT; 2.26 for inhomogeneous distribution in the LS; and 1.07 for inhomogeneous distribution in the UT.

Joshi et al. (2003) applied O_3 changes in the UT in the tropics, UT in the Northern Hemisphere extratropics and globally in the LS. Three very different models were run for each study (UREAD, ECHAM4, and LDM), Efficacies were found to be: 0.71, 0.72 and 0.91 for the three models, respectively, in the UT tropics; 0.63, 1.17, and 0.55, respectively, in the Northern Hemisphere UT; and 1.39, 1.8, and 1.23, respectively, globally in the LS. The difference on stratospheric O_3 efficacies between the models is thought to be due to the different feedback mechanisms of stratospheric water vapor.

Forster and Shine (1999) found that lower stratospheric ozone had a 40% higher climate sensitivity than CO_2, while Joshi et al. (2003) found a 20-80% higher climate sensitivity using three different models. Stratospheric water vapor feedback was included in the stratospheric ozone efficacies for both of these studies and it was determined that this feedback accounts for the large efficacy values. The stratospheric water vapor reaction is already considered in steady-state CTM runs for aircraft emissions, so the efficacies used for radiative forcing should be lower than those found by Joshi et al. (2003).

At this time, it is premature to assign an efficacy with any confidence to stratospheric ozone changes, but the Joshi et al. (2003) and Stuber et al. (2005) results clearly suggest that the efficacy is not the same for UT and LS O_3. Bernsten et al. (2005) also found that ozone perturbations are not linearly additive when O_3 perturbations were tested over Europe and SE Asia separately and combined. The departure from linearity was approximately 8%.

Scattering Aerosols (Direct Effect)

As discussed earlier, aerosols have both a direct and indirect effects on the atmosphere. Cook and Highwood (2004) determined in idealized studies that the direct effect of scattering aerosols is very similar to the effect of changing total solar irradiance (near 1.0). Hansen et al. (2005) found an efficacy of 1.09 for tropospheric sulfates and determined that realistic changes in scattering aerosols had a larger effect at higher latitudes than at lower latitudes. This experiment doubled the current concentrations of sulfates, so it is not clear how relevant this efficacy value is for aircraft emissions near the tropopause. Rotstayn and Penner (2001) have also examined the direct effect of scattering sulfate aerosols. Sulfates in their experiment are distributed in the vertical so that there is an exponential decrease in concentration with height. Direct sulphate efficacy was calculated to be 0.68 for pure forcing (no feedback) and 0.73 for quasi-forcing that included longwave feedback effects. Generally, it is assumed that the direct effect of scattering aerosols has an efficacy between 0.7 and 1.1, with similar efficacies for both stratospheric and tropospheric aerosols. Again, none of these studies directly simulated a change in sulphate emissions by aircraft. In all likelihood, the sulfate effect due to aircraft at the tropopause would be much too small to rise above climate model noise unless the sulfate concentration was multiplied by a large factor.

Absorbing Aerosols (Direct Effect)

Absorbing aerosols are perhaps the most difficult forcing types to infer global mean temperature change from because the linear relationship between RF and temperature change breaks down, and efficacy is not constant for black carbon aerosols (Hansen et al., 1997; Cook and Highwood, 2004; Feichter et al., 2004; Roberts and Jones, 2004; Hansen et al., 2005). For simplicity, the effect of changes in boundary layer black carbon is not discussed here because they are not directly relevant to aircraft issues.

The relative locations of cloud and aerosol layers, along with surface albedo, affect the relationship between RF and temperature (Penner et al., 2003, Cook and Highwood, 2004; Feichter et al., 2004; Johnson et al., 2004; Roberts and Jones, 2004; Hansen et al., 2005). The source of the black carbon also appears to affect the efficacy. Hansen et al. (2005) find efficacies much larger than 1.0 for biomass burning and much smaller than 1.0 for fossil fuel carbon Hansen et al. (2005) found that black carbon had efficacies of 0.5 in the free troposphere to 0.3 in the upper troposphere.

So far, there appears to be no consensus on efficacy for absorbing aerosols. It appears that no simple relationship exists between radiative forcing due to all absorbing aerosols and global mean temperature change. Biomass burning efficacies would not be appropriate for aircraft studies, but the smaller values for fossil fuel carbon may be appropriate. More studies need to be done to gain confidence in these results.

Indirect Aerosol Effects

The indirect effect of aerosols has been examined numerous times in the literature, with recent publications by Rotstayn and Penner (2001), Williams et al. (2001) and Lohmann and Feidhter (2005), but none of these studies relate to emissions in the upper troposphere and resulting effects on cirrus clouds. Rotstayn and Penner (2001) calculate the efficacy for the indirect effect of surface-emitted aerosols to be 0.83 for the first indirect effect (Twomey effect), 0.78 for the second indirect effect (cloud lifetime effect) and 0.86 for the total indirect effect due to sulphate aerosols. Lohmann and Feichter (2005) calculate the efficacy for first indirect effect to be 1.01. Williams et al. (2001) calculated the efficacy for the first indirect effect to be 0.82 and the second indirect effect to be 1.17. The radiative forcings for the first and second indirect aerosol effects do not add linearly.

Contrails

Hansen et al. (2005) and Ponater et al. (2005) find that contrail efficacy is smaller than that for CO_2. Hansen et al. (2005) used 10 times the current contrail value in a GCM experiment to determine the contrail climate sensitivity. The contrail signal did not rise above the model noise level enough for a statistically significant climate sensitivity value to be determined. Ponater et al. (2005) used 20 times the FESG/Fa1 inventory for 2050 aviation contrails in a similar experiment. They calculated the climate sensitivity value for CO_2 and contrails to determine the climate sensitivity in various regions of the world. As expected, there was a larger temperature response over the land than there was over the ocean. The globally averaged efficacy for contrails in this study is 0.6. This value has not been confirmed by any other studies.

Stratospheric Water Vapor

Forster and Shine (1999) determined that the efficacy for stratospheric water vapor is approximately 1.1. Their experiment increased stratospheric water vapor assumed increases in water vapor of 40 ppbv/year in the lower stratosphere and 100 ppbv/year in the upper stratosphere. They noted that it was the change in the lower stratospheric water vapor that contributed most of the radiative forcing. Hansen et al. (2005) also examined stratospheric watervapor, but they only presented efficacy for radiative forcing calculated using a constant sea surface temperature (Fs) and did not present efficacy for stratospheric adjusted radiative forcing (Fa). The efficacy for Fs is 0.96, but it is typically different from that for Fa. Since most of the forcing in the Forster and Shine (1999) scenario was due to lower stratospheric water vapor, this efficacy value is probably appropriate for aircraft studies. Unfortunately there are not enough studies to gain confidence in the value.

Table A. Summary of efficacies found in literature for various forcing agents

Forcing Agent		Efficacy	Source
Long-lived GHGs	All	~1.0 +/- 10%	
	N2O	1.04	Hansen et al., 2005
	CFC (-11 & -12)	1.32	Hansen et al., 2005
	CFC (-11 & -12)	0.94	Forster & Joshi, 2005
	CH4	1.1	Hansen et al., 2005
	CH4	0.95 - 1.08	Bernsten et al., 2005
	CH4	1.18	Ponater et al., 2006
O_3			
	UT (extratropics)	1.07	Stuber et al., 2005
	UT (extratropics)	0.55 – 1.17	Joshi et al., 2003
	UT (tropics)	0.71 – 0.91	Joshi et al., 2003
	LS (extratropics)	2.26	Stuber et al., 2005
	LS (global)	1.8	Stuber et al., 2005
	LS (global)	1.23 – 1.8	Joshi et al., 2003
	LS (global)	1.4	Forster & Shine, 1999
	aviation	1.37-1.55	Ponater et al., 2006
Sulfates (direct)			
	UT	1.09	Hansen et al., 2005
	UT	0.68	Rotstayn & Penner, 2001
	UT	0.73 (w/feedbacks)	Rotstayn & Penner, 2001
Soot (direct)			
	free troposphere	0.5	Hansen et al., 2005
	UT	0.3	Hansen et al., 2005
Sulfates (indirect)			
	1st	0.83	Rotstayn & Penner, 2001
	1st	1.01	Lohmann & Feichter, 2005
	1st	0.82	Williams et al., 2001
	2nd	0.78	Rotstayn & Penner, 2001
	2nd	1.17	Williams et al., 2001
Contrails		0.59	Ponater et al., 2005

References

Andronova, N., M. Schlesinger, S. Dessai, M. Hulme, and B. Li, 2007: The Concept of Climate Sensitivity: History and Development. In *Human-Induced Climate Change: An Interdisciplinary Assessment* . M. E. Schlesinger et al. (eds.), Cambridge University Press.

Berntsen, T. K., J. S. Fuglestvedt, M. M. Joshi, K. P. Shine, N. Stuber, M. Ponater, R. Sausen, D. A. Hauglustaine, and L. Li, 2005: Response of climate to regional emissions of ozone precursors: sensitivities and warming potentials. *Tellus, 57B,* 283-304.

Boer, G. J., and B. Yu, 2003a: Climate sensitivity and climate state. *Clim. Dyn* ., 21, 167-176.

Boer, G. J., and B. Yu, 2003b: Climate sensitivity and response. *Clim. Dyn* ., 20, 415-429.

Bradford, D. F., 2001: Global change: Time, money and tradeoffs. *Nature* 410, 649-650, doi: 10.1038/35070707

Boucher, O., and J. Haywood, 2001: On summing the components of radiative forcing of climate change. *Climate Dynamics, 18,* 297-302.

Chipperfield, M. P., W. J. Randel, G. E. Bodeker, and P. Johnston, 2003: Global ozone: past and future. In: *Scientific assessment of ozone depletion: 2002* : pursuant to Article 6 of the Montreal Protocol on Substances that Deplete the Ozone Layer / World Meteorological Organization., C.A. Ennis (ed.). WMO, Geneva, pp. 4.1-4.90. Chapter 4.

Christiansen, B., 1999: Radiative forcing and climate sensitivity: The ozone experience. *Quart. J. Royal Met. Soc.,* 125, 3011-3035

Chylek, P., G. Lesins, G. Vindeen, J. Wong, R. Pinnick, D. Ngo, and J. Klett, 1996: Black carbon and absorption of solar radiation by clouds. *J. Geophys . Res* ., 101, 23365-23371.

Collins, W. D., et al., 2006: Radiative forcing by well-mixed greenhouse gases: estimates from climate models in the IPCC AR4. *J. Geophys. Res* ., 111, doi:10.1029/2005JD006713.

Cook, J., and E. J. Highwood, 2004: Climate response to tropospheric absorbing aerosols in an intermediate general-circulation model. *Q. J. Roy. Meteor.Soc.* , 130, 175-191.

Cox, S. J., W. C. Wang, and S. E. Schwartz, 1995: Climate response to radiative forcings by sulfate aerosols and greenhouse gases. *Geophys. Res. Lett.,* 22(18), 2509-2512.

Cubasch, U., K. Hasselmann, H. Höck, E. Maier-Reimer, U. Mikolakewicz, B.D. Santer, R. Sausen, 1992: Time-dependent greenhouse warming computations with a coupled ocean-atmosphere model. *Climate Dynamics* , 8, 55-69, DOI 10.1007/BF00209163. URL http://dx.doi.org/10.1007/BF00209163.

Derwent, R.G., et al, 2001: Transient behaviour of tropospheric ozone precursors in a global 3-D CTM and their indirect greenhouse effects. *Climatic Change* , 49, 463-487.

Echaus, R. S., 1992: Comparing the effects of greenhouse gas emissions on global warming. *Energy J* ., 13, 25-35.

Fisher, D. A., C. H. Hales, W.-C. Wang, M. K. W. Ko, N. D. Sze, 1990: Model calculations on the relative effects of CFCs and their replacements on global warming. *Nature* , 344, 513-516.

Forster, P. M. F., and M. J. Joshi, 2005: The role of halocarbons in the climate change of the troposphere and stratosphere. *Climate Change* , 71, 249-266.

Forster, P. M. de F., and K. P. Shine, 1999: Stratospheric water vapor changes as a possible contributor to observed stratospheric cooling, *Geophys. Res. Lett.*, 26, 3309-3312.

Forster, P. M. F. and K. E. Taylor, 2006: Climate forcings and climate sensitivities diagnosed from coupled climate model integrations. *J. Climate*, 19, 6181-6194.

Forster, P. M. F., M. Blackburn, R. Glover, and K.P. Shine, 2000: An examination of climate sensitivity for idealised climate change experiments in an intermediate general circulation model. *Clim. Dyn.*, 16, 833-849.

Forster, P. M. de F., R. S. Freckleton, and K. P. Shine, 1997: On aspects of the concept of radiative forcing. *Clim. Dyn.* 13, 547-560.

Forster, P. M. F., M. Ponater, and W. Y. Zhong, 2001: Testing broadband radiation schemes for their ability to calculate the radiative forcing and temperature response to stratospheric water vapour and ozone changes. *Meteorol. Z., 122,* 1-999.

Forster, P. M. F., K. P. Shine, and N. Stuber, 2006: It is premature to include non- CO_2 effects of aviation in emission trading schemes. *Atmos. Environ.*, 40, 1117-1121.

Freckleton, R. S., E. J. Highwood, K. P. Shine, O. Wild, K. S. Law, and M. G. Sanderson, 1998. Greenhouse gas radiative forcing: Effects of averaging and inhomogeneities in trace gas distribution. *Quart. J. Roy. Met. Soc.*124:2099-2127.

Fuglestvedt, J. S., T. K. Berntsen, O. Godal, R. Sausen, K. P. Shine, and T. Skodvin, 2003: Metrics of Climate Change: Assessing Radiative Forcing and Emission Indices. *Climatic Change*, 58, 267-331.

Fuglestvedt, J. S., T. K. Bernsten, O. Godal, and T. Skodvin, 2000: Climate implications of GWP-based reductions in greenhouse gas emissions. *Geophys. Res. Lett.*, 27, 409-412.

Gauss, M. G. Hyhre, G. Pitari, M. J. Prather, I. S. A. Isaksen, T. K. Bernsten, G. P. Brasseur, F. J. Dentener, R. G. Derwent, D. A. Haughlustaine, L. W. Horowitz, D. J. Jacob, M. Johnson, K.S. Law, L. J. Mickley, J.-F. Muller, P.-H. Plantevin, J. A. Pyle, H. L. Rogers, D. S. Stevenson, J. K. Sundet, M. van Weele, and O. Wild, 2003: Radiative forcing in the 21st century due to ozone changes in the troposphere and lower stratosphere. *J. Geophys. Res.*, 108, NO. D9, 4292, doi:10.1029/2002JD002624.

Gauss, M., I. S. A. Isaksen, D. S. Lee, O. A. Sovde, 2005: Impact of aircraft NOx emissions on the atmosphere – Tradeoffs to reduce impact. *Atmos. Chem. Phys. Discuss.*, 5, 12255-12311.

Godal O., 2003: The IPCC assessment of multidisciplinary issues: The choice of greenhouse gas indices. *Climatic Change*, 58, 243-249.

Godal, O., and J. Fuglestvedt, 2002: Testing 100-year Global Warming Potentials: impacts on compliance costs and abatement profile. *Climatic Change*, 52, 93-127.

Gohar, L. K., G. Myhre, and K. P. Shine, 2004: Updated radiative forcing estimates of four halocarbons. *J. Geophys. Res., 109,* D01107, doi: 10.1029/2003JD004320.

Govindasamy, B., K. E. Taylor, P. B. Duffy, B. D. Santer, A. S. Grossman, and K. E. Grant, 2001b: Limitations of the equivalent CO_2 approximation in climate change simulations. *J. Geophys. Res.*, 106, 22593-22603.

Gregory, J. M., W. J. Ingram, M. A. Palmer, G. S. Jones, P. A. Stott, R. B. Thorpe, J. A. Lowe, T. C. Johns, and K. D. Williams, 2004: A new method for diagnosing radiative forcing and climate sensitivity. *Geophys. Res. Lett.*, 31, L03205, DOI 10.1029/2003GL018747.

Grewe, V., and A. Stenke, 2007: A strategy for climate evaluation of aircraft technology: an efficient climate impact assessment tool – AirClim. *Atmos. Chem. Phys. Discuss.*, 7, 12185-12229.

Hammitt, J. K., A. K. Jain, J. L. Adams, and D. J. Wuebbles, 1996: A welfare-based index for assessing environmental effects of greenhouse-gas emissions. *Nature*, 381, 301-303.

Hammond, A. L., E. Rodenburg, and W. Moomaw, 1990: Commentary in *Nature*, 347, 705-706.

Hansen, J., M. Sato, and R. Ruedy, 1997: Radiative forcing and climate response. *J. Geophys. Res.*, 102, 6831-6864.

Hansen, J., M. Sato, L. Nazarenko, R. Ruedy, A. Lacis, D. Koch, I. Tegen, T. Hall, D. Shindell, B. Santer, P. Stone, T. Novakov, L. Thomason, R. Wang, Y. Wang, D. Jacob, S. Hallandsworth, L. Bishop, J. Logan, A. Thompson, R. Stolarski, J. Lean, R. Willson, S. Levitus, J. Antonov, N. Rayner, D. Parker, and J. Christy, 2002: Climate forcings in Goddard Institute for Space Studies SI2000 simulations. *J. Geophys. Res.*, 107, D18, 4347, doi:10.1029/2001JD001143.

Hansen, J., M. Sato, R. Ruedy, L. Nazarenko, A. Lacis, G.A. Schmidt, G. Russell, I. Aleinov, M. Bauer, S. Bauer, N. Bell, B. Cairns, V. Canuto, M. Chandler, Y. Cheng, A. Del Genio, G. Faluvegi, E. Fleming, A. Friend, T. Hall, C. Jackman, M. Kelley, N. Kiang, D. Koch, J. Lean, J. Lerner, K. Lo, S. Menon, R. Miller, P. Minnis, T. Novakov, V. Oinas, J. Perlwitz, J. Perlwitz, D. Rind, A. Romanou, D. Shindell, P. Stone, S. Sun, N. Tausnev, D. Thresher, B. Wielicki, T. Wong, M. Yao, and S. Zhang, 2005: Efficacy of climate forcings. *J. Geophys. Res.*, 110, D18104, doi: 10.1029/2005JD005776.

Harvey, L. D. D., 1993: A guide to Global Warming Potenitals (GWPs). *Energy Pol.*, 21, 24-34.

Hasselmann, K., R. Sausen, E. Maier-Reimer, and R. Voss, 1993: On the cold start problem in transient simulations with coupled atmosphere-ocean models. *Climate Dynamics*, 9, 53 - 61, doi: 10.1007/BF00210008.

Hasselmann, K., S. Hasselmann, R. Giering, V. Ocana, and H. V. Storch, 1997: Sensitivity Study of Optimal CO_2 Emission Paths Using a Simplified Structural Integrated Assessment Model (SIAM). *Climatic Change*, 37, 345 - 386, doi: 10.1023/A:1005339625015.

Haywood, J.M., and V. Ramaswamy, 1998: Global sensitivity studies of the direct radiative forcing due to anthropogenic sulfate and black carbon aerosols. *J. Geophys. Res.* 103, 6043- 6058.

Hooss, G., R. Voss, K. Hasselmann, E. Maier-Reimer, and F. Joos, 2001: A nonlinear impulse response model of the coupled carbon cycle-climate system (NICCS). *Climate Dynamics*, 18, 189 - 202, doi:10.1007/s003820100170.

Huang, Y., and V. Ramaswamy, 2006: Temperature dependence of gas absorptivity and its implication for climate change problem. *J. Geophys. Res.*, submitted. IPCC, 1990: *Climate Change 1990. The IPCC Scientific Assessment.* Working Group 1 report, Intergovernmental Panel on Climate Change, WMO and UNEP, Houghton, J.T., G.J. Jenkins and J.J. Ephraums (eds.), Cambridge University Press, Cambridge, U.K., pp 364.

IPCC, 1995: Climate Change 1994. Radiative Forcing of Climate Change and an Evaluation of the IPCC IS92 Emission Scenarios. Cambridge, Cambridge University Press, Cambridge, U.K.

IPCC, 1999: *Aviation and the global atmosphere: a special report of the IPCC Working Group I and III.* J.E. Penner, D.H. Lister, D.J. Griggs, D.J. Dokken, and M. MacFarland (eds.). Cambridge University Press, pp. 373.

IPCC, 2001: *Climate Change 2001: The Scientific Basis. Intergovernmental Panel on Climate Change*, J. T. Houghton, et al., editors. Cambridge Univ. Press, Cambridge, UK.

IPCC, 2007a: *Climate Change 2007 – The Physical Science Basis*. Contribution of Working group I to the Fourth Assessment Report of the IPCC. S. Solomon, et al., editors. Cambridge Univ. Press, Cambridge, UK.

IPCC, 2007b: *Climate Change 2007 – Impacts, Adaptation and Vulnerability*. Contribution of Working group II to the Fourth Assessment Report of the IPCC. M. Parry, et al., editors. Cambridge Univ. Press, Cambridge, UK.

Jacob, D. J., R. Avissar, G. C. Bond, S. Gaffin, J. T. Kiehl, J. L. Lean, U. Lohmann, M. E. Mann, R. A. Pielke, V. Ramanathan, and L. M. Russell, 2005: Radiative forcing of climate change. The National Academy Press, Washington, D.C., 207 pp.

Jain, A. K., B. P. Briegleb, K. Minschwaner, and D. J. Wuebbles, 2000: Radiative forcings and Global Warming Potentials of thirty-nine greenhouse gases. *J. Geophys. Res*., 105, 20773-20790.

Joshi, M., M. Ponater, N. Stuber, R. Sausen, and L. Li, 2003: A comparison of climate response to different radiative forcings in three general circulation models: towards an improved metric of climate change. *Clim. Dyn., 20,* 843-854.

Joos, F., I. C. Prentice, S. Sitch, R. Meyer, G. Hooss, G.-K. Plattner, S. Gerber, and K. Hasselmann, 2001: Global warming feedbacks on terrestrial carbon uptake under the IPCC emission scenarios, Global Biogeochemical Cycles, 15, 891–907.

Kandlikar, M., 1995: The relative role of trace gas emissions in greenhouse abatement policies. *Energy Pol*., 23, 879-883.

Kandlikar, M., 1996: Indices for comparing greenhouse gas emissions: integrating science and economics. *Energy Economics*, 18, 265-282.

Kiehl, J. T., 2007: Twentieth century climate model response and climate sensitivity. *Geophys. Res. Lett*., 34, doi:10.1029/2007GL031383.

Lashof, D. A. and D. R. Ahuja, 1990: Relative contribution of greenhouse gas emissions to global warming. *Nature*, 344, 529-531.

Lee, D. S., and R. Sausen, 2004: Climate responses to aviation NOx and CO_2 emissions scenarios. Aviation, Atmosphere and Climate (AAC). Proceedings of a European Conference, Friedrichschafen, Germany, June 30 to July 3, 2003. R. Sausen, C. Fichter and G. Amanatidis, editors. European Commission, Air Pollution research report 83.

Lee, D. S., and R. C. N. Wit, 2006: Potential methods to include the full climate impact of aviation emissions into the European Trading Scheme and their scientific integrity. Proceedings of an International Conference on Transport, Atmosphere, and Climate (TAC), R. Sausen and A. Blum, editors, Oxford UK, June 26-29, 2006.

Lohmann, U., 2002: Possible aerosol effects on ice clouds via contact nucleation, *J. Atmos. Sci.,* 59, 647–656.

Lohmann, U., and J. Feichter, 2005: Global indirect aerosol effects: A review. *Atmospheric Chemistry and Physics*, 5, 715-737.

Manabe, S., and R.F. Strickler, 1964: Thermal equilibrium of the atmosphere with a convective adjustment. *J. Atmos. Sci*., 21(4), 361-385.

Manabe, S., and R. Wetherald, 1967: Thermal equilibrium of the atmosphere with a given distribution of relative humidity. *J. Atmos. Sci*., 24, 241-259.

Manne, A. S., and R. G. Richels, 2001: An alternative approach to establishing trade-offs among greenhouse gases. *Nature*, 410, 675-677.

Marais, K., S. P. Lukachko, M. Jun, A. Mahashabde, and I. A. Waitz, 2007: Assessing the impact of aviation on climate. *Meteor. Z.,* in press.

Marquart, S., M. Ponater, F. Mager, and R. Sausen, 2003: Future Development of Contrail Cover, Optical Depth, and Radiative Forcing: Impacts of Increasing Air Traffic and Climate Change, *Journal of Climate, 16,* 2890-2904.

Mahashabde, A., K. Marais, M. Jun, and I. Waitz, 2007: *APMT Algorithm Description Document. Benefits Valuation Block – Climate Impact APMT System*. Federal Aviation Administration, document APMT-ADD-20-01.

Meehl, G. A., W. M. Washington, J.M. Arblaster, and A. Hu, 2004a: Factors affecting climate sensitivity in global coupled models. *J. Climate*, 17, 1584-1596.

Meehl, G. A., W. M. Washington, C. M. Ammann, J. M. Arblaster, T. M. L. Wigley, and C. Tebaldi, 2004b: Combinations of Natural and Anthropogenic Forcings in Twentieth-Century Climate. *J. Climate*, 17, 3721-3727.

Mendelsohn, R., W. Morrison, M. E. Schlesinger, and N. G. Andronova, 2000: Country-specific market impacts of climate change. *Climatic Change*, 45, 553-569.

Menon, S., J. Hansen, L. Nazarenko, and Y. Luo, 2002a: Climate effects of black carbon aerosols in China and India. *Science*, 297, 2250-2253.

Mickley, L. J., D. J. Jacob, B. D. Field, and D. Rind, 2004: Climate response to the increase in tropospheric ozone since preindustrial times: A comparison between ozone and equivalent CO_2 forcings. *J. Geophys. Res.*, 109, D05106, doi:10.1029/2003JD003653.

Myhre, G., and F. Stordal, 1997: Role of spatial and temporal variations in the computation of radiative forcing and GWP. *J. Geophys. Res.*, 102, 11181-11200.

Myhre, G., and F. Stordal,, 2001: On the tradeoff of the solar and thermal infrared radiative impact of contrails. *Geophys. Res. Lett.*, 28, 3119-3122.

Myhre, G., A. Myhre, and F. Stordal, 2001: Historical evolution of radiative forcing of climate. *Atmos. Env.*, 35, 2361-2373.

Naik, V., A. Jain, K. O. Patten, and D. J. Wuebbles, 2000: Consistent sets of atmospheric lifetimes and radiative forcings on climate for CFC replacements: HCFCS and HFCs, *J. Geophys. Res.*, 105, 6903-6914.

National Research Council (NRC), 2003: *Estimating Climate Sensitivity: Report of a Workshop*. The National Academies Press, Washington, D. C.

Nordhaus, W. D., 1997: Discounting in economics and climate change. *Climatic Change*, 37, 315-328.

Nordhaus, W. D., and J. Boyer, 2000: *Warming the World: Economic Models of Global Warming*. MIT Press, Cambridge, MA.

O`Neill, B. C., 2000: The Jury is Still Out on Global Warming Potentials. *Climatic Change*, 44, 427-443.

O`Neill, B. C., 2003: Economics, natural science, and the costs of Global Warming Potentials. *Climatic Change*, 58, 251-260.

Pan, L. L., J. C. Wei, D. E. Kinnison, R. Garcia, D. J. Wuebbles, and G. P. Brasseur, 2006: A set of diagnostics for evaluating chemistry-climate models in the extratropical tropopause region. *J. Geophys. Res.*, 112, D09316, doi:10.1029/2006JD007792.

Ponater, M., S. Marquart, R. Sausen, and U. Schumann, 2005: On contrail climate sensitivity. *Geophys. Res. Lett.*, 32, L10706, DOI 10.1029/2005GL022580

Ponater, M., V. Grewe, R. Sausen, U. Schumann, S. Pechtl, E. J. Highwood, and N. Stuber, 2006: Climate sensitivity of radiative impacts from transport systems. Proceedings of an

International Conference on Transport, Atmosphere, and Climate (TAC), R. Sausen and A. Blum, editors, Oxford UK, June 26-29, 2006.

Prather, M., and R. Sausen, convening authors, 1999: Potential Climate Change from Aviation, Chapter 6 in Aviation and the Global Environment, Intergovernmental Panel on Climate Change, J. Penner, et al., editors, Cambridge University Press, Cambridge, UK, 189-215.

Ramanathan, V., R.J. Cicerone, H.B. Singh and J.T. Kiehl, 1985: Trace gas trends and their potential role in climate change. *J. Geophys. Res.*, 90, 5547-5566.

Ramanathan, V., P.J. Crutzen, J.T. Kiehl, and D. Rosenfeld, 2001: Atmosphere: Aerosols, climate, and the hydrological cycle. *Science*, 294, 2119-2124.

Ramanathan, V., L. Callis, R .D. Cess, J. Hansen, and I. Isaksen, 1987: Climate-Chemical Interactions and effects of changing atmospheric trace gases, *Rev. Geophys.,* 25(7), 1441-1482.

Ramaswamy, V., O. Boucher, J.D. Haigh, D.A. Hauglustaine, J.M. Haywood, G. Myhre, T. Nakajima, G.Y. Shi, and S. Solomon, 2001: Radiative forcing of climate change. In: Climate Change 2001: The Scientific Basis. Contribution of Working Group I to the Third Assessment Report of the Intergovernmental Panel on Climate Change, J.T. Houghton, Y. Ding, D.J. Griggs, M. Noguer, P.J. van der Linden, X. Dai, K. Maskell, and C.A. Johnson (eds.). Cambridge University Press, Cambridge, UK, pp. 349-416

Roberts, D.L., and A. Jones, 2004: Climate sensitivity to black carbon aerosol from fossil fuel combustion. *J. Geophys. Res.*, 109, D16202 1-12.

Rodhe, H., 1990: A comparison of the contribution of various gases to the greenhouse effect. *Science*, 248, 1217-1219.

Roe, G. H., and M. B. Baker, 2007: Why Is climate sensitivity so unpredictable? *Science*, 318. 629 – 632, DOI: 10.1126/science.1144735.

Rogers, 1. D., and R. D. Stephens, 1988: Absolute infrared intensities for F-113 and F-114 and an assessment of their greenhouse warming potential relative to other chlorofluorocarbons. *J. Geophys. Res.*, 93, 2423-2428.

Rotmans, J., M. G. J. Den Elzen, 1992: A model-based approach to the calculation of Global Warming Potentials (GWP). *Int. J. Climatol.*, 12, 865-874.

Rotstayn, L. D., and J. E. Penner, 2001: Indirect aerosol forcing, quasi forcing, and climate response. *J. Climate, 14(13),* 2960-2975.

Sausen, R., M. Ponater, and N. Stuber, 2002: Climate response to inhomogeneously distributed forcing agents. Non- CO_2 Greenhouse Gases, Van Ham, Baede, Guicherit and Williams-Jacobse (eds.). Millpress, Rotterdam, pp. 377-381.

Sausen, R., and U. Schumann, 2000: Estimates of the Climate Response to Aircraft CO_2 and NO*X* Emissions Scenarios. *Climatic Change, 44*, 27-58.

Sausen, R., I. Isaksen, V. Grewe, D. Hauglustaine, D.S. Lee, G. Myhre, M. Kohler, G. Pitari, U. Schumann, F. Stordal, and C. Zerefos, 2005: Av iation radiative forcing in 2000: An update on IPCC (1999). *Meteor. Z., 14, No. 4,* 555-561.

Schmalensee, R., 1993: Comparing greenhouse gases for policy purposes. *Energy J.*, 14, 245-255.

Schwartz, S.E., 2004: Uncertainty requirements in radiative forcing of climate change. *J. Air and Waste Manage. Assoc.*, 54, 1351-1359.

Shindell, D. T., G. Faluvegi, N. Bell, and G. A. Schmidt, 2005: An emissions-based view of climate forcing by methane and tropospheric ozone. *Geophys. Res. Lett.*, 32, L04803, doi:10.1029/2004GL021900.

Shine, K. P., and P. M. de F. Forster, 1999: The effect of human activity on radiative forcing of climate change: a review of recent developments. *Global and Planetary Change, 20,* 205-225.

Shine, K. P., J. Cook, E. J. Highwood, and M. M. Joshi, 2003: An alternative to radiative forcing for estimating the relative importance of climate change mechanisms. *Geophys. Res. Lett. 30,* CLM 4, doi: 10.1029/2003GL018141.

Shine, K. P., J. S. Fuglestvedt, K. Hailemariam , and N. Stuber, 2005a: Alternatives to the Global Warming Potential for Comparing Climate Impacts of Emissions of Greenhouse Gases, *Climatic Change*, *68,* 281-302.

Shine, K. P., T. K. Berntsen, J. S. Fuglestvedt, and R. Sausen, 2005b: Scientific issues in the design of metrics for inclusion of oxides of nitrogen in global climate agreements, *Proceedings Nat. Acad. Sciences, 102, No. 44,* 15768-15773.

Shine, K. P., T. K. Berntsen, J. S. Fuglestvedt, R. B. Skeie, and N. Stuber, 2007: Comparing the climate effect of emissions of short- and long-lived climate agents. *Phil. Trans. R. Soc. A*, 365, 1903-1914.

Smith, S. J., and T. M. L. Wigley, 2000: Global Warming Potentials: 1. Climatic implications of emissions reductions. *Climatic Change*, 44, 445-457.

Smith, S. J., and T. M. L. Wigley, 2000: Global Warming Potentials: 2. Accuracy. *Climatic Change*, 44, 459-469.

Sokolov, A., 2006: Does model sensitivity to changes in CO_2 provide a measure of sensitivity to the forcing of different nature? *J. Climate*, 19, 3294-3306.

Stevenson, D. S., R. M. Doherty, M. G. Sanderson, W. J. Collins, C. E. Johnson, and R. G. Derwent, 2004: Radiative forcing from aircraft NO*X* emissions: Mechanisms and seasonal dependence. *J. Geophys. Res., 109,* 13 pp.

Stuber, N., M. Ponater, and R. Sausen, 2001: Is the climate sensitivity to ozone perturbations enhanced by stratospheric water vapor feedback? Geophys. Res. Lett., 28, 2887-2890.

Stuber, N., M. Ponater, and R. Sausen, 2005: Why radiative forcing might fail as a predictor of climate change. *Clim. Dyn*., 24, 497-510.

Sygna, L., J.S. Fuglestvedt, and H.A. Aaheim, 2002: The adequacy of GWPs as indicators of damage costs incurred by global warming. *Mitigation Adaptation Strategies Global Change*, 7, 45-62.

Tett, S. F. B., G. S. Jones, P.A. Stott, D. C. Hill, J. F. B. Mitchell, M. R. Allen, W. J. Ingram, T. C. Johns, C. E. Johnson, A. Jones, D. L. Roberts, D. M. H. Sexton, and M. J. Woodage, 2002: Estimation of natural and anthropogenic contributions to twentieth century temperature change. *J. Geophys. Res.,* 107, 4306, doi:10.1029/2000JD000028.

Tol, R. S. J., 2002a: estimates of the damage costs of climate change. Part I: Benchmark estimates. *Environ. & Resource Econ*., 21, 47-73.

Tol, R. S. J., 2002b: estimates of the damage costs of climate change. Part I: Dynamic estimates. *Environ. & Resource Econ*., 21, 135-159.

Wang, W.; Wuebbles, D.J.; Washington, W.M.; Isaacs, R.G.; Molnar, G., 1986: Trace gases and other potential perturbations to global climate. *Reviews of Geophysics and Space Physics*, 24(1), 110-141.

Wang, W.-C., M. Dudek, and X.-Z. Liang, 1992: Inadequacy of effective CO_2 as a proxy in assessing the regional climate change due to other radiatively active gases. *Geophys. Res. Lett.*, 19(13), 1375-1378.

Wang, W.-C., M. Dudek, X.-Z. Liang, and J. T. Kiehl, 1991: Inadequacy of effective CO_2 as a proxy in simulating the greenhouse effect of other radiatively active gases. *Nature*, 350(6319), 573-577.

Williams, K. D., A. Jones, D. L. Roberts, C. A. Senior, and M. J. Woodage, 2001: The response of the climate system to the indirect effects of anthropogenic sulfate aerosols. *Clim. Dyn.*, 17, 846-856.

Wit, R. C. N., B. H. Boon, A. van Velzen, M. Cames, O. Deuber, and D. S. Lee, 2005: Giving wings to emission trading, Design and impacts, Delft, CE, *05.7789.20*, 1-245.

WMO (World Meteorological Organization), 1985: Atmospheric Ozone: Assessment of Our Understanding of the Processes Controlling its Present Distribution and Change. 3 vol. WMO Report No. 16. Geneva.

Wuebbles, D. J., 1989: Beyond CO_2—the other greenhouse gases. Lawrence Livermore National Laboratory report UCRL-99883; Air and Waste Management Association paper 89-119.4.

Wuebbles, D. J., 1995: Weighing functions for ozone depletion and greenhouse gas effects on climate. *Annu. Rev. Energy Environ.*, 20, 45-70.

Wuebbles, D. J., A. K. Jain, K. E. Grant, and K. O. Patten, 1995: Sensitivity of direct global warming potentials to key uncertainties. *Climate Change*, 29, 265-297.

Wuebbles, D. J., et al. (31 total authors), 2006: Workshop on the Impacts of Aviation on Climate Change: A Report of Findings and Recommendations. Partnership for Air Transportation Noise and Emissions Reduction, Report No. PARTNER-COE-2006-004 (available at http://web.mit.edu/aeroastro/partner/reports/climatewrksp-rpt-0806.pdf).

In: Aviation and the Environment
Editor: Jon C. Goodman
ISBN: 978-1-60692-320-7

Chapter 9

AVIATION AND THE ENVIRONMENT: NEXTGEN AND RESEARCH AND DEVELOPMENT ARE KEYS TO REDUCING EMISSIONS AND THEIR IMPACT ON HEALTH AND CLIMATE*

Statement of Gerald L. Dillingham

WHAT GAO FOUND

Aviation contributes a modest but growing proportion of total U.S. emissions, and these emissions contribute to adverse health and environmental effects. Aircraft and airport operations, including those of service and passenger vehicles, emit ozone and other substances that contribute to local air pollution, as well as carbon dioxide and other greenhouse gases that contribute to climate change. EPA estimates that aviation emissions account for less than 1 percent of local air pollution nationwide and about 2.7 percent of U.S. greenhouse gas emissions, but these emissions are expected to grow as air traffic increases.

Two key federal efforts, if implemented effectively, can help to reduce aviation emissions—NextGen initiatives in the near term and research and development over the longer term. For example, NextGen technologies and procedures, such as satellite-based navigation systems, should allow for more direct routing, which could improve fuel efficiency and reduce carbon dioxide emissions. Federal research and development efforts—led by FAA and NASA in collaboration with industry and academia—have achieved significant reductions in aircraft emissions through improved aircraft and engine technologies, and federal officials and aviation experts agree that such efforts are the most effective means of achieving further reductions in the longer term. Federal R&D on aviation emissions also focuses on improving the scientific understanding of aviation emissions and developing lower-emitting aviation fuels.

Next steps in reducing aviation emissions include managing NextGen initiatives efficiently; deploying NextGen technologies and procedures as soon as practicable to realize

* This is an edited, excerpted and augmented edition from GAO Report GAO-08-706T, dated May 6, 2008.

their benefits, including lower emissions levels; and managing a decline in R&D funding, in part, by setting priorities for R&D on NextGen and emissions-reduction technologies. Challenges in reducing aviation emissions include designing aircraft that can simultaneously reduce noise and emissions of air pollutants and greenhouse gases; encouraging financially stressed airlines to purchase more fuel-efficient aircraft and emissions-reduction technologies; addressing the impact on airport expansion of more stringent EPA air quality standards and growing public concerns about the effects of aviation emissions; and responding to proposed domestic and international measures for reducing greenhouse gases that could affect the financial solvency and competitiveness of U.S. airlines.

Source:
FAA.

Sources of Aviation Emissions.

Why GAO Did this Study

Collaboration between the federal government and the aviation industry has led to reductions in aviation emissions, but growing air traffic has partially offset these reductions. The Federal Aviation Administration (FAA), together with the National Aeronautics and Space Administration (NASA), the Environmental Protection Agency (EPA), and others, is working to increase the efficiency, safety, and capacity of the national airspace system and at the same time reduce aviation emissions, in part, by transforming the current air traffic control system to the Next Generation Air Transportation System (NextGen). This effort involves new technologies and air traffic procedures that can reduce aviation emissions and incorporates research and development (R&D) on emissionsreduction technologies. Reducing aviation emissions is important both to minimize their adverse health and environmental effects and to alleviate public concerns about them that could constrain the expansion of airport infrastructure and aviation operations needed to meet demand.

This testimony addresses (1) the scope and nature of aviation emissions, (2) the status of selected key federal efforts to reduce aviation emissions, and (3) next steps and challenges in reducing aviation emissions. The testimony updates prior GAO work with FAA data, literature reviews, and interviews with agency officials, industry and environmental stakeholders, and selected experts.

Mr. Chairman and Members of the Subcommittee:

I appreciate the opportunity to testify before you on aviation emissions, one of the key sources of concern about the environmental effects of aviation. Over the past 30 years, the federal government, the aviation industry, and other private parties have worked collaboratively to achieve steady reductions in aircraft emissions.[1] Nevertheless, increases in air traffic, which have enhanced the nation's productivity and mobility, have partially offset these reductions, as more flights have produced more emissions and congestion has led to flight delays. According to the Federal Aviation Administration (FAA), this growth in air traffic will continue, with the number of flights increasing 20 percent by 2015 and 60 percent by 2030.[2] In light of these developments, concerns about the environmental effects of aviation emissions have persisted. Moreover, better scientific understanding of the potential health effects of certain aviation emissions and their contribution to climate change has intensified the public's concerns.

To accommodate the expected growth in air traffic, FAA is leading a multipronged, multiagency effort to increase the efficiency, safety, and capacity of the national airspace system. This effort includes transforming the current air traffic control system into the Next Generation Air Transportation System (NextGen)[3] and will require airport and runway expansion. The NextGen initiative incorporates research and development (R&D) on emissions-reduction technologies, alternative fuels, and cleaner and quieter air traffic management procedures. This R&D is necessary both to meet anticipated domestic and international environmental standards and to reduce the environmental impact of aviation. Meeting environmental standards can limit the adverse effects of aviation emissions on air quality and climate, and addressing public concerns about aviation emissions is necessary to avoid constraints on the expansion of aviation operations and airport infrastructure planned under NextGen.[4]

Under the National Environmental Policy Act of 1969, agencies evaluate the likely environmental effects of projects they are proposing using an environmental assessment or, if the projects likely would significantly affect the environment, a more detailed environmental impact statement.[5] FAA typically carries out one of these evaluations for federally financed airport construction projects, including the construction of federally subsidized runways. In addition, under the Clean Air Act's conformity provision, no federal agency may approve or provide financial assistance for any activity that does not conform to an applicable state implementation plan.[6] Therefore, FAA must evaluate whether a proposed federal action associated with an airport project conforms with the applicable state implementation plan before approving or funding the project.[7] In addition, the Clean Air Act mandates standards for mobile sources of emission, such as aircraft and the equipment that service them at airports. EPA sets emissions standards for aircraft and has chosen to adopt international emissions standards for aircraft set by the International Civil Aviation Organization (ICAO).[8]

As requested, my testimony today focuses on aviation emissions. It will address the following questions: (1) What are the scope and nature of aviation emissions? (2) What is the status of selected key federal efforts to address aviation emissions? and (3) What are some next steps and major challenges for the federal government, the aviation industry, and

Congress related to aviation emissions? My statement is based on previous GAO reports[9] updated with a synthesis of recent empirical literature and interviews with officials from FAA, the National Aeronautics and Space Administration (NASA), and the U.S. Environmental Protection Agency (EPA); representatives of aviation industry and environmental associations, and selected aviation emissions experts.[10] We balanced the selection of these experts to capture the views of the many different groups involved in aviation emissions reduction efforts and NextGen. We conducted our work from March to May 2008 in accordance with generally accepted government auditing standards. Those standards require that we plan and perform the study to obtain sufficient, appropriate evidence to provide a reasonable basis for our findings and conclusions based on our study objectives. We believe that the evidence obtained provides a reasonable basis for our findings and conclusions based on our study objectives.

SUMMARY

Currently, aviation contributes a modest proportion of total emissions in the United States, but its share could increase in the future, and aviation emissions can have a detrimental effect on health and the environment. Aircraft are the primary source of aviation emissions, but airport operations, including those of service and passenger vehicles, also produce emissions. Together, aircraft operations in the vicinity of the airport and other airport sources emit nitrogen oxides, which lead to the formation of ground-level ozone (also known as smog), and other substances that contribute to local air pollution, as well as carbon dioxide and other greenhouse gases that rise into the atmosphere and contribute to climate change. Aircraft operations in the upper atmosphere are, however, the primary aviation-related source of greenhouse gas emissions. Currently, according to EPA estimates, aviation emissions account for less than 1 percent of local air pollution nationwide and about 2.7 percent of U.S. greenhouse gas emissions. This proportion is, however, expected to grow with projected increases in air traffic, despite expected improvements in fuel efficiency. Notably, according to FAA, emissions of nitrogen oxides from aviation sources will increase by over 90 percent by 2025 if not addressed. This increase is likely to increase ozone, which aggravates respiratory ailments. Increases in air traffic also mean increases in carbon dioxide emissions and increases in aviation's contribution to climate change, according to the International Panel on Climate Change (IPCC).

Two key federal efforts, if implemented effectively, can help to reduce aviation emissions—near-term NextGen initiatives and R&D over the longer term to fully enable NextGen and reduce aircraft emissions. Some NextGen technologies and procedures, such as satellite-based navigation systems, should allow for more direct routing, which could improve fuel efficiency and reduce carbon dioxide emissions. According to FAA, the full implementation of NextGen could reduce greenhouse gas emissions from aircraft by up to 12 percent by 2025. Federal R&D efforts—led primarily by FAA and NASA and often conducted in collaboration with industry and academia—have achieved significant reductions in aircraft emissions over the last 30 years, and FAA and NASA officials and aviation experts agree that such efforts are the most effective means of achieving further reductions in the longer term. As part of the a national plan for aeronautics R&D, issued by the White House

Office of Science and Technology Policy, the federal government supports a comprehensive approach to R&D on aviation emissions involving FAA, NASA, and other federal agencies that is intended both to improve scientific understanding of the impact of aviation emissions and to develop new technologies, fuels, and air traffic management approaches. Better understanding of the nature and impact of aviation emissions can inform the development of lower- emitting alternative fuels, more efficient air traffic management technologies and procedures, and more fuel-efficient aircraft engines.

Reducing aviation emissions includes steps that FAA and others can take to move the implementation of NextGen forward and to support R&D on NextGen and emissions-reduction technologies, as well as technical, financial, and regulatory challenges facing the federal government, the aviation industry, and Congress. One step for FAA is to ensure the efficiency of NextGen's management by, for example, addressing congressional leaders' and stakeholders' concerns about the program's management structure and authority. Another step for FAA is to further deploy, as soon as practicable, NextGen technologies and procedures, such as the more efficient takeoff and landing procedures now in use at a few airports, to realize their benefits and lower emissions levels. A third step, for FAA and NASA, is managing a decline in federal funding for aeronautics research, the research category that includes work on aviation emissions, new aircraft and engine technologies, and alternative fuels. As a result of this decline, NASA is now sometimes developing technologies to a lower maturity level than in the past, and the technologies are less ready for manufacturers to adopt them. The administration's reauthorization bill for FAA seeks some additional funding for an initiative that could lead to the earlier maturation of certain emissions-reduction technologies, but according to some experts, increased funding of the initiative could increase the probability of success and decrease the time needed to achieve that success. Challenges in reducing aviation emissions for the federal government, the aviation industry, and Congress include designing aircraft that can simultaneously reduce noise and emissions of air pollutants and greenhouse gases; encouraging financially stressed airlines to purchase more fuel-efficient aircraft and emissions-reduction technologies; addressing the impact on airport expansion of more stringent EPA air quality standards and growing public concerns about effects of aviation emissions; and responding to proposed domestic and international measures for reducing greenhouse gases that could affect the financial solvency and competitiveness of U.S. airlines.

AVIATION'S SMALL BUT GROWING PROPORTION OF TOTAL EMISSIONS CONTRIBUTES TO HEALTH AND ENVIRONMENTAL EFFECTS

Aviation-related activities contribute to local air pollution and produce greenhouse gases that cause climate change. Aircraft account for about 70 to 80 percent of aviation emissions, producing emissions that mainly affect air quality below 3,000 feet and increase greenhouse gases at higher altitudes. At ground level, airport operations, including those of motor vehicles[11] traveling to and from the airport, ground service equipment,[12] and stationary sources such as incinerators and boilers, also produce emissions. Together, aircraft operations in the vicinity of the airport and other airport sources produce emissions such as carbon monoxide, sulfur oxides, particulate matter, nitrogen oxides, unburned hydrocarbons, hazardous air pollutants,[13] and ozone[14] that contribute to air pollution. In addition, these sources emit carbon dioxide and other greenhouse gases that contribute to climate change, but aircraft operations in the upper atmosphere are the primary source of aviation-related greenhouse gases. Carbon dioxide is both the primary aircraft emission and the primary contributor to climate change. It survives in the atmosphere for over 100 years. Furthermore, other gases and particles emitted by aircraft— including water vapor, nitrogen oxides, soot, contrails,[15] and sulfate—can also have an impact on climate, but the magnitude of this impact is unknown, according to FAA. Figure 1 illustrates aviation's impact on air quality and climate.

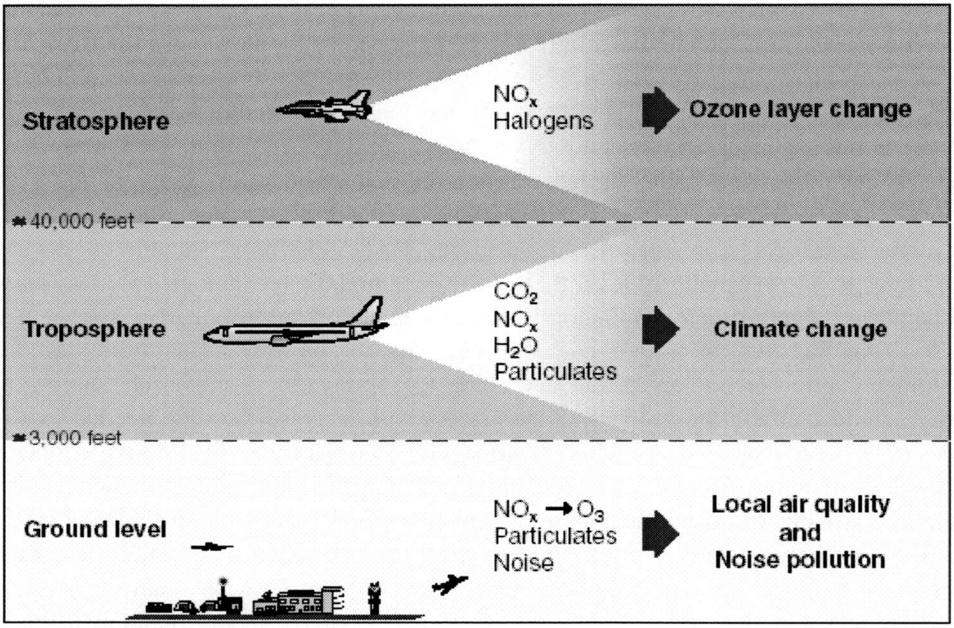

Source: GAO.

Figure 1. Environmental Effects of Aviation Emissions and Noise.

Currently, aviation accounts for a small portion of air pollutants and greenhouse gas emissions. Specifically, aviation emissions represent less than 1 percent of air pollution nationwide, but their impact on air quality could be higher in the vicinity of airports. In addition, aviation accounts for about 2.7 percent of the total U.S. contribution of greenhouse gas emissions, according to the Department of Transportation's Center for Climate Change and Environment. A 1999 study by the United Nations' Intergovernmental Panel on Climate Change (IPCC) estimated that global aircraft emissions generally accounted for approximately 3.5 percent of the warming generated by human activity.[16]

As air traffic increases, aviation's contribution to air pollution and climate change could also grow, despite ongoing improvements in fuel efficiency, particularly if other sectors achieve significant reductions. In addition, aviation's impact on air quality is changing as more fuel-efficient, quieter aircraft engines are placed in service. While new aircraft engine technologies have reduced fuel consumption, noise, and emissions of most pollutants, they have not achieved the same level of reductions in nitrogen oxide emissions, which contribute to ozone formation. According to FAA, nitrogen oxide emissions from aviation will increase by over 90 percent by 2025 without improvements in aircraft emissions technologies and air traffic management, and emissions of other air pollutants will also increase, as shown in figure 2. Additionally, aviation's greenhouse gas emissions and potential contribution to climate change is expected to increase. IPCC has estimated that aircraft emissions are likely to grow by 3 percent per year, outpacing the emissions reductions achieved through technological improvements. Furthermore, as emissions from other sources decline, aviation's contribution to climate change may become proportionally larger, according to FAA. Alternative fuels are not yet available in sufficient quantities for jet aircraft, as they are for some other uses, and therefore aviation cannot yet adopt this approach to reduce its greenhouse gas emissions (see discussion below on U.S. efforts to develop alternative fuels for aviation).

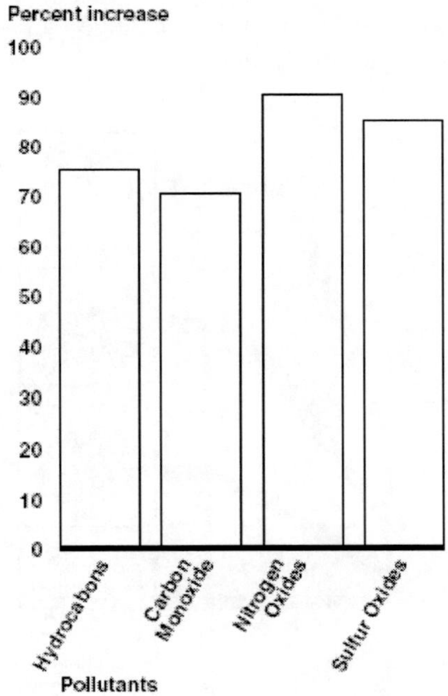

Source: FAA.

Note: According to FAA, the increases in aviation-related pollutants are baseline forecasts that do not account for potential improvements in aircraft technology and air traffic management.

Figure 2. FAA Analysis of Growth in Aviation Related Pollutants by 2025.

Aviation emissions, like other combustible emissions, include pollutants that affect health. While it is difficult to determine the health effects of pollution from any one source, the nitrogen oxides produced by aircraft engines contribute to the formation of ozone, the air pollutant of most concern in the United States and other industrialized countries. Ozone has been shown to aggravate respiratory ailments. A National Research Council panel recently concluded that there is strong evidence that even short-term exposure to ozone is likely to contribute to premature deaths of people with asthma, heart disease, and other preexisting conditions. With improvements in aircraft fuel efficiency and the expected resulting increases in nitrogen oxide emissions, aviation's contribution to ozone formation may increase. In addition, aviation is associated with other air pollutants, such as hazardous air pollutants, including benzene and formaldehyde, and particulate matter, all of which can adversely affect health. Data on emissions of hazardous air pollutants in the vicinity of airports are limited, but EPA estimates that aviation's production of these pollutants is small relative to other sources, such as on-road vehicles. Nevertheless, according to EPA, there is growing public concern about the health effects of the hazardous air pollutants and particulate matter associated with aviation emissions. See appendix I for more detailed information on the health and environmental effects of aviation emissions.

Carbon dioxide and other greenhouse gas emissions from aircraft operations in the atmosphere, together with ground-level aviation emissions that gradually rise into the atmosphere, contribute to global warming and climate change. IPCC's most recent report[17] documents mounting evidence of global warming and projects the potential catastrophic effects of climate change. As figure 6 shows, climate change affects precipitation, sea levels, and winds as well as temperature, and these changes in turn will increasingly affect economies and infrastructure around the world.

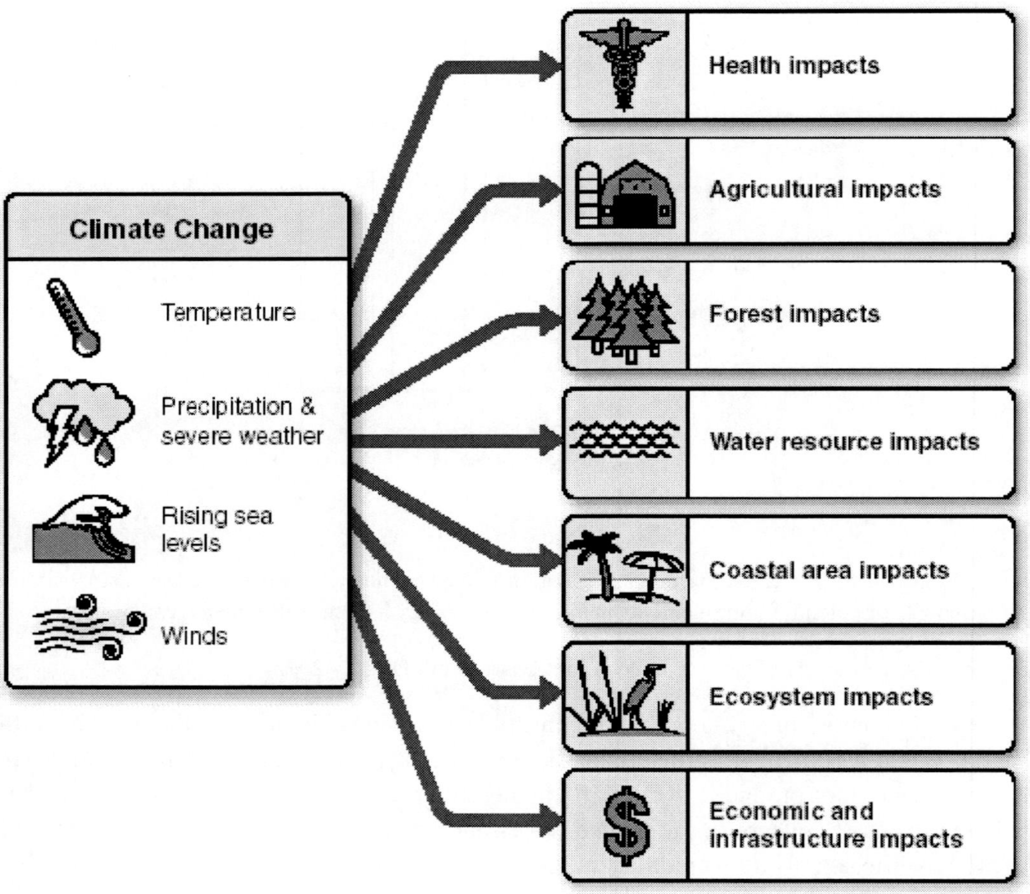

Source: EPA and FAA.

Figure 3. Concerns about the Effects of Climate Change.

KEY FEDERAL EFFORTS TO ADDRESS AVIATION EMISSIONS INCLUDE NEAR-TERM OPERATIONAL CHANGES AND LONGER-TERM R&D INITIATIVES

Two key federal efforts, if implemented effectively, can help to reduce aviation emissions—near-term NextGen initiatives and an array of R&D programs over the longer term to fully enable NextGen and to reduce aircraft emissions. The NextGen initiatives are primarily intended to improve the efficiency of the aviation system so that it can handle expected increases in air traffic, but these initiatives can also help reduce aviation emissions. In addition, the federal government, led by FAA and NASA, has longer-term R&D programs in place to improve the scientific understanding of the impact of aviation emissions in order to inform decisions about emissions-reduction strategies, explore potential emissions-reducing alternative fuels, and develop NextGen and aircraft emissions-reduction technologies.

NextGen Initiatives Have the Potential to Help Reduce Emissions

Technologies and procedures that are being developed as part of NextGen to improve the efficiency of flight operations can also reduce aircraft emissions. According to FAA, the implementation of NextGen could reduce greenhouse gas emissions from aircraft by up to 12 percent. One NextGen technology, considered a centerpiece of NextGen, is the Automatic Dependent Surveillance-Broadcast (ADS-B) satellite aircraft navigation system. ADS-B is designed, along with other navigation technologies, to enable more precise control of aircraft during en route flight, approach, and descent. ADS-B will allow for closer and safer separations between aircraft and more direct routing, which will improve fuel efficiency and reduce carbon dioxide emissions. This improved control will also facilitate the use of air traffic control procedures that will reduce communities' exposure to aviation emissions and noise. One such procedure, Continuous Descent Arrivals (CDA), allows aircraft to remain at cruise altitudes longer as they approach destination airports, use lower power levels, and thereby lower emissions and noise during landings. Figure 3 shows how CDA compares with the current step-down approach to landing, in which aircraft make alternate short descents and forward thrusts, which produce more emissions and noise than continuous descents. A limited number of airports have already incorporated CDA into their operations. For example, according to officials from Los Angeles International Airport, nearly 25 percent of landings at their airport use CDA procedures in one of the airport's standard terminal approaches. In addition, United Parcel Service plans to begin using a nighttime CDA procedure, designed and tested at the Louisville International Airport, for its hub operations.

Two closely associated NextGen initiatives, Area Navigation (RNAV) and Required Navigation Performance (RNP), have the potential to modify the environmental impact of aviation by providing enhanced navigational capability to the pilot. RNAV equipment can compute an airplane's position, actual track, and ground speed, and then provide meaningful information on the route of flight selected by the pilot. RNP will permit the airplane to descend on a precise route that will allow it to avoid populated areas, reduce its consumption of fuel, and lower its emissions of carbon dioxide and nitrogen oxides. [18] See figure 4.

Currently, over 350 RNAV/RNP procedures are available at 54 airports, including Dallas/Fort Worth, Miami International, Washington Dulles, and Atlanta Hartsfield.

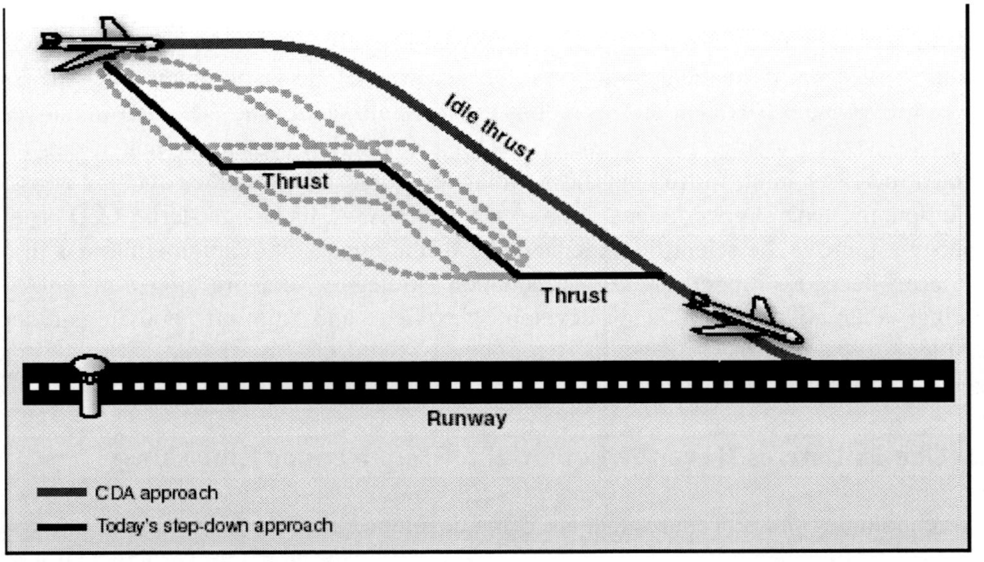

Sources: Naverus and AVTECH.

Note: Continuous Descent Arrivals keep aircraft higher for longer and have them descend at near-idle power to touchdown. Optimal profiles are not always possible, especially at busy airports.

Figure 4. Comparison of CDA and Current Step-Down Approach.

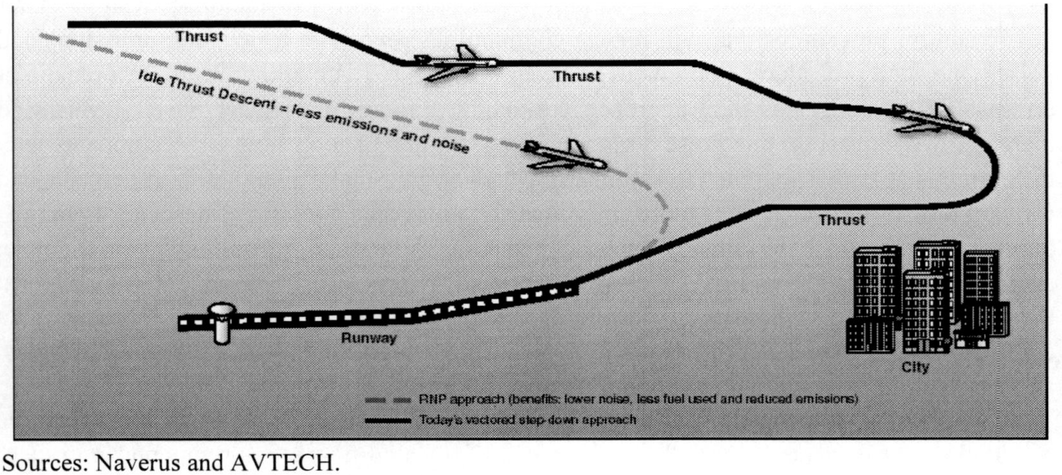

Sources: Naverus and AVTECH.

Note: An RNP approach and path allows for idle-thrust, continuous descent instead of today's step-down approaches with vectors. RNP precision and curved-approach flexibility can shift flight paths to avoid populated areas.

Figure 5. Comparison of RNP and Current Step-Down Approach.

Still another NextGen initiative, High-Density Terminal and Airport Operations, is intended to improve the efficiency of aircraft operations at busy airports, and, in the process, reduce emissions. At high-density airports, the demand for access to runways is high, and arrivals and departures take place on multiple runways. The combination of arrivals, departures, and taxiing operations may result in congestion, which in turn produces delays, emissions, and noise as aircraft wait to take off and land. Under the High-Density Terminal and Airport Operations initiative, which FAA has just begun to implement, aircraft arriving and departing from different directions would be assigned to multiple runways and safely merged into continuous flows despite bad weather and low visibility. To guarantee safe separation, these airports would need enhanced navigation capabilities and controllers with access to increased automation. Under this initiative, aircraft would also move more efficiently on the ground, using procedures that are under development to reduce spacing and separation requirements and improve the flow of air traffic into and out of busy metropolitan airspace. More efficient aircraft movement would increase fuel efficiency and reduce emissions and noise. Although the implementation of this initiative is in the early stages, FAA has identified the R&D needed to move it forward.

Technologies and procedures planned for NextGen should also help improve the efficiency of flights between the United States and other nations, further reducing emissions, particularly of greenhouse gases. A test program scheduled to begin in the fall of 2008, known as the Atlantic Interoperability Initiative to Reduce Emissions (AIRE), sponsored by FAA and the European Commission, Boeing, and Airbus, will involve gate-togate testing of improved procedures on the airport surface, during departures and arrivals, and while cruising over the ocean. Some of the procedures to be tested will use technologies such as ADS-B. A similar effort—the Asia and South Pacific Initiative to Reduce Emissions (ASPIRE)—was launched earlier this year, involving the United States, Australia, and New Zealand.

Federal R&D Focuses on Long-Term Approaches to Addressing Aviation Emissions

We have previously reported[19] that the federal government and industry have achieved significant reductions in some aircraft emissions, such as carbon dioxide, through past R&D efforts, and federal officials and aviation experts agree that such efforts are the most effective means of achieving further reductions in the longer term[20]. As part of the a national plan for aeronautics R&D, issued by the White House Office of Science and Technology Policy, the federal government supports a comprehensive approach to R&D on aviation emissions that involves FAA, NASA, and other federal agencies. According to FAA, this approach includes efforts to improve the scientific understanding of the nature and impact of aviation emissions and thereby inform the development of more fuel-efficient aircraft, of alternative fuels that can reduce aircraft emissions, and of air traffic management technologies that further improve the efficiency of aviation operations. NASA, industry, and academia are important partners in these efforts. Notably, however, the development of breakthrough technologies, such as highly fuel-efficient aircraft engines that emit fewer greenhouse gases and air pollutants, is expensive and can take a long time, both to conduct the research and to implement the new technologies in new aircraft designs and introduce these new aircraft into the fleet. Successfully developing these technologies also requires the support and cooperation of stakeholders throughout the aviation industry.

FAA Supports Research on Improving the Scientific Understanding of Aviation Emissions and on Alternative Fuels

Improving the scientific understanding of aviation emissions can help guide the development of approaches to reducing emissions by improving aircraft manufacturers' and operators' and policy makers' ability to assess the environmental benefits and costs of alternative policy measures. Such an assessment can then lead to the selection of the alternative that will achieve the greatest net environmental benefits. For example, one technology might greatly increase fuel efficiency, but produce higher nitrogen oxide emissions than another, somewhat less fuel-efficient technology. Overall, a cost benefit analysis might indicate that the less fuel-efficient technology would produce greater net benefits for the environment.

FAA currently supports several recent federal efforts to better quantify aviation emissions and their impact through improvements in emissions measurement techniques and modeling capability. One of these efforts is FAA's Partnership for Air Transportation and Emissions Reduction (PARTNER) Center of Excellence.[21] Created in 2003, PARTNER carries on what representatives of airlines, aircraft and engine manufacturers, and experts in aviation environmental research have described as a robust research portfolio. This portfolio includes efforts to measure aircraft emissions and to assess the human health and welfare risks of aviation emissions and noise. For example, researchers are developing an integrated suite of three analytical tools—the Environmental Design Space, the Aviation Environmental Design Tool, and the Aviation Environmental Portfolio Management Tool – that can be used to identify interrelationships between noise and emissions. Data from these three tools, together with the Aviation Environmental Design tool being developed by the Volpe National Transportation Systems Center and others, will allow for assessing the benefits and costs of aviation environmental policy options. Another R&D initiative, the Airport Cooperative Research Program (ACRP),[22] conducts applied research on aviation emissions and other environmental issues facing airports. The program is managed by the National Academies of Science through its Transportation Research Board under a contract with FAA, which provided $10 million for the program in both 2007 and 2008 and is seeking to increase these investments through its reauthorization to specifically focus on aviation environmental issues. Several of the emissions-related projects undertaken through ACRP have concentrated on developing methods to measure particulate matter and hazardous air pollutants at airports in order to identify the sources of these pollutants and determine whether their levels could have adverse health effects. FAA has also developed an Aviation Emissions Characterization roadmap to provide a systematic process to enhance understanding of aviation's air quality emissions, most notably particulate matter and hazardous air pollutants. In addition, FAA, in conjunction with NASA and the National Oceanic and Atmospheric Administration, launched the Aviation Climate Change Research Initiative to develop the scientific understanding necessary for informing efforts to limit or reduce aviation greenhouse gas emissions.

Another effort, the Commercial Aviation Alternative Fuels Initiative (CAAFI),[23] led by FAA, together with airlines, airports, and manufacturers, is intended to identify and eventually develop alternative fuels for aviation that could lower emissions of greenhouse gases, and other pollutants; increase fuel efficiency; and reduce U.S. dependence on foreign oil. CAAFI supports research on low-carbon fuel from sources such as plant oils, algae, and biomass that are as safe as petroleum-based fuel and compare favorably in terms of environmental impact. Part of the research will involve assessing the environmental impact of alternative fuels to determine whether their use could reduce emissions of pollutants that affect climate and air quality. The research will also assess the impact of producing these fuels on the overall carbon footprint. The CAAFI sponsors have set goals for certifying a 50 percent synthetic fuel for aviation use in 2008, a 100 percent synthetic fuel for use by 2010, and a biofuel made from renewable resources such as palm, soy, or algae oils. As part of CAAFI, Virgin Atlantic Airlines, together with Boeing, has tested a blend of kerosene (normal jet fuel) and biofuels in a flight from London to Amsterdam, and Continental, in association with Boeing and jet engine manufacturer General Electric, is planning a similar test in 2009.

NASA Conducts Fundamental Aeronautics R&D in Support of NextGen, Including Efforts That Can Help Lower Emissions

NASA has devoted a substantial portion of its aeronautical R&D program to the development of technologies critical to the implementation of NextGen, as well as new aircraft and engine technologies, both of which can help reduce aviation emissions.

NASA has three main aeronautics research programs – Fundamental Aeronautics, Aviation Safety, and Airspace Systems – each of which contributes directly and substantially to NextGen. For example, the Airspace Systems program supports research on air traffic management technologies for NextGen, and the Fundamental Aeronautics program focuses on removing environmental and performance barriers, such as noise and emissions, that could constrain the capacity enhancements needed to accommodate projected air traffic increases. Appendix II describes in more detail how NASA's aeronautics R&D programs support the implementation of NextGen.

NASA also works with aircraft and aircraft engine manufacturers to increase fuel efficiency and reduce emissions. Their efforts have contributed to a number of advancements in aircraft engine and airframe technology, and NASA's R&D on emissions-reduction technologies continues. NASA has set technology-level goals for reducing greenhouse gases, nitrogen oxides, and noise, which have become part of the U.S. National Aeronautics Plan. For example, the plan includes a goal for developing technologies that could reduce nitrogen oxide emissions during landings and takeoffs by 70 percent[24] below the ICAO current standard. The plan also sets a goal of increasing fuel efficiency (and thereby decreasing greenhouse gases emissions) by 33 percent. These technologies would be incorporated in the next generation of aircraft, which NASA refers to as N+1,[25] by 2015. However, as NASA officials note, these goals must be viewed within the context that each of the goals can be fully met only if it is the only goal. For example, the goal for reducing nitrogen oxides can be fully achieved only at the expense of the goals for lowering greenhouse gas emissions and noise, because it is technologically challenging to design aircraft that can simultaneously reduce all of these environmental impacts.

For the longer term (2020), NASA is focusing on developing tools and technologies for use in the design of advanced hybrid-wing body aircraft, the following generation of aircraft, or N+2. Emissions from these aircraft would be in the range of 80 percent below the ICAO standard for nitrogen oxide emissions during landings and takeoffs, and fuel consumption would be 40 percent less than for current aircraft. The U.S. aircraft and engine manufacturing industry has also set goals for reducing aircraft emissions in the engines the industry plans to produce. According to the Aerospace Industries Association, which represents this industry, its members have set a goal of reducing carbon dioxide emissions by 15 percent in the next generation of aircraft while continuing to significantly reduce nitrogen oxide emissions and noise.

The development of aircraft technologies such as those that NASA is currently working on to reduce emissions can take a long time, and it may be years before the technologies are ready to be incorporated into new aircraft designs. According to FAA, the development process generally takes 12 to 20 years. For example, the latest Pratt and Whitney engine, the geared turbofan, which is expected to achieve significant emissions and noise reductions, took 20 years to develop.

Several Steps Can Be Taken to Help Reduce Aviation Emissions, but Challenges Remain to Be Addressed

Reducing aviation emissions includes steps that FAA and others can take to move the implementation of NextGen forward and support R&D on NextGen and emissions-reduction technologies, as well as technical, financial, regulatory challenges facing the federal government, the aviation industry, and Congress.

Expediting the Implementation of NextGen Can Help Reduce Aviation Emissions

Implementing NextGen expeditiously is essential to handle the projected growth in air traffic efficiently and safely, and in so doing, help to reduce aircraft emissions. Steps to advance NextGen's implementation include management improvements and the deployment of available NextGen components.

Management Improvements Can Move NextGen Forward More Efficiently

Several management actions are important to advance the implementation of NextGen. One such action is to establish a governance structure within FAA that will move NextGen initiatives forward efficiently and effectively. FAA has begun to establish a governance structure for NextGen, but it may not be designed to give NextGen initiatives sufficient priority to ensure the system's full implementation by 2025. Specifically, FAA's implementation plan for NextGen is called the Operational Evolution Partnership (OEP). The manager responsible for OEP is one of nine Vice Presidents who report to the Chief Operating Officer (COO) of FAA's Air Traffic Organization (ATO), who reports directly to the FAA Administrator. While the manager responsible for OEP is primarily responsible for implementing NextGen, other Vice Presidents are responsible for NextGen-related activities in their designated areas. In addition, the FAA managers responsible for airports and aviation safety issues are Associate Administrators who report through the Deputy FAA Administrator to the FAA Administrator. Some of the activities for which these Associate Administrators are responsible are critical to NextGen's implementation, yet there is no direct line of authority between the OEP manager and these activities.

Some congressional leaders and other stakeholders, including aviation industry representatives and aviation experts, view FAA's management structure for NextGen as too diffuse. Some of the stakeholders have called for the establishment of a position or NextGen program office that reports directly to the FAA Administrator to ensure accountability for NextGen results. These stakeholders have expressed frustration that a program as large and important as NextGen does not follow the industry practice of having one person with the authority to make key decisions. They point out that although the COO is nominally in charge of NextGen, the COO must also manage FAA's day-to-day air traffic operations and may therefore not be able to devote enough time and attention to managing NextGen. In addition, these stakeholders note that many of NextGen's capabilities span FAA operational units whose heads are on the same organizational level as the head of OEP or are outside ATO, and they believe that an office above OEP and these operational units is needed. In prior work, we have found that programs can be implemented most efficiently when managers are empowered to make critical decisions and are held accountable for results.[26]

Another management action is needed to help ensure that FAA acquires the skills required for implementation, such as contract management and systems integration skills. Because of the scope and complexity of the NextGen implementation effort, FAA may not have the in-house expertise to manage it without assistance. In November 2006, we recommended that FAA examine its strengths and weaknesses and determine whether it has the technical expertise and contract management expertise that will be needed to define, implement, and integrate the numerous complex programs inherent in the transition to NextGen.[27] In response to our recommendation, FAA has contracted with the National Academy of Public Administration (NAPA) to determine the mix of skills and number of skilled persons, such as technical personnel and program managers, needed to implement the new OEP and to compare those requirements with FAA's current staff resources. In December 2007, NAPA provided FAA with its report on the types of skills FAA will require to implement NextGen, and it has undertaken a second part of the study that focuses on identifying any skill gaps between FAA's current staff and the staff that would be required to implement NextGen.[28] NAPA officials told us that they expect to publish the findings of the second part of the study in the summer of 2008. We believe this is a reasonable approach that should help FAA begin to address this challenge as soon as possible. It may take considerable time to select, hire, train, and integrate into the NextGen initiative what could be a large number of staff.

We have also identified potential approaches for supplementing FAA's capabilities, such as having FAA contract with a lead systems integrator (LSI)–that is, a prime contractor who would help to ensure that the discrete systems used in NextGen will operate together and whose responsibilities may include designing system solutions, developing requirements, and selecting major system and subsystem contractors.[29] However, this approach would require careful oversight to ensure that the government's interests are protected and could pose significant project management and oversight challenges for the Joint Planning and Development Office (JPDO), the organization within FAA responsible for planning NextGen, and for FAA.

Deploying Available NextGen Components Can Demonstrate Their Ability to Operate Together and Achieve Anticipated Efficiencies

Moving from planning to implementing some components of NextGen can begin to demonstrate the potential of the system as well as reduce congestion in some areas of the country, thereby also reducing emissions. Many of the technologies and procedures planned for NextGen are already available, and a few have been implemented individually, such as the CDA procedures in use in Los Angeles and Louisville and ADS-B in Alaska. However, the available technologies and procedures have not yet been deployed simultaneously to demonstrate that they can be operated safely as an integrated suite of technologies and procedures in the national airspace system. Several stakeholders have suggested that FAA consider a gradual rollout of NextGen technologies and procedures in a particular area. For example ADS-B technologies, CDA and RNAV/RNP procedures, and high-density airport operations could be deployed in a defined area of the current system, possibly in sequence over time, to test their combined use and demonstrate the safety of an integrated suite of NextGen advancements. Such a graduated rollout is sometimes referred to as "NextGen Lite." FAA is currently considering a demonstration project in Florida and Georgia, in which it, together with aviation equipment manufacturers and municipalities, would use the NextGen capabilities of ADS-B, RNAV, and RNP for on-demand air taxi fleet[30] operations. As other NextGen capabilities, such as System-Wide Information Management (SWIM),[31] are deployed and as air taxi fleet operations move to other airports and regions, the demonstration will be expanded to include those new capabilities and other airports and regions. According to the airlines and other stakeholders we interviewed, a demonstration of the successful integration of NextGen capabilities and of efficiencies resulting from their use would give the airlines an incentive to equip their aircraft with NextGen technologies. They could then lower their costs by reducing their fuel consumption and decrease the impact of their operations on the environment. The findings from our research indicate that such regional or targeted demonstrations could accelerate the delivery of NextGen benefits while helping to ensure safe operations within the current system. In addition, demonstrations can increase stakeholders' confidence in the overall NextGen initiative.

Resolving Aeronautics R&D Funding Issues Is a Further Step in Addressing Aviation Emissions

Federal funding for aeronautics research, the category that includes work on aviation emissions, has declined over the past decade, particularly for NASA, which historically provided most of the funding for this type of research. NASA's current aeronautics research budget is about half of what it was in the mid-1990s. Moreover, the budget request for aeronautics R&D for fiscal year 2009 is $447 million, or about 25 percent less than the $594 million provided in fiscal year 2007. (See table 1.) According to NASA, about $280 million of the proposed $447 million would contribute to NextGen. In addition, according to NASA officials, a significant portion of the funding for subsonic fixed-wing aircraft is directed toward emissions-related research, and many other research efforts contribute directly or indirectly to potential emissions-reduction technologies.

Table 1. The President's Budget for NASA's Aeronautics Programs for Fiscal Years 2007 and 2008 and Budget Projections for Fiscal Years 2009 through 2013

(Dollars in millions)							
			Fiscal year				
	Enacted		Requeste		Proposed		
Program	2007	2008	2009	2010	2011	20112	2013
Aviation Safety							
Integrated Vehicle Health	30.7	22.2	19.7	19.9	18.8	18.6	19.2
Aging Aircraft	14.9	10.0	10.6	11.3	11.2	12.0	12.4
Integrated Resilient Aircraft Control	22.2	15.3	17.1	18.5	19.0	18.2	18.8
Integrated Intelligent Flight Deck echnologies	19.5	19.3	15.2	16.3	16.0	15.7	16.1
Subtotal	87.3	66.5	62.6	65.9	65.0	64.5	66.5
Airspace Systems							
NextGen – Airspace	85.1	83.3	61.3	56.0	57.3	58.5	60.8
NextGen – Airportal	17.4	16.8	13.3	16.7	16.9	16.9	17.5
Subtotal	102.5	100.1	74.6	72.7	74.2	75.4	78.4
Fundamental Aeronautics							
Subsonic – Rotary Wing	36.1	30.8	25.8	26.6	26.7	26.9	28.0
Subsonic – Fixed Wing	133.9	119.9	108.4	105.3	107.6	109.1	111.5
Supersonics	67.7	53.0	44.0	44.9	44.3	45.2	46.6
Hypersonics	92.8	66.2	57.3	56.4	56.5	57.4	58.4
Subtotal	330.4	269.9	235.4	233.2	235.2	238.6	244.6
Aeronautics Test Program							
Aero Ground Test Facilities	48.5	50.0	48.2	49.4	50.8	51.0	51.0
Flight Operations and Test	25.0	25.1	25.6	26.4	27.2	27.2	27.2
Subtotal	73.5	75.1	73.9	75.8	78.0	78.2	78.2
Total	593.8	511.7	446.5	447.5	452.4	456.7	467.7

Source: NASA.

Note: Most of the research on aircraft emissions reductions that NASA performs is funded through the Fundamental Aeronautics – Fixed Wing program.

As its funding for aeronautics R&D has declined, NASA has emphasized fundamental research, which serves as the basis for developing technologies and tools that can later be integrated into aviation systems, and has focused less on developmental and demonstration work. As a result, NASA is now sometimes developing technologies to a lower maturity level than in the past, and the technologies are less ready for manufacturers to adopt them, resulting in a gap in the research needed to bring technologies to a level where they can be transferred to industry for further development. Failure to address this gap could postpone the development of emissions-reduction technologies.

As a partial response to the gap, the administration has proposed some additional funding for FAA that could be used to further develop NASA's and others' emissions- and noise reduction technologies. Specifically, FAA's reauthorization proposal seeks $111 million through fiscal year 2011 for the CLEEN Engine and Airframe Technology Partnership,[32] which FAA officials said is intended to provide for earlier maturation of emissions and noise technologies while NASA focuses on longer-term fundamental research on noise and emissions. The CLEEN partnership, which is also contained in the House's FAA reauthorization bill,[33] would create a program for the development and maturation of certifiable engine and airframe technologies for aircraft over the next 10 years which would reduce aviation noise and emissions. The legislation would require the FAA Administrator, in coordination with the NASA Administrator, to establish objectives for developing aircraft technology outlined in the legislation. The technology requested to be developed would increase aircraft fuel efficiency enough to reduce greenhouse gas emissions by 25 percent relative to 1997 subsonic jet aircraft technology, and, without increasing other gaseous or particle emissions, reduce takeoff-cycle nitrogen oxide emissions by 50 percent relative to ICAO's standard. Although FAA's reauthorization bill has not yet been enacted, the administration's proposed fiscal year 2009 budget includes $10 million for the CLEEN program.

The CLEEN program would be a first step toward further maturing emissions and noise reduction technologies, but experts agree that the proposed funding is insufficient to achieve needed emissions reductions. While acknowledging that CLEEN would help bridge the gap between NASA's R&D and manufacturers' eventual incorporation of technologies into aircraft designs, aeronautics industry representatives and experts we consulted said that the program's funding levels may not be sufficient to attain the goals specified in the proposal. According to these experts, the proposed funding levels would allow for the further development of one or possibly two projects. Moreover, in one expert's view, the funding for these projects may be sufficient only to develop the technology to the level that achieves an emissions-reduction goal in testing, not to the level required for the technology to be incorporated into a new engine design. Nevertheless, according to FAA and some experts we consulted, the CLEEN program amounts to a pilot project, and if it results in the development of emissions-reduction technologies that can be introduced into aircraft in the near future, it could lead to additional funding from the government or industry for such efforts.

FAA and NASA have identified the R&D that is needed for NextGen, but have not determined what needs to be done first, at what cost, to demonstrate and integrate NextGen technologies into the national airspace system. Completing this prioritization is critical to avoid spending limited funds on lower-priority efforts or conducting work out of sequence. Once the identified R&D has been prioritized and scheduled, cost estimates can be developed and funds budgeted. Prioritizing research needs is an essential step in identifying the resources required to undertake the research.

The European Union is investing substantially in R&D that can lead to fuel-efficient, environmentally friendly aircraft. In February 2008, the European Union announced the launch of the Clean Sky Joint Technology Initiative, with total funding of $2.4 billion over 7 years—the European Union's largest-ever research program. The initiative establishes a Europe- wide partnership between industry, universities, and research centers and aims to reduce aircraft emissions of carbon dioxide and nitrogen oxides by up to 40 percent and aircraft noise levels by 20 decibels. According to FAA, it is difficult to compare funding levels for U.S. and European R&D efforts because of differences in program structures and funding mechanisms, Nevertheless, foreign government investments of such magnitude in R &D on environmentally beneficial technologies could reduce the competitiveness of the U.S. aircraft manufacturing industry, since greater investments are likely to lead to greater improvements in fuel efficiency and keep U.S. aircraft manufacturers competitive in the global economy as well as reducing aviation's impact on the environment.

Reducing the Impact of Aviation Emissions Poses Technical, Financial, and Regulatory Challenges

Reducing aviation emissions will require technological advances, the integration of lower-emitting aircraft and NextGen technologies into airline fleets, and strengthened or possibly new regulations to improve air quality and limit greenhouse gas emissions. Fulfilling these requirements will pose challenges to aviation because of the technical difficulties involved in developing technologies that can simultaneously address air pollutants, greenhouse gases, and noise; constraints on the airline industry's resources to invest in new aircraft and technologies needed to reduce emissions and remain competitive; and the impact that emissions regulations can have on the aviation system's expansion and the financial health of the aviation industry.

Simultaneously Addressing Air Pollutants, Greenhouse Gases, and Noise from Aircraft Presents Technical Challenges

Although the aviation industry has made strides in lowering emissions, more reductions are needed to keep pace with the projected growth in aviation, and achieving these reductions will be technically challenging. NASA's efforts to improve jet engine designs illustrate this challenge: While new designs have increased fuel efficiency, reduced most emissions, and lowered noise, they have not achieved comparable reductions in nitrogen oxide emissions. Nitrogen oxide emissions have increased because new aircraft engines operate at higher temperatures, producing more power with less fuel and lower carbon dioxide and carbon monoxide emissions, but also producing higher nitrogen oxide emissions, particularly during landings and takeoffs, when engine power settings are at their highest. It is during the landing/takeoff cycle that nitrogen oxide emissions also have the greatest impact on air quality. As discussed, nitrogen oxides contribute to ground-level ozone formation. Similarly, as we noted in a report on NASA's and FAA's aviation noise research earlier this year,[34] it is technologically challenging to design aircraft engines that simultaneously produce less noise and fewer greenhouse gas and other emissions. Although it is possible to design such engines, the reductions in greenhouse gases could be limited in engines that produce substantially less noise. NASA and industry are working on technologies to address these environmental trade-offs. For example, the Pratt & Whitney geared turbo fan engine that we mentioned earlier is expected to cut nitrogen oxide emissions in half while also improving fuel efficiency and thereby lowering carbon dioxide emissions. Nevertheless, it remains technologically challenging to design aircraft that can reduce one environmental concern without increasing another.

In a 2004 report to Congress on aviation and the environment,[35] FAA noted that the interdependencies between various policy, technological, and operational options for addressing the environmental impacts of aviation and the full economic consequences of these options had not been appropriately assessed. However, in recent years, FAA has made progress in this area, including its sponsorship of the previously mentioned PARTNER study on the interrelationships between noise and emissions. This study can be used to assess the costs and benefits of aviation environmental policy options.

The Financial Condition of the Airline Industry Creates a Challenge to Implementing Emissions-Reduction Technologies

Most U.S. airlines have stated that they plan to invest in aircraft and technologies that can increase fuel efficiency and lower emissions, but in the near term, integrating new aircraft into the fleet, or retrofitting aircraft with technologies that can improve their operational efficiency, poses financial challenges to the airline industry. Aircraft have an average lifespan of about 30 years, and the airlines can take almost that entire period to pay for an aircraft. The current fleet is, on average, about half as many years old—11 years for wide-body aircraft, and 14 years for narrow- body aircraft—and therefore is expected to be in operation for many years to come. In addition, the financial pressures facing many airlines make it difficult for them to upgrade their fleets with new, state-of-the-art aircraft, such as the Boeing 787 and Airbus A380, which are quieter and more fuel efficient, emitting lower levels of greenhouse gases.[36] Currently, U.S. carriers have placed a small proportion (40, or less than 6 percent) of the over 700 orders that Boeing officials say the company has received for its 787 model. Furthermore, no U.S. carriers have placed orders for the new Airbus 380. These financial pressures also limit the airlines' ability to equip new and existing aircraft with NextGen technologies such as ADS-B that can enable more efficient approaches and descents, resulting in lower emissions levels. FAA estimates that it will cost the industry about $14 billion to equip aircraft to take full advantage of NextGen.

Delays by airlines in introducing more fuel-efficient, lower-emitting aircraft into the U.S. fleet and in equipping or retrofitting the fleet with the technologies necessary to operate NextGen could limit FAA's ability to efficiently manage the forecasted growth in air traffic. Without significant reductions in emissions and noise around the nation's airports, efforts to expand their capacity could be stalled and the implementation of NextGen delayed because of concerns about the impact of aviation emissions. As we previously reported,[37] offering operational advantages, such as preferred takeoff and landing slots, to fuel-efficient, lower-emitting aircraft or aircraft equipped with ADS-B could create incentives for the airlines to invest in the necessary technologies. Similarly, as noted, deploying an integrated suite of NextGen technologies and procedures in a particular region could create incentives for carriers to equip their aircraft with NextGen technologies.

More Stringent Regulatory Standards Pose Challenges for Airport Expansion Projects

Concerns about the health effects of air pollutants have led to more stringent air quality standards that could increase the costs or delay the implementation of airport expansion projects. In recent years, EPA has been implementing a more stringent standard for ozone emissions to better protect the health of people exposed to it, and this standard could require more airports to tighten controls on nitrogen oxides and some types of volatile organic compounds that also contribute to ozone formation. Under the current standard,[38] 122 airports are located in areas that are designated as nonattainment areas. This number includes 43 of the 50 busiest U.S. commercial service airports. In March 2008, EPA further revised the ozone standard, because new evidence demonstrated that exposure to ozone at levels below the level of the current standard are associated with a broad array of adverse health effects.[39] This recent revision to the ozone standard will increase the number of U.S. counties, and hence airports, that will be in nonattainment. EPA estimated that the number of affected counties could potentially grow from 104 to 345 nationwide. While the exact number of airports that will be affected has not been officially determined at this time, FAA estimates that a modest number of commercial service airports in California, Arizona, Utah, Texas, Oklahoma, Arkansas, and along the gulf coast to Florida will be in nonattainment areas for the revised 8-hour ozone standard. According to EPA, any development project beginning in 2011 at these airports would have to conform to the state implementation plan.

As communities gain more awareness of the health and environmental effects of aviation emissions, opposition to airport expansion projects, which has thus far focused primarily on aviation noise, could broaden to include emissions. According to a California air quality official, many of the same communities that have interacted with airports over aviation noise have more recently recognized that they could also be affected by emissions from airport sources. In Europe, concerns about the impact of aviation on air quality and climate change have led to public demands for tighter control over aircraft emissions, and these demands have hindered efforts to expand airports in Birmingham, and London (Heathrow). Moreover, a plan to expand London's Stansted Airport was rejected because of concerns about climate change that could result from additional emissions.

To minimize constraints on the future expansion of airport capacity stemming from concerns about the health and environmental effects of aviation emissions, it will be important for airports; the federal and state governments; and the airline industry to work together to accurately characterize and address these concerns and to take early action to mitigate emissions. As noted, constraints on efforts to expand airports or aviation operations could affect the future of aviation because the national airspace system cannot expand as planned without a significant increase in airport capacity. The doubling or tripling of air traffic that FAA expects in the coming decades cannot occur without additional airports and runways.

Market-Based Initiatives to Reduce Aviation Emissions of Greenhouse Gases Could Pose Challenges for U.S. Airlines by Increasing Their Costs

Concerns about the environmental effects of greenhouse gas emissions have grown steadily over the years, leading to national and international efforts to limit them. In the In the United States, EPA has not regulated greenhouse gas emissions;[40] however, Congress is taking steps to deal with climate change, some of which could include market-based measures that would affect the aviation industry. For example, several bills were introduced in the 110th Congress to initiate cap and trade[41] programs for greenhouse gas emissions[42] None of these bills would include aviation directly in a cap and trade program. However, some could have indirect consequences for the aviation industry by, for example, requiring fuel producers to purchase allowances through the system to cover the greenhouse gas content of the fuel they sell to the aviation sector. The cost of purchasing these allowances could be passed on to fuel consumers, including airlines, raising the cost of jet fuel. Fuel is already the airline industry's largest cost. According to the Air Transport Association, cap and trade programs that significantly increase airline fuel costs could have significant consequences for the industry and such programs could make it more difficult for carriers to pay for aircraft or technologies that would reduce greenhouse gas emissions. As we have previously noted,[43] cap and trade programs can cost-effectively reduce emissions of greenhouse gases such as carbon dioxide, especially when compared with other regulatory programs. However, it is important that the impact of such measures on various sectors of the economy, such as the aviation industry, be thoroughly considered.

Internationally, ICAO has not set standards for aircraft carbon dioxide emissions,[44] but it has been working, with the support of FAA, other government aviation authorities, and the aviation industry, to develop a strategy for addressing the impact of aviation on climate change, among several efforts to address climate change. For example, ICAO published a manual for countries, Operational Opportunities to Minimize Fuel Use and Reduce Emissions. In 2004, ICAO endorsed the development of an open emissions trading system as one option countries might use and endorsed draft guidance for member states on establishing the structural and legal basis for aviation's participation in a voluntary open trading system. The guidance includes information on key elements of a trading system, such as reporting, monitoring, and compliance, while encouraging flexibility to the maximum extent possible. In adopting the guidance last fall at the ICAO Assembly, all 190 Contracting States—with the exception of those in the European Union—agreed that the inclusion of one country's airlines in another country's emissions trading system should be based on mutual consent between governments.

Consistent with the requirement to pursue reductions of greenhouse gas emissions from international aviation through ICAO, some countries that have included the aviation sector in their emissions trading systems or other emissions-reduction efforts have, excluded international flights. Consequently, these countries' efforts will not affect U.S. airlines that fly into their airports. The European Union (EU), however, is developing legislation, which has not been finalized, that would include both domestic and international aviation in an emissions trading scheme.[45] As proposed, the EU's scheme would apply to air carriers flying within the EU and to carriers, including U.S. carriers, flying into and out of EU airports in 2012. For example, under the EU proposal, a U.S. airline's emissions in domestic airspace as well as over the high seas would require permits if a flight landed or departed from an EU airport. Airlines whose aircraft emit carbon dioxide at levels exceeding prescribed allowances would be required to reduce their emissions or to purchase additional allowances. Although the legislation seeks to include U.S. airlines within the emissions trading scheme, FAA and industry stakeholders have argued that U.S. carriers would not legally be subject to the legislation.

While the EU's proposal to include international aviation in its emissions trading system is intended to help forestall the potential catastrophic effects of climate change, according to FAA and airlines, it will also affect the aviation industry's financial health. In particular, according to FAA and airline and aircraft and engine manufacturing industry representatives, the EU's proposal could disadvantage U.S. airlines, which have older, less fuel- efficient fleets than their European competitors. Paying for emissions credits could, according to U.S. airlines, also leave them with less money for other purposes, including investing in newer, more fuel-efficient aircraft and technologies to improve flight efficiency and reduce fuel usage. Furthermore, according to U.S. carriers, the proposed trading scheme unfairly penalizes the aviation sector because it lacks a readily available non-carbon-based alternative fuel, whereas other sectors can use alternative fuels to reduce their emissions.

The governments of many nations, including the United States, oppose the European Union's proposal to unilaterally include international aviation in its emissions trading system because the proposed approach is not consistent with ICAO guidance. Furthermore, such an approach could be inconsistent with international aviation agreements and may not be enforceable. According to FAA, the EU's inclusion of aviation in its emissions trading scheme violates the Chicago Convention on International Civil Aviation[46] and other international agreements. FAA further notes that the EU proposal ignores differences in the U.S. and EU aviation systems[47] and ignores a performance-based approach in which countries decide which measures are most appropriate for goals on emissions. We are currently undertaking for this Subcommittee a study of the EU emissions trading system and its potential impact on U.S. airlines, and other issues relating to aviation and climate change.[48]

Mr. Chairman, this concludes my prepared statement. I would be pleased to respond to any questions that you or other Members of the Subcommittee may have.

APPENDIX I: FEDERAL AGENCY VIEWS ON HEALTH AND ENVIRONMENTAL EFFECTS OF AIR POLLUTION

Pollutant	Heath effects	Environmental effects
Ozone	Lung function impairment, effects on exercise performance, increased airway responsiveness, increased susceptibility to respiratory infection, increased hospital admissions and emergency room visits, pulmonary inflammation, and lung structure damage (long term).	Results from animal studies indicate that repeated exposure to high levels of ozone for several months or more can produce permanent structural damage in the lungs. Ozone is also responsible for several billion dollars of agricultural crop yield loss in the United States each year.
Carbon monoxide	Most serious for those who suffer from cardiovascular disease. Healthy individuals are also affected, but only at higher levels of exposure. Exposure to elevated carbon monoxide levels is associated with visual impairment, reduced work capacity, reduced manual dexterity, poor learning ability, and difficulty in performing complex tasks.	Adverse health effects on animals similar to effects on humans.
Nitrogen oxides	Lung irritation and lower resistance to respiratory infections.	Acid rain, visibility degradation, particle formation. Contributes toward ozone formation, and acts as a greenhouse gas in the atmosphere and, therefore, may contribute to climate change.

Appendix I: (Continued).

Pollutant	Heath effects	Environmental effects
Particulate matter	Effects on breathing and respiratory systems, damage to lung tissue, cancer, and premature death. The elderly, children, and people with chronic lung disease, influenza, or asthma, tend to be especially sensitive to the effects of particulate matter.	Visibility degradation, damage to monuments and buildings, safety concerns for aircraft from reduced visibility.
Volatile organic compounds	Eye and respiratory tract irritation, headaches, dizziness, visual disorders, and memory impairment.	Contribute to ozone formation, odors, and have some damaging effect on buildings and plants.
Carbon dioxide, water vapor, and contrails	None.	Act as greenhouse gases in the atmosphere and, therefore, may contribute to climate change. Contrails and contrail-induced clouds produce warming effect regionally where aircraft fly.
Sulfur dioxide	Breathing, respiratory illness, alterations in pulmonary defenses, and aggravation of existing cardiovascular disease.	Together, sulfur dioxide and nitrogen oxides are the major precursors to acid rain, which is associated the acidification of lakes and streams, accelerated corrosion of buildings and monuments, and reduced visibility.

Sources: EPA and FAA.

Appendix II: Examples of the National Aeronautics and Space Administration's Research and Development Programs Supporting NextGen

NextGen research and development (R&D) needs	NextGen capabilities from the National Aeronautics and Space Administration's (NASA) R&D programs
Safety management procedures that can predict, rather than respond to, safety risks, in a high density, complex operating environment; research to support safety analysis, development of advanced materials for continued airworthiness of aircraft, aircraft system and equipage management; and adaptive aircraft control systems to allow the crew and aircraft to recover from unsafe conditions.	Under its Aviation Safety program, NASA research supports development of Safety Management Systems to provide a systematic approach to manage safety risks; integrates prediction and mitigation of risks prior to aircraft accidents or incidents; and shares safety-related information through programs such as the Aviation Safety Analysis and Information Sharing program.
Improved air traffic management technologies to manage airspace configuration, support increases in volume and complexity of traffic demands, mitigate weather impacts, and maintain safe and efficient operations at airports, decrease runway incursions, and address wake vortex issues.	Under its Airspace Systems program, NASA research supports development of variable separation standards based on aircraft performance levels in the en route environment; trajectory-based operations, traffic spacing, merging, metering, flexible terminal airspace, and expanded airport access; technologies and procedures for safe runway procedures in low-visibility conditions; coordinated arrival/departure management; and mitigation of weather and wake vortex issues.

Management of aviation growth to meet the complexity of operations within the NextGen environment, regulation and certification of new manned and unmanned aircraft, and management of operations in an environmentally sound manner.	Under its Fundamental Aeronautics program, NASA research supports development of improved performance for the next generation of conventional subsonic aircraft, rotorcraft and supersonic aircraft and develops methods for environmental management system to measure and assess reductions in air quality impact, noise, and emissions.

Source: GAO analysis of Joint Planning and Development Office and NASA information.

RELATED GAO PRODUCTS

Aviation and the Environment: FAA's and NASA's Research and Development Plans for Noise Reduction Are Aligned, but the Prospects of Achieving Noise Reduction Goals Are Uncertain. GAO-08-384. Washington, D.C.: February 15, 2008.

Aviation and the Environment: Impact of Aviation Noise on Communities Presents Challenges for Airport Operations and Future Growth of the National Airspace System. GAO-08-216T. Washington, D.C.: October 24, 2007.

Climate Change: Agencies Should Develop Guidance for Addressing the Effects on Federal Land and Water Resources. GAO-07-863. Washington, D.C.: August 7, 2007.

Responses to Questions for the Record; Hearing on the Future of Air Traffic Control Modernization. GAO-07-928R. Washington, D.C.: May 30, 2007.

Responses to Questions for the Record; Hearing on JPDO and the Next Generation Air Transportation System: Status and Issues. GAO-07-918R. Washington, D.C.: May 29, 2007.

Next Generation Air Transportation System: Status of the Transition to the Future Air Traffic Control System. GAO-07-748T. Washington, D.C.: May 9, 2007.

Joint Planning and Development Office: Progress and Key Issues in Planning the Transition to the Next Generation Air Transportation System. GAO-07-693T. Washington, D.C.: March 29, 2007.

Next Generation Air Transportation System: Progress and Challenges in Planning and Implementing the Transformation of the National Airspace System. GAO-07-649T. Washington, D.C.: March 22, 2007.

Next Generation Air Transportation System: Progress and Challenges Associated with the Transformation of the National Airspace System. GAO-07-25. Washington, D.C.: November 13, 2006.

Aviation and the Environment: Strategic Framework Needed to Address Challenges Posed by Aircraft Emissions. GAO-03-252. Washington, D.C.: February 28, 2003.

Aviation and the Environment: Transition to Quieter Aircraft Occurred as Planned, but Concerns about Noise Persist. GAO-01-1053. Washington, D.C.: September 28, 2001.

Aviation and the Environment: Aviation's Effects on the Global Atmosphere Are Potentially Significant and Expected to Grow. GAO/RCED-00-57. Washington, D.C.: February 18, 2000.

Aviation and the Environment: Results from a Survey of the Nation's 50 Busiest Airports. GAO/RCED-00-222. Washington, D.C.: August 30, 2000.

Aviation and the Environment: Airport Operations and Future Growth Present Environmental Challenges. GAO/RCED-00-153. Washington, D.C.: August 30, 2000.

REFERENCES

[1] These emissions include airborne pollutants, which affect air quality, and greenhouse gases, primarily carbon dioxide, which are produced by the combustion of fossil fuel, and contribute to climate change.

[2] These figures are based on a long-range FAA forecast using 2006 as the baseline.

[3] See the list of related products at the end of this statement, especially GAO, *Next Generation Air Transportation System: Progress and Challenges in Planning and Implementing the Transformation of the National Airspace System*, GAO-07-649T (Washington, D.C.: Mar. 22, 2007.)

[4] As we noted in our recent testimony before this Subcommittee, aviation noise has been a greater constraint on airport expansion efforts than aviation emissions, but we are limiting our discussion in this testimony to aviation emissions.

[5] 42 U.S.C. §4332(2)(C).

[6] States are required to submit implementation plans to EPA for reducing emissions in areas that fail to meet the National Ambient Air Quality Standards set by EPA under the Clean Air Act for common air pollutants with health and environmental effects (known as criteria pollutants). Geographic areas that have levels of a criteria pollutant above those allowed by the standard are called nonattainment areas.

[7] 42 U.S.C. §7506(c)(1) (The Conformity Provision).

[8] ICAO is an organization affiliated with the United Nations that aims to promote the establishment of international civilian aviation standards and recommended practices and procedures. FAA, as the U.S. representative to ICAO, in consultation with EPA, works with representatives from other countries to formulate aircraft emissions standards.

[9] See the list of related GAO products at the end of this statement, especially GAO, *Aviation and the Environment: Strategic Framework Needed to Address Challenges Posed by Aircraft Emissions,* GAO-03-252 (Washington, D.C.; Feb. 28, 2003).

[10] We are currently undertaking a study on aviation environmental trends, efforts, and challenges for this Subcommittee and the Subcommittee on Space and Aeronautics, Committee on Science and Technology, House of Representatives.

[11] Motor vehicles include cars and buses for airport operations and passenger, employee, and rental agency vehicles.

[12] Ground service equipment includes aircraft tugs, baggage and belt loaders, generators, lawn mowers, snow plows, loaders, tractors, air-conditioning units, and cargo moving equipment.

[13] Hazardous air pollutants from aviation activities include benzene and formaldehyde.

[14] Ground-level ozone is formed when nitrogen oxides and volatile organic compounds as well as other gases and substances are mixed and heated in the atmosphere.

[15] Contrails are clouds and condensation trails that form when water vapor condenses and freezes around small particles (aerosols) in aircraft exhaust.

[16] Intergovernmental Panel on Climate Change, *Aviation and the Global Atmosphere* (1999).

[17] Intergovernmental Panel on Climate Change, Fourth Assessment Report, *Summary for Policy Makers,* Cambridge University Press, Cambridge, UK, November 2007.

[18] A critical component of RNP is the ability of the navigation system to monitor its achieved navigation performance and to identify for the pilot if an operational requirement is or is not being met during an operation.

[19] GAO-03-252.

[20] Alternatively, some scientists studying options for addressing climate change believe that a price on emissions would represent the most effective means of achieving reductions overall.

[21] FAA Centers of Excellence are FAA partnerships with universities and affiliated industry associations and businesses throughout the country that conduct aviation research in a number of areas, including advanced materials, aircraft noise, and aircraft emissions. PARTNER is a cooperative research organization that includes 10 collaborating universities and approximately 50 advisory board members who represent aerospace manufacturers, airlines, airports, state and local governments, and professional and community groups. NASA, FAA, and Transport Canada are sponsors of PARTNER. The collaborating universities and organizations represented on the advisory board provide equal matches for federal funds for research and other activities.

[22] ACRP was authorized in 2003 as part of Vision 100—Century of Aviation Reauthorization Act, Pub. L. 108-176, Section 712 (Dec 12, 2003).

[23] CAAFI, established in October 2006, is sponsored by the Air Transport Association, the Aerospace Industries Association, and the Airports Council International-North America under the direction of FAA, and involves stakeholders from industry, universities, and other federal agencies, including NASA.

[24] This goal is at a pressure ratio of 30, over the ICAO standard adopted at the Committee on Aviation Environmental Protection's sixth meeting (CAEP 6), with commensurate reductions over the full pressure ratio range.

[25] "N" refers to the current generation of tube-and-wing aircraft entering service in 2008, such as the Boeing 787.

[26] See GAO, *Best Practices: Better Support of Weapon System Program Managers Needed to Improve Outcomes*, GAO-06-110 (Washington, D.C.: Nov. 30, 2005). In this study of private-sector best practices that could be applied to federal programs,[26] we found that program managers at highly successful companies were empowered to decide whether programs were ready to move forward and to resolve problems and implement solutions. In addition, program managers were held accountable for their choices.

[27] GAO, *Next Generation Air Transportation System: Progress and Challenges Associated with the Transformation of the National Airspace System*, GAO-07-25 (Washington, D.C.: Nov. 13, 2006).

[28] NAPA, *Workforce Needs Analysis for the Next Generation Air Transportation System (NEXTGEN): Preliminary Findings and Observations* (Washington, D.C.: December 2007).

[29] GAO-07-25.

[30] Air taxis are small aircraft that can be hired to carry passengers or cargo and are regulated under Part 135 of the Federal Aviation Regulations.

[31] SWIM is information-management architecture for the national airspace system, acting as its "World-Wide Web." SWIM will manage surveillance, weather, and flight data, as

well as aeronautical and system status information and will provide the information securely to users.

[32] CLEEN stands for continuous lower energy emissions and noise.

[33] H.R. 2881.

[34] GAO, *Aviation and the Environment: Impact of Aviation Noise on Communities Presents Challenges for Airport Operations and Future Growth of the National Airspace System,* GAO-08-216T (Washington, D.C.: Oct. 24, 2007).

[35] FAA, *Aviation and the Environment: A National Vision Statement, Framework for Goals and Recommended Actions* (Washington, D.C.: December 2004).

[36] We are currently undertaking a study for this Subcommittee and the House Committee on Transportation and Infrastructure that, among other things, will assess the financial condition of the airlines.

[37] GAO, *Aviation and the Environment: FAA'S and NASA's Research and Development Plan's for Noise Reduction Are Aligned, but the Prospects of Achieving Noise Reduction Goals Are Uncertain,* GAO-08-384 (Washington, D.C.: Feb. 15, 2008).

[38] In 2003, EPA began implementing a new standard that called for concentrations of ozone not to exceed 0.08 parts per million over an 8-hour period. The former standard required concentrations not to exceed 0.12 parts per million over a 1-hour period. The more stringent standard resulted in the designation of more nonattainment areas for ozone. These areas contained 12 airports.

[39] 73 Fed. Reg. 16436 (Mar. 27, 2008). The new standard would lower the allowed concentrations of ozone from 0.08 parts per million in an 8-hour period to 0.075 parts per million during that period.

[40] Recently, however, the Supreme Court ruled that greenhouse gases meet the Clean Air Act's definition of an air pollutant and that EPA has the statutory authority to regulate greenhouse gas emissions from new motor vehicles under the Clean Air Act. *Massachusetts v. Environmental Protection Agency*, 127 S.Ct. 1438, 1459-62 (2008). As a result of this opinion, EPA must take one of three actions: (1) issue a finding that greenhouse gas emissions cause or contribute to air pollution that may endanger public heath or welfare; (2) issue a finding that greenhouse gases do not endanger public health or welfare; or (3) provide a reasonable explanation as to why it cannot or will not exercise its discretion to issue a finding. If EPA makes an endangerment finding, the Clean Air Act requires EPA to regulate greenhouse gas emissions from new motor vehicles. In response to this case, EPA has announced that it will issue an Advance Notice of Proposed Rulemaking on "specific effects of climate change and potential regulation of greenhouse gas emissions from stationary and mobile sources under the Clean Air Act.

[41] Cap and trade programs combine a regulatory limit or cap on the amount of a substance—in this case a greenhouse gas such as carbon dioxide—that can be emitted into the atmosphere with market elements like credit trading to give industries flexibility in meeting this cap. A current example is the cap and trade program for sulfur dioxide under the Clean Air Act. This program includes electric utilities, which are the primary emitters of sulfur dioxide, and established a cap on the utilities' emissions. Sulfur dioxide allowances were primarily given (rather than auctioned) to companies.

[42] S. 28, S, 309, S. 317, S. 485, S. 1168, S. 1177, S. 1201, S. 1554, S. 1766, S. 2191,H.R. 620, H.R. 1590, H.R. 3989, H.R. 4226.

[43] GAO, *Vehicle fuel Economy: Reforming Fuel Economy Standards Could Help Reduce Oil Consumption by Cars and Light Trucks, and Other Options Could Complement These Standards,* GAO-07-921 (Washington, D.C.: Aug. 2, 2007).

[44] According to FAA, the last extensive discussion within ICAO on carbon dioxide emissions from aircraft occurred several years ago. At that time, ICAO's experts agreed that the cost of fuel provided sufficient incentive to minimize fuel consumption – hence carbon dioxide emissions. There was some technical work around 2001 on the development of an aircraft efficiency parameter, which might have been used to target carbon dioxide reductions. However, it failed to identify a parameter that would be able to assess aircraft fleets in their multiple operational environments in an equitable manner.

[45] The emissions trading scheme involves a cap and trade system that sets allowances for greenhouse gas emission for industries and other sources. Parties that pollute below their allowance receive emissions credits, which they can trade in a market to other parties that have exceeded their allowance.

[46] The Chicago Convention on International Civil Aviation of 1944 organized global aviation. According to the Convention, no state may condition the right of transit over or entry into or exit from its territory of any aircraft of another state on their operator's payment of fees, dues, or other charges.

[47] For example, FAA notes that there are considerable differences in the air traffic system efficiencies across the Atlantic, that the United States has a domestic fuel tax while nearly all EU states have none, and that the cost of fuel is about 50 percent more expensive for U.S. airlines because of the dollar's weakness in recent years.

[48] This ongoing work was jointly requested by the Committee on Transportation and Infrastructure, House of Representatives, and the Committee on Science and Technology, House of Representatives.

INDEX

B

C

D

E

M

Q

R

S

T

U